Advances in Intelligent Systems and Computing

Volume 780

Series editor

Janusz Kacprzyk, Polish Academy of Sciences, Warsaw, Poland
e-mail: kacprzyk@ibspan.waw.pl

The series "Advances in Intelligent Systems and Computing" contains publications on theory, applications, and design methods of Intelligent Systems and Intelligent Computing. Virtually all disciplines such as engineering, natural sciences, computer and information science, ICT, economics, business, e-commerce, environment, healthcare, life science are covered. The list of topics spans all the areas of modern intelligent systems and computing such as: computational intelligence, soft computing including neural networks, fuzzy systems, evolutionary computing and the fusion of these paradigms, social intelligence, ambient intelligence, computational neuroscience, artificial life, virtual worlds and society, cognitive science and systems, Perception and Vision, DNA and immune based systems, self-organizing and adaptive systems, e-Learning and teaching, human-centered and human-centric computing, recommender systems, intelligent control, robotics and mechatronics including human-machine teaming, knowledge-based paradigms, learning paradigms, machine ethics, intelligent data analysis, knowledge management, intelligent agents, intelligent decision making and support, intelligent network security, trust management, interactive entertainment, Web intelligence and multimedia.

The publications within "Advances in Intelligent Systems and Computing" are primarily proceedings of important conferences, symposia and congresses. They cover significant recent developments in the field, both of a foundational and applicable character. An important characteristic feature of the series is the short publication time and world-wide distribution. This permits a rapid and broad dissemination of research results.

More information about this series at http://www.springer.com/series/11156

Daniel N. Cassenti
Editor

Advances in Human Factors in Simulation and Modeling

Proceedings of the AHFE 2018 International
Conferences on Human Factors and Simulation
and Digital Human Modeling and Applied Optimization,
Held on July 21–25, 2018, in Loews Sapphire Falls Resort
at Universal Studios, Orlando, Florida, USA

 Springer

Editor
Daniel N. Cassenti
U.S. Army Research Laboratory
Aberdeen Proving Ground, MD, USA

ISSN 2194-5357   ISSN 2194-5365   (electronic)
Advances in Intelligent Systems and Computing
ISBN 978-3-319-94222-3   ISBN 978-3-319-94223-0   (eBook)
https://doi.org/10.1007/978-3-319-94223-0

Library of Congress Control Number: 2018947366

Printed on acid-free paper

This Springer imprint is published by the registered company Springer International Publishing AG
part of Springer Nature
The registered company address is: Gewerbestrasse 11, 6330 Cham, Switzerland

Advances in Human Factors
and Ergonomics 2018

AHFE 2018 Series Editors

Tareq Z. Ahram, Florida, USA
Waldemar Karwowski, Florida, USA

9th International Conference on Applied Human Factors and Ergonomics and the Affiliated Conferences

Proceedings of the AHFE 2018 International Conferences on Human Factors and Simulation and Digital Human Modeling and Applied Optimization, Held on July 21–25, 2018, in Loews Sapphire Falls Resort at Universal Studios, Orlando, Florida, USA.

Advances in Affective and Pleasurable Design	*Shuichi Fukuda*
Advances in Neuroergonomics and Cognitive Engineering	*Hasan Ayaz and Lukasz Mazur*
Advances in Design for Inclusion	*Giuseppe Di Bucchianico*
Advances in Ergonomics in Design	*Francisco Rebelo and Marcelo M. Soares*
Advances in Human Error, Reliability, Resilience, and Performance	*Ronald L. Boring*
Advances in Human Factors and Ergonomics in Healthcare and Medical Devices	*Nancy J. Lightner*
Advances in Human Factors in Simulation and Modeling	*Daniel N. Cassenti*
Advances in Human Factors and Systems Interaction	*Isabel L. Nunes*
Advances in Human Factors in Cybersecurity	*Tareq Z. Ahram and Denise Nicholson*
Advances in Human Factors, Business Management and Society	*Jussi Ilari Kantola, Salman Nazir and Tibor Barath*
Advances in Human Factors in Robots and Unmanned Systems	*Jessie Chen*
Advances in Human Factors in Training, Education, and Learning Sciences	*Salman Nazir, Anna-Maria Teperi and Aleksandra Polak-Sopińska*
Advances in Human Aspects of Transportation	*Neville Stanton*

(continued)

(continued)

Advances in Artificial Intelligence, Software and Systems Engineering	*Tareq Z. Ahram*
Advances in Human Factors, Sustainable Urban Planning and Infrastructure	*Jerzy Charytonowicz and Christianne Falcão*
Advances in Physical Ergonomics & Human Factors	*Ravindra S. Goonetilleke and Waldemar Karwowski*
Advances in Interdisciplinary Practice in Industrial Design	*WonJoon Chung and Cliff Sungsoo Shin*
Advances in Safety Management and Human Factors	*Pedro Miguel Ferreira Martins Arezes*
Advances in Social and Occupational Ergonomics	*Richard H. M. Goossens*
Advances in Manufacturing, Production Management and Process Control	*Waldemar Karwowski, Stefan Trzcielinski, Beata Mrugalska, Massimo Di Nicolantonio and Emilio Rossi*
Advances in Usability, User Experience and Assistive Technology	*Tareq Z. Ahram and Christianne Falcão*
Advances in Human Factors in Wearable Technologies and Game Design	*Tareq Z. Ahram*
Advances in Human Factors in Communication of Design	*Amic G. Ho*

Preface

This volume is a compilation of cutting-edge research regarding how simulation and modeling support human factors. The compilation of chapters is the result of efforts by the 9th International Conference on Applied Human Factors and Ergonomics (AHFE), which provides the organization for several affiliated conferences. Specifically, the chapters herein represent the 3rd International Conference on Human Factors and Simulation and the 7th International Conference on Digital Human Modeling and Applied Optimization.

Simulation is a technology that supports an approximation of real-world scenes and scenarios for a user. For example, a cockpit simulator represents the configuration of the inside of a cockpit and presents a sensory and motor experience to mimic flight. Simulations advance research by providing similar experiences to those scenarios that would otherwise be impractical to carry out in the real world for such reasons as monetary cost or safety concerns. Simulations can support numerous goals including training or practice on established skills.

Modeling is a somewhat different tool than simulation, though the two are often used interchangeably as they both imply estimation of real-world scenes or scenarios that bypass practical concerns. The difference in the context of this book is that modeling is not intended to provide a user with an experience, but rather to represent anything pertinent about the real world in computational algorithms, possibly including people and their psychological processing. Modeling may answer questions about large-scale scenarios that would be difficult to address otherwise, such as the effects of economic interventions or smaller-scale scenarios such as the cognitive processing required to perform a task that is otherwise undetectable by measurement devices.

The goal of the research herein is to bring awareness and attention to advances that human factors specialists may make in their field to address the design of programs of research, systems, policies, and devices. This book provides a plethora of avenues for human factors research that may be helped by simulation and modeling.

The book is divided into the following sections:

1. Virtual Environments and Augmented Reality
2. Modeling and Simulation Applications
3. Extreme Environments and Military Applications
4. Cognitive Modeling
5. Applications in Safety and Risk Perception
6. Digital Modeling and Biomechanics

Special thanks to Thomas Alexander and Vincent Duffy for the significant contributions to the conference on Digital Human Modeling and Applied Optimization. All papers in this book were either reviewed or contributed by the members of editorial board. For this, I would like to recognize the board members listed below:

Human Factors and Simulation

Digital Human Modeling and Applied Optimization

Ravi Goonetilleke, Hong Kong
Brian Gore, USA
Rush Green, USA
Lars Hanson, Sweden
Daniel Högberg, Sweden
Bruce Byung Cheol Lee, USA
Zhizhong Li, China
Ameersing Luximon, Hong Kong
Tim Marler, USA
Russell Marshall, UK
Stefan Pickl, Germany
George Psarros, Norway
Sudhakar Rajulu, USA
Zaili Yang, UK

This book is the first step in covering diverse topics in simulation and modeling. I hope this book is informative and helpful for the researchers and practitioners in developing better products, services, and systems.

July 2018

Daniel N. Cassenti

Contents

Virtual Environments and Augmented Reality

Determining the Ecological Validity of Simulation Environments in Support of Human Competency Development

Glenn A. Hodges[✉]

Modeling, Virtual Environments, and Simulation Institute (MOVES)
Naval Postgraduate School, Monterey, CA 93943, USA
gahodgesl@nps.edu

Abstract. This paper discusses the development and use of an analytical assessment methodology that applies Systems Engineering principles, Ecological Affordance Theory, and Human Abilities, to measure the potential of integrated simulation training environments (ITEs) to support the development of competence in the execution of specific military missions. The results of this research include the development and use of the integrated training environment assessment methodology (ITEAM). ITEAM was used to re-evaluate the ecological validity of several ITEs ability to support the development of specific competencies during training.

Keywords: Systems engineering · Affordance theory
Human ability requirements · Simulation · Training · Military
Competency development · Performance

1 Introduction

The United States Military values integrated simulation training environments (ITE) because they provide a relatively safe space, with a certain level of ecological validity (i.e. that replicate important features of the real world), to develop and maintain warfighting competence [1]. During and between conflicts, the Armed Forces rely heavily on ITE. ITE are comprised of various live, virtual, constructive, and game-based training aids, devices, simulators and simulations that allow men and women to practice skills and engrain knowledge necessary to execute their jobs successfully on the battlefield [2]. Developing and operating ITE is extremely resource intensive and ITE are rarely described as lightweight or turnkey. ITE require verification, validation, and accreditation (VVA) just as their analytical counterparts that support budgetary and force structure decisions. The major difference between ITE that support training and other types of simulation is how they are evaluated (i.e. verified and validated).

The most common method of determining ITE effectiveness is through the use of empirical transfer of training (TOT) studies that are expensive and often provide limited or misleading insight into ITE utility. Some researchers have attempted to use non-empirical means to evaluate ITE in an effort to reduce costs and accurately capture

© Springer International Publishing AG, part of Springer Nature (outside the USA) 2019
D. N. Cassenti (Ed.): AHFE 2018, AISC 780, pp. 3–14, 2019.
https://doi.org/10.1007/978-3-319-94223-0_1

positive system attributes [3–6]. Despite their best efforts only a handful of researchers have had their techniques successfully implemented outside of the research arena and of those few have been used more than a handful of times. Most of the analytical techniques developed have not been extensible, user friendly or well documented to facilitate reuse. Additionally, many have used mathematical equations that have not been validated with empirical data. Many have been automated due to their extreme complexity without concern for program documentation making them nearly impossible to implement by others and their focus has been similar to that of empirical attempts.

Until 2012, the Unites States Army (USA) had a logical system for providing analysis of training programs called the Training Effectiveness Analysis (TEA) system. The TEA system, established in 1975, was a Training and Doctrine Command (TRADOC) program focused on the impacts associated with training and hardware costs, hardware development cycles and complexity, training resources, and the overall effectiveness of Army programs to prepare Soldiers for battlefield conditions [7]. Prior to 2012, the TRADOC Analysis Center at White Sands Missile Range (TRAC–WSMR) was the Army's lead agency for providing technical assistance and conducting TEA for training systems. TRADOC Regulation 350-32 governed the TEA program. At the time, Simpson [8] offered that the Army TEA system was the most robustly defined training analysis system that existed. Several system analysts have described the use of TEA studies for the benefit of their respective programs and offered examples of how they conducted TEA studies [9, 10]. Despite this, in the summer of 2012, the USA officially concluded its last TEA study, eliminating both the office responsible for the conduct and oversight of TEA, and the regulation that governed the TEA system [11]. Recognizing a gap, the author conducted research to develop a logical methodology for assessing ITE that requires few resources and focuses more on the affordances provided by the training environment and less on advanced technology or the opinions of trainees.

2 Competence, Human Abilities, and Affordances

2.1 Competence

The military spends a great deal of time and millions of dollars each year educating and training its workforce due to the substantial number of unique occupational specialties it maintains [12] and the missions it must be prepared to conduct. Some authors have attested that training is the most expensive tool used to develop human competency [13, 14]. The military predominantly uses a task-based training approach and evaluation by subject-matter experts to judge soldier proficiency. However, it is often impossible to determine if a soldier's performance has improved as a result of training due the confounding (i.e. chaotic and unpredictable) environments in which they operate and due to lack of any pre-performance baselining of soldier performance. Therefore, some have offered that instead of evaluating tasks we should view performance using mission essential competencies [15]. Others warn that doing so may result in disappointment [12]. Gilbert [14] approaches performance from a different perspective. He argues for "engineering worthy performance" by attending to accomplishments instead of behavior and by using a set of analytical processes and tools to logically guide the determination of incompetence, thus revealing areas where improvements are needed to create worthy performance. Gilbert's view is instantiated in the form of a performance matrix that depicts performance across multiple levels (i.e. philosophical, cultural, policy, strategic, tactical and logistical). Each of the levels is analyzed using (1) models of accomplishment, (2) measures of deficiency and finally (3) methods of improvement to answer questions about what is required and where and how to achieve and measure it. Gilbert's work resonates here simply because it applies a logical systematic process to identify performance gaps and potential for improved performance. ITEAM was developed with the same view albeit tasks were used instead of accomplishments to dissect performance.

2.2   Human Abilities

Human abilities (HA) are viewed as enduring attributes of the human being (i.e., they are the same in the real or virtual world) and they play an important role in the methodology discussed here. Research on HA requirements has been ongoing since the 1960's and has been used as a tool for empirical work investigating training system design, fidelity, and personnel requirements [16, 17]. As early as 1983, the United States Army Research Institute was using ability requirements in support of system acquisition [18]. The Defense Advanced Research Projects Agency (DARPA) sponsored research into HA to assist the military with job placement and training [19] and with the advent of mobile phones, the Army Research Lab used ability requirements to investigate the effects of mobile phone use while driving [20]. The HA body of research has been developed as part of an umbrella taxonomic effort attempting to standardize the way human performance is described. The objective of the ability requirements approach was to identify and define the fewest number of independent ability categories that would be useful and meaningful for describing performance in the widest variety of tasks [21]. HA development is an iterative process intended to produce a list of verified abilities that are empirically derived from patterns of responses to different tasks. The assumption is that specific tasks require certain abilities and that tasks requiring the same types of abilities can be categorized similarly. This assumption allows researchers to discuss task performance in relative terms. The HA project, through experimentation and collaboration with multiple subject matter experts, derived 52 HA with the possibility of adding more. Examples of HA are oral comprehension, deductive reasoning, dynamic strength, peripheral vision, and sound localization. HA are grouped into one of four categories (i.e., physical, sensory, psychomotor and cognitive). Through years of research, Fleishman and his colleagues analyzed various jobs and tasks to ascertain and develop the list of 52 human abilities that can be found throughout various human activities. During this process, they executed numerous task analyses (TA). Through their process of defining ability requirements, they linked information dealing with task characteristics to HA [21–23]. The results of their efforts led to a means of description, understanding and categorization of human activity (i.e., work) based on HA instead of through the use of TA. Today, both the United States Department of Labor and the United States Army

Research Lab use HA as the basis for their O*NET (http://www.onetonline.org) program and Job Assessment Software System (JASS) [18] respectively.

2.3   Affordance Theory

Affordance theory comes from ecological psychology. Gibson [24] coined the term "affordance" to capture the essence of what an environment offers or provides an animal in either a positive or negative fashion. Precedent exists for the use of affordance theory in supporting computer science and human factors research [25–28]. Affordance theory is naturally associated with HA, most notably with human perception. Gibson's theory of affordances has been met with varying degrees of enthusiasm and criticism over the years [29]. As initially described, the concept of affordances was simple, clear and appealing [30]. However, Gibson's later attempts to describe affordances in more detail, resulted in a situation that "makes them seem like impossible, ghostly entities, entities that no respectable scientist (or science worshiping analytic philosopher) could have as part of their ontology" [31]. Attempts at providing clarity and concrete definitions for affordances have been offered and debated [32, 33]. Affordances are used in this research as a means of identifying the qualities and characteristics of an ITE that are absent or present in relation to the HA associated with specific tasks (i.e. help determine the ecological validity of an ITE). Affordances were used as part of this methodology because they provide context and allow for the opportunity to view an ITE unlike any other approach. Using affordances, we are not only able to identify the characteristics of an ITE that support deliberate practice; but also, why those identified characteristics are important to the trainee's execution of the tasks. Through the use of affordances, we are able to determine specific task elements with the highest likelihood of positive training transfer to the real operational environment.

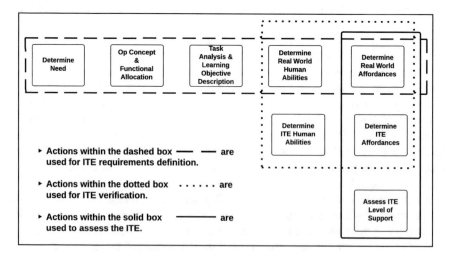

Fig. 1. Processes involved in the ITEAM

3 Integrated Training Environment Assessment Methodology

The Integrated Training Environment Assessment Methodology (ITEAM), is a human-centered, systems engineering approach to ITE analysis. ITEAM was developed based on the lessons learned from the literature and with the recognition that good ITE assessment must first start with a solid understanding of the human ability requirements associated with the necessary performance demands. Considering the pieces of a training program (e.g. humans, requirements, technology) it has been established that computer technology evolves the fastest (i.e. Moore's Law). Requirements determination on the other hand occurs much more slowly. Humans evolve the slowest yet their evolutionary stability is ignored during ITE assessment in favor of focusing on advanced technology. ITEAM counters this by focusing on the affordance support provided by an ITE to the development of human competence.

ITEAM (Fig. 1) was developed as a set of three main logical processes each containing multiple sub-processes. All of the sub-processes are iterative in nature and steps may be abbreviated or skipped depending on the time available and level of detail required. Requirements definition occurs first and proceeds from left to right beginning with determining the need and ending with determining the real world (RW) affordances. Verification follows and builds on requirements definition by determining the ITE HA requirements and ITE affordances. Assessment of ITE support to training happens last and only after the RW and ITE affordances have been identified for comparison.

3.1 Requirements Definition

Proper problem description and analysis are critical to the ITE assessment process. ITEAM groups the activities of determining the need for the ITE, how it will be used, which functions will be performed by the ITE and the human, description of the tasks to be executed during training and the desired learning outcomes, within the boundary of requirements definition. Also included is a list of real world (RW) HA and RW affordance requirements that are necessary to accomplish the training tasks. HA are used to help illuminate the critical aspects (environmental affordances) required of the ITE. Affordances are used to describe the attributes of the ITE that are necessary to support the execution of the desired training.

3.2 Verification

Verification is defined here as "the process of determining that a model or simulation implementation and its associated data accurately represent the developer's conceptual description and specifications" [34]. The sub-processes of ITEAM considered to be useful for verification consist of compiling the identified real world (RW) and system-supported HA as well as the RW and system-provided affordances. During this process, the evaluator uses a task analysis to determine the RW HA and affordance requirements associated with the tasks to be trained. Then, the ITE is investigated to determine what HA it supports and what affordances are available. Comparison of these items provides the basis for an initial judgment on whether or not an ITE will support the execution of the desired training that leads to exemplary performance.

3.3   Assessment

The final process of ITEAM assesses an ITE's ability to support desired training by quantifying ITE affordance resources based on ITE affordance requirements. The quantification of resources provides the customer/stakeholder/user with an estimate of the level of support that the ITE provides. ITE scoring is based on a subject matter expert (SME) evaluator's judgment on the absence or presence of specified affordances using the scale seen in Fig. 2.

Scale Definition

5–Excellent – the ITE contains all but a few (90–100%) of the affordances determined during the analysis
4–Very Good – the ITE contains a significant portion (70–89%) of the affordances determined during the analysis
3–Good – the ITE contains a good portion (50–69%) of the affordances determined during the analysis
2–Fair – the ITE contains some (25–49%) of the affordances determined during the analysis
1–Poor – the ITE contains very few (0–24%) of the affordances determined during the analysis

---

Fig. 2. ITEAM scoring scale definition

Affordance Scoring. Affordances are determined by studying the hierarchical task analysis conducted during the requirements determination phase. Each mission task is decomposed into sub-tasks, which are then evaluated to determine affordance requirements.

Unique Affordances. A unique affordance is identified as one that has not been previously identified, evaluated or accounted for as part of any another subtask evaluation. If a subtask's affordances are unique, then $D = C \div A$, where U = Universal set of affordances; A = subset of U required to execute a subtask in the real world; B = subset of U found in the ITE; $C = A \cap B$, which is the subset of U shared by both A and B; and D = % of support provided by the ITE to the execution of the subtask. The percentage (D) is then compared to the rating scale (Fig. 2) and results in a rating of 1–5 Poor to Excellent.

Affordances Previously Accounted for. ITE support to subtask execution in this situation is treated in the following manner: $D = C \div B$ where U = Universal set of all subtasks; A = set of $\{1, 2, 3, 4, 5\}$ which are the ratings from Fig. 2; B = subset of U for the current evaluation; For each $x \in B$ there is a corresponding A_i; $C = \Sigma A_i$ for all B, which is the sum of the subtask scores for the present analysis; D = the rating number in Fig. 2 when properly rounded (scores containing 0.50 and lower are rounded down).

Affordances Partially Unique and Partially Accounted for. If the affordances for a subtask are partially unique and partially accounted for in other analyses then the calculation is conducted in three steps. *Step one*—Treat previous subtask analyses as a single affordance that is present and unique. *Step two*—Evaluate and account for the presence of all unique affordances associated with the subtask. Once every affordance is accounted for, the calculation for determining the percentage present is conducted as described in *Unique affordances*. The result (rating of 1–5) is temporarily assigned as the subtask score. *Step three*—Obtain the values (scores) for the subtask affordances from the previous analyses (see *Affordances previously accounted for*) and sum them. Add the temporary value for the subtask currently under evaluation. Average this value by the total number of subtasks (including the current one). The derived number represents a number on the scale between 1 and 5 (see Fig. 2) that when rounded appropriately (0.50 and lower round down) provides the qualitative rating for this subtask.

High-Level Task Scoring. High-level tasks are also scored using the scale seen in Fig. 2. The procedure to score a high-level task consists of summing all of the subtask scores and dividing them by the total number of subtasks. The result is a numerical value that is associated with a level of support provided (Poor to Excellent) by the ITE to the deliberate practice of the task.

4 Testing the Methodology

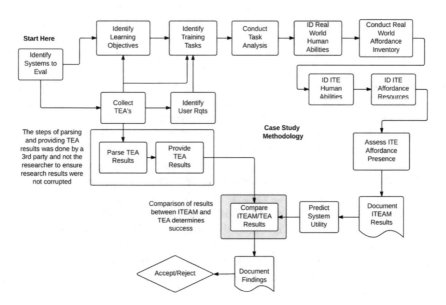

Fig. 3. The process used to validate ITEAM

4.1   Discussion About Studies

ITEAM was used to re-evaluate several existing training environments. Figure 3 depicts the process that was applied. The results for 3 studies were compared to the original empirical assessments to determine whether this analytical process was as effective in determining the utility of the ITEs [35]. In each case, the results indicated that ITEAM answered the questions asked and provided insights that the empirical studies did not. The final research case study plan to determine the ITEAMs validity, consisted of the development, evaluation and pseudo training transfer experiment of a driving environment designed to allow for the deliberate practice of a set of tasks necessary for inclement weather driving. The Inclement Weather Driving Environment (ICWDE) was developed under a consensus belief that California drivers are not well prepared to handle various types of severe weather driving conditions due to the mild climate that occurs throughout most of the state. Two forms of the ICWDE were to be employed during the transfer experiment. To avoid any undue influence or corruption of results, the author was blinded from the development of the experimental environments and was only allowed to view the results of the data analysis after it was conducted. As in the previous case studies, the author began this study with enough information to initiate and execute ITEAM. Only after ITEAM execution was complete and predictions offered was the author provided with the statistical analysis of the performance data from the pseudo transfer study so that comparisons could be made.

Two ITE's developed for this study and experiment were simply referred to as the low fidelity (LF) and high fidelity (HF) environments. The HF environment consisted of a three 11×17 in. computer monitors arranged in a manner to provide an approximate $90°$ field of view (FOV). Logitech G25 driving controls were employed to replicate the driver's controls of a typical vehicle. The controls consisted of the steering wheel that was securely fashioned to a tabletop and the pedal configuration consisting of three pedals of which only two were used. The far-right pedal replicated an accelerator for forward movement and a brake if the vehicle was in reverse. The middle pedal was the brake for forward movement and the accelerator if reverse motion was needed. No gearshift was employed. The driving environment software handled the transmission shifting as if the vehicle was equipped with an automatic transmission. A standard stationary desk chair was used for the vehicle seat. A vibration device (the ButtKicker) was attached to the bottom of the desk chair and linked into the software allowing for limited tactile feedback from the road to the driver. Sound from the driving environment came from a pair of standard Dell desktop speakers.

The software for the driving environment was developed using the Unity Game Engine 4.3.3f1 with several extensions (e.g. Eddie's Vehicle Physics, Boomlagoon JSON parser, Silver Lining). The base scenario of the software contained several different vehicles (e.g. a pickup truck, car, bus), pedestrians, buildings, landscape terrain and roads. Extensions added vehicle physics, realistic skies and weather. Alterations to the vehicle physics (vehicle control) and visibility (weather and skies) were used to create the main differences between the HF and LF environments. The only physical difference between the HF and LF environments was the number and position of the monitors. In the LF environment, only one monitor was used reducing the FOV to approximately 30°.

Pseudo Training Transfer Experiment. The experiment conducted in support of this study was a pseudo training transfer study consisting of tasks performed under conditions of low visibility and low traction to simulate conditions that may be found in inclement weather. All subjects received a basic block of instruction about inclement weather driving from a short video. The subjects were then randomly assigned to one of two groups for practice and testing of skills. Both groups used the same route for initial driving familiarization. Group A used the low fidelity ITE to train and a higher fidelity ITE to test. Group B used the higher fidelity ITE to train and test. After training was conducted, each subject completed a questionnaire requesting their assessment as to how well the ITE supported the training of the specific tasks. Once complete, subjects were asked to drive another route in the ITE. For this second route all subjects used the same version of the ITE. Upon completion of the second route, subjects again filled out a questionnaire asking them to assess how well the ITE supported the execution of the tasks they had been asked to conduct. The tasks that the subjects were asked to execute during the first route were: staying on the road; driving with diminished visibility; following the instructions of a navigator; avoiding stationary objects; driving on a slippery road surface (straight, curved, turns, and intersections). For the second route driven, subjects additionally negotiated oncoming traffic. Group A also had to adjust to a higher fidelity ITE. Each subject drove the same training route after filling out questionnaires and watching a brief instructional video. Half the subjects used one version of the simulator, while half used another in the training task. After filling out a mid-task questionnaire, the subject would complete a second route and fill out a final questionnaire. Both experimental groups used the same version of the simulator in the second task.

ITEAM Assessment of the ICWDE. The front-end analysis of this experiment was consisted of inferring the stakeholder need and intent from the statements above dealing with California drivers needing inclement weather driving exposure. The experimental research assistant (RA) who was responsible for the development of the ITE and running the pseudo transfer study provided general tasks that provided the basis for the ITEAM assessment. Driving domain knowledge was enhanced through the investigation of various authoritative sources such as the National Transportation and Safety Administration (NTSA) and the American Automobile Association (AAA). Additionally, various other sources to include Nationwide insurance and the Virginia state driver education curriculum were consulted. Six high-level tasks were used to evaluate the ICWDE. The core of each task was tied to a focus on visibility, traction and vehicle handling. These tasks provide a representative sample of the types of activities that a driver would need to be proficient in to be a capable driver in inclement weather conditions. Taken at face value the results (Table 1) appear to indicate that the LF ITE should generally provide very good support to the practice of skills necessary for the tasks under investigation. Closer examination of the numbers associated with the scores suggests that the environment may be closer to providing 'Good' overall support versus Very Good. This suggestion is based on the fact that the numerical scores listed in all but one of the 'Very Good' ratings is close to the threshold that would have rounded the score down from 'Very Good' to just 'Good'. Additionally, a summary score of the 6 high-level tasks reveals an overall environmental rating of 'Very Good' at 3.61. This is the score that was used in comparison to the pseudo training transfer experiment.

Table 1. ITEAM assessment results for the low fidelity ICWDE.

High level task	ITEAM score
Avoid obstacles in path of travel	Good 3.33
Turn at intersection from moving vehicle	Very good 3.67
Turn at intersection from stationary vehicle	Very good 3.77
Stop moving vehicle when directed	Good 3.20
Initiate vehicle movement from a running vehicle by accelerating from the halt	Very good 4.00
Control vehicle when driving	Very good 3.67

4.2   Lessons Learned from Failure and Success

After conducting the ITE verification processes on the low fidelity (LF) environment, it quickly became apparent that there would be no differentiation between the scoring of the high fidelity (HF) and LF driving environments. Each contained the same affordances needed to support task execution and the methodology, in its current state of development, does not account for the quality of the affordances during the assessment. From the standpoint of the methodology, the levels of fidelity do not matter only their presence or absence. This realization, that the experimental design would not support evaluation of the methodology, resulted in the discontinuation of the study and the identification of large piece of future work that must be accomplished. The next lesson is that domain experience is necessary in order to use ITEAM effectively. Using properly focused SMEs with domain expertise strengthens the validity and reliability of the ITEAM data and reduces the risk of a false positive with respect to ITE capability. Without domain knowledge and experience, an essential understanding of the necessary ITE affordances does not exist and cannot be determined appropriately. The consistent application of ITEAM mitigates the effects of evaluator bias. Taking the time to draft a study plan, rules for ITE examination and the handling of unique and similar situations all help to mitigate SME bias and strengthen the reliability of the ITEAM results. Additionally, any approach that focuses attention on the tasks needing practice, the HA involved with those tasks, and the necessary ITE affordances, mitigates an overemphasis on the application of advanced technology in the design, development and implementation of ITEs. Taking time to determine the true need for an ITE and a concept for human/ITE interaction is time well spent.

5 Conclusion

The debate over the value of analytical assessment of ITEs continue but we believe that our efforts have shed light on a new way to approach the issue. Implementing a methodical process in assessment efforts forces an accounting of things that the current acquisition process ignores or bypasses. Each process and sub-process of ITEAM unlocks information about the stakeholder's needs and ITE requirements that otherwise might be missed if the methodology was not followed. Furthermore, given that the cost of using the methodology is so small, it will result in savings of time and money in the areas of design, development and manufacturing of ITEs.

Acknowledgements. The work presented here was supported by the Naval Postgraduate School. The opinions expressed in this paper are solely those of the author and should not to be interpreted as an official position of NPS, the USA, or the Department of Defense.

References

1. Salas, E., Rosen, M., Held, J., Weissmuller, J.: Performance measurement in simulation-based training: a review and best practices. Simul. Gaming **40**, 328–376 (2008)
2. Hodges, G., Darken, R., McCauley, M.: An analytical method for assessing the effectiveness of human in the loop simulation environments: a work in progress. In: Proceedings of the 2014 Spring Simulation Multi-conference. The Society for Modeling and Simulation International, Tampa (2014)
3. Tufano, D., Evans, R.: The prediction of training device effectiveness: a review of army models. Technical report, U.S. Army Research Institute for the Behavioral and Social Sciences (1982)
4. Keesling, J., King, J., Mullen, W.: Simulation training strategies for force XXI. Technical report, U.S. Army Research Institute for the Behavioral and Social Sciences (1999)
5. Sticha, P., Campbell, R., Knerr, M.: Individual and collective training in live, virtual and constructive environments–training concepts for virtual environments. Study report, U.S. Army Research Institute for the Behavioral and Social Sciences (2002)
6. Gilligan, E., Elder, B., Sticha, P.: Optimization of simulation-based training systems: user's guide. Technical report, U.S. Army Research Institute for the Behavioral and Social Sciences (1990)
7. Neal, G.: Overview of training effectiveness analysis. In: Proceedings of the Human Factors and Ergonomics Society 26th Annual Meeting, pp. 244–248. SAGE (1982)
8. Simpson, H.: Cost-effectiveness analysis of training in the department of defense. Technical report, Defense Manpower Data Center (1995)
9. Carter, R.J.: Methodologies for evaluating training products and processes. In: Proceedings of the Human Factors and Ergonomics Society Annual Meeting, pp. 258–262. SAGE (1982)
10. Maitland, A.: Training effectiveness analysis: where the operator meets the equipment. In: Proceedings of the Human Factors and Ergonomics Society Annual Meeting, pp. 255–257. SAGE (1982)
11. Drillings, M.: Director's corner. MANPRINT (2013). http://www.manprint.army.mil
12. Kerry, J.: Competency in the military. In: Ford, J. (ed.) Improving Training Effectiveness in Work Organizations. Psychology Press, New York (1997)

13. Salas, E., Cannon-Bowers, J., Kozlowski, S.: The science and practice of training—current trends and emerging themes. In: Ford, J. (ed.) Improving Training Effectiveness in Work Organizations. Psychology Press, New York (1997)
14. Gilbert, T.: Human Competence, Engineering Worth Performance, Tribute edn. Pfeiffer, San Francisco (2007)
15. Bennett, W., Alliger, G., Colegrove, C., Garrity, M., Beard, R.: Mission essential competencies: a novel approach to proficiency-based live, virtual, and constructive readiness training and assessment. In: Ford, J. (ed.) Improving Training Effectiveness in Work Organizations. Psychology Press, New York (1997)
16. Hays, R., Singer, M.: Research issues in training device design. In: Proceedings of the Human Factors Society 27th Annual Meeting, pp. 147–150. SAGE (1983)
17. Napoletano, N.: The eyes have it: simulated sound visualization for testing. In: Proceedings of the Interservice/Industry Training, Simulation & Education Conference, NTSA (2013)
18. Rossmeissl, P., Tillman, B., Rigg, K., Best, P.: Job assessment software system (JASS) for analysis of weapon systems personnel requirements. Technical report, U.S. Army Research Institute for the Behavioral and Social Sciences (1983)
19. Cockayne, W.: Two-handed, whole-hand interaction. Masters thesis, Naval Postgraduate School (1998)
20. Middlebrooks, S., Knapp, B., Tillman, B.: An evaluation of skills and abilities required in the simultaneous performance of using a mobile telephone and driving an automobile. Technical report, U.S. Army Research Laboratory (1999)
21. Fleishman, E., Quaintance, M.: Taxonomies of Human Performance: The Description of Human Tasks. Academic Press, Orlando (1984)
22. Fleishman, E., Mumford, M.: Evaluating classifications of job behavior: a construct validation of the ability requirement scales. Pers. Psychol. 44, 523–575 (1991)
23. Fleishman, E., Bartlett, C.: Human abilities. Ann. Rev. Psychol. 20, 349–380 (1969)
24. Gibson, J.: The theory of affordances. In: The Ecological Approach to Visual Perception, pp. 127–143. Lawrence Erlbaum Associates, Inc., Hillsdale (1986)
25. Bærentsen, K., Trettvik, J.: An activity theory approach to affordance. In: Proceedings of the Second Nordic Conference of Human Computer Interaction, pp. 51–60. Association for Computing Machinery, New York (2002)
26. Chemero, A., Turvey, M.: Gibsonian affordances for roboticists. Adapt. Behav. 15, 473–480 (2007)
27. Lintern, G.: An affordance-based perspective on human-machine interface design. Ecol. Psychol. 12, 65–69 (2000)
28. Rome, E., Paletta, L., Sahin, E., Dorffner, G., Hertzberg, J., Breithaupt, R., Fritz, G., Uğur, E.: The MACS project: an approach to affordance-inspired robot control. In: Rome, E., Hertzberg, J., Dorffner, G. (eds.) Towards Affordance-Based Robot Control, pp. 173–210. Springer, Berlin (2008)
29. Jones, K.: What is an affordance? Ecol. Psychol. 15, 107–114 (2003)
30. Michaels, C.: Affordances: four points of debate. Ecol. Psychol. 15, 135–148 (2003)
31. Chemero, A.: An outline of a theory of affordances. Ecol. Psychol. 15, 181–195 (2003)
32. Stoffregen, T.: Affordances as properties of the animal-environment system. Ecol. Psychol. 15, 115–134 (2003)
33. Turvey, M.: Affordances and prospective control: an outline of the ontology. Ecol. Psychol. 4, 173–187 (1992)
34. Under Secretary of Defense: Modeling and Simulation (M&S) Verification, Validation, and Accreditation (VV&A) Instruction, DoD Instruction 5000-61 (2009)
35. Hodges, G.: Identifying the limits of an integrated training environment using human abilities and affordance theory. Doctoral dissertation, Naval Postgraduate School (2014)

The Use of Immersive Virtual Reality for the Test and Evaluation of Interactions with Simulated Agents

Gabrielle Vasquez[✉], Rhyse Bendell[✉], Andrew Talone[✉],
Blake Nguyen[✉], and Florian Jentsch[✉]

University of Central Florida, Orlando, FL 32826, USA
{gvasquez, rbendell, atalone, bnguyen,
fjentsch}@ist.ucf.edu

Abstract. We aim to better inform the scientific community regarding test and evaluation techniques for validating devices that will potentially be used by individuals interfacing with autonomous robotic teammates (particularly, members of the U.S. Military). Testing within immersive virtual environments (IVRs) similar to those experienced in military operations will be discussed with focus on the use of a commercial gaming engine for task development. Highlights of using commercial gaming engines will be illustrated throughout the paper to emphasize their utility for evaluating future technologies with attention given to testing efficiency and ecological validity. The study of interactions with simulated agents and future communication devices will be described in the context of the Robotics Collaborative Technology Alliance (RCTA) research program.

Keywords: Human-robot interaction · Immersive virtual reality
Simulation

1 Introduction

1.1 Purpose

Overall, our goal is to study human-robot interaction (HRI) topics (situation awareness, usability, trust) relevant to human-robot teaming involving interactions with highly autonomous unmanned ground vehicles (UGVs) and the use of tasks which incorporate principles of serious games and gamification within immersive virtual reality (IVR) environments for studying those topics. Specifically, we are focusing on the use of simulated agents and virtual representations of novel communication interfaces. Important benefits of experimentation in IVR will be emphasized with particular attention given to participant experience (immersion, engagement, learning, motivation), data collection scope (biomechanics, gaze, interaction times, perception reaction time, etc.), and the ability to replicate environments and scenarios that would otherwise be inaccessible.

© Springer International Publishing AG, part of Springer Nature 2019
D. N. Cassenti (Ed.): AHFE 2018, AISC 780, pp. 15–25, 2019.
https://doi.org/10.1007/978-3-319-94223-0_2

1.2 Robotic Collaborative Technology Alliance (RCTA) – Transitioning Robots from Tools to Teammates

As members of the Robotics Collaborative Technology Alliance (RCTA)-a consortium of university, industry, and government researchers seeking to enhance the state of the art in human-robot teaming-we have sought to identify useful methods for testing and evaluating human-robot communication interface designs intended for use with highly autonomous UGVs. The RCTA is primarily focused on advancing four key areas relevant to producing UGVs that move beyond simple tools to useful, autonomous teammates. These areas include: perception, artificial intelligence, mobility and manipulation, and human-robot interaction. If we seek to transition UGVs from their current use as tools to a new role as teammates, then advances in these key areas is essential to achieving this goal.

One key RCTA initiative is to investigate the utility of a multimodal interface (MMI) for optimizing communication and interactions with UGVs (specifically autonomous robotic teammates). Currently, human interactions with robotic teammates are characterized by: (a) a team of dedicated robot operators or "handlers", (b) a substantial amount of time spent "heads-down" viewing a visual display or graphical user interface (GUI), and (c) sustained attention needed to teleoperate, supervise, or manage the robot [1]. Taken together, these characteristics of current human-robot interactions with UGVs require substantial human resources that could be better utilized for other purposes (e.g., completing other critical tasks) [2]. The RCTA's vision for future human-robot teaming is one that involves UGVs communicating with humans in a manner that is similar to how human-human teammates communicate. Characteristics of human-human team communications include: natural, speech based communications, use of non-verbal signals (e.g., gestures), and utilization of both implicit and explicit communication. These characteristics can be incorporated into human-robot communication through the implementation of an MMI. MMIs provide users with the means to interact with a system via more than one sensory modality (e.g., visual and auditory, visual and tactile) [3].

1.3 MMI Evaluation Considerations

Our team is interested in investigating human interactions with robots that have capabilities far beyond current robotic technologies and within environments not easily accessible to the research team. It is for these reasons that we employ other methods for investigating HRI topics. Wizard of Oz techniques are particularly useful, and involve creating illusions in place of some elements of experimental tasks which would otherwise require developing complex systems. This technique was implemented by our team in previous HRI research which used a scale Military Operations in Urban Terrain (MOUT) facility. The MOUT facility served to research HRI in the context of active, on-the-move military scenarios which are relevant for future use of robotics in the field [4]. Primarily, the MOUT facility was created to mimic the way UGVs and unmanned aerial vehicles (UAVs) were used by the military at the time; however, developing a system by which participants could interact with autonomous UVs during experimental tasks was not feasible. Rather than devote time and resources to developing actual autonomous robots, a Wizard of Oz solution was implemented by having one (or more) confederates (out of participant line-of-sight) control the UVs in response to participant commands. This technique proved to be straightforward and reliable which allowed the research effort to progress rapidly. Wizard of Oz methods do not always require an active confederate as the goal is simply to have participants believe that the events of the experiment are real. Given the case of interacting with remote robotic teammates, the illusion could be accomplished with something as low level as playing pre-recorded audio at specific times and letting participants believe that what they are hearing is being generated by an actual teammate completing a task. Our current work investigating human-robot interactions will employ a more complex manifestation by representing a virtual version of a MMI that will allow participants to receive and respond to messages purportedly generated by their remote teammate. By programming the behavior of the MMI and letting participants believe that the data it presents is from an active teammate we can create the illusion of dynamic interactions while maintaining experimental control and repeatability. Additionally tasks will take place in IVR to help engage participants and support the illusion.

As IVR has advanced rapidly in the last decade, consideration should be given to its utility for research in domains that seek to transfer findings into real world practice (the ultimate goal of fields such as HRI). What needs to be established is the ecological validity that using immersive simulation technologies can offer. Because it can be difficult to replicate certain functionalities and capabilities within a live simulation, an IVR environment paired with built-in Wizard of Oz techniques presents an ideal method for investigating human-robot (HR) teaming in a low cost, reliable, and ecologically valid manner.

2   Validity in IVR

2.1   Ecological Validity

Ecological validity describes the relevance of research findings to the real world, and is vital to consider for IVR studies [5]. Studies lacking ecological validity cannot transfer findings to practical applications, and are thereby useless to fields such as HRI (though not necessarily to basic research or the study of cognitive abstractions). It is possible to achieve adequate ecological validity with IVR studies by attending to elements of simulations such as visual realism, physics accuracy, interaction/input familiarity, locomotion, etc.; augmentation of the meaningfulness of findings in these studies may be achieved by additionally incorporating principles of serious games and gamification into experimental tasks. Consideration of such guidelines can help engage and motivate users in tasks so as to elicit reactions and behaviors that are similar to their tendencies in the real world (e.g. ecologically valid performance).

2.2   Gamification and Serious Games

Gamification is a general term used to describe the incorporation of task elements which are enjoyable and interesting to participants. The goal of implementing principles of gamification is to help participants behave as they would naturally, and not as a result of being in a laboratory completing a task (particularly when compensation is involved). An important aspect of gamification is the ability to elicit intrinsic motivation in participants by providing them with a reason to continue exerting effort and maintaining performance. An example of this is the utilization of badges for the Boy Scouts of America: this organization has been able to motivate young children to obtain a mastery of goals, reputation, and identity with the assistance of the extrinsic reward system provided by badges and ceremonies [6]. Additionally, Denny (2013) applied principles of gamification (i.e. rewarding students with badges) in an online, undergraduate course for completion of extra quizzes; this motivated students to go beyond the goals of the course to achieve a higher grade [7, 8]. These cases differ somewhat from experimental tasks as they are less constrained. Gamified IVR tasks must be minutely controlled for the maintenance of internal validity, and a balance must be struck between fun/interest and the production of meaningful results. Achieving the latter is the focus of tasks such as serious games which aim to maximize the applicability of outcomes to real world training, learning, and design [8–12]. Serious games have been shown in past research to help promote learning and knowledge acquisition [13], and elicit engagement, flow, immersion, and a sense of presence [14]. Normally, this is accomplished by gamifying aspects of training tasks to support the immersion and engagement of users [14]; this could be done with the use of badges, points, etc. By implementing dynamic feedback, rewards, into test and evaluation using an IVR simulation, we aim to maintain ecological validity. An example of how we have incorporated these elements is the implementation of a modified signal detection task: performance on this particular task will be automatically calculated during trials (i.e. correct hits and false alarm rates) to give participants an overall understanding of their performance; this will serve to provide immediate feedback and promote motivation in achieving the task objective. Our simulation studies will also achieve validity by replicating real world environments as accurately as possible, and implementing gamified tasks within those virtual realities. The development of the environments for instance will be accomplished by gathering images of urban and rural environments and using the features in the Unreal Engine (a commercial game engine) to recreate them.

3   Immersive Virtual Reality Research Benefits

The principles of gamification may be applied to any simulation that requires human interaction; however, it is likely that incorporation of those principles will be particularly effective in IVR due to shared support of user engagement and motivation. Experiment design rarely takes participant experience into account beyond the consideration of the degree to which it aids the provision of accurate data. Accordingly experiments are designed to be rigid and repeatable, and tend to lose any semblance of enjoyable or familiar tasks that one might encounter in the real world. IVR tasks provide the ability to more accurately mimic complex activities experienced in the real world, and paired with the guidance of gamification can not only elicit real world behaviors from participants but can also promote that behavior by generating an intrinsic desire to continue. It is important for experimental research that participants remain engaged with administered tasks as boredom and disinterest can confound results as strongly as poor training or technical malfunctions.

3.1 Enhanced Presence and Engagement

Commercial hardware designed to present virtual realities can provide enough meaningful feedback to allow users to perform cognitive and manual tasks similarly to how they are performed in the real world. Currently feedback comes primarily in the form of high quality visuals (with high fidelity head tracking to allow for the illusion of motion), a simulation of 3D audio, and a limited amount of vibrotactile information via handheld controllers. While the sensory data provided by simulation tools is a reduced form of what the real world has to offer, it may be enough for most participants to develop a sense of presence and experience immersion in a virtual environment [15–17]. Presence can be considered a measure of the perception one has of physically being in an environment, real or virtual [18–20]. That perception is important for engaging participants and improving performance in IVR studies, and is most effectively created when the virtual reality accurately represents expected reality [21]. Accuracy is sometimes discussed in terms of breadth and depth of sensory stimulation; breadth describes the quantity of sensory modes and depth describes the resolution provided to each sense by virtual reality devices. The depth of current virtual realities is sufficient for creating many engaging tasks, and stimulators such as haptic gloves and displays that can match the resolving power of the human eye can be used to expand the depth range of current technologies. Breadth on the other hand would need to be addressed by more complex systems such as olfactory displays and virtual motion platforms. Interactions with virtual realities has also been shown to improve sense of presence and immersion, and more complex interaction types are an important factor in helping participants behave naturally in IVR. Additionally control over interactions through range of motion and input modes can be implemented at any level from having participants statically view and respond to scenarios to allowing them to move through and interact with a dynamic world. Advancements are being made regularly in the pursuit of improved immersive virtual realities, but their current state is already sufficient for supporting engagement as well as motivation [22].

3.2 Data Collection Benefits

Virtual reality devices both support the performance of natural behaviors and the collection of biomechanic, temporal, and some cognitive data regarding the execution of those behaviors. For example, the information required to match a user's motions to the associated representations in virtual reality is necessarily provided to the devices creating those simulations. That data may be combined with simulation output, and analyzed to reveal tendencies in the motion, input, gaze, task performance metrics, perception reaction time, etc. displayed by participants. One example of why this might be significant for researchers is shown by object identification and signal detection. Foremost the decision a participant makes regarding the identity of a particular object may be recorded as reliably as traditional present-absent signal detection trials, but that data alone may not be enough for researchers to determine which object features and environmental factors were most important in guiding that decision. IVR data collection can allow researchers to analyze the amount of time that participants spent looking at a particular object, which features they fixated on, how much time passed between their first fixation on the object and their decision input action, and even whether they displayed any hesitation when registering their input (provided some degree of motion is required as opposed to only a button press). An object in motion may also have its exact orientation, velocity, world location, and lighting data recorded for quantification of its projection on a participant's eye at every nanosecond of the identification process.

3.3   Real World Scenario Accessibility

Participants' actions in IVR are taken in response to whatever scenario is presented to them, and is not limited by the practicality of running an experiment in the real world. While exciting possibilities exist for basic perceptual research (presenting scenarios with altered physical laws, creating visual illusions in three dimensions, testing audio source localization ability), this capability is extremely relevant for human interaction research as many of the scenarios that are of real significance simply cannot be replicated for testing purposes. A bomb squad searching buildings in a real town for instance has no resources to spare for participation in a research study, and is an overly hazardous and volatile situation for research. Data may be passively collected in special cases for analysis after missions are complete, but repeatability of conditions and events cannot be achieved. IVR can present identical search-and-dispose missions to as many participants as needed and replicate experiences for each one. Though the effects of being in a life-threatening situation may not all manifest due to participants' awareness that they are in a simulation, risk may be induced through gamification and adequate story-telling for generating motivation. The limitations of imposing real risk may be outweighed by the opportunity to conduct tests in dynamic environments requiring large numbers of personnel and technologies such as robotic teammates: conducting research in such scenarios would be impractical for a single participant let alone with sample sizes large enough to reveal meaningful effects.

A particularly useful aspect of using simulations for administering experimental tasks is the ability to present prototypes of technologies that have only been conceptualized. Modeling functionalities and designing a technology's visual appearance are the major requirements for informing a virtual representation that hasn't been physically prototyped. For test and evaluation purposes functionalities and resulting capabilities for future robotic teammates could be developed within a virtual reality simulation, and would not require having to physically build a robot with those features already functioning. For example, experiencing a robot call and respond to reports given by a human could be accomplished with an input button that is programmed to execute the appropriate behavior in the simulation without having to solve the problem of implementing complex hardware.

Because developing new technology takes time and effort, especially in the test and evaluation stage, developing them in simulation can reduce cost and time; developers can determine and implement functionalities and capabilities that can be adjusted as designs are iterated. Such test and evaluation methods can reduce the time and cost that would otherwise be required for fabricating real world prototypes. The field of robotics is a prime example of how simulated prototypes reduce resource requirements: the cost of fabricating a novel robot (excluding the costs of producing designs) includes sensors, computing systems (and typically proprietary programming), locomotion actuators, chassis components, batteries, custom mechanical parts (especially for robots with manipulators), etc. as well as the time and effort of a build team that is skilled enough to assemble parts and problem solve when real world issues arise. The combined costs of manifesting a novel robotic system that is usable for field testing can be tens of thousands of dollars, but the cost of representing that same system in virtual reality can range from under a hundred to several hundred dollars depending on developer pay. Additionally whereas assembling (and debugging) a novel robotic system may take weeks or months, virtually representing a physical appearance and vital functionalities of a robot can be accomplished in a single week by a capable 3D modeler and a VR programmer. Therefore, IVR testing lends support toward the iterative prototyping process that is employed in the development of technologies as it can augment the scope of early model iterations, inform the design of real world interactions, and minimize the resources required for the prototyping phase.

4 HRI Research in IVR

4.1 HRI Relevant Scenarios

Natural behaviors and meaningful performance data can only be elicited from participants if the virtual environment they are experiencing is adequately realistic. This stipulation is necessarily vague as user requirements are not consistent across features of immersive virtual realities. Visual realism for instance has been shown to reach adequate levels of realism for supporting performance far before photorealism or even accurate texturing are achieved [23]. Users tend to expect more realism from functionalities on the other hand: if a lever does not move as anticipated, an item seems like it can be picked up but cannot, or a virtual tool malfunctions or does not behave according to physics users quickly lose their sense of immersion and reject the virtual reality. Achieving adequate realism with simulated robotic agents requires identifying and optimizing the features that users are most likely to require for maintaining engagement and natural interactions. The required features, and their level of realism, will change for any given task set depending on the interaction modes available to users. Given a scenario in which participants are asked to interact with a remote robotic teammate, it is likely that the physical appearance, appearance of locomotion, and essentially all visual elements regarding the physical robot do not require attention. While the stipulation that the teammate is remote (or at least out of line-of-sight) exempts consideration of visual factors, representations of the data and output allegedly generated by the actions of the robot are still necessary. Determination of the form communications should take may be informed by research questions and the relevance of particular data types to the scenario that participants will experience. Examples of the sorts of considerations that may be appropriate for HRI research are discussed below in the context of our current research.

4.2   Use Case: RCTA Experiments

The testing environment we have implemented in IVR is based on a rural, Middle Eastern setting that is meant to replicate a setting similar to a possible modern military operation. This was chosen for the sake of maintaining ecological validity while taking into consideration ability of our chosen development engine, Unreal Engine 4 (UE4), to realistically render such environments. Due to hardware limitations it was determined to be preferable to present relatively simple, uncluttered environments and avoid introducing latency or visual lag to the participant experience. Within the rural environment, participants will be performing tasks that are similar to those that might be executed by dismounted soldier teams. Our primary interests with the study currently in development is the effect of situational awareness and mental workload (with attention to multiple resource theory) on the performance of mission critical tasks. These factors are of interest to the future of human-robot team interactions as effective utilization of autonomous robots implies not interfering with other duties.

Situational awareness describes one's awareness of surroundings, events, objectives, and the changes that occur within relevant environments over time [24]. Given our experimental scenario this includes awareness of what is occurring during a visual search task as well as of the whereabouts, activities, and decisions of remote robotic teammate. The measurement of situational awareness is not consistent in the literature, so we have chosen to distribute verbal SA probes throughout the mission to measure the abilities of users to stay aware of their teammate's actions.

Mental workload is an important consideration for tactical scenarios as overloading can result in serious performance decrements and mission failure. It is considered in the context of mental resources that may be drawn upon for the processing of information and completion of tasks [25]. Multiple Resource Theory (MRT) is a useful method of pinpointing the contribution of various tasks and task elements to experienced workload as it considers the processing capacity and demands on particular sensory modalities [26]. Because there will be an element of time -sharing and a dual-task performance in our experimental design, MRT will allow us to closely investigate the impact of loading (and overloading) audio and visual pathways as well as resulting effects on task performance and situational awareness.

It is very likely that workload impacts situational awareness, and the interaction between the two constructs must be quantifiable for rigorous analysis of either. A simulated commanding officer will be communicate with participants at planned intervals, and inquire about the state of the participant's environment, task, and teammate; responses to these probes will accurately represent a participant's situational awareness during the mission and can be quantified by coding for stratified correct and incorrect responses. While situational awareness will be quantified by the accuracy and speed of participant responses to awareness probes, quantification of workload will be accomplished by detailed performance metrics and validated post trial workload measures. In the mission, the user will be required to perform a signal detection task that will involve selecting potential threats (characters carrying weapons) entering an area, and performance on this task will be considered in tandem with workload measure outcome. As shown in Fig. 1, UE4 is used to create the immersive virtual environment which participants experience through an HTC Vive head mounted display;

participants will be using the HTC Vive controllers to select potential targets for the signal detection task, and provide input to communications from their robotic teammate.

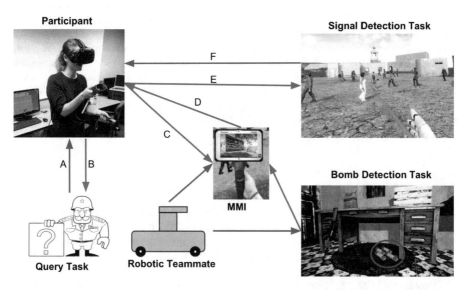

Fig. 1. The three tasks that will be completed by participants for the evaluation of cognitive workload and situational awareness are a signal detection task, a bomb detection task, and a query task. The signal detection task will require participants to view an environment (F) and make input (E) to identify characters who are carrying guns, whereas the query task will require listening to (A) and verbally responding to (B) pre-recorded situational awareness probes. Answers to the probes (B) will derive from information sent to the participant by their simulated robotic teammate (D) via the MMI using visuals or synthetic speech. The bomb identification task will require viewing images (D) and providing input (C) through the simulated MMI. The Wizard of Oz method will be used to generate the robotic teammate, bomb detection, MMI loop that ostensibly feeds into the data sent to participants (C) and reacts to their input (D).

5 Conclusion

The goal with our current projects is to continue using simulations to test HRI topics along with evaluating aspects and functionalities of the MMI and to validate its use for field studies and further research. Others interested in studying HRI topics may considering utilizing simulations developed similarly to the task set described above in order to:

- Ensure their research is able to maintain ecological validity
- Take into account engagement and immersion of users
- Control variables of interest and maintain internal validity

With technology continuously changing and more state-of-the-art technology being released, test and evaluation tools can potentially be implemented in IVR for users to better evaluate the usefulness and efficiency of emerging and near future technologies.

Acknowledgments. The research reported in this document was performed in connection with Contract Number W911NF-10- 2-0016 with the U.S. Army Research Laboratory. The views and conclusions contained in this document are those of the authors and should not be interpreted as presenting the official policies or position, either expressed or implied, of the U.S. Army Research Laboratory, or the U.S. Government unless so designated by other authorized documents. Citation of manufacturers or trade names does not constitute an official endorsement or approval of the use thereof. The U.S. Government is authorized to reproduce and distribute reprints for Government purposes notwithstanding any copyright notation hereon. Support for this endeavor was additionally provided by the University of Central Florida's Office of Research and Commercialization in the form of financial assistance via Mr. Bendell's ORC fellowship.

References

1. Yanco, H.A., Drury, J.L.: Rescuing interfaces: a multi-year study of human-robot interaction at the AAAI robot rescue competition. Auton. Robots **22**(4), 333–352 (2007)
2. Chen, J.Y.C., Haas, E.C., Barnes, M.J.: Human performance issues and user interface design for teleoperated robots. IEEE Trans. Syst. Man Cybern. Part C: Appl. Rev. **37**(6), 1231–1245 (2007)
3. Perzanowski, D., Schultz, A.C., Adams, W., Marsh, E., Bugajska, M.: Building a multimodal human-robot interface. IEEE Intell. Syst. **16**(1), 16–21 (2001)
4. Evans III, A.W., Hoeft, R.M., Rehfeld, S.A., Feldman, M., Curtis, M., Fincannon, T., Ottlinger, J., Jentsch, F.: Advancing robotics research through the use of a scale MOUT facility. Proc. Hum. Factors Ergon. Soc. Ann. Meet. **49**(6), 742–746 (2005)
5. Rizzo, A., Buckwalter, J.: Virtual reality and cognitive assessment and rehabilitation: the state of the art. Stud. Health Technol. Inf. **44**, 123–145 (1997)
6. Deterding, S.: Gamification: designing for motivation. Interactions **19**(4), 14–17 (2012)
7. Denny, P.: The effect of virtual achievements on student engagement. In: Proceedings of CHI 2013: Changing Perspectives, pp. 763–772 (2013)
8. Landers, R.N., Armstrong, M.B.: Enhancing instructional outcomes with gamification: an empirical test of the technology-enhanced training effectiveness model. Comput. Hum. Behav. **71**, 499–507 (2017)
9. Michael, D.R., Chen, S.: Serious games: games that educate, train, and inform. Muska & Lipman/Premier-Trade. Thomson Course Technology, Boston (2006)
10. Abt, C.C.: Serious Games. University Press of America, Boston (1987)
11. Landers, R.N.: Developing a theory of gamified learning: linking serious games and gamification of learning. Simul. Gaming **45**(6), 752–768 (2014)
12. Roth, S.: Serious gamification: on the redesign of a popular paradox. Games Cult. **12**(1), 100–111 (2017)
13. Connolly, T.M., Boyle, E.A., MacArthur, E., Hainey, T., Boyle, J.M.: A systematic literature review of empirical evidence on computer games and serious games. Comput. Educ. **59**(2), 661–686 (2012)
14. Boyle, E., Connolly, T.M., Hainey, T.: The role of psychology in understanding the impact of computer games. Entertain. Comput. **2**(2), 69–74 (2011)

15. Heeter, C.: Being there: the subjective experience of presence. Presence: Teleoper. Virtual Environ. **1**(2), 262–271 (1992)
16. Snow, M.P., Williges, R.C.: Empirical modeling of perceived presence in virtual environments using sequential experimentation techniques. In: Proceedings of the Human Factors and Ergonomics Society, 41st Annual Meeting, pp. 1224–1228 (1997)
17. Witmer, B.G., Singer, M.J.: Measuring presence in virtual environments (Tech. Rep. No. 1014). U.S. Army Research Institute, Washington, DC (1994)
18. Barfield, W., Sheridan, T., Zeltzer, D., Slater, M.: Presence and performance within virtual environments. In: Barfield, W., Furness, T.A. (eds.) Virtual Environments and Advanced Interface Design, pp. 473–513 (1995)
19. Barfield, W., Weghorst, S.: The sense of presence within virtual environments: a conceptual framework. In: Salvendy, G., Smith, M. (eds.) Human–Computer Interaction: Software and Hardware Interfaces, pp. 699–704 (1993)
20. Steuer, J.: Defining virtual reality: dimensions determining telepresence. J. Commun. **42**, 73–93 (1992)
21. Bowman, D., McMahan, R.: Virtual reality: how much immersion is enough? Computer **40**, 36–43 (2007)
22. McMahan, R.P., Bowman, D.A., Zielinski, D.J., Brady, R.B.: Evaluating display fidelity and interaction fidelity in a virtual reality game. IEEE Trans. Vis. Comput. Graph. **18**, 626–633 (2012)
23. Slater, M.: Place illusion and plausibility can lead to realistics behaviour in immersive virtual environments. Philos. Trans. R. Soc. B **364**, 3549–3557 (2009)
24. Endsley, M.R.: Design and evaluation for situation awareness enhancement. Proc. Hum. Factors Soc. Ann. Meet. **32**, 97–101 (1988)
25. Moray, N.: Mental Workload: Its Theory and Measurement. Plenum, New York (1979)
26. Wickens, C.D.: Multiple resources and mental workload. Hum. Factors **50**, 449–455 (2008)

Beyond Anthropometry and Biomechanics: Digital Human Models for Modeling Realistic Behaviors of Virtual Humans

Thomas Alexander$^{(\boxtimes)}$ and Lisa Fromm

Department of Human Factors, Fraunhofer-FKIE, Zanderstr. 5, 53177 Bonn,
Germany
{Thomas.Alexander,Lisa.Fromm}@FKIE.Fraunhofer.de

Abstract. Spatial layout of workplaces and geometric analysis of future products is a prominent application of Digital Human Models (DHMs). DHMs describe characteristics of body dimensions, body shape and motions of the future workers and users. Further applications of DHM are animated, computer-generated and photo-realistic figures for populating computer games and movies. In contrast to these applications, other areas of human modeling, e.g. human behavior modeling or cognitive modeling, have not been applied broadly.

These models address different levels of human behavior, including human information processing. This paper presents several of these models and uses them as a basis for generating the idea of a comprehensive digital human model. Such a model is applicable for optimizing complex work processes as well as populating virtual environments. It is concluded that there will be no single solution for modeling and simulation all variations and aspects of human behavior, but that there is a need for a reference architecture as a generic interface between the different models.

Keywords: Digital human modeling · Modeling · Simulation
Human behavior

1 Introduction

Modelling and simulation provide well-established methods, means and technologies for a broad spectrum of applications. This includes education and training of personnel (e.g. driving and flight simulators), design of technical systems (e.g. computer-aided design, finite-element simulation), and optimization of production processes and organizations (e.g. material or energy flow in industrial planning). Nowadays, there are many modelling and simulation tools available, which are successfully used and applied in many applications and offer valid results of various kinds.

In general, simulation describes the

> *"... process of designing a model of a real or imagined system and conducting experiments with this model to understand the behaviour of the system or to evaluate strategies for its operation. Assumptions are made about this system and algorithms and relationships are derived to describe these assumptions"* [1].

© Springer International Publishing AG, part of Springer Nature 2019
D. N. Cassenti (Ed.): AHFE 2018, AISC 780, pp. 26–33, 2019.
https://doi.org/10.1007/978-3-319-94223-0_3

Because most environments and, thus, situations in the real world are far too complex for a comprehensive understanding and description, a simulation is always limited to relevant aspects. Therefore, a simulation is not as complete and comprehensive as reality.

The same is true for DHMs. They are a part of the simulation and refer to relevant aspects of human characteristics and capabilities. DHMs are closely linked to the application: In case of spatial layout, they refer to human body shape and dimensions; in case of performance, they refer to human performance, measured e.g. in response times, and error.

Consequently, the application area of DHMs is as diverse as the application area for modeling and simulation itself. It ranges from designing new products and workplaces to computer-generated entities in training environments, movies or games. Different DHMs can be used in different phases throughout a total lifecycle of a new system. Not all of them are addressed by a single DHM. Instead, multiple DHMs are used for their special target application.

2   Spatial and Physical Design

Ergonomic design of new products and production systems requires thorough consideration of human anthropometric and biomechanical characteristics. Standardized data collection and presentation provided a valuable base for a systematic design, which was soon followed by the introduction of first anthropometric drawing templates [2]. These templates were useful for technical drawings. With the increase of computer-aided design (CAD), the first ergonomic digital human modeling (DHM) systems were developed and applied shortly afterwards. Modern DHMs make a spatial and physical design for a broad population of future users possible.

2.1   Anthropometric DHMs

Anthropometric DHMs offer modeling functions for considering human body dimensions, movements, and maximum strength in designing new products, systems or personal equipment [3, 4]. In addition, they provide a broad database of body dimensions for different nations, gender and age groups. Moreover, they provide statistics for modeling inter- and intraindividual variance. Several DHMs have been integrated into the industrial system lifecycle or even become part of CAD-packages. Among others, the most common models are RAMSIS, JACK and Virtual Human [5–7]. Most of the anthropometric models provide a photorealistic appearance and offer functionality for product and workplace design.

2.2   Biomechanic DHMs

Biomechanical models address human movement and motion. Usually, they are based on the laws of technical mechanics and apply them to the human body. A simple biomechanical model understands the human body as a complex mechanical system (including simple joints of different degrees-of-freedom with static connections between them). Applying mathematic cost functions, including e.g. forces, torques, and load, facilitates a subsequent movement simulation in the simulated environment [8–11]. Biomechanical models either simulation self-induced motion or externally induced motion, e.g. for virtual crash tests.

The original focus of biomechanics was to calculate maximum loads of postures for workplace design, to optimize motion for better performance and to reduce occupational workload during energetic work or to reduce injuries for driver's safety. Today's complex models, e.g. AnyBody, have been introduced as biokinematic models and allow insights into internal properties during movements.

3 Virtual Environments

Virtual humans for virtual simulation, movies and games are another important application area for DHMs. Most of these DHMs were implemented with a background in computer graphics and animation. The main foci of these models are photorealism and subjectively perceived realism. They look and appear very realistic. It is often not possible to differentiate between computer-generated and real images.

A simulation of realistic behavior is not trivial and requires a lot of off-line, manual programming. Motion-capture of real actors is often used for this. In this case, an actor controls the virtual entity as a representative.

Realistically appearing motion is critical because observers and users are very sensitive to inaccuracies or differences from their anticipated actions. This effect has been called the "Uncanny Valley" [12]. This is because our lifelong experience allows us to notice even small inconsistencies instantly. Furthermore, we often use motion for inferring emotional states, intentions, and goals of the acting entity. Accordingly, slight inaccuracies in motion modelling might therefore easily lead to incorrect inferences about its future actions and goals.

4 Human Performance Models

Performance models are used for resource planning and optimization. Human performance is not necessarily included in process models. Human performance models model human behavior outputs, such as error or time to finish a task. They usually consist of special modules for perception, attention, cognition, memory and motor reaction. By combining them, results about complete human-machine-system's reliability or performance can be estimated. Performance models are widely used to optimize complex processes.

Human performance models refer to goal selection and goal generation and, thus, model behavior at a higher level. With sufficient information and data, it is possible to use them for modeling human decision making [13].

A functional task analysis of a process includes a description of all subprocesses and the interrelations between them. Visualization tools focus on a visual analysis – usually in temporal order. Functional analysis goes one step further and allows for a statistical or mathematical analysis regarding workload, time effort, or other variables. Usually, modeling frameworks build the basis for function and task allocation.

An example for discrete event simulations that can be used for human performance modeling is Micro Saint Sharp. It is applied for modeling and simulation of human performance and processes in the health sector [14]. Another example is *Integrated Performance Modeling Environment* (IPME), which has been specially designed to analyze human performance. This model is based on the Micro Saint Sharp simulation engine. It provides a discrete event simulation environment and a realistic representation of humans in complex environments [15].

MIDAS (*Man-Machine Integration Design and Analysis Systems*) is an integrated suite of software components developed to aid in the design of complex human-machine systems. The modeling goal was to develop an engineering environment, which contains tools and models to assist in the conceptual phase of crew station development, and to anticipate training requirements [16].

The "Editor for Manual Work Activities" (ema) is another example for independent software used to simulate industrial work processes in the digital factory [17]. The goal of ema is to simulate activities of the human worker. This allows a more holistic planning method, which carries out self-initiative work instructions. A library of typical activities included in the software makes it possible to compile a parametrized description of an activity by specifying the general framework (e.g. objects to handle or target positions). The description of an activity is initially analyzed, checked for plausibility and finally geometrically simulated. The results show the necessary postures for the digital human model and further parameters for indicative time analyses and ergonomic analyses [18].

5 Human Information Processing and Cognition

Cognition is a complex domain and therefore places special demands on computational modeling. Most cognitive models require a formal analysis. They are based on a syntactic and semantic rule system [19–21]. Recently, probabilistic models have gained importance. They serve as a software tool for solving problems and are based on fundamental knowledge from psychological research. Cognitive models often simulate probabilities of special states and transitions between the states. For instance, working on different tasks sequentially defines a unique order of the tasks. Subjects might choose different orders based on their working strategy. The probabilities of the transitions from one stage to another can be used to model these strategies expressed in Bayesian terms or Hidden Markov Models.

One of these models is the Atomic Components of Thought – Rational or ACT-R [22, 23]. It serves as a framework for modeling different tasks in a special programming language. This specific model can rely on general assumptions, which are provided by ACT-R or can be specified by the author. ACT-R has been used successfully for producing user models in many applications. To some extent, an application of such a model for goal generation and goal selection might be possible. In this case, such a model could be applied to control the goal of a motion of a low-level movement model.

6   A Modeling Structure of Intelligent Appearing Behaviors

There are several different ways to model a large variety of possible behaviors. A simple, first approach is to differentiate between a pure physical motion on the one hand and intelligently controlled motion and, thus, behaviors on the other. Physical motion refers to simple movements of dumb, passive objects and low-level movements of animated, active objects. Intelligent motion focuses on the control behind the movement.

6.1   A Procedural Approach

A hierarchical, procedural approach serves as the basis for a first understanding. This approach is built on the geometry of static objects. They provide information about their spatial positions, their type and other properties. Dynamics is then introduced on the next higher level. It refers to changes of position during time. Both levels refer to the visible output. The next level, i.e. the cybernetic level, introduces a more complex behavior and adds motion control. The approach continues and adds higher levels of behavior. This includes goal selection (teleotic level) and goal synthesis (geneotic level) [24].

According to this approach, behavior is considered as a subset of subordinate levels, including goal generation, goal selection, comparison of goal and state, and single movements itself. The approach enables detailed modeling and simulation for a realistic appearing visual output. A model based on this approach is highly adaptable and would not depend on a specific environment. Current motion models have not been able to address this whole spectrum. They are specialized either on modeling lower, i.e. movements, or higher, i.e. behavioral, level motion. Most of them are limited to special applications and require major modifications for others.

Nonetheless, the general shortcoming of this procedural approach is that its base is purely mechanic/cybernetic. It does not consider the perception of motion and considers behaviors as a set of separate items. Simulating relationships like causality gets complicated. For simulating social relations or team performance it gets even more complicated and complex.

6.2   An Alternative Approach Introducing Rules and Relationships

Relationships and links between different motions are essential for motion perception and understanding of complex intelligent behavior. Physical motion is just one aspect of behaviors, but causal and intentional movements create the perception and illusion of intelligent behavior. Therefore, the original procedural approach may take explicitly the links and relationships between events and motions into account.

One possibility is to relate it to linguistics and semiotics. In this case, a behavior is understood as a media for communication. Such a compositional approach focuses on the compositional character of motion and its role in communication. In this approach, motion is referred to as a concept of hierarchically constituted entities and patterns, like words and sentences in language. Motion patterns are stored and simply retrieved when needed following syntactic rules (Fig. 1).

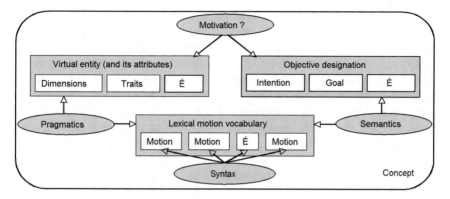

Fig. 1. A compositional, linguistic alternative for structuring human (goal-directed) behaviors

The lowest level refers to basic movements and their relation to static and geometric attributes and properties of the entity or the DHM. Kinematics are either calculated by biomechanical functions or retrieved from a databank. The lexical level also describes the link and connections between the different entities. This level considers movements and simple behaviors as a baseline vocabulary, similar to words.

On the syntactic level, the different single behaviors are combined for more complex behaviors. It defines the relationship between the isolated behaviors and is based on traits of the acting virtual entity. Several single units are connected to form a more complex behavior. Transitions between simple behaviors have to be determined to assure continuity.

At the semantic level, which relates to the designation, behavioral pattern are correlated to the meaning or overall goal of the motion. The relation between them and the goal of the action are specified. By this, an observer is able to infer intention and understand even more complex situations.

The highest level is called the conceptual level. It addresses constructs like intelligence and autonomy. This level provides functions for goal selection and generation. The conceptual level therefore models behavior on a higher level. It relates to external factors, like environmental stimuli, as well as internal factors, like motivation and traits of the acting entity.

This approach for structuring motion is based more on the relationship between elements of motion and the overall communicative purpose of the system. It considers motion as a further media to enhance communication between the system and the user.

7 Conclusions and Outlook

There are many different human models available, which offer different functionalities for different applications. Several of them are already used daily. They have been found to be beneficial for the design and optimization of new workplaces, products and processes. But most of them address a single aspect of a human entity only. Therefore, they are not sufficient for a comprehensive overview. It was also discovered that a single, comprehensive model is not (and will not be) available [25]. Most DHMs are not connected to each other and there are very few interfaces between them today. This leads to a time- and labor-intensive restart of the design work for each model. Instead, it would make sense to define standards for interfaces between the different models.

These standards should address the combination of different human behavior models during the system design lifecycle. A process model can be used to model the operational background and to perform a first analysis of functionality gaps. Detailed models, e.g. anthropometric models, can be applied in order to test the geometric design. In parallel, cognitive models can be used to analyze the information presentation und information interaction. Finally, virtual simulation models can be applied to carry out detailed analyses and evaluate the final design. They can also be used for education and training.

Such an interoperability of different DHMs requires the definition of interfaces. However, most of the DHMs are too different from each other so that a pure definition of interfaces on a technical level is no sufficient solution. Instead, a reference architecture for a consequent integration of different DHMs into a common simulation framework is required. First steps into this direction are currently investigated [25].

References

1. Reilly, E.D.: Simulation. In: Reilly, E.D. (ed.) Concise Encyclopedia of Computer, pp. 690–695. Whiley, West Sussex (2004)
2. Alexander, T.: Methoden der anthropometrischen Cockpitgestaltung. In: Gärtner (ed.) Anthropometrische Cockpitgestaltung. DGLR, Bonn (1995)
3. Chaffin, D.B.: Human motion simulation for vehicle and workplace design. Hum. Factors Ergon. Manuf. **17**(5), 475–484 (2007)
4. Muehlstedt, J., Kaussler, H., Spanner-Ulmer, B.: The software incarnate: digital human models for CAx and PLM systems. Zeitschrift fuer Arbeitswissenschaft **62**(2), 79–86 (2008)
5. Human Solutions: RAMSIS NextGen 1.1, Ergonomics User Guide. Kaiserslautern (2015)
6. Siemens: Jack product information. https://www.plm.automation.siemens.com/en_us/products/tecnomatix/manufacturing-simulation/human-ergonomics/jack.shtml
7. Dassault: V5 DMU for Human Simulation product information (2018). https://www.3ds.com/products-services/enovia/mid-market/v5-digital-mockup/product-design/human-movement/
8. Andersson, A., Nordgren, B., Hall, J.: Measurement of movements during highly repetitive industrial work. Appl. Ergon. **27**(5), 343–344 (1996)
9. Cerveri, P., Pedotti, A., Ferrigno, G.: Evolutionary optimization for robust hierarchical computation of the rotation centres of kinematic chains from reduced ranges of motion the lower spine case. J. Biomech. **37**(12), 1881–1890 (2004)
10. Rohmert, W., Laurig, W., Philipp, U., Luczak, H.: Heart rate variability and work load measurement. Ergonomics **16**(1), 33–44 (1973)
11. Winter, D.A. (ed.): Biomechanics of Human Movement. Wiley, New York, Chichester, Brisbane, Toronto (1997)
12. Mori, M., MacDorman, K.F., Kageki, N.: The uncanny valley. IEEE Rob. Autom. Mag. **19**, 98–100 (2012)
13. Wickens, C.D.: Multiple resources and performance prediction. Theor. Issues Ergon. Sci. **3**(2), 159–177 (2002)

14. Swain, J.J.: Simulation software survey. OR/MS Today, vol. 38, no. 5. Informs, Baltimore (2011)
15. Archer, S., Headley, D., Allender, L.: Manpower, personnel, and training integration methods and tools. In: Booher, H.R. (Hg.) Handbook of Human Systems Integration. Wiley-Interscience (Wiley Series in Systems Engineering and Management), Hoboken (2003)
16. Gore, B.F.: Man-machine integration design and analysis system (MIDAS) v5: augmentations, motivations, and directions for aeronautics applications. Technical report, NASA Ames Research Center, Moffett Fileds, CA (2011)
17. Leidholdt, W.: Der "Editor menschlicher Arbeit - EMA" - ein Planungsinstrument für manuelle Arbeit: 2. Symposium Produktionstechnik innovativ und interdisziplinär - im Fokus des Automobil- und Maschinenbaus. Zwickau (2007)
18. Leschner, K.: Benutzerhandbuch ema V5: Version 1.4.1.0: imk automotive GmbH (2014)
19. Chomsky, N.: Syntactic Structures. De Gruyter, Berlin (2002)
20. Johnson-Laird, P.N.: Mental models in cognitive science. Cogn. Sci. 4(1), 71–115 (1980)
21. Samuelson, W., Zeckhauser, R.: Status quo bias in decision making. J. Risk Uncertain. 1(1), 7–59 (2004)
22. Anderson, J.R., Matessa, M., Lebiere, C.: ACT-R: a theory of higher level cognition and its relation to visual attention. Hum.-Comput. Interact. 12, 439–462 (1997)
23. Bothell, D.: ACT-R 7 Reference Manual (2015). http://act-r.psy.cmu.edu/wordpress/wp-content/themes/ACT-R/actr7/reference-manual.pdf
24. Ellis, S.R.: Preface, Conference Proceedings of the Symposium on Intelligent Motion and Interaction in Virtual Environments. NASA CP, Moffett Fields (2005)
25. Gunzelmann, G., Gaughan, C., Huiskamp, W., van den Bosch, K., de Jong, S., Alexander, T., Bruzzone, A.G., Tremori, A.: In search of interoperability standards for human behaviour representation. In: Proceedings of the I/ITSEC, Orlando, FL, 1–4 December 2014. National Training and Simulation Association, Arlington (2014)

Determining the Effect of Object-Based Foveated Rendering on the Quality of Simulated Reality

Varun Aggarwal$^{(\boxtimes)}$, Denise Nicholson$^{(\boxtimes)}$, and Kathleen Bartlett$^{(\boxtimes)}$

Soar Technology, Inc., Orlando, FL, USA
varunneilaggarwal@gmail.com, {Denise.Nicholson,
Kathleen.Bartlett}@soartech.com

Abstract. Rendering virtual environments for simulated reality applications often proves to be a challenge due to the high demands of simulations capable of imparting enough detail to appease the human eye. Traditional simulated environments typically require hardware capable of providing enormous graphical processing power in order to render entire scenes in high detail, limiting simulations to high end computers. However, most of this detail is wasted, as human eyes are only capable of perceiving detailed information in a small central field of view. In the peripheral regions of human vision, the majority of the final image that is seen by the human is filled in by the brain, based on context and the minimal detail provided by the eyes. This three-phase research effort attempts to identify whether reducing the level of detail of objects in the peripherals of a virtual reality simulation affects the perceived quality of the simulation.

Keywords: Virtual environments · Latency · Three dimensional models

1 Introduction

The goal of this research is to reduce the amount of processing power required to render a virtual environment. This discussion summarizes phase I efforts and explains the phase II approach of a three-year plan to design a method of rendering virtual environments with increased efficiency, by reducing the detail of objects in the peripherals. Research has been conducted on decreasing the resolution of the parts of the environment seen only by peripheral vision, but a method of foveated rendering based on the level of detail of individual objects, rather than the scene as a whole, does not exist. The underlying goal of this project is to determine how the presence of less detailed objects in the peripheral vision affects the quality of a simulation. In the first phase of the plan, the researchers determined the point at which six specific objects have the minimum possible detail and are still recognizable on a two-dimensional screen [1]. In the second phase, a simulation will be created where the objects in the peripherals are brought down to this point, in order to test whether the decreased detail is noticeable to the user. If the decreased detail has no effect on the quality of the simulation, then the new method of rendering can be applied to any virtual

© Springer International Publishing AG, part of Springer Nature 2019
D. N. Cassenti (Ed.): AHFE 2018, AISC 780, pp. 34–44, 2019.
https://doi.org/10.1007/978-3-319-94223-0_4

environment, in order to increase efficiency and allow simulations to be run using less-powerful hardware.

1.1 History of Simulated Reality

The idea of creating simulated environments with the goal of mimicking real ones has been around for hundreds, if not thousands, of years. The earliest examples of simulated reality date back to the early 19th century, when artists began to paint 360-degree murals on the walls of circular rooms. The goal of these panoramic paintings was to make a viewer standing inside the room feel as though they were actually inside the painting. These paintings were a poor substitute for reality, but they paved the way for future advancements in simulated reality.

In 1838, researcher Charles Wheatstone demonstrated a simple but incredibly important concept. He proved that showing the right and left eyes a slightly different picture, adjusted for their relative position to the head, created an image that looked three-dimensional. He went on to create the stereoscope, a device that used this principle to allow people to view pictures as three-dimensional scenes in front of them [2].

Over 100 years later, the stagnating field of simulated reality was brought back from the brink of collapse by a few major developments. In 1960, the very first simulated reality head-mounted display (HMD) was created, named the Telesphere Mask by its inventor, Morton Heilig. In 1961, the Philco Corporation developed the first HMD with head tracking, which allowed operators to manipulate cameras from afar, giving a sense of presence in areas that were too dangerous to send humans. In 1968, Ivan Sutherland's Sword of Damocles connected an HMD to a computer for the very first time, providing a basic virtual environment that could show the user any scene that the computer was capable of processing. In 1987, the term virtual reality was coined by Jaron Lanier, founder of the Visual Programming Lab, which sold the very first consumer VR HMD, the EyePhone 1. The EyePhone 1 cost over $9,000 and required a further $9,000 in peripherals, dooming it to failure before it ever hit the market. However, the system succeeded in generating public interest in simulated reality. In 1995, Nintendo released the Virtual Boy, the first VR video game console. A year later, the Virtual Boy was discontinued due to poor sales, likely a result of the uncomfortable headset design and lack of color graphics [2].

After the failure of the virtual boy, very little major progress was made in consumer VR, until 2012, when Oculus VR began developing the Oculus Rift. In 2013, Oculus released their first developer kit, the DK1, selling over 50,000 headsets despite being an unfinished prototype. In 2014, Oculus released the DK2, which improved the DK1's resolution and positional tracking technology. The DK2 was a huge success, with over 100,000 headsets sold [3]. On November 27, 2015, the Gear VR mobile VR headset was released by Oculus, becoming the first successful consumer VR headset. In 2016, the Oculus Rift, HTC Vive, Playstation VR, and Daydream View were released, and in 2016, sales for these headsets reached a combined total of over 6 million units, proving the consumer viability of VR [4].

1.2   Current State of Simulated Reality

In the field of simulated reality, two major markets drive the industry. The first is to provide education and entertainment to the public. With VR, people can freely travel around the world and even back in time, experiencing the world in a way that would not be possible otherwise. Students can attend classes on other continents, or even bring the top professors in the world into their own homes or classrooms. Another growing market is in the area of providing effective and realistic training programs for doctors, soldiers, and other professionals. In a virtual environment, a doctor can practice performing difficult operations without risking the safety of a patient, while a soldier can gain the combat experience necessary to stay alive in a fight [5].

In order to accomplish both of these goals, simulated environments need to be as realistic as possible. However, as these environments, and the digital models they are composed of, become more realistic, they also become increasingly difficult to render. The increased level of detail requires larger files with more data to be processed. This, in turn, requires more powerful hardware to run the necessary calculations. The level of graphical processing capability required to operate a VR headset that meets the current consumer standard is far beyond that of any computer that is readily available to the public. The monetary resources required to build or purchase such a machine makes widespread, consumer-ready VR impossible at the moment. In time, this problem will become even more pronounced, as advancements in the technological capability of VR headsets will require even more detailed simulated environments [6].

Due to demand for realism and high-fidelity experiences by today's users, rendering virtual environments within acceptable latency limits proves to be a computational challenge for lightweight computing platforms (e.g., mobile devices). Traditional simulated environments typically use all the processing power available to render the entire scene in high detail, limiting simulations to higher-end computers. One approach to optimize processing power and reduce latency for three-dimensional models is to use varying, decreased, level of detail (LOD) for distant representations [7].

This research attempts to optimize future resources by expanding the adaptive LOD approach based on the object's location in the field of view (FOV) in addition to the object's distance. Such FOV adaptation would take advantage of state-of-the-art head and gaze tracking capabilities. With the development of new VR headsets that are capable of tracking eye movements, it is now feasible for rendering engines to use a process called foveated rendering, where objects outside of the fovea are not rendered at full detail [8].

2   Purpose

2.1   Long-Term Goal

The overarching goal of this three-year project is to determine how the presence of less detailed objects outside of the user's central gaze affects the quality and perceived realism of a simulation. The research done during the first phase of the project helped to identify the point at which specific objects have the minimum possible detail and are still recognizable [1]. If the decreased detail has no effect on the quality of a simulation, then the new method of rendering can be applied to any virtual environment, in order to increase efficiency and allow simulations to be run using less-powerful hardware. The primary impact of this development would be a reduction of financial barriers to consumer VR technology, expanding the user base of VR beyond enthusiasts to average consumers.

2.2   Short-Term Engineering Goals Achieved in Phase 1

The initial engineering goal of this project was to create three-dimensional models of six specific objects at the minimum level of detail required for them to be recognized.

- The experimental models had a maximum of 80% of the detail of the initial models.
- There were no statistically significant differences between the time and accuracy of recognition of the experimental models and the initial models.

A simulation was designed that presented randomized sets of objects of various LOD. Subjects were asked to choose an object from the group based on an on-screen prompt. The speed and accuracy of each subject's response was recorded to determine the LOD at which there was no difference in recognition from the full-detail objects.

2.3   Object Collection

All of the objects used in phase I of this research were taken from the Unity 3D Asset Store (see Fig. 1). Using the "Free" and "Most Popular" filters, the most commonly used objects that were available for public use were identified, and the top fifty distinct types of objects were recorded and downloaded. In order to prevent duplicates that would confuse the participants and interfere with the results, objects of the same type,

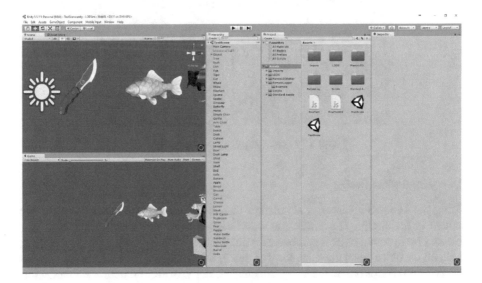

Fig. 1. Creation of a simulation in unity

such as two different variations of a "car" object, were removed, and replaced with objects further down the list. After the identification of the top fifty objects, six were chosen as experimental models, with the remaining forty-four serving as background objects. These six objects, which included an apple, desk, street light, tiger, tree, and vase, were chosen based on three factors. They each represented a unique shape with varying feature complexity, a unique category of object, and were the most used objects in their categories.

After all of the objects had been chosen and sorted into experimental and background categories, the researchers took the objects composed of multiple meshes and replaced them with similar objects composed of a single mesh, in order to reduce the complexity of the objects and ensure accurate results. The final fifty objects were then loaded into the Unity 3D Engine, for use in simulations. Each object was run through the Mantis Integrated Unity Level of Detail (LOD) tool five times, a software tool that reduces detail by collapsing polygons in an approximately linear scale, to create six different copies of each object. As shown below in Fig. 2 for a Fish, the copies each had a different level of detail, from zero percent detail (LOD Group 0) to one hundred percent detail (LOD Group 5), in increments of twenty percent. These three hundred objects were sorted into six different groups based on their level of detail, so that group zero had fifty objects with zero percent detail, and group five, the control group, had fifty objects with one hundred percent detail.

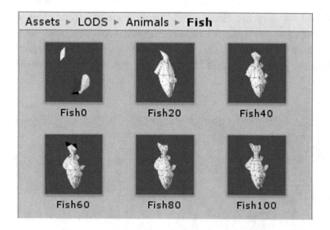

Fig. 2. Levels of detail for "fish" object

2.4 Simulation Design

Using the Unity 3D Editor, a simulation that displayed random sets of objects was written in JavaScript. The simulation was designed to display a set of nine randomly selected background objects and one experimental object in a 5 × 2 grid. At any given time, all ten objects had the same level of detail, with each level of detail being used six times, in a random order. Each experimental object was also be used six times, in an order determined by the level of detail, so that each level of detail for each object was be displayed exactly once, for a total of thirty-six sets of objects. The simulation included a prompt at the top of the screen, identifying the experimental model by name and asking users to click on it.

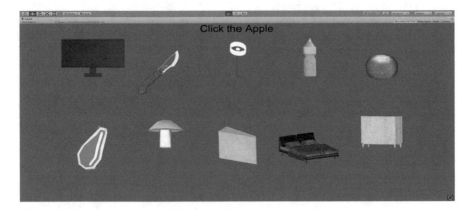

Fig. 3. Final simulation

The amount of time elapsed before the participant clicked and the accuracy of the selection were written to the debug log, along with the name and level of detail of the experimental object. Using preexisting open source code for c# and Google Scripts, a logging system that took all entries from the debug log and wrote them to a google sheets file was created. After the completion of the simulation (Fig. 3), it was compiled as a web application using Unity's built-in compiler and uploaded to GitHub, where it was hosted on GitHub Pages as a web applet accessible to anyone with the link.

2.5   Data Collection and Analysis

In order to collect the large amounts of data required for statistically significant results, a Human Intelligence Task (HIT) was posted on Amazon Mechanical Turk, a platform for crowdsourcing data, asking workers to fill out a demographics questionnaire and give consent before running the simulation on a personal computer. In addition to these workers, emails containing the same set of instructions were sent out to some of the researcher's contacts, who had expressed interest in taking part. After data was collected from one hundred different participants, a single spreadsheet containing each data point was compiled. Due to the low upload speeds experienced by some participants, a portion of the data was not uploaded correctly and had to be deleted. After ensuring that all remaining data was valid, the raw data from the spreadsheet was compiled and entered into the Minitab statistical analysis software. A binary logistic regression comparing each data set to the control was run in order to determine the minimum level of detail required for recognition accuracy on par with that of the control, for each of the six experimental models. Additionally, an ANOVA test with a Tukey pairwise comparison was run to determine the minimum level of detail required for recognition speed on par with that of the control, for each of the six models (Fig. 4).

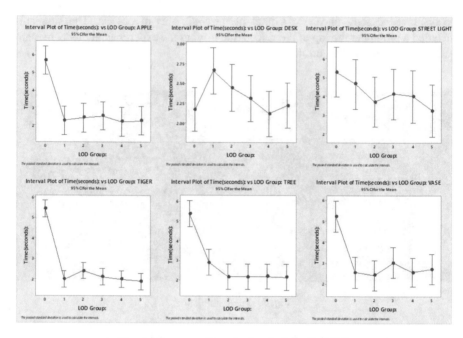

Fig. 4. Response times by level of detail (LOD)

As shown below in Table 1, the speeds of response of LOD Groups 1 and higher all had P-value >/=0.05 to relation to the speed of response of the control, LOD Group 5. In addition, the Desk and Street Light objects response times were similar even when reduced to minimum detail in Group 0.

Table 1. Tukey test P-value results for speed of response compared to control

Object	All	LOD 0 vs. LOD 5	LOD 1 vs. LOD 5	LOD 2 vs. LOD 5	LOD 3 vs. LOD 5	LOD 4 vs. LOD 5
Apple	0.000	0.000	1.000	1.000	0.998	1.000
Desk	0.098	1.000	0.279	0.891	0.998	0.997
Street Light	0.341	0.271	0.679	0.996	0.943	0.972
Tiger	0.000	0.000	0.999	0.449	0.969	0.999
Tree	0.000	0.000	0.618	1.000	1.000	1.000
Vase	0.000	0.000	1.000	0.993	0.992	1.000

Table 2 shows the accuracy of recognition of each group, and compares each to the accuracy with which participants recognized the objects at 100% detail in LOD Group 5. The cells highlighted in yellow represent an accuracy of above 90% of the accuracy of Group 5, while the cells highlighted in green represent accuracies above 95%.

Table 2. Accuracy of recognition by LOD regression prediction

Object	LOD 0	LOD 1	LOD 2	LOD 3	LOD 4	LOD 5
Apple	0.428	0.670	0.847	0.938	0.976	0.991
Desk	0.806	0.855	0.893	0.923	0.944	0.960
Street Light	0.409	0.544	0.673	0.780	0.859	0.913
Tiger	0.339	0.762	0.952	0.992	0.998	0.999
Tree	0.624	0.778	0.881	0.940	0.971	0.986
Vase	0.606	0.751	0.855	0.920	0.958	0.978

= Predicted accuracy is 90% of Predicted accuracy at LOD 5

= Predicted accuracy is 95% of Predicted accuracy at LOD 5

The results of the phase I study showed that the minimum level of detail (LOD) required for speed and accuracy was identical to that of the control, assuming a required match rate of 90%, as follows:

- Apple: 60% (LOD Group 3)
- Desk: 40% (LOD Group 2)
- Street Light: 80% (LOD Group 4)
- Tiger: 40% (LOD Group 2)
- Tree: 60% (LOD Group 3)
- Vase: 60% (LOD Group 3)

When the required match rate was increased to 95%, the following levels of detail were required:

- Apple: 80% (LOD Group 4)
- Desk: 60% (LOD Group 3)
- Street Light: 100% (LOD Group 5)
- Tiger: 40% (LOD Group 2)
- Tree: 60% (LOD Group 3)
- Vase: 80% (LOD Group 4)

3 Phase I Conclusions

The results of the Phase I study demonstrated that it was possible to significantly reduce the level of detail required to recognize virtual objects, with polygon reductions of up to 60% in some cases. This means that object-based foveated reality has the potential to reduce the processing cost of virtual environments by a similar margin, if future studies verify similar recognition results when the object is outside of the central fovea gaze. Using the data gathered in this research, the overall hypothesis of this project can be tested, showing the potential for object-based foveated reality to reduce the hardware requirements of virtual environments and lower the cost barrier to consumer virtual reality.

4 Phase II Study Plan

The next phase of this research will build on the phase I findings, but will use a simulated visual task, observation of a blue cube, as the main visual focus, while providing objects in the simulation's background for which the level of detail has been reduced. This investigation will help identify the accuracy of peripheral, not-attentional vision recognition of objects with reduced detail.

4.1 Materials/Procedure/Simulation Design

Using the Unity3D engine, a virtual reality simulation featuring a room with three tables, arranged in a U shape, will be designed for the HTC Vive virtual reality headset. Three of the objects found in the previous study will be placed in a random location on top of each of the two parallel tables. A small blue cube of side length one inch will be placed in the middle of the center table.

After activating the simulation, the participant will have one minute to walk around freely, examining the room and growing accustomed to it. During this time, all six objects on the two side tables will be rendered in full detail.

After one minute is up, the participant will be moved to the center of the room, between the two side tables, facing the middle table. At this point, the participant will be prompted to look at the blue cube, and determine how many times it changes color.

The cube will randomly change color between dark blue and black, forcing the participant to focus on the cube with their foveal vision.

Five seconds after the cube begins to change color, the six objects on the side tables will change. If the participant is a member of the control group the objects will continue to be rendered in full detail, but if the participant is part of the experimental group, each object will be reduced down to its individual minimum level of detail.

4.2 Participant Gathering

The number of participants necessary will be determined by conducting a pilot study using the researchers who were not involved in creating the simulation and determining the expected variation in the data. Participants will be recruited from graduate student classes at the University of Central Florida. Each participant will fill out a questionnaire about the quality of their eyesight. If a participant's vision is not 20/20 and the participant does not use a vision correction tool with the correct prescription, a new participant will take their place.

4.3   Testing

Once the number of participants is determined, half of the participants will be placed in the control group, and half will go to the experimental group. Before entering the simulation, participants will be told that their objective is to observe color changes in the cube, and that the six experimental objects are simply background clutter. Participants will not be told anything about foveated rendering or reduced levels of detail.

Although the color changing cube is simply a device designed to grab attention, the number of times it changes color will be logged for each participant, in order to ensure that this does not influence the results.

After each participant has exited the simulation, they will be asked to fill out a questionnaire, requiring them to rate the comfort and realism of the simulation on a scale from 1 to 10, with 1 being completely uncomfortable and unrealistic, and 10 being completely comfortable and realistic.

The scores for both comfort and realism will be recorded for each participant, and the data from the experimental group will be compared to that of the control to determine whether there is any statistically significant difference between them.

4.4   Risk/Safety

The safety risks of virtual reality are minimal. In the event that someone in a simulation experiences any negative symptoms, such as eye strain or motion sickness, they are immediately taken out of the simulation and given time to recover.

4.5   Data Analysis

All data will be recorded both electronically and physically in data tables, and all final results will be organized into bar graphs to compare the experimental group and the control. A chi square analysis will be used on all data to ensure that it is significant and not due to random chance.

5   Possible Conclusions

This work, if successful, will help simulated reality programs run on computers that would otherwise have been incapable of handling them. This could allow virtual reality to spread to consumers at a reduced cost, by removing the need for an expensive machine, which is the largest financial burden of consumer virtual reality at the moment.

6 Future Application

The next step in this research beyond phase II involves the creation of a rendering tool that integrates object-based foveated rendering techniques natively into the rendering of simulated environments. An application add-on will be created for Unreal Engine, a leading simulation design tool, that allows developers to add foveated models for objects in their simulations, to be used in the peripherals. In phase III, the rendering engine will be tested using various methods of changing the models in order to determine the most efficient method of rendering peripheral objects at reduced detail while ensuring that foveal objects are not compromised in the event of sudden movement or latency. The software will be released to the public with instructions for developers interested in using foveated rendering for their simulations and resource guides for those wishing to recreate or expand upon the research presented in the first two phases of this study.

References

1. Aggarwal, V., Nicholson, D.: Optimization of computer generated three dimensional models for decreased latency in virtual environments. In: Interservice/Industry Training, Simulation and Education Conference (I/ITSEC), Orlando, FL (2017)
2. History of Virtual Reality: Virtual Reality Society, 10 January 2015. https://www.vrs.org.uk/virtual-reality/history.html. Accessed 15 Feb 2017
3. Hayden, S.: Oculus reveals more than 175,000 rift development kits sold. Road to VR, 12 June 2015. https://www.roadtovr.com/oculus-reveals-175000-rift-development-kits-sold/
4. Durbin, J.: VR Sales Analysis: Rift and Vive to Sell Under 500,000 in 2016, PS VR to Top 2 Million. UploadVR, 14 November 2016. https://uploadvr.com/superdata-headset-sales-analysis/. Accessed 15 Feb 2017
5. Nicholson, D., Schmorrow, D., Cohn, J.: Virtual environments for training and education: VE components and training technologies. Praeger Security International, Westport, CT (2009)
6. Cohn, J., Nicholson, D., Schmorrow, D.: Virtual environments for training and education: integrated systems, training evaluations, and future directions. Praeger Security International, Westport, CT (2009)
7. Sik, L.L., Pattanaik, S.: Rendering large scale synthetic urban terrain. In: Interservice/Industry Training, Simulation and Education Conference (I/ITSEC), Orlando, FL (2011)
8. Perceptually-Based Foveated Virtual Reality. Nvidia Corporation. Accessed 6 Feb 2017

A Test Protocol for Advancing Behavioral Modeling and Simulation in the Army Soldier Systems Engineering Architecture

Joan H. Johnston[1]([✉]), Samantha Napier[2], Clay Burford[1],
Shanell Henry[2], Bill Ross[3], and Colleen Patton[4]

[1] United States Army Research Laboratory Human Research and Engineering
Directorate (ARL HRED), Orlando, FL, USA
{joan.h.johnston.civ, clayton.w.burford.civ}@mail.mil
[2] ARL HRED, Aberdeen Proving Ground, MD, USA
{Samantha.j.napier.civ, shanell.l.henry.civ}@mail.mil
[3] Cognitive Performance Group, Orlando, FL, USA
bill@cognitiveperformancegroup.com
[4] University of Maryland, College Park, MD, USA
patton.colleen.e@gmail.com

Abstract. When developing military capabilities, it is essential to determine the effect that equipment, tasks, and training will have on Soldier and unit readiness. Human factors processes require technology enablers that streamline the systems engineering design process. Toward this end, a collaboration between the U.S. Army Science and Technology Objectives for the Soldier Systems Engineering Architecture (SSEA) and Training Effectiveness for Simulations (TEfS) was established in 2015. An operational SSEA is envisioned to enable analysts to implement systems engineering procedures to achieve more effective results in predicting best system designs. Best practices from TEfS is critical to improving the predictive analyses. In this paper we describe best practices from a squad training use case and propose how the training method could be applied as a test protocol in SSEA SaaS modeling.

Keywords: Behavioral modeling · Systems Engineering Architecture
Team performance · Training effectiveness

1 Introduction

In 2016 the U.S. Army Training and Doctrine Command-Army Capabilities Integration Center identified the Big 6+1 key capabilities and systems that will allow the Army to close capability gaps in the context of the Army Operating Concept; they also provide a framework to focus future force development and prioritize research, development and acquisition activities [1]. The Big 6 are: future vertical lift; combat vehicles; cross-domain fires; advanced protection; expeditionary mission command/cyber-electromagnetic; and robotics and autonomous systems. The Plus 1 cross-cutting capability is Soldier and team performance and overmatch which emphasizes human-centered design of weapons and equipment and enables the Army to make investments

© Springer International Publishing AG, part of Springer Nature (outside the USA) 2019
D. N. Cassenti (Ed.): AHFE 2018, AISC 780, pp. 45–55, 2019.
https://doi.org/10.1007/978-3-319-94223-0_5

in research on cognitive, social and physical development, as well as making training and leader-development activities a priority. Consequently, in 2017, the US Army Warfighting Challenges (WC) raised a number of "Plus 1" human-centered questions in response to the Big 6. For example, WC#11 asks: how can remote sensor and robotic technology be developed and integrated to support combined arms reconnaissance, and WC#13 asks: how will advanced manned and unmanned air/ground capabilities be integrated to facilitate collection, development, and dissemination of actionable combat information for rapid targeting across domains? [2].

These hard questions indicate human-centered engineering will need to be involved early in the acquisition process for the Big 6, such as modeling and tradeoff analyses, to make better informed decisions. Critical to combat effectiveness is developing an understanding of Soldier performance in the context of the operational environment and the Soldier as a System (SaaS) (e.g., crew, squad, and platoon) so that the potential tradeoffs can be accurately modeled and assessed [3]. This sentiment has been voiced by many subject matter experts on human factors engineering in complex socio-technical systems [4, 5].

In anticipation of this need for SaaS behavioral modeling and prediction technologies the Army Research Laboratory established the Soldier Systems Engineering Architecture (SSEA) Science and Technology Objective (STO) in 2015 to streamline and accelerate the systems engineering design process [3]. An operational SSEA is envisioned that enables analysts to effectively implement systems engineering procedures to achieve faster, more effective results in predicting best system designs, and determining the learning requirements for achieving an effective level of operational performance within the new systems, including understanding how the requirements extend to the squad and above. Developers can then make design decisions that can minimize learning requirements and determine the most effective training strategies that will be needed [6].

The purpose of this paper is to describe a collaborative effort between researchers leading the US Army's SSEA and Training Effectiveness for Simulations (TEfS) STOs. The goal of the TEfS STO is to demonstrate the benefits of tools and methods for improved training outcomes and an improved training technology investment strategy. The payoff from both efforts will be increased levels of Soldier proficiency and readiness. In this paper we describe best practices from a recently completed TEfS squad training use case and propose an initial concept for how its training development and implementation protocol could be leveraged as a test protocol for future use in SSEA SaaS modeling research.

2   Soldier – Equipment – Task Framework

It is envisioned that SSEA will enable analysts to use automated systems engineering procedures utilizing behavioral modeling technologies to achieve faster, more effective results in predicting best system designs and learning requirements. Toward this end the SSEA STO is developing technologies, tools, and data that enable development and use of a human-centered systems engineering architecture to address known Soldier/Tactical unit requirements and gaps. A principle-based Soldier architecture and framework is envisioned that will enable system level tradeoff analyses and create the foundation for design parameters for next generation Soldier systems and subsystems based on human performance capabilities, the full complement of equipment, and mission tasks [3]. The payoffs will be increased efficiencies and optimized performance of the SaaS. The Soldier-Equipment-Task (SET) framework consists of three fundamental components: the Soldier (human capabilities), the equipment required to execute the mission (technological advancements); and the operational context to include mission essential tasks, the tactics, techniques, and procedures, and the operational environment (mission accomplishment). The SET framework extends from the SaaS, to the squad, which is a collection of systems (or System of Systems) [3]. A key SSEA STO objective is to create a standard set of processes, models, and techniques towards the creation of a collaborative environment centered on Soldier performance, which can drive systematic development of Soldier centric capabilities, enable system level tradeoff analyses, and predict changes in the SaaS in operationally relevant environments.

An example of a system level tradeoff analysis is determining the impact that new robotic resources would have on Soldier physical and cognitive load and understanding this impact on the SaaS (e.g., team and squad) and then using this information to inform technology interface and task design. Therefore, a main focus of the SSEA STO is developing an expanded set of behavioral and cognitive models to more accurately depict SaaS performance considerations that would inform end-user tradeoff analysis studies.

The SSEA operating concept is to capture "data elements" (e.g., behaviors) for the Soldier, Equipment, and Task within a database, capture relationships among data elements (such as within a Model Based Systems Engineering language such as SysML), provide access to data repositories, and provide novel query capabilities to enable an end user (e.g., analyst or researcher) to search for relevant data, highlight requirement issues, include learning requirements, and potentially launch modeling and simulation services. The SSEA envisions an end user in a particular community as being able to utilize one or more SSEA capabilities to conduct tradeoff analyses on materiel modifications in functionality that optimize SaaS performance, to include levels of proficiency based on learning requirements. The envisioned approach is to assess the variation in the condition sets and parameters that would indicate in qualitative and quantitative form that the cognitive, physiological, attitudinal, and social impacts of their study would recommend a portfolio of training capabilities (e.g., instruction, tutoring, virtual simulation, and or live training exercises) that could enable the accommodation of the engineering changes. Or, the envisioned SSEA capability set could recommend that the instruction and training solutions are so complex that the engineering changes are a no-go all together. Indicators of design effectiveness may decrease as the analysis moves from the Soldier to the squad and higher because the demands on the Soldier employing a new system (e.g., higher levels of stress, greater demands on situation awareness and teamwork), may negatively impact the performance of the squad. Whereas, for example, redesign and/or selecting effective team simulation-based training solutions may reduce the overall negative effects, thereby satisfying the design recommendations.

To achieve the vision for better human systems integration tools the underlying behavioral modeling capability is the key to an effective and accurate SSEA. Currently, the Improved Performance Research Integration Tool (IMPRINT) is the Department of Defense's human performance modeling and cognitive workload prediction tool used by human factors practitioners across the Services to assess and compare system designs and their effects on operator performance [7]. An IMPRINT model is based on a detailed representation of the mission in the form of a task network containing the tasks that Soldiers are likely to perform using system capabilities according to conceptual system designs. The task network enumerates the tasks performed and are connected according to the order in which tasks are performed in the field, and can represent concurrently-executed tasks. Each task is then annotated with a time of execution or probability distribution of execution times, values for the workload incurred while performing the task, and other information. Alternative designs are often built into the IMPRINT models reflecting independent variables of an experimental design. These alternatives can be modifications to equipment used by the Soldier reflected in task properties, or allocations of tasks to operators reflected in changes to the task network or operator assignments. The result is a functional simulation of different experimental conditions that produce predictive performance data. IMPRINT allows analysts to predict quantitative performance metrics including execution time and cumulative and resource-specific workload values during task execution. While the IMPRINT stressor parameters allow the analyst to specify certain conditions such as heat, cold, or noise which can affect the output metrics no option is currently built into IMPRINT to account for a holistic stress level experience under high-arousal contexts and adjust the predicted values. However, the tool does include a flexible plug-in architecture that allows for the adjustment of predicted values, which could be calculated based on an independent model of stress effects on performance.

To advance the IMPRINT capability toward increased accuracy in predicting Soldier utilization of new systems and system designs, it is important to understand how they will use them in varying conditions, particularly stressful situations, so that the relationships between the Soldier, Equipment, and Task can be better defined [7]. Understanding how Soldiers complete their tasks under realistic and stressful conditions is an appropriate starting point. The components needed to create this methodology are a way to relate the Soldier, Equipment and Task; quantitatively measure stress in a way that allows it to be related to performance metrics; and augment scenario events to provide empirical data to combine the first two components. To address this issue, SSEA and TEfS STO scientists have collaborated over the past year to develop an initial test method that could better define the behavioral models required by both STOs. The programs are leveraging data collection being conducted with Soldiers during controlled simulation training and live training exercises.

The approach draws from the US Navy's Tactical Decision Making Under Stress (TADMUS) and Manning and Affordability (M&A) programs to optimize Sailor performance through better training and combat ship design [8, 9]. The TADMUS program was an operational research program that engaged experienced US Navy officers and senior enlisted personnel throughout cycles of development, test, and experimentation. A stress exposure training protocol was developed with event-based scenarios and measures. This protocol was initially tested in a team training simulation lab environment with active duty Navy personnel as participants, then moved to training experiments and demonstrations at high fidelity combat team training installations, to pier-side, embedded shipboard training facilities, and then in deployed combat ship training exercises. Since then, research findings resulted in numerous training effectiveness guidelines for conducting scenario-based training in a variety of task domains (e.g., aviation, medical, and space exploration) [10]. Lessons learned from the TADMUS research enabled implementation of the M&A program which conducted extensive studies to develop individual and team behavioral models in order to develop new shipboard Multi-Modal Watch Station (MMWS) concepts for optimized manning in the combat information center [9, 11, 12]. The program leveraged the TADMUS scenario-based training protocol to compare 8-person watchstanding teams using the standard ship air defense warfare (ADW) system with 4-person ADW teams interacting with the prototype MMWS. Researchers manipulated the task protocol with controlled periods of low- and high-information stressors to determine if workload and performance could be sustained with reduced crew sizes. One study found ADW teams using the MMWS had greater situation awareness and performed more accurate tactical communications than teams using the standard ADW system [12]. The next section describes the Army Squad Overmatch (SOvM) training program that leveraged the TADMUS protocol, and like the M&A initiative approach we describe how the SOvM scenario-based training protocol could be adapted to conduct SaaS behavioral modeling studies.

3 Squad Overmatch

Starting in 2013, the SOvM research program leveraged the TADMUS scenario-based training approach to train dismounted squads in tactical decision making under stressful, urban combat conditions, with a focus on advanced situation awareness (ASA) and resilience and performance enhancement (RPE) skills [13]. The ASA skills use pattern recognition, predictive analysis and anticipatory thinking skills for assessing high risk contexts. The RPE skills involve using attention and concentration, and physiological control skills to manage and reduce distracting negative thoughts and physical reactions experienced under stress. These skill areas were identified first from the TADMUS findings, and then verified after thorough research literature reviews and interviews with SMEs as the key task stressors and cues that trigger performance problems in terms of psychomotor and cognitive processes. Initially, the SOvM team investigated how to improve existing TADSS and new training technologies to provide combat realistic exercises and experiences. Then, working with the Army Maneuver Center of Excellence (MCoE) in 2014, the study team developed and evaluated a SOvM stress exposure training demonstration and generated requirements for building ASA and RPE skills. Classroom instruction, simulation-based training, and live training exercises were developed and implemented over a two day period.

Then in 2015, the SOvM team adapted the training to focus on developing Tactical Combat Casualty Care (TC3) skills that would enable squad members to effectively conduct care under fire, tactical field care, casualty collection point care, and casualty evacuation [13]. Currently, medics and combat life savers receive their training independent of the unit they are assigned to assist. No Army squad training is conducted with embedded medics to learn TC3 skills which places tremendous stressors on the squad to adapt as a team when these events happen in combat. By adding an additional team member to the squad we created a more realistic set of battlefield conditions, but were expecting the stress exposure training to offset the added stress of learning how to work with the additional member. The SOvM for TC3 team expanded the training to include Team Development (TD), and Integrated After Action Reviews (IAARs). The TD skills involve helping team members adapt to high stress and reduce errors by exchanging critical information in a timely manner, providing priorities to focus decision making, proactively monitoring each other for signs of stress, providing backup and support, and taking corrective actions without having to be asked. The IAAR skills involve facilitating an event-based review, asking squad members to discuss the impact the squad's actions had on tactics, TC3, ASA, RPE, and TD. In fall of 2015, the SOvM team successfully conducted a training demonstration at the MCoE that integrated a medic into each of four US Army squads, and a Corpsman into each of three Marine Corps squads. Then in 2016 a training effectiveness experiment was conducted at the MCoE with eight US Army squads. Instruction, simulation, and live exercises were conducted over a four-day period, with multiple measures taken for learning, cognitions, attitudes, physiology, and performance [14].

4   A Scenario-Based Assessment Protocol

Lessons learned from the SOvM effort presented a unique opportunity to explore how its scenario-based training protocol could be adapted to create an assessment collection protocol for SSEA behavioral modeling studies. This approach could create the needed realistic, event-based conditions for collecting behavioral measures during periods of low and high levels of cognitive, physiological, and performance load, and in the context of a dynamic, adaptive team. In the next section we describe the steps in developing the SOvM 2016 scenario-based experimental protocol and then later discuss this approach in the context of conducting SSEA behavioral modeling studies.

4.1   Mission Tasks and Objectives

Scenario development was accomplished through an Integrated Research Team (IRT) of cognitive scientists, simulation developers, and military subject matter experts who possessed recent, relevant experience with dismounted infantry operations in urban environments. Then a series of steps were performed starting with defining the performance objectives in the context of TC3 beginning with the overarching goal of focusing on how infantry squads plan and conduct combat patrols in dynamically complex environments where they must manage operational and emotional stressors that affect decision making and problem solving.

4.2   Tactical Tasks and Soldier Warrior Leader Tasks

First, squad tactical tasks were identified from the Army Mission Essential Task List (METL). The METL is a composite, comprehensive list of the specific tasks necessary for a unit to master in order to complete whatever missions they may be assigned. The following squad tactical tasks for SOvM were defined: apply Troop Leading Procedures (TLP) to plan, organize and prepare for missions; determine the pattern of life baseline; recognize changes in the pattern of life; assess changes in the pattern of life; use cues and indicators to make sense of tactical situations; interact with civilian populations; minimize casualties; and defeat the enemy.

4.3   Flow Diagram of Key Events

An outline of key events and associated team behaviors in chronological order was developed for two simulation training scenarios, and three live training scenarios (M1, M2, and M3). The chronological list of events for M2 is:

1. Establish a listening post (LP)/observation post (OP)
2. Depart LP/OP
3. Civilian interactions in village
4. Key leader engagement and tactical questioning with high value target
5. Proxemic push as village civilians move away from the central square
6. Interview civilian woman
7. Sniper fire results in Soldier receiving Gun Shot Wound (GSW) to arm and a civilian woman receiving GSW to chest
8. Squad conducts movement toward sniper locations
9. Soldier receives GSW to chest at Sniper Location

   Next, a visual diagram of events was created that enabled the IRT to discuss, expand, and revise the logical flow of events, plan expected squad actions, and to manipulate the general timing for each event. For example, the flow diagram for scenario M2 had the squad mission objective of conducting a zone reconnaissance in order to conduct key leader engagement; exploit intelligence; confirm location of a suspected arms cache; and, exploit the site, if able. Each live training scenario was designed to take the squad about 45 min to complete.

4.4   Network Model Storyboard

Using concept mapping tools, the IRT developed an event network model that included the mission thread and assumed task branches off of the thread. Each model identified tactical tasks, decision points, information flows, situational awareness cues and stressors. The storyboard enabled developing a more detailed flow of events, decision points, communications, and linkages between squad tasks. The flow of events graphically depicts a sequence of events and indicators that are part of the situation. Each scenario storyboard was used to depict the squad's performance steps that make up the training event. Each situation was linked to battle drills, warrior leader tasks, and stressors. The storyboard had color codes with: blue indicating a tactical task; yellow indicating a cue or indicator; pink indicating an assessment; and green indicating information exchange with higher or adjacent unit.

Next the research team prepared a written narrative of the network model storyboard that included a synopsis, a context, and a situation. The synopsis identified the mission, major performance steps, and defined the skills that were to be practiced or trained as the anticipated end-state that was expected. The synopsis described the situation from the perspective of the platoon leader, the enemy, and the civilian population. Then a detailed description of the problem was developed from the squad leader's perspective. The context explained the conditions that the Soldiers would experience in terms of critical cues, factors and indicators that comprise their operating environment. The situation described the problem set and required tactical thinking skills; situational variables that should be perceived; and combat stressors that would be observed or experienced during the training.

4.5   Support Package and Detailed Master Scenario Event List

Following this, a support package was prepared to create the conditions for scenario execution. A single platoon Operation Order (OPORD) described the basic framework for the mission and places the learners in the tactical mindset necessary for problem solving and decision making. A Fragmentary Order (FRAGO) was developed that identified a mission requirement within the framework of the OPORD. The FRAGO enabled the squad leader to apply TLPs for planning and preparation before each scenario. Each squad reviewed the TLP to prepare a summary of the order along with a 5 by 8 in. mission card. Each situation was supported by an update to intelligence picture which reinforced priority intelligence requirements and provided linkages between scenarios. All of the materials up until this point were then used to develop a detailed Master Scenario Event List (MSEL) in a spreadsheet format to provide detailed information on the who, what, where, how and why of each decision trigger. Each MSEL was organized with a set of decision triggers that were in chronological order and numbered sequentially by scenario.

4.6   Event-Based Performance Objectives

The MSEL was then used to develop the event-based performance objectives for ASA, TD, and TC3 for each scenario. Table 1 presents a sample of the 51 event-based performance objectives identified for ASA, TD, and TC3 mapped to the nine scenario events in scenario M2. Many objectives were repeated across events and some events had many more performance objectives than others to ensure the scenarios had sufficient levels of stressors. For example, Event 7 had sniper fire resulting in a Soldier receiving a GSW to his arm and a civilian woman receiving a GSW to her chest. This was expected to elicit such TD behaviors from the squad as "provides complete and accurate medical reports" (see Table 1, item 3 under TD), and "squad leader and team leaders provide guidance and state priorities regarding roles for continuing mission" (see Table 1, item 5 under TD). Also, such TC3 behaviors were expected as "waits for suppressive fire or other cover before retrieving casualty" (see Table 1, item 3 under

TC3) and "squad leader directs team leaders to suppress enemy to maintain tactical focus" (see Table 1, item 4 under TC3).

Table 1. Sample of scenario M2 event-based performance objectives.

ASA	M2 events
1. Establishes geographic points of interest (avoidance or common use of an area)	1
2. Establishes atmospheric details (information that is or is not in line with baseline from intelligence)	1
3. Establishes that groups of civilians are engaging in mimicry, adoration, directing attention, or are part of an entourage	1
4. Positively identifies key leader	1
5. Employs guardian angel/geometries of observation	2, 4, 6
TD	
1. Communicates to team members when groups of people are engaging in mimicry, adoration, directing attention, or are part of an entourage	1
2. Corrects errors in information repeated on radio	1
3. Provides complete and accurate medical reports	7, 9
4. Squad asks platoon leader for guidance in further care of civilian casualty	7
5. Squad leader and team leaders provide guidance and state priorities regarding roles for continuing mission	8
TC3	
1. Provides MANDoWN report to squad leader	7, 9
2. Establishes security/provides cover after injury occurs	7, 9
3. Waits for suppressive fire or other cover before retrieving casualty	7, 9
4. Squad leader directs team leaders to suppress enemy to maintain tactical focus	7, 9
5. Sends up first 5 lines of 9-line report	7, 9

4.7 Prepare Supporting Materials for Implementing Protocol

Once performance objectives for each scenario had been vetted, the team prepared mission cards and high value target packages. Mission cards are tactical job aids that contain all mission essential information for the squad leader. Target packages were prepared for each high value individual and facilitated the identification of key leaders or threat personnel who were introduced through the intelligence process. The pattern of operations within the scenario was established and followed for each patrol starting with mission orientation (receive an order and analyze the mission); planning and preparation (apply TLP and prepare for action); and combat patrolling (movement to objective; actions on the objective, and change of mission).

4.8 Measures

Individual TC3, ASA, and TD performance checklists were developed for the five scenarios based on performance objectives in Table 1. Other measures collected in the

2016 experiment were assessments of knowledge acquisition, confidence in learning, cognitive workload, coping strategies, perceived stress, team perceptions of performance, action, efficacy, and cohesion, shared situation awareness, and physiological measures of heart rate variability and salivary alpha amylase (physical stress) [14].

5   A Test Protocol for Application to SSEA Modeling Studies

As was demonstrated by the U.S. Navy TADMUS and M&A programs, we propose the SOvM training effectiveness best practices can be adapted to use as a test protocol to develop realistic, multi-dimensional behavioral models for SSEA studies. The vision is for these models to be used to understand the impact of new designs on the SaaS and the training requirements for learning how to use new systems [6]. The first goal is to establish models of individual, team, and unit proficiency against standardized scenarios created with the test protocol. An initial study should be conducted that establishes a baseline of patterns and relationships of Soldier cognitions, attitudes, behaviors, and physiology under varying levels of realistic stressors, first in the context of a 4-person fire team and then in the 9-person squad. The measures listed in Sect. 4.8 would be a good starting point. The event-based performance objectives would enable collecting the types of behaviors that will be affected because the timing of stressors and the behavioral demands are known. For example, a casualty event has multiple performance objectives as indicated in Table 1. So, we would predict that Soldier's cognitive load is high during this event because of increased communications with other squad members; decision making for combat lifesaving activities; continuing the mission tactical actions; cognitive stress reactions to the event; and the physical load of moving the casualty to a casualty collection point. If the Soldier performs poorly in the casualty event this would also affect the ability of the fire team and the squad to adapt to addressing the TC3 requirements. For example, the fire team members may be unable to communicate their status to their team leader, and consequently, the team leader is then unable to communicate the team's status with the squad leader. Communication delays diminish the squad's ability to move quickly to address mission task requirements and address casualty needs. The behavioral data from this baseline would be collected under controlled conditions using the protocol and would then be converted to behavioral models to be tested as part of design modeling research.

   To illustrate we present a hypothetical example. The Army is developing Device X to improve the efficiency of communications among squad members about the status of casualty injuries during TC3 events. Designers create a simulated software prototype of Device X using the SSEA modeling tools. Designers then employ a simulation of the TC3 test protocol scenario to determine how implementation of Device X would affect the behavioral models for the Soldier and unit in the scenario. It would be expected that Device X would change the behavioral models by improving communication performance. If Device X interferes with communications then the designers can diagnose problems, explore and develop modifications, and test it again. This iterative testing phase using the SSEA modeling and simulation tools would also determine the most effective learning and training requirements for employing Device X.

In summary, when developing and testing existing or new military capabilities it is of the utmost importance to understand the impact that equipment and tasks will have on Soldier performance in the context of the SaaS so that the tradeoffs can be accurately modeled and assessed. In this paper we described a scenario-based team assessment protocol used in training effectiveness research and then proposed how it could be adapted to conduct SSEA behavioral modeling studies in the future to address the Army WCs in response to the Big 6+1 capabilities.

References

1. Chassé, C.: Training and doctrine command's big 6+1 capabilities. Armor: Mounted Maneuver J. **CXXVIII**, 60–63 (2017)
2. United States Army Warfighting Challenges, February 2017
3. Auer, R.J., Burford, C.W., Gallant, S., McDonnell, J.: A soldier system engineering architecture modeling and simulation application. In: MODSIM World Conference Proceedings (2015)
4. Savage-Knepshield, P., Martin, J., Lockett III, J., Allender, L. (eds.) Designing Soldier Systems: Current Issues in Human Factors (2012)
5. Stowers, K., Oglesby, J., Sonesh, S., Leyva, K., Iwig, C., Salas, E.: A framework to guide the assessment of human–machine systems. Hum. Factors **59**, 172–188 (2017)
6. Hawley, J.K., Mares, A.L.: Training and testing revisited: a modest proposal for change. ITEA J. **30**, 251–257 (2009)
7. Napier, S., Best, C., Patton, D., Hodges, G.: Using an augmented training event to collect data for future modeling purposes. In: International Conference on AugCog, pp. 421–430 (2016)
8. Cannon-Bowers, J.A., Salas, E.E. (eds.): Making decisions under stress: implications for individual and team training. American Psychological Association (1998)
9. Pharmer, J.A., Hildebrand, G., Campbell, G.E.: Measuring the impact of human-centric design on human performance in a combat information center. In: International Symposium on Optical Science and Technology, pp. 221–227 (2000)
10. Salas, E.: Team Training Essentials: A Research-Based Guide. Routledge, Abingdon (2015)
11. Freeman, J.T., Campbell, G.E., Hildebrand, G.: Measuring the impact of advanced technologies and reorganization on human performance in a combat information center. In: Proceedings of the HFES, vol. 44, pp. 642–645 (2000)
12. Osga, G.A., Van Orden, K.F., Kellmeyer, D., Campbell, N.L.: Task managed watchstanding: concepts for 21st century naval operations. In: Proceedings of the HFES, vol. 44, pp. 457–460 (2000)
13. Milham, L.M., Phillips, H.L., Ross, W.A., Townsend, L.N., Riddle, D.L., Smith, K.M., Butler, P.V., Wolf, R.J., Irizarry, D.J., Hackett, M.G., Johnston, J.H.: Squad-level training for tactical combat casualty care: instructional approach and technology assessment. J. Def. Model. Simul.: Appl. **14**, 345–360 (2017)
14. Johnston, J., Gamble, P., Patton, D., Fitzhugh, S., Townsend, L., Milham, L., Riddle, D., Phillips, H., Smith, K., Ross, W., Butler, P., Evan, M., Wolf, R.: Squad overmatch for tactical combat casualty care: phase II initial findings report. Program Executive Office Simulation, Training and Instrumentation, December 2016

Human Factors for Military Applications of Head-Worn Augmented Reality Displays

Mark A. Livingston$^{(\boxtimes)}$, Zhuming Ai, and Jonathan W. Decker

Naval Research Laboratory, 4555 Overlook Avenue SW,
Washington, DC 20735, USA
{mark.livingston, zhuming.ai,
jonathan.decker}@nrl.navy.mil

Abstract. Research into the human factors of augmented reality (AR) systems goes back nearly as far as research into AR. This makes intuitive sense for an interactive system, but human factors investigations are by most estimates still relatively rare in the field. Our AR research used the human-centered design paradigm and thus was driven by human factors issues for significant portions of the development of our prototype system. As a result of early research and more recent prototype development, mobile AR is now being incorporated into military training and studied for operational uses. In this presentation, we will review human factors evaluations conducted for military applications of mobile AR systems as well as other relevant evaluations.

Keywords: Augmented reality · Head-worn displays · Human factors
Military applications

1 Introduction

In the defining paper for augmented reality (AR) and virtual environments, Sutherland [1] described several aspects of the user experience with the virtual world. Given the interactive nature of the system he envisioned, this makes intuitive sense. Yet multiple reviews [2–5] have found a low frequency of human factors evaluations for AR, although there is a recognized trend of an increasing number and percentage of papers that include some form of evaluation. This has been noted recently for AR construction management [6] and safety-critical systems [7]. We note the difference that many make between AR, which may be defined [8] as an interactive system, whereas a mixed reality (MR) system may not be interactive. Reviews of the literature on MR, which to many includes AR, also notes the need for evaluation and validation of medical MR systems [9, 10].

There are many potential reasons for the low frequency of evaluations. Many of these reasons fall into the category of practical challenges, many of which have little to do with AR and its construction with a specific set of technology products. Hardware for AR, and mobile AR in particular, has evolved significantly in the past decade. In such a scenario, an evaluation of a particular piece of hardware may not be valid for very long, limiting the value of the test. Another practical challenge is simply the expense of conducting a test; this is especially true in mobile AR, where a proper test

© Springer International Publishing AG, part of Springer Nature (outside the USA) 2019
D. N. Cassenti (Ed.): AHFE 2018, AISC 780, pp. 56–65, 2019.
https://doi.org/10.1007/978-3-319-94223-0_6

may require being outdoors, sometimes with immature hardware that is not ready for use in inclement weather or by people who are not aware of the fragility of experimental hardware. Certainly, the expense in time can also be a major factor. This can be compounded by the fact that individual differences can be a significant factor in the variance of user performance. Thus a test may need to recruit a large pool of volunteers, have each volunteer conduct multiple tests, or accept lower significance in the results.

Our response to these challenges was the adoption of an iterative development process [11] including tests at multiple levels. Perceptual tests helped us understand the requirements the application was placing on certain hardware components. Cognitive tests helped us understand the user performance that the system supports and (especially early in the development cycle) identify limiting factors that we needed to address. We explore case studies from our development experience and examples from elsewhere in the literature relevant to our efforts to develop military AR applications.

2 Related Work

Although the consistent conclusion drawn is that relatively few papers include reports of formal or informal evaluations of AR technology or applications, there have been four surveys of such evaluations [2–5] since 2005.

Some consistent themes emerge from these surveys. First is that relatively few studies were being reported in early literature, but that this number is growing, both in raw counts and in relative frequency within the literature identified. This raises one issue with tracking user-based evaluations of AR: finding the relevant literature. Because AR is a technology that has been shown to have wide-ranging applications, some of the studies are reported in technology-centric venues, whereas other studies appear in application-centered publications. While it would be intuitive to presume that technology-focused evaluations appear in publications that are considered to be technology-focused, the evidence from a survey that focused on one venue [3] supports the conclusion that at least some application-level evaluations appear in what is considered to be a technology-focused conference. It should be noted that application papers are solicited for this conference. The studies also consistently find that perceptual studies are one of (if not the) most commonly reported type of evaluation. We argue that this makes sense when the technology is rapidly evolving and requirements for that evolving hardware are being rewritten for the expanding pool of applications that researchers and developers are using AR to provide.

Bai and Blackwell [4] noted many challenges for evaluating AR and MR; of these, we note that two particularly affect the ability to conduct an evaluation of mobile AR systems. First is the tolerance of the system to error; in particular, we focus on the error of the registration subsystem in aligning the graphics with the desired objects or locations in the real world. This is a well-known source of error in expressing the relationship between the real and virtual portions of the AR experience. The quality of tracking systems to provide a sufficiently accurate estimate of a mobile user's position and orientation in an outdoor setting has historically been a challenge; some recent systems have made significant advances (e.g. [12]), but any evaluation of a system that exhibits high registration error (by whatever definition of "high" is appropriate for the application) risks the evaluation merely revealing that the registration accuracy is insufficient to truly evaluate the usability of the system.

The second challenge noted [3] that strongly affects mobile AR evaluation is the lack of control of the environment. Outdoor illuminance (and its variation), weather conditions, background noise, shadows, and wind may all create confounding factors during evaluation of a system. These may affect evaluations aimed at a specific feature differently than an evaluation of an entire system, but it is important to note these conditions and try to either reduce them as confounding factors or balance them as factors sufficiently for valid statistical testing of hypotheses related to them.

McKendrick et al. [13] used functional near infrared spectroscopy (fNIRS) to measure the workload experienced by a user in an AR outdoor navigation system. They found the fNIRS was a better measure than subjective post-hoc evaluation of workload and that AR led to a reduced mental workload for the navigation task.

Aaltonen and Laarni [14] evaluated a wearable soldier navigation system using subjective post-hoc evaluation. They found users were generally receptive to the technology, but had a list of requirements of robustness to outdoor conditions, comfort, and ease of use that needed to be addressed. They also found that multiple modalities (visual, audio, and tactile) had distinct advantages and recommended exploration of a multimodal system for guiding navigation.

3   Case Studies in Human Factors Evaluations

3.1   Depth Perception and X-ray Vision

The idea of studying depth perception and the concept of X-ray vision began for us with a meeting with our client, a U.S. Marine acting as a subject matter expert (SME). He described urban environments as presenting a challenge for keeping track of where friendly forces were once they disappear behind urban infrastructure. We offered that an AR system, with its see-through view, could help convey known location information to troops in a usable fashion, without needing to look away from the environment to a hand-held display (which was and is the current state of the art). This was judged by our SME to be an important problem to solve, and thus we began to design visual representations of hidden people and objects within urban environments.

The original purpose was see-through view (a capability to which the term "X-ray vision" has been attached for over two decades, although its specific origin is unclear). It became clear that we were also dealing with the perception of depth, and we expanded the research to depth of virtual objects within the visible region of the environment. (That is, virtual objects at locations in which a real object would have been visible.) Because this was a specific capability within a large set of scenarios, we were able to abstract the task away from the application down to a purely perceptual experience for the user.

This abstraction served us well in many ways. We were able to avoid a requirement to have SME participants in a human factors study, which made conduct of a series of studies easier than had we needed time from SMEs on a regular basis. We did not need any prior experience with AR or military procedures to study the issues. We did include military symbols in our visual representations.

We summarize findings reported previously [15]. We conducted a series of studies focusing on visual representations that worked for depth perception and X-ray vision for outdoor scenarios. These studies examined distances ranging from a few meters up to (in separate studies) over a thousand meters. We found that drawing a virtual copy of the ground plane, drawing a virtual wall over a virtual object that matched the physical number of intervening surfaces, and drawing a virtual tunnel that also indicated the number of intervening surfaces were among the best methods. We also studied labeling distances explicitly, though this presumes legibility of the display, which is not always achievable (see Sect. 3.3), especially in outdoor environments. Other authors found contradictory results, and we noted the use of hand-held versus head-worn displays, differing visual representations, differing distances, and other potentially confounding factors. The issue is far from resolved. The conclusion for the purposes of this survey is not specific results, but the ecological method by which we identified the problem, connected potential solutions to the underlying science, and sought solutions that attempted to satisfy both the need for application usability and scientific discovery.

3.2 Mobile AR for Military Training

The major challenge of evaluating the usability of applications built for military training or operations was how to measure the success of the system. For our purposes, we felt that subjective evaluation by the participants would be of limited value. Familiarity of the technology has increased over the past decade, and the novelty of the technology has decreased in a corresponding fashion. When we began our evaluation, we believed the novelty would have been an impediment to getting useful data from subjective post-hoc surveys of the participants. This is not to say that this method does not generate useful feedback; many authors have used this (e.g., [14]), and the lessons learned can be valuable in a similar way to how expert evaluation is valuable within the human–centered design process. However, we were at a point in our development process when we felt it was necessary to move beyond subjective expert evaluation. The particular application was room clearing; that is, ensuring that there are no active threats in the rooms of a building. We designed our study to use the type of training (AR, non-AR) as a factor in performance of the small-scale simulation of this task, conducted in our laboratory. We set up our system to track users through the building and determine whether avatars to "clear" from a room would appear during their training. A team either saw no avatars at all during multiple training scenarios, or multiple avatars in multiple training scenarios.

We engaged a SME who had worked as a training instructor for the U.S. Marine Corps and asked him to evaluate the performance of our participants (in two-person teams) on the same criteria he would have evaluated Marines during training. While his criteria are in a plain sense subjective, the SME had no knowledge of which type of training the team had undergone, so his attribution of performance could not be to the type of training. His criteria were aggressiveness, movement, security, communication between team members, and coordination between team members. These were judged based on his experience, although our novice participants generally fared poorly on them. Objective measures were also employed, counting whether the participants, hostile forces, and non-combatant actors survived according to the laser-tag weapons, per a formula designed by the SME to reflect military priorities of survival and achievement of the mission objective. We also were able to passively record data about the movement, including complete tracking of body and head motion. In the end, the most convincing measure of the effect of the system was the tracking of "room sweep." Each team member is responsible for looking at a portion of each room entered; we were able to measure (to within the accuracy of our tracking system) what portion of this space the team member actually viewed. The result that was most convincing in the potential of the system was that room sweep increased with increasing repetition when participants had trained against AR avatars in the rooms; room sweep decreased when participants had trained against empty rooms. Only after this evaluation did we conduct a similar study with SME participants [16].

Objective measures are relatively easy to devise in navigation applications, which are a common task for military personnel. In comparing visual displays with tactile and audio display (plus hybrids of these three), Elliott et al. [17] found that navigation was fastest (objective measure) with the visual head-worn display, but (subjective) workload was lowest with the hybrid visual-tactile configuration. Using a hand-held navigation device resulted in lower situation awareness (a measure of awareness of the surrounding physical environment) than with the tactile or visual modalities. Tactile was preferred by virtue of not requiring visual attention (presumably, meaning not taking it away from something else). We note that objective measures of workload were found to be better than subjective measures [13] and would argue for these objective measures to be applied when practical.

Maintenance and repair tasks are common for military personnel, and a prototype system for assisting in this was evaluated against an analog of the hand-held display then in use [18]. The AR system allowed mechanics to locate tasks more quickly. The AR system was also subjectively judged to be intuitive to use.

While both subjective and objective measures are important in determining the success or limitations of an AR system, in our experience, purely subjective measures are not sufficient to convince a user community of the value of the system. It is also often reported (e.g. [19]) that users will prefer a system that does not lend itself to the best performance. These considerations certainly are important for mobile AR.

3.3 Basic Perception in Head-Worn AR Displays

We define basic perception as low-level discernment of objects in the view of an environment (avoiding specifying real or virtual). Being able to recognize symbols on an AR display or in a real environment while wearing an AR display is a prerequisite to reading information conveyed in either portion of the merged environment. Similarly, interpreting a variety of military symbols overlaid on an environment or that exist in an environment seen through an AR display requires being able to discern these symbols. Note the dual nature of the task: seeing both the real and the virtual with sufficient fidelity to make use of the information conveyed.

There have been several attempts to quantify features of basic perception [20]. We report here some new data of contrast sensitivity for a head-worn display, as well as

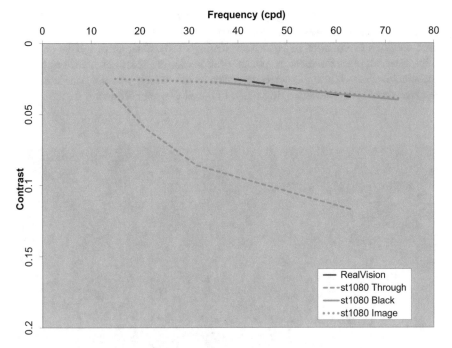

Fig. 1. Measured contrast sensitivity function for our st1080 head-worn display. Under the curve is the visible region for the given view condition.

(objective) color distortion and (subjective) color perception in that display. We draw graphs in the style of previous results [20] to facilitate comparison.

A small study was conducted with the st1080 display from silicon micro display[1]. Eight subjects viewed modified Landolt-C targets with varying gap size and contrast [20]. We then computed a threshold for reliable discernment of the location (which side of the square) contained the gap. This threshold identified the boundary between visible and non-visible in the canonical space of frequency (inverse of object size) and contrast in order to reliably see an object. The result is graphed in Fig. 1. Note that the background behind the graphics made little difference to our users; the lines for the graphics-on-black background ("st1080 Black") and graphics-on-realistic scene background ("st1080 Image") are basically identical where both could be detected. Moreover, they are competitive with the threshold determined without any intervening AR display ("RealVision"). On the other hand, the optics of the display and any light created by it, even when displaying a black screen to become see-through ("st1080 Through"), clearly interfered a little with users' contrast sensitivity. To put the result in context, the minimum contrast for an optometric exam is 0.8, so the "eye chart" presented in this work had much lower contrast than a clinical evaluation.

---

[1] http://www.siliconmicrodisplay.com/st1080.html.

Another (slightly overlapping) pool of participants performed a color matching task in similar conditions as the contrast sensitivity study. This study was again designed to mimic previous evaluations of head-worn displays [20]. Qualitative results are shown in Fig. 2, which plots the mean response locations for the color patches along with the color specification. While the correspondence is not always obvious, one can get the sense that there is a qualitative difference in the display conditions.

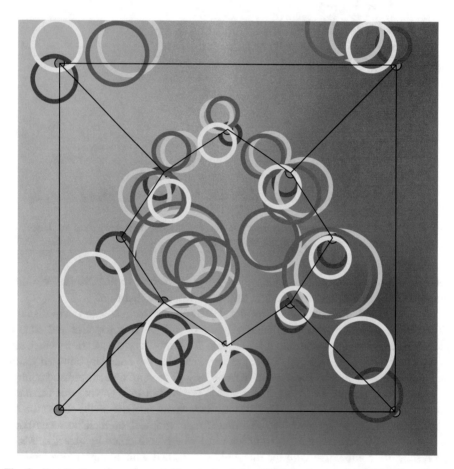

Fig. 2. Results from the color matching study of the st1080 head-worn display. Users matched color patches with no intervening display (black circles), graphics-on-black background (magenta), graphics-on-white background (cyan), see-through (yellow). The dark circles connected in the mesh indicate the color specification of the patches. Note that it is hard to determine the correspondence between the subjective match and the color specification. The background shows CIELAB color space at the L = 65 slice (the displayed intensity). In the blue (bottom) half, error in color matching is especially high for the graphics conditions, whereas users had some success in compensating in the see-through condition and in the baseline (no display) condition. The size of the circle is the standard deviation of the ΔE metric (mean match versus color specification), taken across all subjects for each color patch.

The baseline condition of matching with no intervening display (black circles in Fig. 2) was statistically better than all other display conditions under the ΔE error metric relative to the color specification, $t(7) = 6.07$, $p < 0.001$. Additionally, the see-through condition (yellow circles) was statistically better than either of the graphics conditions, $t(7) > 3.23$, $p < 0.01$. The graphics-on-white background yielded lower ΔE than graphics-on-black by a marginally significant amount, $t(7) = 1.90$, $p < 0.10$.

One key driver in researching the basic perception was feedback from participants in a user study of the training application. User complained about the quality of the display, saying that they felt they had difficulty perceiving depth and even see the walls in the physical environment. This prompted our research, although as described in our previous survey [20], other researchers were finding reasons to investigate these aspects of AR displays.

4 Discussion and Conclusions

We argue that developers of AR systems must be aware of opportunities for useful evaluations during the development process. We employed a strategy of iterating between perceptual and application evaluations, between novice and SME participants in studies. Specific, detailed evaluations of hardware performance (or user performance that is dependent on a feature of the AR hardware) may be turned into specifications for required hardware performance. System-wide evaluation of user performance on application goals can indicate what the limiting factors are and drive new research to improve the system performance. We reaffirm our belief in this iterative strategy and present our success in pushing the military applications forward as evidence of the success. Both training [21] and operational [22] aspects of the system continue to be studied for their usefulness to military personnel.

We have also argued previously [23] that those considering human factors evaluations in AR need to consider the perceptual and/or cognitive theory that underpins the results. One could easily extend this argument to a suggestion for interdisciplinary teams of evaluators, to including perceptual or cognitive psychologists.

Human factors of augmented reality are a critical aspect of the technology becoming useful for end users. Despite significant attention to a few key issues and ample notice that more complete evaluation of the human factors is lacking, research on these issues seems to be only recently gaining interest. If nothing else, this review should convince more researchers to incorporate these issues into their research.

Acknowledgements. This research was supported by the Army Research Laboratory.

References

1. Sutherland, I.E.: The ultimate display. In: Proceedings of IFIP Congress (1965). Information Processing
2. Swan, J.E., Gabbard, J.L.: Survey of user-based experimentation in augmented reality. In: Proceedings of 1st International Conference on Virtual Reality (2005)

3. Dünser, A., Grasset, R., Billinghurst, M.: A survey of evaluation techniques used in augmented reality studies. In: ACM SIGGRAPH Asia Courses (2008)
4. Bai, Z., Blackwell, A.F.: Analytic review of usability evaluation in ISMAR. Interact. Comput. **24**, 450–460 (2012)
5. Dey, A., Billinghurst, M., Lindeman, R.W., Swan II, J.E.: A systematic review of usability studies in augmented reality between 2005 and 2014. In: IEEE International Symposium on Mixed and Augmented Reality (ISMAR-Adjunct) (2016)
6. Chu, M., Matthews, J., Love, P.E.D.: Integrating mobile building information modelling and augmented reality systems: an experimental study. Autom. Constr. **85**, 305–316 (2018)
7. Grabowski, M., Rowen, A., Rancy, J.-P.: Evaluation of wearable immersive augmented reality technology in safety-critical systems. Saf. Sci. **103**, 23–32 (2018)
8. Azuma, R.T.: A survey of augmented reality. Presence: Teleoper. Virtual Environ. **6**(4), 355–385 (1997)
9. Kersten-Oertel, M., Jannin, P., Collins, D.L.: The state of the art of visualization in mixed reality image-guided surgery. Comput. Med. Imaging Graph. **37**(2), 98–112 (2013)
10. Chen, L., Day, T.W., Tang, W., John, N.W.: Recent developments and future challenges in medical mixed reality. In: Proceedings of International Symposium on Mixed and Augmented Reality (2017)
11. Livingston, M.A., Swan II, J.E., Julier, S.J., Baillot, Y., Brown, D., Rosenblum, L.J., Gabbard, J.L., Höllerer, T.H., Hix, D.: Evaluating system capabilities and user performance in the battlefield augmented reality system. In: Proceedings of NIST Workshop on Performance Metrics for Intelligent Systems (2004)
12. Zhu, Z., Branzoi, V., Sizintsev, M., Vitovich, N., Oskiper, T., Villamil, R., Chaudhry, A., Samarasekera, S., Kumar, R.: AR-weapon: live augmented reality based first-person shooting system. In: IEEE Winter Conference on Applications of Computer Vision (2015)
13. McKendrick, R., Parasuraman, R., Murtza, R., Formwalt, A., Baccus, W., Paczynski, M., Ayaz, H.: Into the wild: neuroergonomic differentiation of hand-held and augmented reality wearable displays during outdoor navigation with functional near infrared spectroscopy. Front. Hum. Neurosci. **10**, Article 216 (2016)
14. Aaltonen, I., Laarni, J.: Field evaluation of a wearable multimodal soldier navigation system. Appl. Ergon. **63**, 79–90 (2017)
15. Livingston, M.A., Dey, A., Sandor, C., Thomas, B.H.: Pursuit of 'X-ray vision' for augmented reality. In: Human Factors in Augmented Reality Environments. Springer, New York (2012). Chap. 4
16. Livingston, M.A., Brown, D.G., Julier, S.J., Schmidt, G.S.: Mobile augmented reality: applications and human factors evaluations. In: Proceedings of NATO Human Factors and Medicine Panel Workshop on Virtual Media for Military Applications (2006)
17. Elliott, L.R., van Erp, J.B.F., Redden, E.S., Duistermaat, M.: Field-based validation of a tactile navigation device. IEEE Trans. Haptics **3**, 78–87 (2010)
18. Henderson, S., Feiner, S.: Exploring the benefits of augmented reality documentation for maintenance and repair. IEEE Trans. Vis. Comput. Graph. **17**(10), 1355–1368 (2010)
19. Nielsen, J., Levy, J.: Measuring usability: preference vs. performance. Commun. ACM **37**(4), 66–75 (1994)
20. Livingston, M.A., Gabbard, J.L., Swan II, J.E., Sibley, C.M., Barrow, J.H.: Basic perception in head-worn augmented reality displays. In: Human Factors in Augmented Reality Environments. Springer, New York (2012). Chap. 3
21. Squire, P., Muller, P.: Making augmented reality...a reality. Future Force, Naval Sci. Technol. Mag. **1**(3), 45–49 (2014)

22. Gans, E., Roberts, D., Bennett M., Towles, H., Menozzi, A., Cook, J., Sherrill, T.: Augmented reality technology for day/night situational awareness for the dismounted soldier. In: Display Technologies and Applications for Defense, Security, and Avionics IX; and Head- and Helmet-Mounted Displays XX, Proceedings of SPIE, vol. 9470 (2015)
23. Livingston, M.A.: Issues in human factors evaluations of augmented reality systems. In: Human Factors in Augmented Reality Environments. Springer, New York (2012). Chap. 1

Latent Heat Loss of a Virtual Thermal Manikin for Evaluating the Thermal Performance of Bicycle Helmets

Shriram Mukunthan[1]([✉]), Jochen Vleugels[1], Toon Huysmans[2,3],
and Guido De Bruyne[1,4]

[1] Product Development, Faculty of Design Sciences, University of Antwerp,
Ambtmanstraat 1, 2000 Antwerp, Belgium
{shriram.mukunthan, jochen.vleugels,
guido.debruyne}@uantwerpen.be
[2] Vision Lab, Department of Physics, University of Antwerp (CDE),
Universiteitsplein 1, 2610 Antwerp, Belgium
[3] Applied Ergonomics and Design Department of Industrial Design,
Delft University of Technology, Landbergstraat 15,
2628 CE Delft, The Netherlands
t.huysmans@tudelft.nl
[4] Lazer Sport NV, Lamorinierestraat 33-37 bus D, 2018 Antwerp, Belgium

Abstract. Thermal performance of three bicycle helmets for latent heat loss was evaluated through a virtual testing methodology using Computational fluid dynamics (CFD) simulations. The virtual thermal manikin was prescribed with a constant sweat rate of 2 g/h and a constant sweat film thickness of 0.3 mm. The simulations were carried out at 6 m/s until convergence was achieved. The results from steady state simulations show heat loss of 158 W from manikin without helmet and approximately 135 W with helmets. However, the thermal performance of helmets with a sweating manikin has been reduced from 89–93% to 84–87%. These results imply that evaporative/latent heat loss plays a significant role in thermal performance of helmets. Therefore, thermal performance tests for helmets should also include testing of helmets for evaporative heat loss.

Keywords: Thermal manikin · Evaporative heat transfer
Convective heat transfer · Cooling efficiency · Turbulence models
CFD · Thermal performance

1 Introduction

Cycling is popular. It is healthy and environmental friendly. Unfortunately, it is also the third most dangerous form of transport resulting in injuries and mortalities. Head injury is reported as the direct cause of death in about 69% of these cycling fatalities [1, 2]. Bicycle helmets reduce the risk of head injuries, caused by accidents. The usefulness of bicycle helmets for reducing risks of head injury has been shown [3–5]. This has resulted in substantial research on improving the safety features of the helmet. However, helmet usage among cyclists is low when compared to other types of protective

© Springer International Publishing AG, part of Springer Nature 2019
D. N. Cassenti (Ed.): AHFE 2018, AISC 780, pp. 66–78, 2019.
https://doi.org/10.1007/978-3-319-94223-0_7

head gear usage. This dislike towards helmets can be attributed to physical and thermal discomfort among users [6–8]. For example, motorcycle helmet usage rates in Italy are reported to be 93% and 60% for Northern Italy and Southern Italy respectively. A difference that can at least be partially explained through headgear induced thermal discomfort in warm environments [9–11]. Though helmets can provide thermal benefits in cold conditions [12] helmets receive an adverse reaction from users in warm conditions. Thermal comfort or discomfort is a product of heat transfer in the volume between head and helmet or lack thereof. This is also influenced by other factors like environmental conditions, body temperature, helmet design etc. Local discomfort on the head in warm conditions is sufficient to cause whole body discomfort [13] and hence this local discomfort resulting from sweating and skin-wettedness needs to be addressed. Headgear increases head insulation which results in heat entrapment and sweating especially in warm conditions and can easily affect comfort perception as heat loss differences of 1 W can be sensed by helmet users [14]. The most common approach to improve comfort is to increase ventilations thereby enhancing convective cooling and sweat evaporation [15, 16] but this cannot be considered as the only solution since the underlying factors that cause thermal discomfort are plentiful.

To understand the influence of thermal comfort on users, it is important to understand the mechanisms and metrics pertaining to thermal comfort. This has paved way for a number of studies on thermal aspects and thermal parameters of headgear. The study of thermal aspects begins with understanding the thermo-physiological processes involving human head and heat transfer. Human head is a part of human body that shows a high heat-sensitivity [17]. Due to high heat sensitivity and unique biology, under forced convection, head can contribute up to one-third of total metabolic body heat dissipation [8]. Heat transfer from head mainly happens through convection (natural and forced), evaporation and radiation. Therefore, it is paramount to understand each mode of heat transfer and its influence on thermal comfort in detail. Studies on convection cooling of head have been carried out by multiple agencies to evaluate and validate helmet designs. These studies use one of the two methodologies namely subject studies and object studies. Subject studies or user trials allow gaining insight into the global or local heat loss of test persons. More specifically, they allow quantifying thermo-physiological parameters such as heat storage, core body temperature, skin temperature and sweat production [18, 19]. Several studies have monitored the effects of sweat on heat transfer and in turn on thermal comfort. Quantification of sweat rates using ventilated capsules [19, 20] and methodologies using absorbent pads [21] have helped in understanding the evaporation mechanism and the existence of a correlation between local and global sweat rate [19, 21]. Taylor and Machado-Moreira [20] developed predictive equations for sweat rate in head at rest or exercising. User trials have also emphasized the importance of insight into latent heat loss for optimizing headgear for thermal performance. Results of user trails are unfortunately not always straightforward for generalization towards helmet improvement due to high individual differences in test subjects.

Hence object studies or bio-physical methods have gained importance. Biophysical methods have focused on thermal manikin head studies. Thermal manikin studies induce a heat source onto the surface of a manikin head. Then, the applied heat is kept constant and surface temperatures are recorded or controlled and the applied heat load is recorded in different segments of the manikin head. Different thermal manikin heads have already been developed and investigations on dry heat loss in headgear with the use of a physical thermal manikin head were performed [15, 22]. Additionally, the effect of solar radiation [23] and hair [24] was also investigated. Studies on thermal perception resulting from ventilation changes in motorcycle helmets using thermal manikin was analyzed. None of the above-mentioned studies have analyzed the efficiency of helmets with respect to removal of sweat and heat loss through evaporation. Therefore, analyzing the latent heat loss or heat transfer through evaporation is important to evaluate the helmet design. Evaluation of helmet performance for thermal comfort can be performed using numerical models which have been used to evaluate aerodynamic performance [25] and ventilation performance [26] previously. These numerical models used in evaluating thermal performance of helmets can also be coupled with thermo-physiological models. Several thermo-physiological mathematical models [27, 28] have been developed with different levels of accuracy and complexity to provide the numerical models with realistic values to model thermoregulatory response of the human head and play a crucial role in helmet evaluation. Each of the studies allows gaining insight into the complex phenomenon of local heat loss on the human head, but fail to provide a concrete methodology that allows optimizing bicycle helmets for sensible and latent heat loss. In this research, analysis of heat transfer through evaporation is carried out using virtual models. This paper describes the work carried out to evaluate the ventilation efficiency of three helmet designs with a commercial computational fluid dynamics (CFD) program using Eulerian wall film methodology to simulate sweat production and skin-wettedness on the virtual thermal manikin head. The helmet efficiency values obtained from the virtual experiments will be compared with the efficiency values obtained from sensible heat loss (convection cooling) tests [29].

2   Heat Transfer Between User and Environment

To understand thermal comfort in detail, the thermo-physiological response of human body must be understood. The core temperature of human body is maintained approximately around 37 °C and from this phenomenon we can infer that there exists a heat transfer mechanism between body and environment such that heat is dissipated when in surplus and retained when in deficit. This relationship between heat generation in the body and transfer to environment is dynamic resulting in a heat balance which is given by heat balance equation [30]:

$$S = M - W - (E_{res} + C_{res} + E_{sk} + C_{sk} + K + R) \tag{1}$$

where M is metabolic rate and W is work rate, quantifying the rate of heat production. Heat loss/gain is defined by evaporation (E), convection (C), conduction (K) and radiation (R). Evaporation and convection occurs through respiration (res) and skin (sk) and S is heat storage in the human body which is expressed in kJ/h or W or W/m^2. The human head exhibits unique heat transfer properties owing to its lack of vaso-constriction responses. In addition to this, lack of clothing or insulation that protects other parts of the body results in high transfer of heat from the head. Heat transfer from the head is non-homogenous and results in different parts of the head exhibiting different heat transfer characteristics. For example, the nose exhibits higher cooling rate than the face and cheeks [31]. The opposite of this is also true resulting in an increase in temperature when core temperature of the body increases [31]. Thus, modelling and testing of multiple modes of heat transfer can be a complicated task. For bicycle helmets, having a layer of insulating expanded polystyrene, heat loss is predominantly defined by convective heat loss and evaporative heat loss. Convective heat transfer is defined by:

$$\dot{Q}_{cs} = h_c A (T_h - T_a) \tag{2}$$

where, $\dot{Q}_{cs}$ is convection heat transfer per unit time, A is surface area of the object, h_c is convective heat transfer coefficient, T_h is Temperature of head surface, T_a is Temperature of air/fluid. Thermal performance of bicycle helmets is quantified using cooling efficiency. The evaporation heat dissipation is given by

$$\dot{Q}_{ls} = h_e A (P_{sk} - P_a) \tag{3}$$

where, $\dot{Q}_{ls}$ is evaporation heat transfer per unit time, A is surface area of the object, h_e is evaporative heat transfer coefficient, P_{sk} is vapour pressure of skin, P_a is vapour pressure of air/fluid.

Cooling efficiency of a helmet is thus defined as the ability of a helmet to dissipate the heat from the head to the environment, relative to a nude head.

$$\text{Cooling efficiency} = \frac{\text{Heat transfer with helmet}}{\text{Heat transfer (nude head)}} \times 100 \tag{4}$$

3   Materials and Methods

3.1   Computational Domain and Geometry

The methodology used in virtual testing of helmets for latent heat loss follows the same procedure as the previous research paper [29]. Digital models of three helmets to be tested were obtained with high resolution 3D scanning. These three helmets are chosen because the thermal performance values of these helmets for sensible heat loss, determined in the previous study [29] can be compared with the results from this study. The digital model of head was generated using 3D shape modelling of human head. This is achieved using the mathematical model of human head [32] that estimates the 3D shape of an arbitrary human scalp from one dimensional anthropometric measurements on the external surface of a head. 3D scans of human head from multiple subjects were used in developing the mathematical model. The surface details of head and helmet were smoothed to ease grid generation. The digital model of head is split into five measurement zones (scalp, face, forehead, ears and neck) to understand the zonal/local behavior. The assembly between head and helmet is done such that there is realistic contact and no penetration between head and the helmet mesh (Fig. 1). The zones of head and helmet that come in contact were stitched together. The head and the helmet were placed in a computational domain according to practice guidelines.

The size of the computational domain was $6.5 \times 2 \times 2$ m^3. The test object was placed 1.5 m away from the inlet plane to avoid pressure gradients (Fig. 2). Grid sensitivity analysis was carried out for different mesh sizes to determine influence of cell size on the results.

3.2   Resolving Boundary Layer

Prisms were generated in the layer between head and fluid to model the boundary layer in the solid-fluid interface (Fig. 3(b)). Prism layers provide a high resolution for resolving the thin boundary layer at the head using a low-Reynolds number modeling technique. The boundary layer is resolved into three layers depending on the distance from the wall surface. A laminar sublayer near the wall where no slip condition is applied and dominated by viscous forces and characterized by laminar flow. This laminar region is followed by a buffer region where the flow transitions from laminar to turbulent. The final layer of the flow region is the turbulent region where the inertial forces dominate the flow characteristics.

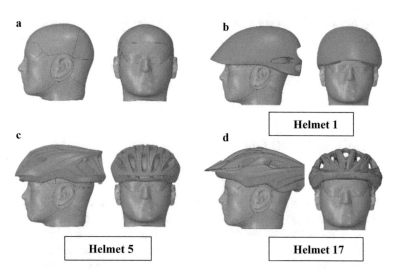

Fig. 1. Digital version of thermal manikin with (a) measurement zones: (b) closed channel helmet (helmet 1): (c) and (d) open channel helmets (helmet 5 and helmet 17)

Resolving the boundary layer is paramount because the flow separation, if not modelled accurately, can result in unrealistic results. The boundary layer that comprises of a viscous layer and a turbulent layer influences the flow on the boundaries and heat transfer. As the velocity of incoming air increases, the Reynolds number (Re) around the object increases resulting in a decrease in viscous sublayer thickness. This necessitates high resolution of the grid close to the walls/test object and is controlled by the dimensionless parameters $y*$ or y^+ for which the condition is $1 \le y*/y^+ \le 5$. The dimensionless quantity $y*$ is used instead of y^+ because $y*$ provides more information on the grid as well as on stagnation and reattachment points. The size of the cell ranges from 0.5 mm near the head surface to 100 mm near the domain walls. Tetrahedral cells (Fig. 3) were used to create the volume mesh with ortho-skewness values and aspect ratio within the allowable limits. The final mesh of the domain contained about 15–23 million cells depending on the helmet geometry and design.

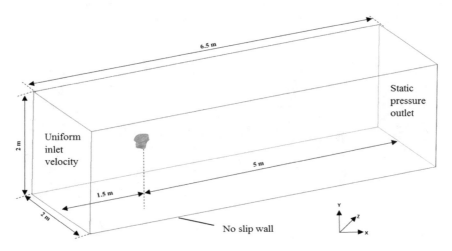

Fig. 2. Computational domain and boundary conditions for thermal manikin with helmet

3.3 Boundary Conditions

The boundary conditions used in the virtual testing of latent heat loss is same as the boundary conditions used in sensible heat loss tests [29]. At the inlet, uniform velocity of 6 m/s (21.6 kph) was imposed with default turbulent intensity value of 5% to assign a median relative air movement while cycling. The outlet of the domain is imposed with ambient static pressure. The surface of the head and helmet was modelled as a no-slip boundary wall with zero roughness. Temperature of incoming air was set to 20 °C (294 K) and the surface temperature of head zones was set to 30 °C (304 K). Each measurement zone was defined with a mass flux value of 0.00033 kg/m² s which approximately equals to a sweating rate of 2 g/h on each surface and liquid film (sweat) of constant thickness 0.3 mm was defined on the headform zones.

This research used Eulerian wall film methodology to define and analyze sweat layer that was modelled on the headform surface. This methodology is a built-in module available in Fluent that provide the user with information about thin liquid films on the wall surface. Latent heat transfer that is controlled mainly by the properties of the sweat layer was simulated using Eulerian wall film method for films with minimum thickness and the flow of film was assumed to be parallel to the surface. Mass, Momentum and Energy equations were solved separately for the film using second order discretization. This method also provides extensive details on film parameters and properties which can serve as a tool for identifying regions of low or high heat transfer.

3.4   Turbulence Models and Simulation Settings

The numerical simulations were performed using commercial CFD module Fluent which is based on finite-volume method. The realizable k-ε model with enhanced wall method for boundary layer resolution using one equation Wolfstein model was used to solve Reynolds-averaged Navier-Stokes (RANS) equations.

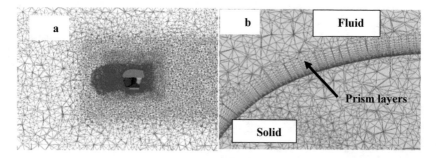

Fig. 3. Computational grid of manikin head and helmet: (a) grid showing different zones with different cell size; (b) prism layers on the surface (solid-fluid interface) to resolve boundary layer

The choice of realizable k-ε model was made based on experiments and existing literature which showed that realizable k-ε model was more accurate when compared to other models like standard k-ε, RNG k-ε, SST k-ω model or one equation model like Spalart-Allmaras model. SIMPLE algorithm was used to couple pressure and velocity. Second order discretization schemes were used for pressure, energy and liquid film. The convection terms and the viscous terms of the governing equations were solved using second order equations. The gradients used in discretizing convection and diffusion terms were computed using least square cell based method since it is computationally inexpensive and provides comparable results to other node-based methods. A steady state calculation was carried out with convergence of residuals monitored every iteration. The change in residual values determined the number of iterations and as the residual values of continuity (10^{-4}), momentum (10^{-6}), energy (10^{-6}), turbulence dissipation rate (10^{-4}) reach their specified values, the simulation was terminated. In addition to residual values, heat flux values on each measurement zone on the head were recorded for every iteration and change in these values for subsequent iterations was verified. The change in total heat transfer rate and film heat transfer rate for every iteration was also recorded and used as a parameter for deciding the number of iterations.

4   Numerical Simulations: Results

4.1   Thermal Manikin Without Helmet

Thermal manikin used in this latent heat loss testing is similar to that of the thermal manikin used in sensible heat loss testing [29] in terms of size and geometry. This is confirmed by comparing the surface area of measurement zones used in both tests as shown below in Table 1. This comparison of surface area proves that the results from the two tests can be compared. The surface area of neck region of both the manikins differ. However, this will not influence the final results since the neck region is not included in heat transfer calculations as it serves as a buffer zone between the headform and the base on which it is placed.

Table 1. Surface area of thermal manikins used in convective and evaporative heat loss virtual testing

Zone	Surface area (m²)	
	Latent heat loss manikin	Sensible heat loss manikin
Scalp	0.0638	0.0676
Forehead	0.0096	0.0141
Face	0.0258	0.0324
Ear-right	0.0068	0.0052
Ear-left	0.0069	0.0051
Neck	0.0392	0.0208

Table 2. Latent/evaporative heat loss from thermal manikin at 6 m/s

Zone	Area (m²)	Latent heat loss (W)
Scalp	0.0638	83.3
Forehead	0.0096	15.8
Face	0.0258	40.7
Ear-right	0.0068	9.3
Ear-left	0.0069	9.4
Total		**158.5**

Table 3. Latent/evaporative heat loss from manikin at 6 m/s with helmets

V = 6 m/s	Surface area (m²)	Nude head	Helmet 5	Helmet 17	Helmet 1
Scalp	0.064	83.3	69.2	71.6	73.0
Forehead	0.010	15.8	13.9	13.7	13.5
Face	0.026	40.7	34.2	34.2	34.2
Ear-right	0.007	9.3	8.4	8.5	8.4
Ear-left	0.007	9.4	8.6	8.8	8.6
Total (W)		**158.5**	**134.3**	**136.8**	**137.6**
Cooling efficiency (%)			84.7	86.3	86.8

Heat transfer values from virtual thermal manikin with the surface temperature maintained at 30 °C and prescribed with a constant sweat rate of 2 g/h on each measurement zone is shown in Table 2. Heat transfer from headform at 6 m/s prescribed with a constant sweat rate is 158.5 W. Heat transfer from scalp region with the largest surface area is 83 W followed by face with 41 W.

This is in direct correlation with sensible heat loss study where the heat loss from scalp and face regions were maximum [29].

4.2   Thermal Manikin with Helmets

The heat transfer values from individual measurement zones of manikin are tabulated in Table 3. The cooling efficiency which is a metric used to evaluate the thermal performance of helmets is also tabulated. The results from the simulations show a decrease in heat transfer from the headform with a helmet. This is as expected since helmets hinder the heat transfer by acting as an insulating layer. The scalp region of head which is covered by the helmet shows a decrease in heat transfer in the order of 12–16%. The forehead and face also shows decrease in heat transfer in the order of 14% and 16% respectively.

The thermal performance of the helmets with and without sweat can be evaluated by comparing the cooling efficiency values of the helmets with and without sweat as shown in Table 4. The efficiency values of helmets without sweat were obtained from a study on the similar topic [29].

Table 4. Comparison of cooling efficiency of helmets with and without sweat

V = 6 m/s	Cooling efficiency (%)	
Helmet	Convective cooling (no sweat)	Evaporative cooling (with sweat)
5	91.9	84.7
17	89.3	86.3
1	92.4	86.8

5   Discussion

5.1   Thermal Manikin Without Helmet

The heat transfer from headform measurement zones with a prescribed sweat rate is higher than heat transfer from the head tested without sweat layer (dry heat loss). This is as expected because the sweat layer which is modelled as a liquid water film with the same surface temperature (304 K) facilitates high heat transfer. These values of heat transfer are comparable to the values predicted by a study [31], which is in the order of 200–250 W. The difference in total heat loss value between this virtual test and the study by Rasch et al. [33] can be attributed to the following reasons. The sweat rate defined in this testing is of a constant value whereas in the study, the heat loss was measured from a human subject where there is a possibility of sweat rate varying considerably at different measurement regions. Another major influence on the results would be the thickness of sweat film. The thickness of the sweat film which was set to 0.3 mm for this virtual testing was obtained from running multiple simulations with varying film thickness (0.1 mm, 0.2 mm, 0.3 mm, 0.5 mm) and arriving upon the thickness value that matched best with the results of physical tests. This film thickness values can also be defined specific to a particular zone for more accurate results.

Presence of hair on the head in the test subjects from the study can also influence the heat transfer between the user and environment. However, this difference in heat transfer values between the physical testing of real subjects and virtual testing using thermal manikin will not influence the goal of this study, since this study focusses mainly on the thermal performance of helmets for evaporative heat loss using the same thermal manikin.

5.2   Thermal Manikin with Helmets

From the results (Table 3), it is evident that the helmets tested virtually are capable of removing the heat through evaporation to certain extent. The decrease in heat transfer from scalp region between 12–16% can be attributed to the insulation provided by the helmet. The flow pattern on front of the forehead shows a flow restriction by the front of helmet and hence a decrease in heat transferred from that region. The face region, in spite of zero interaction with the helmet shows decrease in heat transfer. Flow separation could be considered as one reason. However, the heat transfer from face region is constant (34.2 W) for all three helmets and hence overlooked. Since this study focusses mainly on heat loss through sweating and evaporation, it is important to establish the impact of the presence of a sweat layer of constant thickness (0.3 mm) on thermal performance of helmets. This can be achieved by comparing the efficiency of helmet with and without sweat as shown in Table 4. From the results, it is evident that the thermal performance of helmets with sweat decreases marginally when compared with the respective thermal performance values without sweat [29]. Sweat layer of constant thickness on the surface of head acts as an entrapment of heat generated from head during cycling. From this research it can be inferred that the helmet design must be evaluated for evaporation cooling in addition to pure convection cooling during design validation.

5.3   Limitations and Further Research

To the best knowledge of the authors, a CFD study about the influence of latent heat loss/sweating on the thermal performance of helmets has not yet been published. The present study is based on this assumption and hence a new methodology has been proposed to model the effects of sweat on thermal performance of helmets. However, the study is also subjected to some limitations briefly mentioned below.

A main limitation of the study is that the virtual testing performed using commercial CFD codes was not verified experimentally. Experimental verification with a thermal manikin that can simulate sweating will be performed in the near future. Several minute details on the helmet (e.g., visors) were not included in the model to simplify the computational mesh. Such geometrical features like visors, chin straps and straps to control fit should be included in the analysis. The sweat rate and the sweat film thickness can be improved by researching on individual measurement zones' sweat rates and applying zone specific sweat rates at predefined source locations. Furthermore, head sweat rate prediction equations developed by Taylor and Machado-Moreira [20], can be integrated into the model using user defined functions (UDF) to make virtual testing more accurate. This study was carried out for only three helmets due to computation and time constraints. More helmets with different vent profiles and designs should be tested to understand the extent to which this model can be used to validate thermal performance.

6 Summary and Conclusion

Thermal performance of three helmets for evaporative heat loss through sweating was analyzed using Computational Fluid Dynamics (CFD) simulations. Simulations were carried out at windspeed 6 m/s (21.6 kph) inside a wind tunnel using a thermal manikin modelled from a 50th percentile head shape of a western population that is derived from an anthropometric head shape model with individual measurement zones. The individual zones were maintained at 30 °C at an ambient temperature of 20 °C. The manikin remained vertical and no inclination was applied. The study shows that the helmets evaluated allow removing 84–87% of the heat from the head in the presence of sweat. These values are less as compared to 89–93%, cooling efficiency derived from sensible heat loss only [29]. This study clearly shows that the presence of sweat on the surface of head impacts the thermal performance of helmets.

References

1. Fife, D., Barancik, J.I., Chatterjee, B.F.: North-Eastern Ohio Trauma study: II, injury rates by age, sex, and cause. Am. J. Publ. Health **74**, 473–478 (1984)
2. Wood, T., Milne, P.: Head injuries to pedal cyclists and the promotion of helmet use in Victoria. Aust. Accid. Anal. Prev. **20**, 177–185 (1988)
3. Thompson, R.S., Rivara, F.P., Thompson, D.C.: A case control study of the effectiveness of bicycle safety helmets. N. Engl. J. Med. **320**, 1361–1367 (1989)
4. Attewell, R.G., Glase, K., McFadden, M.: Bicycle helmet efficacy: a meta-analysis. Accid. Anal. Prev. **33**(3), 345–352 (2001)
5. Olivier, J., Creighton, P.: Bicycle injuries and helmet use: a systematic review and meta-analysis. Int. J. Epidemiol. **46**, 278–292 (2017)
6. Sacks, J.J., Kresnow, M., Houston, B., Russell, J.: Bicycle helmet use among American children. Inj. Prev. **2**, 258–262 (1996)
7. Villamor, E., Hammer, S., Martinez-Olaizola, A.: Barriers to bicycle helmet use among Dutch pediatricians. Child Care Health Dev. **34**, 743–747 (2008)
8. Bogerd, C.C., Aerts, J.M., Annaheim, S., Bröde, P., De Bruyne, G., Flouris, A.D., Kuklane, K., Mayor, T.S., Rossi, R.M.: Thermal effects of headgear: state-of-the-art and way forward. Extrem. Physiol. Med. **4**(1), A71 (2015)
9. Servadei, F., Begliomini, C., Gardini, E., Giustini, M., Taggi, F., Kraus, J.: Effect of Italy's motorcycle helmet law on traumatic brain injuries. Inj. Prev. **9**, 257–260 (2003)
10. Orsi, C., Stendardo, A., Marinoni, A., Gilchrist, M.D., Otte, D., Chliaoutakis, J., Lajunen, T., Özkan, T., Pereira, J.D., Tzamalouka, G., Morandi, A.: Motorcycle riders' perception of helmet use: complaints and dissatisfaction. Accid. Anal. Prev. **44**, 111–117 (2012)
11. Papadakaki, M., Tzamalouka, G., Orsi, C.: Barriers and facilitators of helmet use in a Greek sample of motorcycle riders: which evidence? Transp. Res. F Traffic Psychol. Behav. **18**, 189–198 (2013)

12. Lehmuskallio, E., Lindholm, H., Koskenvuo, K., Sarna, S., Friberg, O., Viljanen, A.: Frostbite of the face and ears: epidemiological study of risk factors in Finnish conscripts. BMJ **311**, 1661–1663 (1995)
13. Zhang, H., Arens, E., Huizenga, C., Han, T.: Thermal sensation and comfort models for non-uniform and transient environments, part III: whole-body sensation and comfort. Build. Environ. **45**, 399–410 (2010)
14. Brühwiler, P.A., Ducas, C., Huber, R., Bishop, P.A.: Bicycle helmet ventilation and comfort angle dependence. Eur. J. Appl. Physiol. **92**, 698–701 (2004)
15. Liu, X., Holmer, I.: Evaporative heat transfer characteristics of industrial safety helmets. Appl. Ergon. **26**, 135–140 (1995)
16. De Bruyne, G., Aerts, J.M., Vander Sloten, J., Goffin, J., Verpoest, I., Berckmans, D.: Quantification of local ventilation efficiency under bicycle helmets. Int. J. Ind. Ergon. **42**, 278–286 (2012)
17. Nadel, E.R., Mitchell, J.W., Stolwijk, J.A.J.: Control of local and total sweating during exercise transients. Int. J. Biometeorol. **15**, 201–206 (1971)
18. De Bruyne, G., Aerts, J.M., Van der Perre, G., Goffin, J., Verpoest, I., Berckmans, D.: Spatial differences in sensible and latent heat losses under a bicycle helmet. Eur. J. Appl. Physiol. **104**, 719–726 (2008)
19. De Bruyne, G., Aerts, J.M., Vander Sloten, J., Goffin, J., Verpoest, I., Berckmans, D.: Transient sweat response of the human head during cycling. Int. J. Ind. Ergon. **40**, 406–413 (2010)
20. Taylor, N.A.S., Machado-Moreira, C.A.: Regional variations in transepidermal water loss, eccrine sweat gland density, sweat secretion rates and electrolyte composition in resting and exercising humans. Extrem. Physiol. Med. **2**, 4 (2013)
21. Bain, A., Deren, T., Jay, O.: Describing individual variation in local sweating during exercise in a temperate environment. Eur. J. Appl. Physiol. **111**, 1599–1607 (2011)
22. Brühwiler, P.A.: Heated, perspiring manikin headform for the measurement of headgear ventilation characteristics. Meas. Sci. Technol. **14**, 217–227 (2003)
23. Brühwiler, P.A.: Radiant heat transfer of bicycle helmets and visors. J. Sports Sci. **26**, 1025–1031 (2008)
24. Bogerd, C.P., Brühwiler, P.A.: The role of head tilt, hair and wind speed on forced convective heat loss through full-face motorcycle helmets: a thermal manikin study. Int. J. Ind. Ergon. **38**, 346–353 (2008)
25. Blocken, B., Defraeye, T., Koninckx, E., Carmeliet, J., Hespel, P.: CFD simulations of the aerodynamic drag of two drafting cyclists. Comput. Fluids **71**, 435–445 (2013)
26. Pinnoji, P.K., Haider, Z., Mahajan, P.: Design of ventilated helmets: computational fluid and impact dynamics studies. Int. J. Crashworthiness **13**, 265–278 (2008)
27. Stolwijk, J.A.: A Mathematical Model of Physiological Temperature Regulation in Man, Washington, USA (1971)
28. Fiala, D., Psikuta, A., Jendritzky, G., Paulke, S., Nelson, D.A., van Marken Lichtenbelt, W. D., Frijns, A.J.H.: Physiological modeling for technical, clinical and research applications. Front. Biosci. **S2**, 939–968 (2010)
29. Mukunthan, S., Kuklane, K., Huysmans, T., De Bruyne, G.: A comparison between physical and virtual experiments of convective heat transfer between head and bicycle helmet. In: Proceedings of AHFE, pp. 517–527 (2017)
30. Blatteis, C., Boulant, J., Cabanac, M., Cannon, B., Freedman, R., Gordon, C.J., Hales, J.R. S., Horowitz, M., Iriki, M., Janský, L., Jessen, C., Kaciuba-Uscilko, H., Kanosue, K., Kluger, M.J., Laburn, H.P., Nielsen-Johannsen, B., Mercer, J.B., Mitchell, D., Simon, E., Shibata, M., Szekely, M., Szelenyi, Z., Werner, J., Kozyreva, T.: Glossary of terms for thermal physiology. Jpn. J. Physiol. **51**, 245–280 (2001)

31. Gavhed, D., Mäkinen, T., Holmér, I., Rintämaki, H.: Face temperature and cardio-respiratory responses to wind in thermoneutral and cool subjects exposed to −10 °C. Eur. J. Appl. Physiol. **83**, 449–456 (2000)
32. Danckaers, F., Lacko, D., Verwulgen, S., De Bruyne, G., Huysmans, T., Sijbers, J.: A combined statistical shape model of the scalp and skull of the human head. In: Proceedings of AHFE, pp. 538–548 (2017)
33. Rasch, W., Cabanac, M.: Selective brain cooling is affected by wearing headgear during exercise. J. Appl. Physiol. **74**, 1229–1233 (1993)

Modeling of 3D Environments
for Collaborative Immersive Applications
Scenarios

Alinne Ferreira[1($\boxtimes$)], Jordan Rodrigues[2], Anselmo Paiva[2($\boxtimes$)],
Ivana Maia[1($\boxtimes$)], and João Leite[1($\boxtimes$)]

[1] Departamento Acadêmico de Desenho, Campus São Luís Monte Castelo,
Instituto Federal de Educação, Ciência e Tecnologia do Maranhão (IFMA),
São Luís, Brazil
alinnemartins_@outlook.com, ivana.maia@ifma.edu.br,
jpleite284@gmail.com
[2] Núcleo de Computação Aplicada – Universidade Federal do Maranhã
(UFMA), São Luís, MA 65080-805, Brazil
anselmo.c.paiva@gmail.com

Abstract. Collaborative immersive environments are an important way for communication and interaction between people located in distant areas. This paper presents the development of virtual worlds for this class of 3D applications. It focuses at their use in a collaborative virtual classroom for the improvement of the teaching methods. The paper includes a review of virtual worlds and 3D modeling studies, it discusses the role of ergonomics, with the purpose of designing 3D models that follow ergonomics guidelines and principles and the technical specifications. The developed worlds are expected to convey a realistic and immersive experience to the user.

Keywords: Virtual reality · Immersive environment
Physical and cognitive ergonomic

1 Introduction

Virtual Reality refers to the technology that provides maximum sensation of reality to individuals who uses it. This occurs with the use of immersive devices or virtual reality devices, which allow the immersion of the individual in a virtual environment. According to Pimentel [1], Virtual Reality (VR) is the use of high technology to convince the user that he finds himself in another reality, provoking his involvement completely.

Although it has been available for many decades, this technology has been a hit Brazil only recently. This fact is a consequence of the complexity in acquiring the devices, considering the high cost that they had. However, these resources are currently more accessible and are being adopted by companies and institutions as a means to stimulate and facilitate the teaching and execution of works.

In this way, an immersive classroom environment has been developed with the aim of allowing connected individuals to perform the same tasks performed in a physical

© Springer International Publishing AG, part of Springer Nature 2019
D. N. Cassenti (Ed.): AHFE 2018, AISC 780, pp. 79–85, 2019.
https://doi.org/10.1007/978-3-319-94223-0_8

classroom. According to Fuks [2], this interaction model aims at collaboration in three dimensions: communication, coordination and cooperation. In this way, Pimentel [3] defines that in a collaborative environment individuals need to exchange information (communication) and organize (coordination) so that they can work together in a shared environment (cooperation). This work contributes to remote communication based on the three dimensions cited.

2 Objective

The objective of this work was the development of a collaborative immersive environment, more specifically: a classroom, in which an individual connects to the environment and can perform the same tasks as the ones performed in a physical classroom, such as: slide presentation, videos, graphics, drawings and other activities. The real location of the "immersed" individuals is irrelevant. This fact is possible with the use of immersive devices such as RIFT glasses, prioritized in this project.

Thus, this work aims to promote a diversified interaction, which tends to stimulate and facilitate the teaching-learning process covering several areas of knowledge.

3 Methodology

In order to develop this project, several planned steps were established and accomplished with the purpose of assuring the acquisition of knowledge that enabled the good progress of the work.

In this way, the methodology consisted initially in a systematic analysis of scientific articles, dissertations and books whose subjects deal with studies on graphical interfaces, immersive environments, three-dimensional modeling and physical and cognitive ergonomics.

After the collection of information and in-depth studies on these topics, some technical norms and ergonomic standards were selected and used as guidelines during the modeling process of the furniture and classroom. Hence, this research based on the normative determinations of the Brazilian Association of Technical Norms.

4 Results

After the execution of the procedures mentioned above, in the section 'Methodology', the expected results of the project were obtained: an immersive classroom environment, which reflects an environment conducive to teaching because it has been adapted to the norms and pre-established standards.

4.1 3D Models

During the development process of the three-dimensional models (chairs, tables, place), the relevance of the study of ergonomics was observed, considering that, when applied correctly, it allowed the design of ideal models that actually comply with established norms in relation to the dimensions appropriate to school furniture, and thus, the immersive environment of the classroom has become extremely similar to the physical environment conducive to teaching. In addition to the furniture, place and other objects, characters were also modeled to be used as avatars by users.

This similarity between the real and the virtual/artificial, causes the sensation of reality to be driven in the individual connected to the environment, providing a better experience and ensuring the achievement of the objective of this work.

It should be noted that in this project, the norms established by the Brazilian Association of Technical Standards (ABNT) were followed, that is, all models behaved according to NBR 14006/2008 [4] (Brazilian standard), which specifies the appropriate dimensions of school furniture for an individual student (Figs. 1, 2, 3, 4, 5, 6, 7, 8 and 9).

Fig. 1. 3D model of the student's chair.

Fig. 2. 3D model of the teacher's chair.

Fig. 3. 3D model of the student's desk.

Fig. 4. 3D model of the teacher's desk.

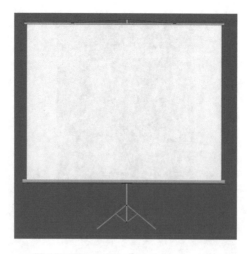

Fig. 5. 3D model of projection screen.

Fig. 6. 3D model of the classroom.

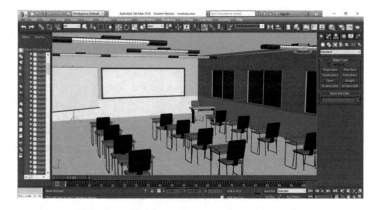

Fig. 7. 3D model of the classroom displayed in 3DS Max Software.

Fig. 8. 3D model of the avatar.

Fig. 9. 3D model of the avatar.

The avatar models confirm the anthropometric characteristics of mesomorphs suggested in Iida [5].

5 Discussion

The virtual reality is very useful considering that immersive collaborative environments are being increasingly required in several areas of knowledge and in the industrial field, as a means to optimize teaching processes, training and experiences. Industries, for example, opt for this technology in order to train and instruct employees in a safe way, avoiding work accidents, highlighting the effectiveness of virtual reality and how it makes numerous operations and practices easier and safer. As virtual reality offers a distinct method of interaction, the learning process tends to be intensified.

Educational institutions, by adhering to educational methodologies that encompass immersive collaborative environments, make a class more stimulating and productive, improving the students' performance proportionally.

Considering that the cost of virtual reality devices currently on the market is more affordable, the acquisition of these devices has become more common to a larger audience, spreading society's knowledge about this technology and fomenting the interest of big institutions.

6 Conclusion

Collaborative immersive environments enable interaction between physically distant people. From this fact, the scientific merit of the presented work is based on the contribution to the improvement of techniques of modeling and optimization of the development of three-dimensional environments. Through the research carried out, the relevance of the study of ergonomics and applications of its standards became evident, even in virtual models. With the design of ergonomically correct models, the virtual classroom starts to expose characteristics that make up a real classroom inducing the maximum sensation of reality in the connected individual.

In this work a collaborative immersive environment of a classroom was developed, which allows the connected user to visualize and interact performing the tasks commonly observed in a teaching place, demonstrating that there are other viable ways to promote the teaching- learning experience.

References

1. Pimentel, K., Texeira, K.: Virtual Reality – Through the New Looking Glass, 2nd edn. McGraw-Hill, New York (1995)
2. Fuks, H., Raposo, A.B., Gerosa, M.A., Pimentel, M., Filippo, D., Lucena, C.J.P.: Teorias e modelos de colaboração, pp. 16–33. Sistemas Colaborativos. Ed. Campus (2011)
3. Pimentel, M., Gerosa, M., et al.: Modelo 3C de Colaboração para o desenvolvimento de Siste-mas Colaborativos. Anais do III Simpósio Brasileiro de Sistemas Colaborativos – SBSC, pp. 58–67 (2006)
4. Norma Brasileira: ABNT 14006, Móveis escolares – Cadeiras e mesas para conjunto aluno individual, vol. 2, Rio de Jeneiro (2008)
5. Iida, I., Buarque, L.: Ergonomia, Projeto e Produção, 3 edn, Blucher, São Paulo (2016)

Modeling and Simulation Applications

Manned-Unmanned Teaming: US Army Robotic Wingman Vehicles

Ralph W. Brewer II[1(✉)], Eduardo Cerame[2], E. Ray Pursel[3],
Anthony Zimmermann[4], and Kristin E. Schaefer[1]

[1] United States Army Research Laboratory,
Aberdeen Proving Ground, MD, USA
{ralph.w.brewer.civ,
kristin.e.schaefer-lay.civ}@mail.mil
[2] US Army TARDEC, Warren, MI, USA
eduardo.j.cerame.civ@mail.mil
[3] Naval Surface Warfare Center, Dahlgren, VA, USA
eugene.pursell@navy.mil
[4] DCS Corp., Alexandria, VA, USA
anthony.j.zimmermann2.ctr@mail.mil

Abstract. Manned-unmanned teaming is the synchronization of Soldiers, manned and unmanned vehicles, and sensors that may improve situational understanding, greater lethality, and improved survivability during military operations. However, since unmanned vehicle autonomy capabilities are constantly advancing, it is difficult to integrate the human team and assess the performance of the team during early design. This work provides an overview of the US Army Wingman program and the human factors integration and assessment capabilities that support improved manned-unmanned teaming performance during joint gunnery operations. The discussion culminates with human integration and team assessment capabilities for interaction with respect to both fielded and software-in-the-loop simulation systems.

Keywords: Wingman · Autonomy · Human factors
Manned-unmanned teaming · Warfighter Machine Interface
Software-in-the-loop · Qualification

1   Introduction

The US Army seeks to identify current and emerging technologies and projections of technology-enabled concepts that could provide significant military advantage during operations in complex, contested, or congested environments between now and 2028. These include development of advanced technologies that support integration of joint manned-unmanned teaming (MUM-T) initiatives. As unmanned technologies advance from traditional teleoperation to more interdependent operations with advanced decision-making capabilities, it is essential to develop appropriate collaboration between the team members [1]. For effective teaming to occur, throughout the development life cycle of the technology, it must include human team members to advance the potential for trusted team development. A driving reason for this focus on the

© Springer International Publishing AG, part of Springer Nature (outside the USA) 2019
D. N. Cassenti (Ed.): AHFE 2018, AISC 780, pp. 89–100, 2019.
https://doi.org/10.1007/978-3-319-94223-0_9

human element in MUM-T operations is that effective teaming and appropriate use of the technology is dependent on the human's understanding of the system, its behaviors, and the reasoning behind those behaviors [2]. If human expectations do not match system behaviors, people will question the accuracy and effectiveness of the system's action [3–5] which can lead to degradations in trust which can be directly linked to misuse or disuse of the system, even if it is operating effectively [6, 7].

For this work, the US Army Wingman program provides a real-world MUM-T example for understanding some of these processes for integrating the human into the manned-unmanned team throughout system development. It further supports the US Army Asymmetric Vision/Decide Faster Initiative in three ways. First, there is a requirement for advanced training and simulation technology that will allow the Warfighter the capability to use decide faster skills when working with intelligent agents leading to advanced team performance. Second, development of tools should be synergistic and optimized for mission planning during multiple phases of operation. Third, technology that supports situation awareness (SA) is key to providing battlespace awareness during both day and night operations. The following sections of the paper provide an overview of the US Army Wingman program including an overview of the simulation and field assessment components (Sect. 2), human integration and assessment protocols (Sect. 3), and implications for future work (Sect. 4).

2   US Army Wingman Program

The goal of the US Army Wingman program is to provide robotic technological advances and experimentation to increase the autonomous capabilities of joint manned and unmanned ground combat vehicles. The team currently consists of a manned command vehicle (MCV) with a 5-man crew (right side Fig. 1) consisting of the manned vehicle driver, vehicle commander (VC), a long-range advanced scout surveillance system (LRAS3) operator, robotic vehicle operator (RVO), and robotic vehicle gunner (RVG), paired with a single unmanned weaponized robotic ground

Fig. 1. Real-world Wingman prototype vehicles

vehicle (left side Fig. 1). In the future, the single MCV will work cooperatively with multiple unmanned vehicles supporting MUM-T in complex, uncertain environments.

While there have been a number of research efforts that have looked at the processes for increasing the span of control of human-agent teams [8–10] and calibrating appropriate team size ratios [11, 12], much of the research is still in its infancy. Wingman's uniqueness is in looking at the human factors associated with MUM-T between a 5-man crew and a weaponized robotic vehicle conducting gunnery operations. Three of these roles in particular (VC, RVO, RVG) need to have a clear understanding of the system state and autonomous decision-making capabilities due to the very nature of this high-risk task. More specifically, the design of the system has a teaming aspect and requires oversight of the unmanned system with respect to human input for both mobility and gunnery operations.

Mobility operations are being developed to support multiple levels of autonomy including control-by-wire or teleoperation, waypoint finding, semi-autonomous driving via defined go-no go zones, and full autonomy. In line with most research discussions on levels of autonomy [13], it is unlikely that the RVG will maintain only a single type of control authority throughout a gunnery mission. Therefore, a goal for effective teaming is to assess the capability of the operator to appropriate toggle between control authority modes with respect to team or mission needs. What we aim to limit is inappropriate changes in control authority, or misuse of the system, due to a degradation in trust. To reach this end, the operator must have accurate and appropriate task, mission, and environmental information from the other human team members as well as from the Wingman vehicle.

Unlike mobility operations, there has been very little research in unmanned gunnery operations. Gunnery operations require a direct interaction between the RVG, VC, fire control system, and weapon. The autonomy-enabled features include finding potential targets within the weapon's field of view, tracking user-selected targets and keeping the weapon trained on those targets while applying lead based on range as well as user-applied adjustments. The human team members are still responsible for the global decision-making associating with firing on a target. As such, efficient team communication is required for target engagement. The VC ultimately authorizes engagement of a target, and the gunner is in-the-loop for the actual trigger pull. Due to the limitations in the current research, there is a specific need for research to characterize the complexities in finding, identifying, and engaging targets revolving around sensor and networking delays and the limited operator SA inherent in unmanned weapon systems.

Further, performance is a direct result of the MUM-T interoperability, where the manned vehicle is often located at a remote location from the Wingman vehicle outside of direct line of sight. Therefore, in support of prior research [2, 14], a technical solution for providing shared SA was key. Accomplishment if this goal rested with the development of the Warfighter Machine Interface (WMI), which provided interactive customizable displays for the VC, RVG, and RVO (see Fig. 2). Each Wingman WMI has access to shared SA data, categorized by subsystem across the bottom of each display, including major subsystems such as map, sensor, alerts, etc. The map screen provides an interactive aerial image, MIL-STD-2525B symbols, mobility plans, sensor fields-of-view, and grid reference lines. The sensor screen provides live video feeds with overlays providing SA such as azimuth, elevation, heading, and field-of-view. The VC and RVG use the sensor feeds to positively identify potential targets for engagement. Each WMI also has SA data available in a common toolbar and prioritized alerts visible as pop-ups at the top of the screen.

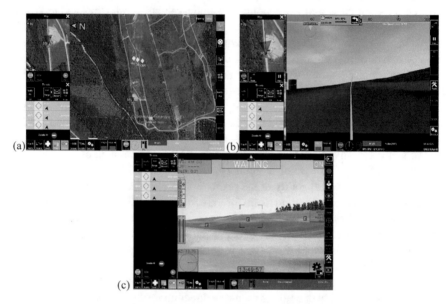

Fig. 2. WMI Screens whereby (a) is the VC screen with battle space objects represented by yellow clover leaf icons marking potential targets; (b) is the RVO screen where the green line on the driving camera marks the waypoint plan; and (c) is the RGV screen with red boxes denote system identified targets.

2.1   Assessing Real-World Gunnery Operations

Field testing the real-world system is important to improve technology integration of hardware and software sets, as well as test the system and team operations on a live gunnery test course. Real-world performance offers the collection of much needed valuable data. It is also possible to test technical challenges that are hard to simulate, such as intermittent communication between the robotic and command vehicles, physical hardware failures due to wear and tear, weather, or hardware specific software bugs, and safety limitations. In general, a much smaller portion of the data collected supports improvements in terms of the WMI usability and understanding how a war-fighter would execute the mission with the current capabilities. This is partially due to difficulties in getting operators other than engineers who are very familiar with the system, and limitations on the number of variations of missions we can execute. By the end of a test event, operators are effectively doing a rehearsed performance rather than a true combat exercise.

2.2 Development of the Simulation SIL

A software-in-the-loop (SIL) simulation supports assessment of MUM-T operations that were limited in field testing, as well as addresses a number of research-specific assessment needs. For example, using a simulation eliminates the safety issues and supply issues that come with real-world tests, especially when operating with live ammunition. Working with Warfighters then allows us to collect key feedback on how they would actually use these systems in a combat scenario including its usability, identification of key features that need to be streamlined, and the need for additional features to facilitate share SA and engender trust. Simulation also allows for execution of a wider range of scenarios than on traditional gunnery ranges. In general, the SIL allows for testing with the same autonomy software as the vehicle but with a focus on the operator's interaction with the system rather than a focus on the integration and performance of the autonomy software and hardware.

Alongside the human factors data collection capabilities, there are other valuable areas of use for the SIL with operator training, supporting software development, and as a presentation tool. Prior to test events involving robotic vehicles, training robotic operators on how to use the system properly is paramount. Normally, this consists of a brief presentation of the interface and then spending time on the range practicing to use the interface and learning the system. With the SIL, we reduce the need for the range time dedicated to training, and can even train additional operators while the real system is in use. For the software developers working on the autonomy algorithms, the SIL provides an additional method of testing new behaviors without the need of a physical asset. This reduces the amount of wasted time when assets are overbooked or undergoing repair. Finally, the complex nature of the Wingman system integration can make it difficult to convey to someone outside the program everything that goes on during a robotic gunnery exercise. The SIL provides an easier way for people who are new to the program so see and interact with the system without having to attend one of the infrequent test events.

3 Human Integration and Assessment

Integrating and assessing the human elements of teaming is a difficult process. For gunnery operations, Soldiers must be tactically and technically proficient in employing their weapon systems. The Army provides guidelines as defined in Training Circular (TC) 3-20.31 titled Training and Qualification, Crew [15] for commanders to train and evaluate all of their direct fire ground system crews. The manual breaks it down into three categories main gun equipped, air to ground missile, and mounted machine gun (MMG) platforms where a crew is defined as, any direct fire platform, wheeled or tracked, manned by multiple Soldiers to effectively employ that platform [15]. The Wingman crew is unlike any crew manned by just Soldiers. Instead, the addition of an autonomous platform with an autonomous remote engagement system (ARES) to the manned crew aboard the MCV transitions a manned gunnery team to MUM-T. Unfortunately, there are no standards set forth in the TC for MUM-T. Instead, we start by using the standards for the remote weapon station (RWS) gunnery from the MMG category. Using current action, conditions, and standards for remote weapon station gunnery exercises the SIL allows crews to progress through a series of increasingly more difficult gunnery exercises culminating in qualification during Table VI exercises (Table 1). The research and testing will drive the development of gunnery standards with MUM-T.

Table 1. Crew tables

Crew qualification		TC 3-20.21, training and qualification, crew
Prerequisites	Table I	Gunnery skills test
	Table II	Simulations
	Table III	Proficiency
Live training	Table IV	Basic
	Table V	Practice
Standard	Table VI	Qualification

MUM-T integration and assessment includes the capability to define training and qualification standards, conduct crew ratings, and identify additional training requirements. Those standards are ones taken directly from MMG qualification requirements (Table 2). The communication between a regular crew is through the vehicle intercom system. With this MUM-T, the typical audio communication is now digital between the two vehicles. Training of the vehicle crew evaluators to score a Wingman team will include fire commands, crew duties in response to a fire command, safety, and the timing of the engagements to determine a score.

Table 2. Table VI qualification standards

Table VI, qualification	
Action	Engage and destroy stationary and moving targets placed in a tactical array, during day and night and limited visibility from a stationary or moving vehicle using full caliber ammunition
Condition	Given the following • An operational wingman vehicle with weapon system • Appropriate 7.62 mm ammunition • A certified vehicle crew evaluator • Full scale targets to meet the scenario requirements
Standard	The crew must • Score a minimum of 700 of 1000 points overall • Score 70 points or more on all targets presented on seven of ten engagements • Qualify at least one night engagement

Specifically, assessment of human factors provides a means to assess the additions of the RVO and RVG to the traditional gunnery team. These efforts help to iteratively define and decrease the gap between autonomous vehicle control and required level of human collaboration. DIDEA is an acronym for detect, identify, decide, engage, and assess. Specific communication between team members is integral to the performance score. This includes identification of threat as friend or foe, classification of level of threat per target, order selection of targets, and firing commands. Trained vehicle crew evaluators will time the engagement process and provide feedback based on how quickly and safely the crew destroys the target(s). Assessment of penalties by the evaluators are for safety violations as well as incorrect crew duties in the DIDEA engagement process.

3.1 Human Integration

The WMI is the primary means by which MUM-T integration occurs. It supports functional integration for mobility and target engagement, as well as support for advanced situation awareness.

Mobility Engagement. At present, the WMI supports two main mobility modes. The most basic is teleoperation, where the user is directly controlling the vehicle's gear, throttle, brake, and steering, either through on screen controls or with an external gamepad. The other mode is waypoint following, where the operator creates a way-point plan through a dedicated planning screen, and then commands the vehicle to follow that plan. The operator has a range of options when generating the plan, including the type of waypoint placed, the maximum vehicle speed for each waypoint, the radius of the circle around the waypoint the vehicle must reach to register as hitting the waypoint, and how long the robot should wait a hold point in the plan. The type of waypoint chosen determines the type of path planner the robot uses, either an exploratory planner designed for sparse waypoints where the robot finds its own exact path, or a route following planner where the robot follows closely a much more densely populated waypoint plan. The operator through the record route option often generates these dense waypoint plans. This recorded route is savable and it is loadable along with manually generated waypoint plans for future operations. During a mission, if needed the robotic operator can pause the vehicle along its waypoint plan if necessary due to a safety issue with the robotic vehicle or if the vehicle command orders the vehicle to stop. If necessary, the operator can also modify the waypoint plan, either by changing parameters or locations of waypoints, and then reload the plan to the vehicle for it to take effect.

Target Engagement. The RVG has positive control of the ARES by using the WMI and a control grip controller or gamepad controller. The controller provides palm, trigger, and sensor controls such as focus and zoom. The WMIs primary use is target designation and engagement. Designation is by an external entity, such as via LRAS3 cues or ARES image recognition; or via RVG WMI actions, either by click or click-and-drag gestures on the ARES video or by adding a BSO to the tactical map screen. Targets appear in the RVG's threat portal. The prefix of the threat name shows the source of generated threats; a prefix of "LRAS3" indicates LRAS3-generated threats, a prefix of "SAF-T" (Supervised Autonomous Fires Technology) indicates ARES-generated threats, and no prefix (number-only) represents RVG map generated threats. Sorting of threats may be done by different criteria, such as range, name, type, etc. This helps the RGV determine the order to select targets. Upon receipt of a threat, the RVG may then tap "accept" or "decline" on the WMI to slew the turret for engagement. The RVG, when authorized by the VC, would be able to engage the threat using the controller, once positively identified as an intended target.

Improving SA. To support both mobility and target engagement, a series of tactical graphics are in development for the WMI (Fig. 3). These graphics allow the sharing of battlefield hazards and engagement tactics among the WMI users. Access to tactical graphics helps to develop a user's understanding of the battlefield, and WMI users are more likely to adapt to the mission in the presence of relevant SA data.

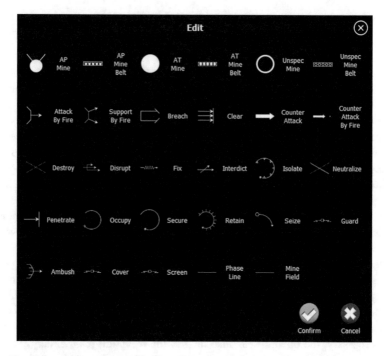

Fig. 3. Tactical graphics menu will allow the team to better understand the shaping of the battlefield

3.2   Recording and Assessing MUM-T Operations

Current manned vehicle gunnery evaluations record the speed at which crews engage targets during each engagement as major part of the scoring criteria. Vehicle crew evaluators sit in the gunnery range tower manually recording the actions of the crews during runs. To accomplish this within the SIL required the development of a number of processes to collect data to support MUM-T assessment from setting up customizable engagements to assessing interactions and performance.

Setting Up the Engagements. The design of a dynamic graphical user interface (GUI) incorporated the ability to set up scenarios and for adjusting scenarios during run-time to control target engagements (Fig. 4). Having the full complement of targets available in the GUI to schedule during run-time enables experimenters to control the pace and content of an experiment run and allows developers to test features that require specific locations and types of targets or sequence of targets. An additional capability was to load a scenario from a configuration file at run-time that allows a demonstration to be set, rehearsable, and easily reproduced. The specific GUI for each of these applications may be different in content, but the mechanisms that enable these capabilities are nearly identical. While being able to standardize and reproduce scenarios may enable training transfer of the simulation training device, the ability to vary starting conditions and dynamically adjust scenarios should reduce the learning effects of executing several experimentation runs using the same simulation.

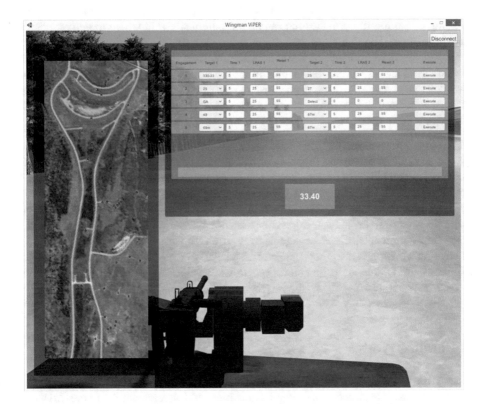

Fig. 4. Target selection GUI created in Unity to set up and maintain customizable engagements allowing the experimenter to choose from a set of possible targets and associated target features.

Assessing Interaction with Gunnery Operations. The underlying robot operating system (ROS) architecture supports both the platform and weapon system. Using a rosbag [16] allows for recording and playback of topics. Topics are buses over which nodes in the ROS environment exchange messages. The bag file collected in the SIL, for example, contains data from the "/dsat_weapon_payload" topic. This topic contains data on the weapon and weapon sight orientation, status of the weapon, status and commands (i.e. palm grip triggered, trigger pulled, commanded rotation percentage, etc.) sent to the weapon, status of the sight, and the vehicle's position and orientation.

Assessing Performance in the SIL. Performance data supports assessment of the Table VI qualification. As the Unity simulation component of the Wingman SIL presents and manages the targets and simulates the rounds fired, it is the authoritative source for the data concerning those elements. Events such as weapon firings, sensor message transmission, and target manipulations enter into an event queue and the simulation assigns a time of execution. As each event executes, the code writes a timestamped data entry to the log file. Similarly, when a projectile impacts an item and the identity of the item it struck is determined (target, ground, vehicle, etc.), the simulation records another timestamped data entry reflecting that event. Post processing of the log file provides the salient measurements with which to derive metrics and measures. Those metrics can range from number of shots fired per successful engagement, time from sensor cue to first shot, to time from sensor cue to first hit, and so on. The logging feature makes note of all the times targets go up, targets go down, weapons start and end firing times, and destruction of target times. These times are crucial at determining the score of each engagement coupled with an evaluation of each member's crew duties.

Assessing Interactions with the WMI. All VC, RVO, and RVG interactions with the WMI, as well as generated threat information, is available using the delivered WMI Logger executable. WMI interactions include cursor position, gesture type (as defined by Qt GUI framework), as well as timestamp; and is saved in a comma-separated value (.csv) format in the "logs" folder. The WMI uses logging to record each user interaction with the screen. Relevant user interactions include button presses, drag gestures, sensor play/stop commands, threat engagements, and any updates to displayed symbology. Threat information is also timestamped and labeled to match the threat portal, with the threat's latitude, longitude, and elevation. These logs provide an understanding of how a user interacts with the WMI, identifying possible efficiency improvements for future releases.

4   Future Work

The US Army Wingman program provides an interesting use case where we are able to study and explore aspects of human integration and assessment in future MUM-T operations. While this work may not be able to provide answers to all open teaming research questions, it does provide insight into some possible ways to assess human and team performance in both field and simulation testing. There are four main areas of future research relating to integration with autonomy.

First, different weapon firing patterns are required for engagement for different types of targets. For example, point target firing uses deliberate and accurate fire, while area targets use burst firing in a 'Z' pattern. Therefore autonomous identification of targets (e.g., vehicle or point target versus troops or area target) would assist in improved situation awareness and increased performance.

The second area of future research relates to human-to-autonomy communication throughout the firing engagement. Natural language processing would help to provide additional recognition to the ARES system so that upon a 'fire and adjust' command, the system could autonomous adjust to a target based on human feedback of where rounds are landing, or to autonomously slew to the next viable target following completion of the previous target. Making adjustments based on sensing and feedback could greatly reduce time and increase the accuracy of the firing platform and crew.

The third area of integration through the SIL is the development of the LRAS3 operator station and manned vehicle driver. Future integration will include incorporating the existing LRAS3 simulator interface (joystick and handgrips). This station will provide a better link to the real world, which is a means to provide better support for target detection and identification. The other position not currently modeled is the driver of the command vehicle. Addition of this station will include an outside-the-vehicle viewpoint that allows the driver to assist in target identification and control of the command vehicle movement within the simulation. The expansion of the outside-the-vehicle viewpoint will also include the vehicle commander's station. Lastly, the goal is to integrate all five manned vehicle stations on the actual HMMWV, allowing simulation while operators are in appropriate CV positions.

The fourth area of future research is to assess the human as a sensor using psycho-physiological measurement. Using wearable technologies, we will be able to calculate changes in behavior and psycho-physiological state associated with changes in stress, workload, and trust. By fusing this data, we will be able to quantify specific times when either changes in control authority should or could take place, as well as when to communicate additional information to improve shared SA and engender appropriate trust in the team.

5 Conclusion

The development of the SIL for the Wingman program provides advanced training through simulation allowing the Warfighter the capability to use decide faster skills to advance MUM-T. The WMI is a tool which is synergistic and optimized for mission planning during all phases of operation. Wingman project supports SA providing battlespace awareness currently just day operations with additional technologies in the future to conquer night operations.

Acknowledgments. The views and conclusions contained in this document are those of the authors and should not be interpreted as representing the official policies, either expressed or implied, of the Army Research Laboratory or the U.S. Government. The U.S. Government is authorized to reproduce and distribute reprints for Government purposes notwithstanding any copyright notation herein.

References

1. Phillips, E., Ososky, S., Grove, J., Jentsch, F.: From tools to teammates: toward the development of appropriate mental models for intelligent robots. Proc. Hum. Fact. Ergon. Soc. **55**(1), 1491–1495 (2011). https://doi.org/10.1177/1071181311551310

2. Chen, J.Y.C., Procci, K., Boyce, M., Wright, J., Garcia, A., Barnes, M.: Situation awareness-based agent transparency. Army Research Laboratory (2014)

3. Bitan, Y., Meyer, J.: Self-initiated and respondent actions in a simulated control task. Ergonomics **50**, 763–788 (2007)

4. Seppelt, B.D., Lee, J.D.: Making adaptive cruise control (ACC) limits visible. Int. J. Hum.-Comput. Stud. **65**, 192–205 (2007)

5. Stanton, N.A., Walker, G.H., Young, M.S., Kazi, T., Salmon, P.M.: Changing drivers' minds: the evaluation of an advanced driver coaching system. Ergonomics **50**, 1209–1234 (2007)

6. Lee, J.D., See, K.A.: Trust in automation: designing for appropriate reliance. Hum. Fact. J. Hum. Fact. Ergon. Soc. **46**, 50–80 (2004)

7. Schaefer, K.E., Straub, E.R.: Will passengers trust driverless vehicles? Removing the steering wheel and pedals. In: 2016 IEEE International Multi-Disciplinary Conference on Cognitive Methods in Situation Awareness and Decision Support (CogSIMA), pp. 159–165. IEEE (2016)

8. Chen, J.Y.C., Durlach, P., Sloan, J., Bowens, L.: Human-robot interaction in the context of simulated route reconnaissance mission. Mil. Psychol. **20**(3), 135–149 (2008). https://doi.org/10.1080/08995600802115904

9. Crandall, J.W., Goodrich, M.A., Olsen Jr., D.R., Nielsen, C.W.: Validating human-robot interaction schemes in multitasking environments. IEEE Trans. Syst. Man Cybern. Part A Syst. Hum. **35**(4), 438–449 (2005). https://doi.org/10.1109/TSMCA.2005.850587

10. Wang, H., Lewis, M.K., Velagapudi, P., Scerri, P., Sycara, K.: How search and its subtasks scale in N robots. In: Proceedings of the International Conference on Human Robot Interaction, pp. 141–148. ACM, New York (2009). https://doi.org/10.1145/1514095.1514122

11. Burke, J.L., Murphy, R.R.: RSVP: an investigation of remote shared visual presence as common ground for human-robot teams. In: ACM/IEEE Human-Robot Interaction, pp. 161–168. ACM, New York (2007)

12. Murphy, R.R., Griffin, C., Stover, S., Pratt, K.: Use of micro air vehicles at Hurricane Katrina. In: Proceedings of the International Workshop on Safety, Security, and Rescue Robots, p. 27. IEEE Press, Gaithersburg (2006)

13. Parasuraman, R., Sheridan, T.B., Wickens, C.D.: A model for types and levels of human interaction with automation. IEEE Trans. Syst. Man Cybern. Part A Syst. Hum. **30**(3), 286–297 (2000)

14. Schaefer, K.E., Straub, E.R., Chen, J.Y.C., Putney, J., Evans, A.W.: Communicating intent to develop shared situation awareness and engender trust in human-agent teams. Cogn. Syst. Res.: Spec. Issue Situat. Aware. Hum.-Mach. Interact. Syst. (2017). https://doi.org/10.1016/j.cogsys.2017.02.002

15. US Army Training and Doctrine Command: Training and qualification, crew. Training Circular No.: TC 3-20.31. Department of the Army (US), Washington (DC), 17 March 2015

16. Rosbag. http://wiki.ros.org/rosbag

Monitoring Task Fatigue in Contemporary and Future Vehicles: A Review

Gerald Matthews[1($\boxtimes$)], Ryan Wohleber[1], Jinchao Lin[1],
Gregory Funke[2], and Catherine Neubauer[3]

[1] Institute for Simulation and Training,
University of Central Florida, Orlando, FL, USA
{gmatthews, rwohlebe, jlin}@ist.ucf.edu
[2] Air Force Research Laboratory, Wright-Patterson Air Force Base,
Dayton, OH, USA
gregory.funke.1@us.af.mil
[3] U.S Army Research Laboratory, University of Southern California,
Los Angeles, CA, USA
catherine.neubauer@gmail.com

Abstract. This article reviews advancements in methods for detection of task-induced driver fatigue. Early detection of the onset of fatigue may be enhanced by spectral frequency analysis of the electrocardiogram (ECG) and analysis of eye fixation durations. Validity may also be improved by developing algorithms that accommodate driver sleep history assessed using mobile actigraphic methods. Challenges to development of fatigue indices include ensuring that metrics are valid across the range of task demands encountered by drivers. Future autonomous vehicles will place novel demands on the driver, and research is needed to test the applicability of current fatigue metrics.

Keywords: Driver fatigue · Safety · Autonomous vehicles · Actigraphy
Electrocardiogram · Eye tracking · Subjective stress

1 Introduction

Driver fatigue reflects multiple factors including lack of sleep, circadian rhythms, and the direct impact of the cognitive demands of driving ("task fatigue"). The neuroscience of sleep and circadian effects is quite well understood, but task fatigue has attracted less attention. It may be expressed as both active fatigue, reflecting prolonged cognitive overload, and passive fatigue, resulting from underload and monotony [1]. Different forms of task fatigue may have differential impacts on driver attention and behavior. The introduction of new vehicle technology may interact with the safety impacts of task fatigue. Studies have shown that automated vehicle operation rapidly leads to passive fatigue, and deficits in driver alertness persist following reversion to manual control [2]. This article reviews new methodological developments that may contribute to enhancing fatigue detection in both current and future vehicles.

Figure 1 shows an outline conceptual model that distinguishes fatigue factors which may be associated with inattention, and hence with performance and safety impairment.

© Springer International Publishing AG, part of Springer Nature 2019
D. N. Cassenti (Ed.): AHFE 2018, AISC 780, pp. 101–112, 2019.
https://doi.org/10.1007/978-3-319-94223-0_10

Distal factors are those that influence attention indirectly, though raising the driver's vulnerability to fatigue states which are the more direct and proximal influence.

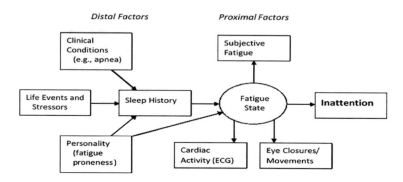

Fig. 1. Conceptual model of multiple driver fatigue factors.

Distal factors include sleep history; shortened or disturbed sleep may elevate the likelihood of experiencing fatigue during driving. Sleep history is influenced, in turn, by various other factors including clinical conditions, life events, and stressors that lead to disrupted sleep. The remaining distal factor shown in Fig. 1 is personality. In addition to general traits such as dispositional boredom, mind-wandering, and impulsivity, some individuals appear to be especially prone to fatigue in the specific context of driving [3]. Certain styles of appraisal and coping may promote mental disengagement from the driving task.

The fatigue state that develops during driving is the critical influence on inattention, and hence safety. Rating scales such as the Stanford Sleepiness Scale (SSS) [4] capture the progressive increase in sleepiness that eventually leads to behavioral impairment, e.g., from feeling foggy, to fighting sleep, to imminent sleep onset. SSS ratings of four (somewhat foggy) and above are associated empirically with higher crash risk [5]. However, for drowsiness detection it is important to identify relatively low levels of state fatigue that may represent the first sign of danger, prior to the driver actually experiencing substantial sleepiness. The fatigue state can be detected psychophysiologically [6], through well-known metrics such as increasing spectral power in the lower frequency bands of the electroencephalogram (EEG) [7] and the percentage of eye closure time (PERCLOS) [8]. Sensors can detect both current states changing over periods of minutes or so, and acute sleepiness episodes (microsleeps). There are numerous psychophysiological measures of fatigue, including EEG, but many are currently too intrusive to incorporate into an in-car monitoring device. Here, we focus on two of the less intrusive methods, the electrocardiogram (ECG) and eye tracking.

The remainder of this article reviews selected recent developments in fatigue assessment that may contribute to early detection of fatigue and drowsiness during driving. Of the distal factors shown in Fig. 1, we focus especially on objective assessment of sleep history using actigraphy. Wristband devices linked to a smartphone may be used for this purpose, facilitating practical application. We also review the three indices of fatigue state shown in the figure. Recent work on multidimensional models of stress and fatigue may inform measurement of subjective states. Traditionally, ECG studies have utilized decreased heart rate as a measure of lower arousal, and hence of fatigue [7]. More recent work has applied spectral frequency analysis and non-linear indices derived from the ECG to develop indices that are more sensitive to fatigue onset. Similarly, eye tracking methodologies are moving beyond PERCLOS to develop indices based on temporal and spatial patterns of eye fixation. We conclude by highlighting some challenges for these new fatigue assessment methodologies, including those associated with automation technology.

2 Assessment of Sleep History: Actigraphy

Actigraphy involves the measurement of movement over time with a wearable, usually portable, device. Sleep is inferred from a lack of movement, using algorithms that typically evaluate the amount of movement within a given time epoch (typically 1 min) in the context of movement in epochs before and after the epoch [8]. In newer algorithms, epoch activity is recorded typically at a rate of one data point every 2 s. Following processing to eliminate artifacts, a weighted score is then calculated based on epoch activity in the context of the surrounding epochs and a comparison is made with a threshold to decide if the epoch should be classified as "sleep" or "awake" [9]. These weights are typically determined by the fit they provide to data from polysomnography (PSG), a method considered the gold standard in sleep evaluation.

Following classification of epochs, several sleep parameters can be assessed including Sleep Latency (SL; amount of time from intention to sleep – or "lights out" – to sleep); Total Sleep Period (TSP; amount of time from onset of sleep to final waking); Total Sleep Time (TST; actual time spent asleep); Intermittent Awakenings/Wake After Sleep Onset (IA/WASO; total time spent awake after initial sleep onset); Midsleep Awakenings (MSA; number of wakenings between initial sleep onset and final wakening); and Sleep Efficiency (SE%; proportion of time between initial sleep onset and final awakening spent asleep).

Actigraphic algorithms have traditionally shown high rates of agreement with established EEG-based sleep assessment methods [10]. However, while agreement across sleep and wakeful epochs is often high, actigraphy is more apt to identify sleep correctly than wakefulness, particularly in transitions between sleep and wakefulness. Agreement between actigraphy and PSG is evaluated in terms of "sensitivity" – agreement in sleep classification – and "specificity" – agreement in awake classification, and accuracy. A typical finding [11] was that while actigraphy was a highly sensitive measure (97–99%), it had poor specificity (34–44%). That is, actigraphy tends to overestimate sleep latency, total sleep time, and sleep efficiency, but underestimates awakenings. In consequence, actigraphy tends to perform better in normal compared to clinical populations [10].

Actigraphy provides more accurate measurement of sleep history than subjective scales and daily sleep diaries [12]. A review [10] concluded that despite some methodological challenges, actigraphy was a cost-effective method useful for, among other things, gauging sleep patterns and circadian rhythms and identifying insomnia, sleep apnea, and transient and chronic sleep schedule disorders.

Recent work has focused especially on the validity of wearable sleep-tracking technologies available to consumers via a range of commercial products, and controlled via smartphone. General reviews of these products have typically been rather skeptical of their validity [13] but some may be suitable for research purposes. Fitbit devices, similar to the lab actigraphs reviewed above, show high sensitivity in detecting sleep but only moderate specificity in detecting wakefulness [14, 15]. These findings support the use of Fitbit products in applied research, provided their limitations are recognized.

In driving research, actigraphy supports selecting participants, monitoring compliance with instructions, and monitoring the impacts of manipulations. In a study comparing performance impacts of normal sleep vs. a two-hour sleep reduction [16], 7day actigraphic monitoring was used prior to the manipulation nights as a screener (participation required an average $\geq 85\%$ sleep efficiency), in addition to assessing sleep history in the different experimental conditions. The design allowed the researchers to demonstrate effects of sleep restriction that were distinct from those of subjective fatigue. Another study [17] employed actigraphy to verify compliance with a 7-day sleep schedule leading up to a simulated drive designed to test the combined influence of low doses of alcohol and restricted sleep on driving performance. Results indicated that even legal levels of alcohol consumption before driving can be dangerous when sleep is restricted. A simulated driving study [18] used actigraphy along with selfreported sleep to gauge the effect of a night on call and resulting sleep deprivation. Investigators were able to verify decreased sleep and quantify the relative sleep amounts between on and off call nights. In another naturalistic investigation [19], commercial driver sleep amounts were investigated to verify previously self-reported values and to investigate the impact of experiencing a "critical event" on sleep habits. Key findings include that increasing sleep requirements under the new regulations indeed resulted in more overall sleep levels, and that involvement in incidents was typically preceded by low driver sleep in the 24 h prior to the incident.

3 Subjective States: Multidimensional Models

Researchers have commonly treated subjective fatigue as a unitary state that can be measured with single ratings of sleepiness such as the SSS [4]. Recent work on the subjective states experienced during task performance suggests a multivariate approach may be more appropriate. Experimental and psychometric evidence identifies three fundamental state dimensions that define the person's experience of task performance [20]. These are task engagement (energy, task motivation, concentration), distress (tension, unhappiness, lack of confidence) and worry (self-focus, low performance selfesteem, intrusive thoughts about task and personal concerns). The multidimensional model has been validated in studies showing that states correlate with both psychophysiological indices and objective performance measures [20, 21].

Low task engagement, i.e., tiredness, demotivation and distractibility, is central to fatigue states [22]. However, depending on the task context, fatigue may also be associated with distress, or with elevations in worry associated with mind-wandering. It is thus important to evaluate the multidimensional pattern of response during fatiguing drives. A study that manipulated active and passive fatigue confirmed different response patterns [2], consistent with theory [1]. Active fatigue was induced by forcing the driver to compensate for frequent wind-gusts; it produced moderate magnitude declines in task engagement, coupled with increased distress. By contrast, passive fatigue induced by vehicle automation elicited large-magnitude loss of task engagement, and little change in distress, relative to a control condition. These different patterns of fatigue response also had behavioral implications; speed of response to an emergency event was slowed only with passive fatigue [2]. The impacts of different forms of phone use in the vehicle may be assessed similarly; texting appears to be associated primarily with elevated distress [23], whereas a phone conversation raised task engagement, relative to no phone use [24].

Subjective state measures have limited practical utility for fatigue monitoring, but they are valuable for research on objective measures because they can define the context within which objective assessments may be diagnostic. For example, diagnosticity of psychophysiological measures may vary according to whether passive or active fatigue is induced [6, 25]. Profiling the subjective response in an experimental study defines the fatigue state experienced by drivers. Similarly, high levels of worry in the study indicate that mind-wandering may be a threat to safety, and diagnostic monitoring can be designed accordingly.

4   ECG: New Metrics

Traditionally, heart rate is interpreted as an autonomic arousal index, reflecting sympathetic and parasympathetic influences. Fatigue and drowsiness are low-arousal states, and thus heart rate should decrease [7]. Studies have confirmed that heart rate declines when drivers become fatigued [26]. Heart rate variability (HRV) has also been proposed as a fatigue index based on a study of prolonged real driving [27], which showed that HRV increased markedly with driving time, but dropped when events which re-alerted drivers occurred. As sensor technology provides increasing scope for unobtrusive heart rate monitoring, it is timely to consider whether novel metrics for fatigue may improve over traditional ones.

4.1   Spectral Frequency Metrics

The power spectral density (PSD) of the ECG is analyzed using the Fast Fourier transform to calculate power in different frequency. Fatigue researchers distinguish high frequency (HF: 0.15–0.4 Hz) and low frequency (LF: 0.04–0.15 Hz) bands. Some studies also analyze very low-frequency (VLF: 0.003–0.04 Hz) and ultra-low frequency bands (ULF: 0.0–0.0033 Hz). HF and LF bands may reflect parasympathetic and sympathetic control of heart rate, respectively [28]. Drowsiness without stress is then represented by high HF and low LF, supporting the use of the LF/HF ratio as a measure of drowsiness, i.e., as low values for the ratio. The ratio is also interpreted as a measure of sympathovagal balance reflecting autonomic readiness. From this perspective, "fatigue" may be defined as a combination of stress and sleepiness that may occur if the driver is actively trying to counter sleepiness and stay awake [28]. High HF/low LF might thus represent passive fatigue, with high HF/high LF corresponding to active fatigue. These interpretations are based on a simplified view of cardiac physiology that ignores interaction between sympathetic and parasympathetic branches and other issues. However, even if spectral power indices are not pure measures of cardiac functions, they may still be useful in the applied context.

Several studies have confirmed the prediction that LF/HF ratio should be lower in fatigue states. A study of 12 truck drivers [29] asked participants to sleep two hours less than usual and perform a 15-min "active" simulated drive designed to maintain alertness, followed by a monotonous drive of up to 2 h. The LF/HF ratio was significantly lower in the fatigued state (1.2) than in the alert state (1.7). In another study [30], participants drove for 80 min on a simulated monotonous highway. For each individual, the drive was divided into an initial alert phase followed by a drowsy phase time marked at the point the driver showed behavioral signs of sleepiness such as long eye blinks and yawning. For individuals exhibiting sleepiness, HF power was higher in the drowsy phase, LF was lower, and the LF/HF ratio was lower. There was no significant effect on mean heart rate.

Other studies have shown moderator effects, i.e., the ratio index only appears to be diagnostic of fatigue in certain circumstances. In one such study, 15 participants performed monotonous simulated drives at different times of day [31]. Lower LF/HF was diagnostic of drowsy state only during night time hours. In another investigation [32] participants performed three 15-min simulated drives with different landscapes. With an open, presumably monotonous, landscape, the LF/HF ratio declined after about 10 min, suggesting possible sensitivity to rather minor levels of fatigue. An experiment on exercise effects [33] had two groups of 20 participants perform a 120-min simulated drive. One group had a physical exercise break half way through. The control group, but not the exercise group, showed decreases in LF/HF ratio at the end of the drive.

Taken together, these studies suggest robust effects of drowsiness and fatigue on LF/HF ratio, especially when simulated drives are designed to be monotonous. However, some studies reported contrary findings. One study [34] had participants perform five relatively short drives intended to provoke different mental states. A tiring drive elicited subjective sleepiness, lowered heart rate, and increased HRV. However, both LF and HF power increased, and the ratio was little changed. Another relatively largescale study of 60 drivers performing on a 90-min simulated drive found that LF/HF ratio *increased* at the end of the drive, as did HF, contrary to expectation [35]. A seat vibration manipulation increased the magnitude of these effects. The article states that participants were required to keep a small red dashboard light illuminated by maintaining accelerator pressure within prescribed limits. Possibly, the sustained attention requirements of this task required alertness and suppressed parasympathetic control of variability.

There is a rather separate research literature in which LF/HF is used as an index of stress or overload in various contexts including vehicle driving [36]. In addition, the 0.1 Hz component of the ECG, coinciding with LF, has quite a long history as a workload index [37].

4.2 Non-linear Metrics

A few recent studies have utilized additional metrics based on non-linear analyses. A study [38] had 12 participants perform for 120 min on a simulator that required "active engagement" (frequent control movements). They calculated two non-linear measures from the ECG, Lempel-Ziv complexity and sample entropy. Both measures declined substantially over time. By contrast, another simulator study [39] found that a different measure of entropy, Approximate Entropy (ApEn), was higher at the end of a fatiguing drive than at the beginning. Finally, there is a report that the energy of the ECG signal was higher in a drowsy state, and that this fatigue index was more reliable across different times of day compared with LF/HF [31].

5   Eye Tracking: Beyond PERCLOS

Common eye tracking metrics used to gauge sleepiness, drowsiness, and fatigue relate to blink rate and eye lid closure duration. The most prominent of these is PERCLOS [8], which is considered a standard drowsiness gauge by many [40]. PERCLOS is the proportion of time that a person's eyes are more than 80% closed and, at its conception, served as the definitional measure of drowsiness [8]. Other research has since used a more liberal 70% closed version of the metric [40]. The seminal work on PERCLOS [8] stipulated that PERCLOS of 7.5% or less indicated that a person was "awake", the 7.5% to 15% range was questionable, and PERCLOS greater than 15% indicated the person was "drowsy". While PERCLOS has been successful for predicting performance on vigilance tasks, more so even than subjective sleepiness ratings [40], it does have some drawbacks. These include limited diagnosticity of performance impairment, sensitivity to individual differences [41], and insensitivity to certain types of fatigue state such as "driving without awareness" (DWA) or "highway hypnosis" [42]. In this state, drivers are able to keep their lane, but cannot respond adequately to unexpected events.

5.1 Fixation Duration and Drowsiness

An alternative eye tracking method for gauging fatigue utilizes fixation duration. Eye movements consists of frequent, quick movements called saccades, interspersed with periods of steady gaze called fixation [43]. During fixations, perception and cognitive activity occur and extended fixations can indicate difficulty in extracting information. Specifically, as a person struggles to maintain focus and attention with fatigue, fixations lasting 150–900 ms, which are associated with cognitive processing, decrease. Fixations longer than 900 ms, indicative of staring, and less than 150 ms, which may relate to low level unconscious control but not deep processing, increase [41]. Mean fixation duration relates more to workload than to fatigue [43]. Eye tracker recording at 5001000 Hz is typically recommended for analyses of fixation duration.

5.2   Recurrence Quantification Analysis (RQA)

Another, non-linear, eye tracking metric, recurrence quantification analysis (RQA), may be useful for quantifying the amount and characteristics of stereotyped behavior, like the eye movements said to accompany DWA. RQA is a method of quantifying the amount of patterning and dynamic structure in a time series [44]. The rationale for RQA stems from the dynamical systems conception that the behavior of any complex system, which is variable but generally stable, is a consequence of interactions between highly variable underlying components and processes [45]. The influence of these many components and their interactions on the collective behavior of the system make it possible for information about the underlying dynamics of a potentially multivariate system to be gleaned from a single scalar time series. As mentioned above, severe fatigue can induce a driving hypnosis or DWA state that is characterized by eye movement stereotypy [42]. Thus, finding high determinism in a time series may indicate an early onset of fatigue. Theoretically, heightened determinism in behavioral variance is indicative of rigid over-control, or high constraint, which can produce stable behavior in a predictable environment, but lacks the flexibility to allow the system to adjust to unpredictable environmental features [45].

6   Methodological Challenges

Thus far, we have identified a range of novel measures that may be incorporated into studies of task-induced driver fatigue and drowsiness to enhance fatigue detection. However, various methodological challenges must be met to optimize fatigue diagnosis in empirical studies.

Variation in Study Design. Various methods are used for comparing response in fatigued and non-fatigued conditions. Each has its own strengths and weaknesses, but insufficient effort has been devoted to establishing convergence between different methods. Commonly, time-on-task is used as a proxy for fatigue, e.g., by having participants perform a monotonous, long-duration simulated drive for a time period up to 120 min [33, 39]. This method has the strength that build-up of fatigue can be continuously. A weakness is that it does not accommodate individual differences in rate of fatigue development. In addition, other factors such as boredom and motivation may change during the drive, potentially affecting responses.

An alternative approach is to generate a contrast between qualitatively different states of fatigue and alertness, comparing fatigue responses in drives of different characteristics [29, 34]. This may be done within-subjects. Researchers may also define a behavioral criterion for fatigue such as yawning, and apply it to determine the transition between alertness and fatigue in a single drive [30]. Fatigue states may also be defined over short time intervals. The ECG can be used to identify short, time-limited episodes of sleepiness [46]. Generally, these methods may provide stronger contrasts between states than in other methods, but there is also a risk that definition of states may be imperfect, given inter-individual variability in response.

Control of Task Demands. Most experimental studies tend to utilize undemanding, low workload drives that are probably conducive to fatigue induction. Conversely, elevated workload or stress, whether or not intended, may threaten the validity of certain indices, such as the LF/HF ratio from the ECG [28], through generating sympathetic arousal. Similarly, with eye tracking metrics, it is important to be able to distinguish changes in general state from changes that are driven by the traffic environment. For example, if we see a decrease in gaze dispersion during driving, does that reflect a general state of fatigue or a strategic decision to pay less attention to roadside objects? Future research requires more systematic exploration of the role of task demands, including workload transitions: threats to safety may arise when low workload is interrupted by events that raise workload.

Generalization to Real Driving. Laboratory-based simulation studies are important for controlling cognitive demand and stress factors, but of course lack realism and ecological validity. However, by using non-intrusive measures using actigraphy, ECG and eye tracking, research may proceed by building diagnostic models based on carefully-designed simulation studies, and then checking for generalization to realdriving. A challenge is that on-road studies often use rather small sample sizes, which may also lead to issues of participant selection and representativeness.

Development and Construction of Algorithms. Fatigue diagnosis may be enhanced by developing algorithms that integrate information from multiple sensors [31]. Given that different fatigue indices are not always highly correlated, this appears to be a wellfounded strategy. Algorithms can also be individualized to weight the specific responses that are most diagnostic in the individual driver. One challenge is that much work of this kind is highly data-driven, rendering algorithms vulnerable to chance outcomes, and of questionable generality. Also, such studies sometimes fail to report basic findings on how fatigue influenced the individual measures, which makes it hard to understand the underlying neuroscience basis.

Application to Next-Generation Vehicles. The challenge of accommodating fatigue diagnosis to different levels and types of task demand is especially acute because of the changing nature of driving, as vehicles become more autonomous. For example, fatigue may impact the driver's ability to take over manual control of the vehicle following an interval of automation. A diagnostic index of fatigue should predict performance following both voluntary manual take-over, and involuntary take-over, as when the automation recognizes an emerging traffic situation it cannot handle. In addition, full automation will increase the scope for drivers utilizing in-car information and entertainment systems during periods of automated vehicle control. It is unclear how these novel activities will impact both fatigue itself [22] and the diagnosticity of psychophysiological fatigue indices for attentional impairment following reversion to manual control. Next generation technology also provides additional scope for fatigue countermeasures. Currently, in-car fatigue monitors do no more than warn the driver, but future vehicles may have an adaptive automation capability, i.e., the automation could take control when driver fatigue is detected [6].

7 Conclusions

Existing driver fatigue research has been quite successful in identifying diagnostic markers for states of moderate-to-severe sleepiness that can be incorporated into vehicle driver status monitoring devices. Technological advances in acquiring and processing data from non-intrusive and/or non-contact sensors promise to improve on current methods. However, several methodological challenges remain. First, it remains important to detect fatigue at an early stage to pre-empt possibly dangerous levels of inattention. This article has reviewed how developments in ECG and eye tracking methods may contribute to enhancing sensitivity in fatigue detection. Second, interindividual variability in response to fatigue is a known issue for fatigue diagnosis. Use of actigraphy via consumer devices to assess sleep history and personalization of algorithms provide a possible solution. Third, the nature of the driving task will change radically as increasingly autonomous vehicles are introduced. The challenge of ensuring that fatigue detection methodologies remain valid has yet to be addressed.

Acknowledgments. Gerald Matthews and Ryan Wohleber gratefully acknowledge research support from DENSO Corporation.

References

1. Desmond, P.A., Hancock, P.A.: Active and passive fatigue states. In: Hancock, P.A., Desmond, P.A. (eds.) Stress, Workload and Fatigue, pp. 455–465. Lawrence Erlbaum, Mahwah (2001)
2. Saxby, D.J., Matthews, G., Warm, J.S., Hitchcock, E.M., Neubauer, C.: Active and passive fatigue in simulated driving: discriminating styles of workload regulation and their safety impacts. J. Exp. Psychol. Appl. **19**, 287–300 (2013)
3. Matthews, G.: Towards a transactional ergonomics for driver stress and fatigue. Theor. Issues Ergon. Sci. **3**, 195–211 (2002)
4. Hoddes, E., Zarcone, V., Smythe, H., Phillips, R., Dement, W.C.: Quantification of sleepiness: a new approach. Psychophysiology **10**, 431–436 (1973)
5. Connor, J., Norton, R., Ameratunga, S., Robinson, E., Civil, I., Dunn, R., Jackson, R.: Driver sleepiness and risk of serious injury to car occupants: population based case control study. Br. Med. J. **324**, 1125–1128 (2002)
6. Wohleber, R.W., Matthews, G., Funke, G.J., Lin, J.: Considerations in physiological metric selection for online detection of operator state: a case study. In: Schmorrow, D., Fidopiastis, C. (eds.) Foundations of Augmented Cognition: Neuroergonomics and Operational Neuroscience. Springer, Cham (2016)
7. Borghini, G., Astolfi, L., Vecchiato, G., Mattia, D., Babiloni, F.: Measuring neurophysiological signals in aircraft pilots and car drivers for the assessment of mental workload, fatigue and drowsiness. Neurosci. Biobehav. Rev. **44**, 58–67 (2014)
8. Wierwille, W.W., Wreggit, S.S., Kirn, C.L., Ellsworth, L.A., Fairbanks, R.J.: Research on Vehicle-Based Driver Status/Performance Monitoring; Development, Validation, and Refinement of Algorithms for Detection of Driver Drowsiness (No. HS-808 247 VPISU ISE 94-04) (1994)
9. Mullaney, D.J., Kripke, D.F., Messin, S.: Wrist-actigraphic estimation of sleep time. Sleep **3**, 83–92 (1980)

10. Sadeh, A., Hauri, P.J., Kripke, D.F., Lavie, P.: The role of actigraphy in the evaluation of sleep disorders. Sleep **18**, 288–302 (1995)
11. de Souza, L., Benedito-Silva, A.A., Pires, M.L.N., Poyares, D., Tufik, S., Calil, H.M.: Further validation of actigraphy for sleep studies. Sleep **26**, 81–85 (2003)
12. Monk, T.H., Buysse, D.J., Rose, L.R.: Wrist actigraphic measures of sleep in space. Sleep **22**, 948–954 (1999)
13. Ko, P.R.T., Kientz, J.A., Choe, E.K., Kay, M., Landis, C.A., Watson, N.F.: Consumer sleep technologies: a review of the landscape. J. Clin. Sleep Med. **11**, 1455–1461 (2015)
14. de Zambotti, M., Baker, F.C., Willoughby, A.R., Godino, J.G., Wing, D., Patrick, K., Colrain, I.M.: Measures of sleep and cardiac functioning during sleep using a multi-sensory commercially-available wristband in adolescents. Physiol. Behav. **158**, 143–149 (2016)
15. Kang, S.G., Kang, J.M., Ko, K.P., Park, S.C., Mariani, S., Weng, J.: Validity of a commercial wearable sleep tracker in adult insomnia disorder patients and good sleepers. J. Psychosom. Res. **97**, 38–44 (2017)
16. Philip, P., Sagaspe, P., Moore, N., Taillard, J., Charles, A., Guilleminault, C., Bioulac, B.: Fatigue, sleep restriction and driving performance. Accid. Anal. Prev. **37**, 473–478 (2005)
17. Vakulin, A., Baulk, S.D., Catcheside, P.G., Anderson, R., van den Heuvel, C.J., Banks, S., McEvoy, R.D.: Effects of moderate sleep deprivation and low-dose alcohol on driving simulator performance and perception in young men. Sleep **30**, 1327–1333 (2007)
18. Ware, J.C., Risser, M.R., Manser, T., Karlson Jr., K.H.: Medical resident driving simulator performance following a night on call. Behav. Sleep Med. **4**, 1–12 (2006)
19. Hanowski, R.J., Hickman, J., Fumero, M.C., Olson, R.L., Dingus, T.A.: The sleep of commercial vehicle drivers under the 2003 hours-of-service regulations. Accid. Anal. Prev. **39**, 1140–1145 (2007)
20. Matthews, G.: Multidimensional profiling of task stress states for human factors: a brief review. Hum. Fact **58**, 801–813 (2016)
21. Matthews, G., Reinerman-Jones, L., Abich IV, J., Kustubayeva, A.: Metrics for individual differences in EEG response to cognitive workload: optimizing performance prediction. Pers. Indiv. Differ. **118**, 22–28 (2017)
22. Matthews, G., Neubauer, C.E., Saxby, D.J., Wohleber, R.W., Lin, J.: Dangerous intersections? A review of studies of fatigue and distraction in the automated vehicle. Accid. Anal. Prev. (in press)
23. Neubauer, C.E., Matthews, G., Saxby, D.J.: The effects of cell phone use and automation on driver performance and subjective state in simulated driving. In: Proceedings of the Human Factors and Ergonomics Society, vol. 56, pp. 1987–1991 (2012)
24. Neubauer, C.E., Saxby, D.J., Matthews, G.: Fatigue in the automated vehicle: do games and conversation distract or energize the driver? In: Proceedings of the Human Factors and Ergonomics Society Annual Meeting, vol. 58, pp. 2053–2057 (2014)
25. May, J.F., Baldwin, C.L.: Driver fatigue: the importance of identifying causal factors of fatigue when considering detection and countermeasure technologies. Transp. Res. F Traffic Psychol. Behav. **12**, 218–224 (2009)
26. Jap, B.T., Lal, S., Fischer, P., Bekiaris, E.: Using EEG spectral components to assess algorithms for detecting fatigue. Expert Syst. Appl. **36**, 2352–2359 (2009)
27. O'Hanlon, J.F.: Heart rate variability: a new index of driver alertness/fatigue (No. 720141). SAE Technical Paper (1972)
28. Vicente, J., Laguna, P., Bartra, A., Bailón, R.: Drowsiness detection using heart rate variability. Med. Biol. Eng. Comput. **54**, 927–937 (2016)
29. Patel, M., Lal, S.K., Kavanagh, D., Rossiter, P.: Applying neural network analysis on heart rate variability data to assess driver fatigue. Expert Syst. Appl. **38**, 7235–7242 (2011)

30. Awais, M., Badruddin, N., Drieberg, M.: A hybrid approach to detect driver drowsiness utilizing physiological signals to improve system performance and wearability. Sensors **17**, 1991 (2017)
31. Sahayadhas, A., Sundaraj, K., Murugappan, M.: Drowsiness detection during different times of day using multiple features. Australas. Phys. Eng. Sci. Med. **36**, 243–250 (2013)
32. Zhao, X., Wei, Z., Li, Z., Zhang, Y., Feng, X.: Threshold research on highway length under typical landscape patterns based on drivers' physiological performance. Discret. Dyn. Nat. Soc. 1–15 (2015)
33. Liang, W.C., Yuan, J., Sun, D.C., Lin, M.H.: Changes in physiological parameters induced by indoor simulated driving: effect of lower body exercise at mid-term break. Sensors **9**, 6913–6933 (2009)
34. Schmidt, E., Decke, R., Rasshofer, R.: Correlation between subjective driver state measures and psychophysiological and vehicular data in simulated driving. In: 2016 IEEE Intelligent Vehicles Symposium (IV), pp. 1380–1385. IEEE (2016)
35. Jiao, K., Li, Z., Chen, M., Wang, C., Qi, S.: Effect of different vibration frequencies on heart rate variability and driving fatigue in healthy drivers. Int. Arch. Occup. Env. Health **77**, 205–212 (2004)
36. Muñoz-Organero, M., Corcoba-Magaña, V.: Predicting upcoming values of stress while driving. IEEE Trans. Intell. Transp. Syst. **18**, 1802–1811 (2017)
37. Nickel, P., Nachreiner, F.: Sensitivity and diagnosticity of the 0.1-Hz component of heart rate variability as an indicator of mental workload. Hum. Fact. **45**, 575–590 (2003)
38. Wang, L., Wang, H., Jiang, X.: A new method to detect driver fatigue based on EMG and ECG collected by portable non-contact sensors. PROMET-Traffic Transp. **29**, 479–488 (2017)
39. Zhao, C., Zhao, M., Liu, J., Zheng, C.: Electroencephalogram and electrocardiograph assessment of mental fatigue in a driving simulator. Accid. Anal. Prev. **45**, 83–90 (2012)
40. Dinges, D.F., Mallis, M.M., Maislin, G., Powell, IV, J.W.: Evaluation of Techniques for Ocular Measurement as an Index of Fatigue and the Basis for Alertness Management (Monograph No. DOT HS 808 762). National Highway Traffic Safety Administration, Washington, DC (1998)
41. Schleicher, R., Galley, N., Briest, S., Galley, L.: Blinks and saccades as indicators of fatigue in sleepiness warnings: looking tired? Ergonomics **51**, 982–1010 (2008)
42. Briest, S., Karrer, K., Schleicher, R.: Driving without awareness: examination of the phenomenon. In: Gale, A. (ed.) Vision in Vehicles XI, pp. 89–141. Elsevier, Amsterdam (2006)
43. Poole, A., Ball, L.J.: Eye tracking in HCI and usability research. In: Encyclopedia of Human Computer Interaction, vol. 1, pp. 211–219 (2006)
44. Russell, S.M., Funke, G.J., Flach, J.M., Watamaniuk, S.N., Strang, A.J., Miller, B.T., Dukes, A., Menke, L., Brown, R.: Alternative indices of performance: an exploration of eye gaze metrics in a visual puzzle task. Technical report (No. AFRL-RH-WP-TR-2014-0095), Air Force Research Laboratory, Wright-Patterson Air Force Base (2014)
45. Kloos, H., Van Orden, G.: Voluntary behavior in cognitive and motor tasks. Mind Matter **8**, 19–43 (2010)
46. Furman, G.D., Baharav, A., Cahan, C., Akselrod, S.: Early detection of falling asleep at the wheel: a heart rate variability approach. In: Computers in Cardiology, pp. 1109–1112. IEEE (2008)

Estimating Human State from Simulated Assisted Driving with Stochastic Filtering Techniques

Gregory M. Gremillion[✉], Daniel Donavanik,
Catherine E. Neubauer, Justin D. Brody, and Kristin E. Schaefer

United States Army Research Laboratory,
Aberdeen Proving Ground, MD, USA
{gregory.m.gremillion.civ, daniel.donavanik.ctr,
catherine.e.neubauer2.ctr, justin.d.brody.ctr,
kristin.e.schaefer-lay.civ}@mail.mil

Abstract. This work proposes a process for formulating a model and estimation scheme to predict changes in decision authority with a simulated autonomous driving assistant. The unique component of this modeling approach is the use of direct estimation of governing mental decision states via recursive psychophysiological inference. Treating characteristic quantities of the environment as inputs, and behavioral and physiological signals as outputs, we propose the estimation of intermediate or underlying psychological states of the human can be used to predict the decision to engage or disengage a driving assistant, using methods of stochastic filtering. Such a framework should enable techniques to optimally fuse information and thereby improve performance in human-autonomy driving interactions.

Keywords: Autonomous driving assistants · Stochastic filtering
Decision-making · Simulation

1 Introduction

A major research gap associated with human-agent teaming is the development of appropriate collaboration and joint decision-making with autonomous agents. While there have been a number efforts to advance collaboration, especially in the field of artificial intelligence, the difficulty lies in that most fielded systems are still traditionally teleoperated or preprogrammed. Through advancements to perception, mobility, and decision-making algorithms, it is possible for intelligent systems to transition to integrated and collaborative tasking [1]. With that role change, the frequency and manner in which humans engage with and control these systems will need to evolve.

One purpose for integrating autonomy is to help relieve some of the task demand placed on the human to better utilize their capabilities. But, without appropriate collaboration or effective joint decision-making, the integration of an autonomous agent may elicit poorer performance and control by reducing task engagement, increasing user drowsiness [2] and even decreasing situation awareness [3]. Additionally, the type of autonomy or level of transparency may cause the human team member to feel less in control, thereby decreasing situation awareness and in turn reducing trust in the system [4]. Furthermore, any effort to manage these interactions that does not monitor the mental state of human counterparts will be less effective. In order to avoid these negative consequences, which can lead to misuse or disuse of the system [5], this research must look toward understanding the process for integration and collaboration, building on the strengths of both the human and autonomous agents. To advance teaming effectiveness, three major topics specific to joint decision-making must be addressed: determining decision authority, managing trust-based decisions, and developing a mental decision model. This is motivated by the need to *calibrate trust* in autonomy-enabled systems in the presence of complex contexts and changing psychological states.

1.1   Decision Authority

Collaboration requires the development of shared mental models of both individual task elements and larger team goals [6–8] to support appropriate decision authority. It is essential to understand which team member will maintain the primary authority to make decisions under which set of constraints or conditions. While some of this authority may be task driven, appropriate changes can be hindered by individuals' unwillingness to relinquish control to an intelligent agent, misunderstanding of agent intent, or miscalibration of trust leading to either over- or under-reliance in the agent [9, 10].

Here the paradigm of leader-follower driving presents specific advantages toward more simply and tractably characterizing decisions to toggle authority. This task may be characterized with a compact, time-continuous state and action space where performance may be quantified by metrics of tracking accuracy to desired lane and headway, or range, reference values. Given knowledge of the capabilities of the human and autonomous agent to regulate these states, an optimal toggling of steering and throttle authority can be determined, such that the more proficient counterpart is in control.

Assuming such an optimal desired authority state may be computed, enacting it can remain challenging. Safety, usability, and task requirements might warrant a system in which the human retains the ability to toggle authority between themselves and the autonomous agent. In such a case, providing a feedback mechanism that *recommends* rather than *enforces* this optimal authority state may still effect improvements in human-autonomy interaction. This methodology relies on the human to maintain some level of compliance with the recommendation. The expected proficiency of each of these agents and likelihood of compliance are also dependent upon environmental and human psychological factors.

1.2   Managing Trust-Based Decisions

A key element in optimizing trust-based decision authority is appropriately quantifying the human team member's psychological and physiological state during a task. This is critical because the psychological and emotional complexities of an individual within performance-based settings can be understood in terms of a transactional model of stress and emotion [11], which treats stress as an interaction between personal (e.g., individual differences) and environmental factors (e.g., high or low task demands). Cognitive processes, including appraisal of task demands, and choice of coping strategy, play a key role in mediating the interaction between external demands and stable patterns of behavioral response [12] making a compact quantitative representation of the psychological and physiological state of the human essential.

The goal of arbitrating authority of the vehicle commands is to drive this authority state toward optimality, meaning the most proficient driver holds driving authority at any given moment. In the context of control theory, this is analogous to tracking a reference signal (see Fig. 1 for a block diagram). The uncontrolled system is represented by the human driver arbitrating authority in response to only the environment and their own internal state. The analogous control-theoretic block diagram representation for this would be a system without any input control signal (Fig. 1, grey dashed). The introduction of an authority recommendation that is based only on relative performance between the human and autonomous agents, without incorporating the mental state of the driver, is analogous to open loop control (Fig. 1, grey solid). Finally, the addition of the mental decision state as feedback used to modulate this authority recommendation is analogous to closed loop control (Fig. 1, black solid).

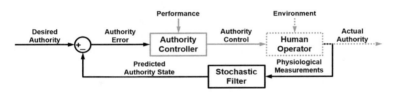

Fig. 1. Task authority control block diagram for uncontrolled (grey dashed), open-loop controlled (grey solid), and closed-loop controlled (black solid) human-agent team

Without closed loop feedback from a model-based estimate of the underlying state of a system, it is difficult if not impossible to ensure that the system will track a desired output. Therefore, an estimate of the likelihood of an authority toggle is desired to modulate the recommendation based on this likelihood, further encouraging compliance with the optimal recommendation and improving overall performance. Generating this likelihood requires prediction of human behavior, and therefore necessitates formation of a model of the internal dynamics of the human driver's mental decision process.

1.3   Current Work: Modeling and Stochastic Filtering

Previous efforts to predict human interaction with an autonomous counterpart as it relates to trust have focused on abstract or data driven methods, which are limited in generality or explanatory power. [13] utilized a post hoc approach to model interactions by inferring trust, incrementing or decrementing its value based on the frequency of human interventions, which lacks predictive capability and a standardizing framework for quantifying accuracy. [14] employs machine learning algorithms to map directly from a superset of input features to a very small set of target output classes typically associated with a behavioral response, such as labeling an image or intervening with an autonomous counterpart. These methods are able to account for nonlinearities in the human-autonomy system through densely connected neural network models. Given sufficient data, this approach can yield relatively high accuracy predictions, but can be difficult to generalize to new task domains, provide no guarantee of optimality, and lack explanatory power over the system model. Probabilistic estimation of human mental states from behavioral cues was demonstrated by [15]. Facial and oculography measurements were used in a Bayes network to infer fatigue. This framework, however, did not incorporate a process model, as with a *dynamic* Bayes network, to propagate information about the mental state forward in time, as is the case in recursive methods. By instead casting inference of these interactions in terms of estimation via stochastic filtering, as is used for many physical systems, the estimate gains statistical optimality while preserving an interpretable, generalizable model.

The driver's decision to shift authority between themselves and their autonomous counterpart is dependent on their current psychological state, the task, and the environment itself. External variables that drive the propagation of psychological states can be thought of as the inputs to this mental dynamic system. Conversely, output measurement signals such as electro-dermal activity (EDA), electro-oculography (EOG), heart rate (HR), heart rate variability (HRV), and electro-encephalography (EEG), map to the subject's state through an observation or measurement model. Psychophysiological measurement is proving to be a promising means of providing insight into quantifying changes in mental state, and more specifically trust-based decision authority [16]. However, the specific features associated with trust and how data can be used for enhancing human-agent teaming are still an open research question.

Given identified process and measurement models, which map external environmental inputs (e.g. road and traffic conditions) to time varying internal psychological states (e.g. workload and stress) to external physiological outputs (e.g. skin conductance and pupil diameter), it is possible to optimally infer and predict the decision to toggle driving authority through recursive probabilistic filtering. This is, in effect, applying the mathematical frameworks of stochastic filtering state estimation typically used on physical or robotic systems to the domain of psychophysiological inference and behavioral prediction.

Extended Kalman filter (EKF) based solutions to recursive state estimation address multiscale heterogeneous sensor fusion chiefly by maintaining the algorithmic independence of the individual sensor models. This allows each incoming sensor measurement across the suite of sensors to asynchronously update the filter, allowing the system to cope with time-varying sensor models. This *loosely coupled* paradigm [17] allows individually-derived estimators specific to each modality to generate a low-dimensional estimate of the system state with respect to each modular sensor subcomponent, which is then propagated through the estimator in a computationally efficient manner. However, loosely coupled stochastic estimators may suffer from a challenging problem of initialization and sensor calibration, in part because they do not directly exploit covariances between conditionally dependent sensor models.

Finally, because we are dealing with physiological signals whose baseline performance changes within and across subjects and whose process and measurement models are expected to be highly nonlinear, the first order linearization used in the traditional EKF is likely inappropriate. In this case the more versatile unscented Kalman filtering (UKF) and particle filtering approaches may be used, which allow more flexibility in defining the system models while maintaining the modular sensor independence described above. Each has been successfully applied to probabilistic inference of physical systems to form optimal stochastic estimates, but each comes with trade-offs. [18] demonstrated the application of a particle filter, a simple method of propagating the multimodal distribution of postulated states but is computationally demanding and relatively less accurate. [19] applied the UKF, a stochastic filter with greater computational efficiency but relies upon the existence of normal Gaussian estimate distributions and can suffer from catastrophic deteriorations of the sample covariance.

2 Methodology

2.1 Simulated Leader-Follower Driving Task

A simulated leader-follower driving task was created in SimCreator^{TM} for use with a 6 degree-of-freedom motion platform (Fig. 2a). Drivers operated a simulated vehicle on a two-lane closed circuit roadway, where they had to make decisions as to whether or not to toggle driving authority to an autonomous assistant to maintain lane position and headway to the lead vehicle (Fig. 2b and c), while performing a secondary pedestrian classification button press task. Data was collected for four experimental conditions where participants interacted with two different driving assistants (one with only throttle control and one with both steering and throttle control) each of which had a high and a low performance condition. Simultaneous recordings of EDA, HR, oculography, pupillometry, and EEG, were collected. Full methods including number of participants, task descriptions, and procedure can be found in [16, 20]. From a control theoretic perspective, this experiment corresponds to the open loop (Fig. 1, grey dashed), where the subject's decision to toggle the driving assistant was guided only by their mental state.

Fig. 2. Simulated leader-follower driving experiment (a) motion platform, (b) example simulation view, and (c) roadway course map

From a mathematical modeling and system identification perspective, autonomy-assisted leader-follower driving crucially provides a constrained, low dimensional, and continuous state space, which is fully defined by ownship position with respect to the lane of travel and the lead vehicle, and the time derivatives of those two quantities. The action space is limited to only steering and throttle commands. Driving is a continuous, time-critical task, in which momentary decisions must be made within varying environmental conditions. This requires the subject to regularly assess the decision to toggle driving authority, encoding temporally rich information in the physiological output measurements. However complex the internal mental process, it is ultimately mapped onto a binary behavioral state in the form of the authority toggle itself. Finally, driving is a relevant, real-world task, in which a large fraction of the population has prior domain knowledge. These characteristics are well-posed for a parsimonious representation of human-autonomy interaction that is suited for modeling stochastic filtering, in contrast to more abstract and open-ended tasks a human-agent team might perform.

2.2 Process for Stochastic Filtering State Estimation

Stochastic filtering state estimation requires identifying and partitioning quantities that parsimoniously describe the generalized system's inputs, states, and outputs, defining their relations via a model structure, and estimating the model's parameters. The first of these requirements is accomplished through a priori selection of physiological cues for mental state. A data driven, information-theoretic approach is used to select features from a superset of simulation environment data and physiological measurement signals. This is itself a challenge, as illustrated by [21], as there may be a many-to-one or many-to-many mapping from psychological to physiological features. Carefully designed derivative features can reduce this model complexity to one-to-one mappings.

Specifically, eye tracking data can provide indications of what information the subject might find important during an interaction, which can be used to support trust calibration. Repeated gaze fixation on an area of interest, such as an area on a user display, suggests that the information being displayed in that area may be useful to task [22, 23], whereas increased fixation duration might indicate that the subject finds the information difficult to understand [23]. Further, the pattern and targets of gaze fixation may indicate the direction of an impending decision [23, 24], and can be used to estimate the level of cognitive effort being expended while processing information [25]. Pupillometry can provide an informative measure of workload [26, 27]. Workload has been directly associated with trust and is a significant factor in decision-making for both driving tasks and human-autonomy interactions [28].

Two physiological measures that have been associated with cognitive effort and decision-making capabilities within the domain of human-robot trust are EDA and HRV. EDA is a sensitive measure of emotional arousal such that levels increase during periods of anxiety and cognitive effort [29], as well as engagement [30], e.g. tonic and phasic EDA should increase with levels of anxiety or frustration in response to errors by an autonomous counterpart. HRV is often used in conjunction with EDA measures to infer the cognitive and affective effect of a stimulus, and has been associated with orienting behavior [31], or to infer levels of workload and trust [32, 33], e.g. states of high trust should correspond to high HRV, low HR, and low tonic EDA levels.

In addition to these peripheral physiological modalities, EEG provides a very rich, though complex mechanism for monitoring psychological states [34, 35]. In conjunction with measurements of pupil diameter or eye blinks, EEG can be used to estimate workload [36] or stress [35], respectively. Driving-specific antecedents in EEG measurements have also be shown, and thus provide fruitful avenue for predictive measures at high time resolutions [37, 38].

A complementary approach to determining features of interest that predictively encode trust-based decisions is information-theoretic analysis. To reach this end, machine learning based feature selection algorithms and data reduction techniques are used to identify signals within the dataset that are informative of these authority toggles from the leader-follower simulated driving dataset, leveraging *relevance* and *redundancy* criteria derived from mutual information (MI) metrics, which describes how much information one variable contains about another. For example, the value of variable *A* may potentially be uninformative given examination of variables *B* and *C*.

Given identified features of interest from a priori and data driven methods, the next task is defining a state representation and model structure that is consistent with stochastic filtering which captures the essential inputs, states, and measurements of the human-autonomy system. The control theoretic definitions for these quantities in Sect. 1.3 are used to define the input, output, and state vectors, u, y, and x, for the human-autonomy system. This process of defining the system representation is based on separating those quantities that parsimoniously describe the system: (1) those sensory inputs the human may perceive, (2) those sensory outputs the human generates, and (3) the quantities needed to sufficiently relate each of these quantities across time. Effectively this will be a process of pruning and sorting of the potentially relevant features.

Having constructed an appropriate system representation, the parameters of the process and observation model must be estimated. When using the particle filter or UKF, these relationships may be stochastic or nonlinear, but are assumed to be *locally linear* for implementation purposes. The constraints on the parameters of the process and observation models are chosen based on the a priori relationships described above, and their values were adjusted using stepwise regression to fit the linear parameters values. For the preliminary results presented here, a particle filter is used, due to its ease of implementation and ubiquity in physical systems research, to demonstrate estimation of the trust-based decision to toggle the autonomous driving assistant from subject data.

3 Results

3.1 Feature Selection

Our ultimate goal in this project is to construct robust models that can make predictions about a driver's trust-based decisions from a range of psychophysiological and environmental data. Here we present preliminary results to identify quantities of interest, form a model representation, and perform stochastic filtering of the authority decision state. To demonstrate the procedure and structure of the modeling and estimation process for the human-autonomy assisted driving system, we posit system input, state, and measurement vectors. Furthermore, we postulate locally linear models relating each of these quantities across subsequent time steps, such that recursively filtering each will yield an optimal estimate of the decision state. We expect that the amount of potential features of interest available to our system will far surpass its computational limits of real-time estimation. This is denoted as the "curse of dimensionality" that often hinders high-dimensional system representations. To alleviate this, it will be necessary to work with only a tractable subset of the available data. As previously described, techniques of machine learning and mutual information were applied to accomplish this.

Multiple machine learning algorithms were applied to the simulated leader-follower driving dataset. Specifically, several neural networks (NN) and support vector machines (SVM) were trained using a standard 80–20 cross validation to assess performance. This performance was compared to that of the same algorithms using two transformations of the data. Principal component analysis (PCA) was applied as a dimensionality reduction technique to replace the original 412 base physiological and simulation environment variables with 100 transformed variables. Each derived variable is an expertly chosen linear combination of raw variables. We then applied an MI-based techniques to select an optimal set of 40 of the new variables. Our analysis was performed using *scikit-learn* [39] for the machine-learning and *scikit-feature* for the variable selection [40]. The results of this process are summarized in Table 1.

Table 1. Neural network error metric

	NN (all variables)	NN (40 MI-based variables)	SVM (all variables)	SVM (40 MI-based variables)
412 base variables	33.1	38.7	21.5	93.2
100 PCA variables	69.2	69.9	21.9	94.1

Note: Values are Euclidean norm of the difference between the prediction and the target vectors.

For each algorithm, the error was computed by aggregating the differences between the ground truth in the testing set and the prediction made by the machine learning algorithm. The data in this table suggest: (1) that selecting a subset of variables based on MI is possible with minimal impact on NN performance, since error rates with the selected subset of variables are comparable to those using all of the variables; (2) performance of SVM is more likely to be impacted, since error rate for these classifiers increased significantly when fewer variables were used; and (3) the dimensionality reduction associated with PCA has a detrimental impact on classifier performance when used with NN, since the error for the PCA-based dataset is higher than for the full dataset. It should be emphasized, however, that all three of these conclusions are preliminary and warrant further investigation.

3.2   Model Structure

The *state vector* of the current model includes dynamic, partially unobservable psychological values, and specifically comprises the subject's continuous authority decision state, stress, and workload. The first of these states is included by design, as it is the target quantity we aim to estimate, and is defined as a scalar value on the interval [0, 1] where lower values indicate a high likelihood to disengage the driving assistant. The *input vector* consists of externally observable quantities that affect the propagation of those states. In our current model, these are defined as the task score decrements, lane position error, inverse time-to-contact, headway, road curvature, and ownship speed. The *measurement vector* is defined as observable variables that are affected by the states of the system, which include physiological and behavioral outputs. For our current model, this comprises the binary driving authority, phasic EDA, tonic EDA, HR, HRV, pupil diameter, and horizontal EOG variance. The relationship between each of these vectors is displayed graphically as a recursive network (Fig. 3), where the edges between nodes correspond to process and observation models, and missing edges imply no direct relationship was assigned. Linear model parameters were fit with stepwise regression with sign convention constraints for positive or negative relationships (respectively, solid or dashed in Fig. 3). Temporal processes between states equivalent to integration or memory are shown (dotted).

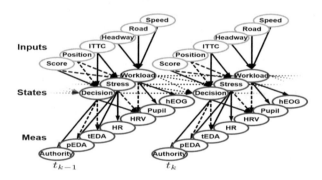

Fig. 3. Network representation for recursive model structure.

3.3 Filter Implementation

Using the process and observation model structure described in Sect. 3.2, a particle filter was implemented to estimate the 3-state mental system for a single subject from the simulated leader-follower driving dataset described in Sect. 2.1, to demonstrate the operation of the filter. Figure 4 displays the results of this estimation for a 150 s portion of the task in which two disengagements of the driving assistant occurred.

Approximately 80 s from the start of the timetrace, input signals u (Fig. 4a) drive the estimated decision state x_1 (Fig. 4b), toward 0, indicating increased likelihood to disengage the autonomy. This corresponds to the actual authority disengagement shown in y_1 (Fig. 4c). The decision state estimate increases as the subject engages the driving assistant. Additionally, approximately 40 s from the start of the timetrace, the estimate of the decision state precipitously decreases, but then increases before reaching a value that would presumably correspond to an authority toggle. This "sub-threshold" response, where a mental state builds toward a behavioral output but

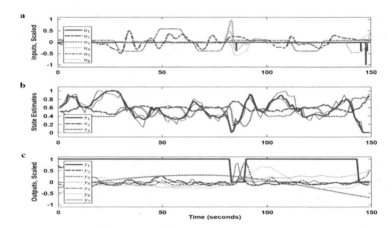

Fig. 4. Particle filter state estimation for a single simulated leader-follower driving subject.

ceases prior to reaching some intrinsic threshold, is precisely the type of process this work aims to predict. Estimates of such phenomena can be used to preemptively manage a potentially detrimental behavior. Thereby, human autonomy interactions indicative of miscalibrated trust can be mitigated more effectively. The spread of the particles for the decision state estimate are shown (Fig. 4b, thin solid) are shown, which qualitatively indicate the confidence of the mental state estimate. While these are preliminary results for only a single subject trial, they are illustrative of ability of the filter to recursively infer hidden states through which the inputs map to the outputs.

4   Summary and Future Directions

This work outlines a motivation and methodology for estimating unobservable mental states from observable environmental inputs and physiological and behavioral outputs, using stochastic filtering in the context of simulated leader-follower driving. These efforts can be leveraged to inform optimal fusion of joint decisions, assist with calibrating trust, and improve human-autonomy collaboration and performance.

The process described herein will be applied to the existing dataset for all subjects and experimental conditions. Current efforts are also focused on developing a longitudinal experimental paradigm, with the goal of collecting a sufficiently large dataset to form a model that has robust predictive power across a range of inter- and intra-subject variations, with a common model structure. Additionally, more complex and accurate nonlinear or probabilistic models and more sophisticated methods of parameter estimation will be utilized, such as approximate Bayesian computation Monte Carlo methods. Parameter estimation can also be improved by recording subject self-reports of likelihood to toggle authority during the task, as a target signal for fitting the state estimate.

Acknowledgements. The authors thank Jason Metcalfe, Amar Marathe, Jamie Lukos, Justin Estepp, Kim Drnec, Victor Paul, Benjamin Haynes, Corey Atwater, and William Nothwang for the simulated driving dataset. The views and conclusions contained in this document are those of the authors and should not be interpreted as representing the official policies, either expressed or implied, of the Army Research Laboratory or the U.S. Government. The U.S. Government is authorized to reproduce and distribute reprints for Government purposes notwithstanding any copyright notation herein.

References

1. Phillips, E., Ososky, S., Grove, J., Jentsch, F.: From tools to teammates: toward the development of appropriate mental models for intelligent robots. In: Proceedings of the Human Factors and Ergonomics, pp. 1491–1495 (2011)
2. Hancock, P.A., Verwey, W.B.: Fatigue, workload and adaptive driver systems. Acc. Anal. Prev. **29**(4), 495–506 (1997)
3. Stanton, N.A., Young, M.S.: Driver behaviour with adaptive cruise control. Ergonomics **48** (10), 1294–1313 (2005)
4. Koo, J., Kwac, J., Ju, W., Steinert, M., Leifer, L., Nass, C.: Why did my car just do that? Explain-ing semi-autonomous driving actions to improve driver understanding, trust, and performance. Int. J. Interact. Des. Manuf. **9**(4), 269–275 (2015)
5. Parasuraman, R., Manzey, D.H.: Complacency and bias in human use of automation: an at-tentional integration. Hum. Fact.: J. Hum. Fact. Ergon. Soc. **52**(3), 381–410 (2010)
6. Adams, J.A., Deloach, S.A., Scheutz, M.: Shared mental models for human-robot teams. In: Proceedings of AAAI, Stanford, pp. 99–105 (2014)
7. Awais, M., Henrich, D.: Human-robot collaboration by intention recognition using probabilistic state machines. In: Proceedings of International Workshop on Robotics in Alpe-Adria-Danube Region (RAAD), pp. 75–80. IEEE (2010)
8. Ososky, S., Schuster, D., Jentsch, F., Fiore, S., Shumaker, R., Lebiere, C., Kurup, U., Oh, J., Stentz, A.: The importance of shared mental models and shared situation awareness for transforming robots from tools to teammates. In: Proceedings of SPIE - The International Society for Optical Engineering, p. 838710 (2012)
9. Chen, J.Y.C., Barnes, M.J.: Human–agent teaming for multirobot control: a review of human factors issues. IEEE Trans. Hum.-Mach. Syst. **44**(1), 13–29 (2014)
10. Schaefer, K.E., Brewer, R., Putney, J., Mottern, E., Barghout, J., Straub, E.R., Orlando, F.L.: Relinquishing manual control: collaboration requires the capability to understand robot intent. In: Proceedings of the IEEE 9th International Workshop on Collaborative Robots and Human Robot Interaction, pp. 359–366 (2016)
11. Lazarus, R.S.: The cognition-emotion debate: a bit of history. Handb. Cogn. Emot. **5**(6), 3–19 (1999)
12. Matthews, G.: Towards a transactional ergonomics for driver stress and fatigue. Theoret. Issues Ergon. Sci. **3**(2), 195–211 (2002)
13. Xu, A., Dudek, G.: Trust-driven interactive visual navigation for autonomous robots. In: IEEE International Conference on Robotics and Automation (ICRA), pp. 3922–3929 (2012)
14. Takahashi, M., Kubo, O., Kitamura, M., Yoshikawa, H.: Neural network for human cognitive state estimation. In: Advanced Robotic Systems and the Real World Intelligent Robots and Systems 1994, pp. 2176–2183 (1994)
15. Ji, Q., Zhu, Z., Lan, P.: Real-time nonintrusive monitoring and prediction of driver fatigue. IEEE Trans. Veh. Technol. **53**(4), 1052–1068 (2004)

16. Metcalfe, J.S., Marathe, A.R., Haynes, B., Paul, V.J., Gremillion, G.M., Drnec, K., Atwater, C., Estepp, J.R., Lukos, J.R., Carter, E.A., Nothwang, W.D.: Building a framework to manage trust in automation. In: Proceedings of SPIE 10194, Micro- and Nanotechnology Sensors, Systems, and Applications IX, p. 101941U (2017)

17. Donavanik, D., Hardt-Stremayr, A., Gremillion, G.M., Weiss, S., Nothwang, W.D.: Multi-sensor fusion techniques for state estimation of micro air vehicles. In: SPIE Defense and Security, p. 98361V. International Society for Optics and Photonics (2016)

18. Rosencrantz, M., Gordon, G., Thrun, S.: Decentralized sensor fusion with distributed particle filters. In: Proceedings of the Nineteenth conference on Uncertainty in Artificial Intelligence, pp. 493–500 (2002)

19. Van Der Merwe, R., Wan, E., Julier, S.: Sigma-point Kalman filters for nonlinear estimation and sensor-fusion: applications to integrated navigation. In: AIAA Guidance, Navigation, and Control Conference and Exhibit, p. 5120 (2004)

20. Gremillion, G.M., Metcalfe, J.S., Marathe, A.R., Paul, V.J., Christensen, J., Drnec, K., Haynes, B., Atwater, C.: Analysis of trust in autonomy for convoy operations. In: Proceedings of International Society for Optics and Photonics Defense+Security, p. 98361Z (2016)

21. Cacioppo, J.T., Tassinary, L.G.: Inferring psychological significance from physiological signals. Am. Psychol. **45**(1), 16 (1990)

22. Just, M.A., Carpenter, P.A.: Eye fixations and cognitive processes. Cogn. Psychol. **8**(4), 441–480 (1976)

23. Poole, A., Ball, L.J.: Eye tracking in HCI and usability research. In: Encyclopedia of Human Computer Interaction, vol. 1. pp. 211–219 (2006)

24. Glaholt, M.G., Reingold, E.M.: Eye movement monitoring as a process tracing methodology in decision making research. J. Neurosci. Psychol. Econ. **4**(2), 125 (2011)

25. Marshall, S.: Measures of attention and cognitive effort in tactical decision making. Decis. Mak. Complex Environ. **321**, 332 (2007)

26. Beatty, J.: Pupillometric measurement of cognitive workload. TR-22, Department of Psychology, California University Los Angeles (1977)

27. Schwalm, M., Keinath, A., Zimmer H.D.: Pupillometry as a method for measuring mental workload within a simulated driving task. Hum. Fact. Assist. Autom. 1–13 (2008)

28. Parasuraman, R., Sheridan, T., Wickens, C.: Situation awareness, mental workload, and trust in automation: viable, empirically supported cognitive engineering constructs. J. Cogn. Eng. Decis. Mak. **2**(2), 140–160 (2008)

29. Shi, Y., Ruiz, N., Taib, R., Choi, E., Chen, F.: Galvanic skin response (GSR) as an index of cognitive load. In: CHI 2007, Extended Abstracts on Human Factors in Computing Systems, pp. 2651–2656 (2007)

30. Bethel, C.L., Burke, J.L., Murphy, R.R., Salomon, K.: Psychophysiological experimental design for use in human-robot interaction studies. In: International Symposium on Collaborative Technologies and Systems (2007)

31. Figner, B., Murphy, R.O.: Using skin conductance in judgment and decision making re-search. In: Handbook of Process Tracing Methods for Decision Research, pp. 163–184 (2011)

32. Matthews, R., McDonald, N.J., Trejo, L.J.: Psycho-physiological sensor techniques: an over-view. In: 11th International Conference on Human Computer Interaction (HCII), pp. 22–27 (2005)

33. Montague, E., Xu, J., Chiou, E.: Shared experiences of technology and trust: an experimental study of physiological compliance between active and passive users in technology-mediated collaborative encounters. IEEE Trans. Hum.-Mach. Syst. **44**(5), 614–624 (2014)

34. Schaefer, K.E., Scribner, D.R.: Individual differences, trust, and vehicle autonomy: a pilot study. In: Proceedings of the Human Factors and Ergonomics Society Annual Meeting, vol. 59, no. 1. pp. 786–790 (2015)
35. Haak, M., Bos, S., Panic, S., Rothkrantz, L.J.M.: Detecting stress using eye blinks and brain activity from EEG signals. In: Proceeding of the 1st Driver Car Interaction and Interface, pp. 35–60 (2008)
36. Rozado, D., Dunser, A.: Combining EEG with pupillometry to improve cognitive workload detection. Computer **48**(10), 18–25 (2015)
37. Haufe, S., Treder, M.S., Gugler, M.F., Sagebaum, M., Curio, G., Blankertz, B.: EEG potentials predict upcoming emergency brakings during simulated driving. J. Neural Eng. **8** (5), 056001 (2011)
38. Lin, C.T., Chuang, C.H., Huang, C.S., Tsai, S.F., Lu, S.W., Chen, Y.H., Ko, L.W.: Wireless and wearable EEG system for evaluating driver vigilance. IEEE Trans. Biomed. Circuits Syst. **8**(2), 165–176 (2014)
39. Pedregosa, F., Varoquaux, G., Gramfort, A., Michel, V., Thirion, B., Grisel, O., Blondel, M., Prettenhofer, P., Weiss, R., Dubourg, V., Vanderplas, J.: Scikit-learn: machine learning in Python. J. Mach. Learn. Res. **12**, 2825–2830 (2011)
40. Li, J., Cheng, K., Wang, S., Morstatter, F., Trevino, R.P., Tang, J., Liu, H.: ACM feature selection: a data perspective. Comput. Surv. **50**(6), 94 (2017)

Translating Driving Research from Simulation to Interstate Driving with Realistic Traffic and Passenger Interactions

Jean M. Vettel[1,2,3(✉)], Nina Lauharatanahirun[1,3],
Nick Wasylyshyn[1,3], Heather Roy[1], Robert Fernandez[4],
Nicole Cooper[3], Alexandra Paul[3], Matthew Brook O'Donnell[3],
Tony Johnson[4], Jason Metcalfe[1], Emily B. Falk[3],
and Javier O. Garcia[1,3]

[1] US Army Research Laboratory,
Aberdeen Proving Ground, Adelphi, MD, USA
jvettel@gmail.com
[2] University of California, Santa Barbara, Santa Barbara, CA, USA
[3] University of Pennsylvania, Philadelphia, PA, USA
[4] DCS Corporation, Alexandria, VA, USA

Abstract. In this driving study, participants were assigned to a driver-passenger dyad and performed two drives along Interstate-95 in normal traffic conditions. During the driving session, the driver had to safely navigate the route while listening and discussing news stories that were relayed by the passenger. The driver then performed a set of memory tasks to evaluate how well they retained information from the discussion in a multitask context. We report preliminary analyses that examined subjective factors which may influence success in social communication, including trait and state similarity derived from questionnaires as well as physiological synchrony from implicit state measurements derived from brain activity data. Although this dataset is still in collection, these initial findings suggest potential metrics that capture the contextual complexity in naturalistic, multitask environments, providing a rich opportunity to study how successful communication reflects shared social and emotional experiences.

Keywords: Interstate driving · Social network structure · State questionnaires
EEG · Neural synchrony · Communication · Individual differences

1 Introduction

For the majority of Americans, driving serves as an essential component of life activities, providing a means for commuting to work, attending social gatherings, and transporting goods from stores to home [1, 2]. Thus, driving has become a task that consumes a large amount of time for many, and the automotive industry has sought technological innovations that improve both the comfort and safety of driving. An impressive suite of technologies have parameterized core components of driving,

© Springer International Publishing AG, part of Springer Nature (outside the USA) 2019
D. N. Cassenti (Ed.): AHFE 2018, AISC 780, pp. 126–138, 2019.
https://doi.org/10.1007/978-3-319-94223-0_12

including collision-avoidance sensors, lane-keeping technology, adaptive cruise control, and voice-activated controls [3–5]. As these assistive features have improved, several self-driving cars have been approved for on-road testing. Waymo and Uber have autonomous vehicles driving along normal commute routes in Silicon Valley, Pittsburgh, and Austin (to name a few). While the timeframe for a full conversion from human drivers to automated drivers is unknown (for an in-depth prediction, see [6]), the success of self-driving cars amidst human-driven cars suggest that the nature of driving may rapidly evolve [7, 8]. Soon, drivers may need to spend less effort safely navigating their vehicle; instead, drivers may need to balance the basic oversight of autonomous driving while they engage in other tasks, such as social communication.

In our driving study, we still rely on human drivers for controlling the vehicle, but driving along Interstate-95 is concurrent with a communication task between a driver and passenger. This route was chosen for minimal risk driving conditions with clear lane markings, minimal navigation decisions, and calm traffic patterns. These conditions approximate the level of engagement that drivers may need to oversee the performance of near-term self-driving cars [9], so here, we use it as a proxy for studying how communication dynamics may be influenced in a naturalistic, multitask context. Navigating the interstate with real traffic dynamics carries risk for injury if the driver does not maintain sufficient engagement with the primary driving task [10]. This task hierarchy provides a context to study how a multitask environment influences performance on a secondary task, namely communication with an in-car passenger.

Our experimental design, however, investigates additional layers of contextual complexity that may modulate performance on the communication task. Successful communication inherently involves implicit and explicit processing of information between two or more individuals, and previous research has shown that increased synchrony between people correlates with successful transfer of information [11–13]. Furthermore, recent extensions suggest that concurrent activity between brains may represent abstract cooperative efforts (i.e. hyperscanning: [14]), including rhythmic tapping [15] and musicians performing [16]. Complementary results have been observed in social domains where individuals with similar neural activity during social exclusion demonstrate similar susceptibility to peer influence [17] and have similar real-life social network structures [18].

Our core hypothesis posits that successful communication depends upon shared social experiences and similar emotional states that facilitate joint understanding of information and interest in comprehending another's perspective on a topic. In our driving study, we collected several metrics about a participant's social interactions, including their real-life social network structure as well as their interactions with their dyad partner outside of the study. Similarly, a participant's communication performance will likely be heavily influenced by their current state, e.g., emotional, physiological, and cognitive states. We collected both explicit estimates of state, indexed by self-report questionnaires, as well as implicit estimates from physiological data from brain (EEG) and body (HRV, GSR). Consequently, our experimental design allows us to examine how these various contextual factors influence a driver's performance on a communication task that is embedded in a multitask, driving context. Here, we present a set of preliminary analyses on only a small subset of these individual difference measures. Although this data is still in collection, our initial findings indicate the promise of similarity and synchrony metrics to capture trait and state influences on performance in a naturalistic, multitask context.

2 Methods

Participants. The present study used data from twenty-eight adults (68% male) between the ages of 21 and 55 (M = 38.02; SD = 11.44) who participated as part of an ongoing longitudinal experiment aimed at investigating the communication dynamics between driver-passenger dyads during interstate driving under naturalistic conditions. All study volunteers provided informed consent in accordance with study approval from the accredited Institutional Review Board at U.S. Army Research Laboratory and in accordance with the U.S. Army Research Laboratory Human Research Protection Program (32 CFR 219 and DoDI 3216.01). Participants were recruited either from the U.S. Army Research Laboratory (Aberdeen Proving Ground) or DCS Corporation (Alexandria, VA and Abingdon, MD locations) to ensure that they received liability insurance in the event of a car accident (none occurred). Inclusion criteria consisted of being at least 21 years of age, having normal or corrected to normal visual acuity, and possession of an unrestricted driver's license for a minimum of two years. Participants were excluded if they had medical conditions that prevented normal driving (e.g., seizures) or motion sickness in cars. All criteria were assessed through self-report.

Experimental Design. An overview of the experimental design for the 9–15 week longitudinal study is depicted in Fig. 1. During a 40-min intake session, participants received an actigraphy watch (Readiband Actigraph SBV2; Fatigue Science, Vancouver, BC) to monitor sleep and physical activity throughout the course of the study, provided a cell phone number to receive daily text messages during the study, and completed a one-time set of trait assessments. In these preliminary analyses, we only report trait data from the Social Network Information questionnaire.

Participants were then assigned to driver-passenger dyads based on their schedule availability for drives. Each dyad completed two driving sessions: one where they were the passenger and one as the driver. Each driving session took approximately 2.5 to 3 h to complete. Traffic, weather, and vehicle conditions were assessed prior to each session, and drives were only conducted when both participants and the experimenters agreed that conditions met minimal risk criteria.

The driving session occurred in an all-wheel drive, 2016 Ford Fusion Titanium instrumented with an Ergoneers D-Lab data acquisition system. As depicted in Fig. 1B–D, each driving session consisted of three segments: pre-drive, on-drive, and post-drive. The D-Lab recorded time-synchronized multi-sensor vehicle environment data for all three segments, including MobilEye and On-Board Diagnostics (OBD) data, audio, and four channels of video. One camera recorded the external environment out the front windshield, one out the back windshield, one angled at the driver seat in the car, and one angled at the passenger seat. The MobilEye provided driving performance data by monitoring the vehicle position and elements of the external environment, such as lane markings and other traffic. Additional driving performance can be ascertained from OBD data, including steer angle, speed, acceleration, and braking. None of these data are reported in these preliminary analyses, but the recorded dialogue in the audio file and facial/body gestures in the videos serve as the basis for planned analyses about successful communication during the driving session.

During the pre-drive segment, each participant separately completed a series of state assessments while they were outfitted with a set of multimodal physiological sensors to measure brain activity, respiration, heart rate, and galvanic skin responses. Each participant then completed pre-drive tasks based on their role for the driving session. The driver reviewed the route and safety procedures outside the vehicle, while the passenger sat in the vehicle and watched 16 videos of unique news stories on a tablet. The passenger was asked to remember as many news stories as possible to share them with the driver in conversation during the on-drive segment. The dyad then jointly watched an instructional video about their responsibilities for the drive and provided 3 min of baseline physiological data. Once in the vehicle, the physiological sensors were paired to recording devices and synchronized with the D-Lab system data through the use of a common reference signal (engine RPM) recorded to all data logs using OBD splitters; this common reference ensured time synchronization across participants and vehicle/task events during the driving session. In these preliminary analyses, we only report physiological data from the brain sensors as well as state data from three questionnaires, the Perceived Stress Scale (PSS), the Motivation Visual Analog Scale (MVAS), and the Driver-Passenger Social Interaction questionnaire (DPSI).

The on-drive segment consisted of three subcomponents. During the first 10 min, the dyad jointly listened to a 2-min podcast about the importance of sleep or physical activity for healthy living (one topic was assigned to each drive; counterbalanced between dyads) and then discussed their opinions about the health information. Next, they drove 40 miles on Interstate-95, turning around at 20 miles, and the passenger communicated details about the news stories with the driver in two sequential memory tasks. The first was an open recall task where the passenger had 5 min to share as many news stories as they could remember and engage the driver in a discussion about their opinions on the topics. The second memory task was a cued recall where the passenger saw a visual cue for each of the 16 unique news stories, and the dyad discussed the topic for 1 min each. After the dyad exited the interstate, they spent the final 10 min jointly listening to a second podcast on the same health topic as the start of the drive, and they freely discussed their opinions on the additional health information.

Finally, in the post-drive segment, each participant separately completed a series of survey assessments as well as a 32-item recognition memory task about fine-grained details from the news stories (e.g., change tax burden to 25% or to 32%, where one is accurate and the other is a lure). However, the driver completed two additional memory tasks without the passenger present. The first was a 3-min open recall to recount as many news stories and conversation details as possible, and the second was a cued recall task where they had 40 s to talk aloud about each of the 16 cued news clips, including details about the opinions discussed with their dyad partner during the drive. In short, across these two additional tasks, the driver reported details that captured how successfully the passenger communicated the news stories as well as the success of the dyad's information exchange in the discussion itself.

After all of the tasks in the driving session were completed, the experimenters removed physiological sensors and debriefed the participants. The participants then completed the daily text tasks while wearing the actigraphy watch for 2–3 weeks before returning for their second drive. The flow of drive 2 mirrored drive 1, except that the participants changed roles (i.e., driver in drive 1 is passenger in drive 2) and they discussed a different, unique set of the 16 news clips.

2.1   Trait Similarity Metric

Social Network Analysis. During the intake session, participants completed a web-based application [19] to characterize their real-life social networks. In this task, participants identified up to 10 people with whom they have communicated in the past week in each of four communication types: face-to-face conversation, voice call, text messaging, and online social media. The maximum number of unique individuals possible to include was 40, but many participants had communicated with the same subset of people across the media or had communicated with fewer than 10 people through at least one medium in the past week, so all social networks consisted of fewer than 40 individuals. For each unique individual listed, the participant then indicated the strength of their relationship ("closeness"), subjective estimates of their friend's preferences ("driving riskiness" and "political interest"), and which of their friends knew each other. From the latter responses, an undirected graph of their most recent contacts was computed, where each node corresponds to a friend and an edge connects every pair of people who know each other. The size of the social networks ranged from 13 to 33 people (Mean = 21.0; SD = 5.0).

We calculated the density of each social network by taking the number edges and dividing by the number of possible edges (i.e., the number of edges in a fully connected graph where every friend of the participant is also a friend with each other). Thus, a higher density corresponds to a social network in which more of the participant's friends know each other, while a lower density indicates that a participant has more distinct groups of friends.

2.2   State Similarity Metrics

Perceived Stress Scale (PSS). During the pre-drive segment, participants provided self-report responses on a 10-item Perceived Stress Scale [20], capturing subjective perceptions of stress. Responses were scored on a 5-point Likert scale (0 = never to 4 = very often) regarding thoughts and feelings experienced within the last month. Sample items included, "In the last month, how often have you been able to control irritations in your life?" and "In the last month, how often have you felt nervous and stressed?" A total score was calculated by summing scores on all items, where higher total scores indicated higher levels of perceived stress.

Motivation Visual Analog Scale (MVAS). During the pre-drive segment, participants were asked to provide subjective ratings of general motivation on a scale from 0 ("I am not motivated") to 100 ("I am very motivated"). The scale was represented as a horizontal line with an initial position of the slider at 50 in the middle. Visual analog scales

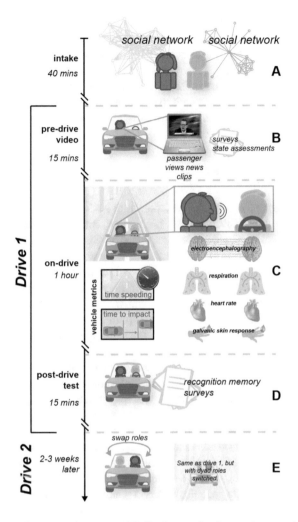

Fig. 1. *Experimental Design Overview.* (A) During an intake session, participants provided social network information as a trait variable. (B) Both driver and passenger completed pre-drive state surveys (e.g., PSS, MVAS, DPSI) and tasks specific to their role in the drive. In particular, the passenger separately watched 16 video news clips that served as discussion topics during the drive. (C) During the on-drive segment, the passenger communicated the news topics and led a discussion with the driver. All vehicle CAN bus data (speed, time to contact, steering, lane position, etc.) and multimodal physiological data (EEG, respiration, heart rate, and galvanic skin responses) were measured and synchronized continually throughout the driving session. (D) In the post-drive segment, participants independently completed memory tasks and post-drive surveys. (E) Several weeks later, the dyad returned for a second drive, but they swapped roles (driver in drive 1 is passenger in drive 2) and discussed a different set of news clips.

are typically used to provide a visual guide for participants to rate their perception of subjective feelings or states on a continuous scale by choosing a location between two extremes [21–24].

Driver-Passenger Social Interaction Questionnaire (DPSI). In order to assess the existing relationship between dyad partners prior to the study session, participants answered 5 items regarding the frequency of interaction as well as the likelihood of accepting advice from their partner for general and work-related concerns. For the present study, we focused on the frequency of casual interaction prior to the study session. During the pre-drive segment, participants indicated frequency using the following scale: 0 = Never, 1 = Past 12 Months, 2 = Past 6 Months, 3 = Past Month, 4 = Past Week, and 5 = Last Day.

2.3   Physiological Synchrony Metric

Electroencephalography (EEG). For this study, we used the ABM B-Alert X24 Electroencephalography (EEG) system (Advanced Brain Monitoring, Inc, Carlsbad, CA) sampled at 256 Hz. The flexible electrode strip consisted of a set of flat electrodes in standard 10–20 scalp locations, and conductive paste was placed on cylindrical foam pads on each electrode to serve as the conductive medium between the scalp and sensor. The electrode strip was then affixed to a headband that was adjusted to fit the head of the participant, and a wireless transmitter that attached to the headband sent EEG signals to a separate device for data capture. The B-Alert system is relatively light, weighing less than 200 g, allowing the participant to freely move during the driving session.

The EEG data were preprocessed to remove nuisance and non-brain signals, such as muscle activity, electrical noise in the car, and vehicle movement, using the PREP approach implemented in EEGlab [25]. The steps included are: (1) line noise removal via a frequency-domain (multi-taper) regression technique to remove 60 Hz and harmonics present in the signal, (2) a robust average reference with a Huber mean, (3) artifact subspace reconstruction to remove residual artifact with the standard deviation cutoff parameter set to 5, (4) band-pass filtering using a Butterworth filter with 2-dB attenuation at 2 and 50 Hz, and (5) an automated independent component analysis-based component removal to specifically target residual muscle and eye-related artifacts that may influence our physiological synchrony metric [26]. This artifact removal procedure has been shown to be robust to high artifact environments [27, 28].

EEG Synchrony. Across the entire duration of the on-drive segment, we estimated EEG synchrony in the alpha band (8–12 Hz) between dyad partners using Matlab (Mathworks, Inc.). For each channel pair, the phase-locking value (PLV) was computed using a Hilbert transform across 2 s EEG epochs in 62 ms steps. PLV estimates the similarity in phase between the two signals [12], where 1 equates with perfect phase locking and 0 with no phase locking. In this preliminary analysis, we focused on alpha since this frequency band has been proposed as a gating mechanism for perceptual information [29] or access controller for semantic knowledge [13], both of which may underlie successful communication via physiological synchrony.

3   Results

Our preliminary analyses focused on identifying and describing variability of subjective factors that may influence social communication between and within driver-passenger dyads. Here, we utilized data from intake and drive 1 sessions from a dataset still in collection, and we examined similarity among both trait and state metrics as well as synchrony for a physiological metric derived from scalp EEG data.

3.1   Real-Life Social Networks Capture Trait Similarity Among Dyads

We first used social network analysis to examine individual differences in trait metrics derived from the participant's real life social interactions. Using an online tool [19], participants reported their most recent interactions and indicated which friends knew one another. From these relationships, we computed the density of the participant's social network. A social structure with high density denotes a network where a large proportion of a participant's friends know one another, while a low density indicates that the participant may have distinct clusters of friends who know one another but little crossover among subgroups of friends.

The social network density for each dyad collected thus far is plotted in Fig. 2: individual participants are indicated by a circle connected to their dyad partner by a line, so that their adjoining line reflects their similarity (short line = high similarity, long line = low similarity). Across dyads, we observe variability in density, including dyad 4 where the participants' network densities are 0.61 and 0.68 and dyad 3 where each participant's density is 0.32. In contrast, we also have dyads where the participants have dissimilar network density scores, such as dyads 8 and 12.

Next, we illustrate the social network structures for several dyad partners as a graph. In each graph, the participant is represented as the red node, friends are black nodes, and edges between nodes represent a direct connection. For participants in dyads 2 and 3, their social network reflects clusters of friends with interconnections, but often the only connection between clusters is through the participant. In contrast, dyad 8 includes one participant who has a similar cluster structure, while the other mostly interacts with friends who also know one another. Thus, within the current participant sample, we observe variability both within and between dyad participants, indicating that this trait measure can be examined as a covariate to account for variablity in communication success in future analyses.

3.2   Dyad Interaction Outside of the Experiment Relate to State Effects at Drive 1

We next examined intra- and inter-dyad variability in subjective reports of motivation and stress at the start of their first driving session. Variability in intra-dyad scores was observed for metrics (Fig. 3 top). Absolute difference scores were calculated to determine the level of intra-dyad similarity for subjective state measures. Specifically, dyads 1, 2, 5, 6, 8, and 10–12 exhibited relatively high levels of intra-dyad similarity for motivation as indicated by lower absolute difference score values (short adjoining lines). In contrast, dyads 3, 4, 7, 9, 13 and 14 displayed relatively low levels of intra-dyad similarity for motivation (long adjoining lines). Across dyads, there was high inter-dyad variability in general motivation (range = 0–40; SD = 15.10).

For perceived stress, dyads 1, 5, 10–12, and 14 exhibited relatively high levels of intra-dyad similarity (short adjoining lines), whereas dyads 2–4, 6–9, and 13 showed relatively low levels of intra-dyad similarity (long adjoining lines). Across dyads, results demonstrated high inter-dyad variability in perceived stress at the start of their first drive session (range: 1–14; SD: 4.04). Interestingly, we observed consistency for the intra-dyad similarity between these two state metrics. Dyads 1, 5, and 10–12 had low absolute difference values for both the motivation and stress state metrics. Notably, dyads 2, 5, and 6 also displayed high levels of similarity for both density of social networks (trait) and general motivation (state).

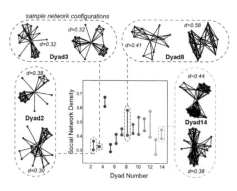

Fig. 2. *Trait Similarity Metric.* Participants' social network information was extracted using Friendly Ocean [19]. Examples from several dyads are depicted. For each subject's social network, density or the number edges divided by the number of possible edges was calculated. Higher density reflected social networks in which participants' friends were acquainted (see dyad 8, right), while lower density within the network is indicative of distinct groups of friends (see Dyad 2).

Finally, we capitalized on the fact that our participants were recruited from two cohesive work environments, and most had interacted with each other prior to the experiment. This analysis examined whether intra-dyad similarity in state metrics related to the frequency of the dyad's real-life interactions. In Fig. 3 (bottom row), the scatter plots relate the two state metrics (motivation left and stress right) with the participant's reported frequency of casual interaction in the DPSI questionnaire done during the pre-drive segment of the driving session. Preliminary results demonstrate a negative trend between general motivation scores and frequency of interaction, such that participants with relatively higher motivation scores are those that interact less frequently (r = −.28). In contrast, there appears to be a positive trend between perceived stress and frequency of interaction in the real-world demonstrating that those who interact more frequently reported higher levels of perceived stress (r = .29). These trends suggest that frequency of interaction with a dyad partner may produce differential effects in subjective state metrics. Individuals may be more highly motivated to make a strong first impression with new colleagues, while they simultaneously experience heightened performance-related stress with partners with whom they interact

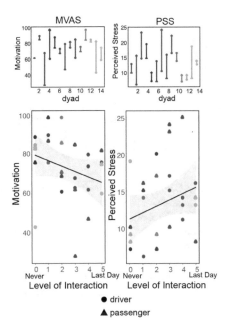

Fig. 3. *State Similarity Metrics.* Top row: state estimates for each participant, organized by dyads, for MVAS and PSS pre-drive surveys are depicted. Lines connecting driver and passenger scores reflect their level of similarity. Bottom row: self-report scores on the MVAS and PSS are plotted as a function of the self-reported level of interaction outside the experiment, where 0 = never, 1 = last 12 months, 2 = last 6 months, 3 = last month, 4 = last week, and 5 = last day.

frequently. However, we acknowledge that these emerging trends are preliminary and may change as we collect additional data.

3.3   Neural Synchrony Captures Time-Evolving Relationships Between Dyad Partners

While the first two analyses investigated similarity in state and trait variables, our final preliminary analysis examined the amount of synchrony between physiological measurements during the drive. In Fig. 4, the PLV synchrony metric is shown for a single dyad for the initial task of the on-drive segment of drive 1. The first time segment (labeled −60 to 0) is the joint listening task to a health podcast, while the second time segment (labeled 0 to 60) is the open discussion about the health information. The heat map at the bottom of Fig. 4 displays the range of PLV across the time interval with a maximum synchrony of PLV = 0.64. The purple line located above the heat map averages the synchrony across channels and displays a complex temporal profile of intervals of synchrony and asynchrony within and across both time segments. Finally, the two topographic plots on the head models at the top of Fig. 4 reveal that the spatial pattern of this physiological synchrony metric is maximal in the back of the scalp, which likely reflects visual processing. This might indicate a jointly perceived change in the environment (e.g., traffic), but future research will relate these metrics and spatial

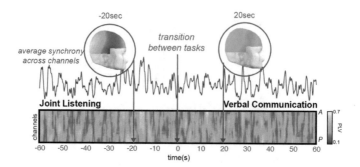

Fig. 4. *Inter-subject Synchrony.* Sample time segment of the PLV synchrony metric for a single dyad during the initial task of the on-drive segment of drive 1. The topographic plots on the head models at the top show the spatial distribution of the PLV across the scalp 20 s before and after the task transition. The purple line plot displays the average synchrony across channels shown in the heat map at the bottom of the figure.

patterns to successful communication to investigate whether neural synchrony can account for moment-to-moment fluctuations in successful communication.

4   Discussion

The present study provides a framework for understanding the subjective factors that may influence social communication variability within a multitask, driving context. Our core hypothesis asserts that successful communication between driver-passenger dyads will likely occur when similarity in shared social experiences or emotional states arise, providing a foundation for engagement and interest in understanding another's perspective [30, 31]. In this paper, we conducted preliminary analysis on a set of subjective factors that operationalized these trait and state relationships. These results demonstrated strong intra- and inter-dyad variability among social network structure in the trait similarity metric. Furthermore, we also observed a trending relation showing that real life interactions between dyad partners accounted for variability in motivation and stress states at the beginning of their first driving session. Finally, we investigated implicit measures of state captured in the physiological synchrony from scalp EEG sensors. Preliminary results identified that alpha activity captures dynamic fluctuations in neural synchrony which can be used as a covariate in future analyses of communication success. Collectively, these explicit and implicit metrics of similarity and synchrony may play critical roles in shaping successful communication between dyad partners.

Our future analyses will examine both the audio and video files collected during the drive to characterize and quantify dyad communication [32]. From the transcript of the driver's post-drive memory tasks, we can determine the number of news stories remembered as well as recall of idiosyncratic details from the conversation, reflecting driver engagement when communicating with their partner. Furthermore, natural language processing [33, 34] will be applied in order to extract sentiment information and quantify attitudes expressed by individuals within and across dyads. We expect that the passenger's valence while communicating the news stories may influence memory and recall of the driver [35]. Similarly, analysis of the videos will focus on extracting gestures and facial expressions that may also capture implicit metrics of connection that drive successful communication [36–38]. Lastly, we plan to investigate how these metrics of social communication are related to driving performance, including how social interactions and bonding can positively impact not only performance in the vehicle but health and well-being more generally.

In summary, our initial results demonstrate promising metrics that can quantify how social and emotional experiences influence performance in naturalistic, multitask environments. By identifying ways to quantify contextual complexity, this translational research can augment our understanding about how to enhance human performance within contexts with increasingly sophisticated technology automation.

References

1. McKenzie, B.: Who drives to work? Commuting by automobile in the United States: 2013. American Community Survey Reports (2015)
2. Triplett, T., Santos, R., Rosebloom, S., Tefft, B.: American Driving Survey: 2014–2015. AAA Foundation for American Safety (2016)
3. Vahidi, A., Eskandarian, A.: Research advances in intelligent collision avoidance and adaptive cruise control. IEEE Trans. Intell. Transp. Syst. **4**(3), 143–153 (2003)
4. Braitman, K.A., McCartt, A.T., Zuby, D.S., Singer, J.: Volvo and infiniti drivers experiences with select crash avoidance technologies. Traffic Inj. Prev. **11**(3), 270–278 (2010)
5. Xiao, L., Gao, F.: A comprehensive review of the development of adaptive cruise control systems. Veh. Syst. Dyn. **48**(10), 1167–1192 (2010)
6. Bansal, P., Kockelman, K.: Forecasting Americans' long-term adoption of connected and autonomous vehicle technologies. Transp. Res. Part A: Policy Pract. **95**, 49–63 (2017)
7. Fagnant, D.J., Kockelman, K.: Preparing a nation for autonomous vehicles: opportunities, barriers and policy recommendations. Transp. Res. Part A: Policy Pract. **77**, 167–181 (2015)
8. Milakis, D., van Arem, B., van Wee, B.: Policy and society related implications of automated driving: a review of literature and directions for future research. J. Intell. Transp. Syst. **21**(4), 324–348 (2017)
9. Reimer, B.: Driver assistance systems and the transition to automated vehicles: a path to increase older adult safety and mobility? Publ. Policy Aging Rep. **24**(1), 27–31 (2014)
10. Medeiros-Ward, N., Cooper, J.N., Strayer, D.L.: Hierarchical control and driving. J. Exp. Psychol. Gen. **143**(3), 953–958 (2014)
11. Hari, R., Himberg, T., Nummenmaa, L., Hämäläinen, M., Parkkonen, L.: Synchrony of brains and bodies during implicit interpersonal interaction. Trends Cogn. Sci. **17**(3), 105–106 (2013)
12. Dumas, G., Nadel, J., Soussignan, R., Martinerie, J., Garnero, L.: Inter-brain synchronization during social interaction. PLoS ONE **5**(8), e12166 (2010)
13. Klimesch, W., Sauseng, P., Hanslmayr, S., Gruber, W., Freunberger, R.: Event-related phase reorganization may explain evoked neural dynamics. Neurosci. Biobehav. Rev. **31**(7), 1003–1016 (2007)
14. Babiloni, F., Cincotti, F., Mattia, D., Mattiocco, M., Fallani, F.D.V., Tocci, A., Bianchi, L., Marciani, M.G., Astolfi, L.: Hypermethods for EEG hyperscanning. In: Engineering in Medicine and Biology Society (2006)
15. Tognoli, E., Lagarde, J., DeGuzman, G.C., Kelso, J.A.S.: The phi complex as a neuromarker of human social coordination. Proc. Natl. Acad. Sci. **104**(19), 8190–8195 (2007)

16. Lindenberger, U., Li, S.-C., Gruber, W., Müller, V.: Brains swinging in concert: cortical phase synchronization while playing guitar. BMC Neurosci. **10**(1), 22 (2009)
17. Wasylyshyn, N., Hemenway Falk, B., Garcia, J.O., Cascio, C.N., O'Donnell, M.B., Bingham, C.R., Simons-Morton, B., Vettel, J.M., Falk, E.B.: Global brain dynamics during social exclusion predict subsequent behavioral conformity. Soc. Cogn. Affect. Neurosci. **13**(2), 182–191 (2018)
18. Schmälzle, R., O'Donnell, M.B., Garcia, J.O., Cascio, C.N., Bayer, J., Bassett, D.S., Vettel, J.M., Falk, E.B.: Brain connectivity dynamics during social interaction reflect social network structure. Proc. Natl. Acad. Sci. **114**(20), 5135–5138 (2017)
19. O'Donnell, M.B., Falk, E.B.: Big data under the microscope and brains in social context: integrating methods from computational social science and neuroscience. Ann. Am. Acad. Polit. Soc. Sci. **659**(1), 274–289 (2015)
20. Cohen, S., Williamson, G.M.: Stress and infectious disease in humans. Psychol. Bull. **109**(1), 5–24 (1988)
21. Aitken, R.C.B.: Measurement of feelings using visual analogue scales. Proc. Roy. Soc. Med. **62**(10), 989–993 (1969)
22. Bond, A., Lader, M.: The use of analogue scales in rating subjective feelings. Psychol. Psychother.: Theory Res. Pract. **47**(3), 211–218 (1974)
23. McCormack, H.M., Horne, D.J., Sheather, S.: Clinical applications of visual analogue scales: a critical review. Psychol. Med. **18**(4), 1007–1019 (1988)
24. Wewers, M.E., Lowe, N.K.: A critical review of visual analogue scales in the measurement of clinical phenomena. Res. Nurs. Health **13**(4), 227–236 (1990)
25. Delorme, A., Makeig, S.: EEGlab: an open source toolbox for analysis of single-trial EEG dynamics including independent component analysis. J. Neurosci. Methods **134**(1), 9–21 (2004)
26. Winkler, I., Haufe, S., Tangermann, M.: Automatic classification of artifactual ICA-components for artifact removal in EEG signals. Behav. Brain Funct. **7**(1), 30 (2011)
27. Bigdely-Shamlo, N., Mullen, T., Kothe, C., Su, K.-M., Robbins, K.A.: The prep pipeline: standardized preprocessing for large-scale EEG analysis. Front. Neuroinform. **9**, 16 (2015)
28. Garcia, J.O., Brooks, J., Kerick, S., Johnson, T., Mullen, T.R., Vettel, J.M.: Estimating direction in brain-behavior interactions: proactive and reactive brain states in driving. NeuroImage **150**, 239–249 (2017)
29. Jensen, O., Mazaheri, A.: Shaping functional architecture by oscillatory alpha activity: gating by inhibition. Front. Hum. Neurosci. **4**, 186 (2010)
30. Higgins, E.T., Echterhoff, G., Crespillo, R., Kopietz, R.: Effects of communication on social knowledge: sharing reality with individual versus group audiences. Jpn. Psychol. Res. **49**(2), 89–99 (2007)
31. Higgins, E.T., Pittman, T.S.: Motives of the human animal: comprehending, managing, and sharing inner states. Annu. Rev. Psychol. **59**, 385 (2008)
32. McLellan, E., MacQueen, K.M., Neidig, J.L.: Beyond the qualitative interview: data preparation and transcription. Field Methods **15**(1), 63–84 (2003)
33. Joshi, A.K.: Natural language processing. Science **253**(5025), 1242–1249 (1991)
34. Chowdhury, G.G.: Natural language processing. Ann. Rev. Inf. Sci. Technol. **37**(1), 51–89 (2003)
35. Pickering, M.J., Garrod, S.: An integrated theory of language production and comprehension. Behav. Brain Sci. **36**(4), 329–347 (2013)
36. Buck, R.W., Virginia, S.J., Miller, R.E., Caul, W.F.: Communication of affect through facial expressions in humans. J. Pers. Soc. Psychol. **23**(3), 362–371 (1972)
37. Cacioppo, J.T., Petty, R.E., Losch, M.E., Kim, H.S.: Electromyographic activity over facial muscle regions can differentiate the valence and intensity of affective reactions. J. Pers. Soc. Psychol. **50**(2), 260–268 (1986)
38. Cacioppo, J.T., Bush, L.K., Tassinary, L.G.: Microexpressive facial actions as a function of affective stimuli: replication and extension. Pers. Soc. Psychol. Bull. **18**(5), 515–526 (1992)

Challenges with Developing Driving Simulation Systems for Robotic Vehicles

Kristin E. Schaefer[1(✉)], Ralph W. Brewer[1], Brandon S. Perelman[1],
E. Ray Pursel[3], Eduardo Cerame[4], Kim Drnec[1], Victor Paul[4],
Benjamin Haynes[4], Daniel Donavanik[2], Gregory Gremillion[2],
and Jason S. Metcalfe[1]

[1] United States Army Research Laboratory,
Aberdeen Proving Ground, MD, USA
{kristin.e.schaefer-lay.civ, ralph.w.brewer.civ,
brandon.s.perelman.ctr, kim.a.drnec2.ctr,
jason.s.metcalfe2.civ}@mail.mil
[2] United States Army Research Laboratory, Adelphi, MD, USA
{daniel.donavanik.ctr,
gregory.m.gremillion.civ}@mail.mil
[3] Naval Surface Warfare Center, Dahlgren, VA, USA
eugene.pursell@navy.mil
[4] United States Army TARDEC, Warren, MI, USA
{eduardo.j.cerame.civ, victor.j.paul2.civ,
ben.a.haynes2.civ}@mail.mil

Abstract. There are a number of reasons to use computer-based simulation in human-robot interaction research. Most predominant is the assessment of humanin-the-loop interactions for robotic technologies that do not yet exist, are in prototype development, or are in early test and evaluation stages of development. In these cases, simulation can provide insight into how the human may interact with the real world robotic vehicle. However, there are a number of challenges to designing and developing these simulations so that findings translate to interaction with real-world systems. This work specifically looks at the incorporation of human factors into driving simulations for robotic military vehicles. Three use cases address a number of these challenges.

Keywords: Autonomy · Driving simulation · Human-robot interaction

1 Introduction

A difficulty with conducting research aimed at high risk contexts, such as during development of autonomous capabilities for military vehicles, is that the physical space requirements to adequately exercise system capabilities are typically large, costly, and may pose real threats to system evaluators or test participants. Moreover, even with the ready availability of adequately controlled and safe spaces, the vehicles themselves must be built and tested to verify safety before any experimentation. Simulation approaches have thus been developed to enable experimentation that may be more tightly controlled, safe, rapidly deployable, and cost effective. Key to the notion of

© Springer International Publishing AG, part of Springer Nature (outside the USA) 2019
D. N. Cassenti (Ed.): AHFE 2018, AISC 780, pp. 139–150, 2019.
https://doi.org/10.1007/978-3-319-94223-0_13

using driving simulation to advance autonomy development is the need for the simulation environment to have enough fidelity that realistic human-autonomy interactions, or at least essential elements thereof, may be evoked and assessed. As such, simulation may need to provide varying levels of visual and motion cues [1] either through the platform or within the virtual environment; have sufficient sub-system realism; and be able to collect the necessary data, including human-in-the-loop (HITL) performance outcomes. Often times, these pieces of the simulation are instantiated in separate software and hardware components, thus creating an underlying requirement for a simulation framework capable of coordinating the precise distribution, execution and timing of all of those simulation components.

This work addresses three main challenges in developing driving simulation for teaming with autonomy-enabled manned or unmanned vehicles in military operations. Section 2 discusses the challenges involved in developing driving simulations that can advance understanding of human interaction with autonomy-enabled vehicles that either do not yet exist or are in early development stages. Section 3 provides insight into how driving simulation may be leveraged to support remote operation of real-world autonomy-enabled vehicles and directly integrate with the vehicle software. Section 4 addresses the conceptual and technical challenges of assessing adaptive capabilities associated with future human-autonomy teaming or joint operation initiatives. Three specific US Army use cases, Applied Robotics for Installations and Base Operations (ARIBO), Wingman, and Privileged Sensing Framework (PSF) Autonomy Research Pilot Initiative (ARPI), provide some possible solutions to the technical and conceptual challenges that arise in the development of driving simulation for human-in-the-loop (HITL) operations. Taken together, these use cases provide a basis for addressing both the challenges and opportunities available through use of driving simulation in science and technology efforts aimed at future military operations, including development of next generation crew station technologies.

2   Interaction with Systems that Do Not Yet Exist

Simulation is paramount in assessing human factors challenges with robots that do not exist or are in early design. It is often the only way to attain some early insight into the potential HITL interactions that could occur with a robotic vehicle. The main difficulty with this type of simulation is an increased number of unknowns associated with the capabilities and autonomy-enabled features of the vehicle. In the likely case of integrating and planning for such a complex system where detailed operations of certain components are held as proprietary intellectual property, inherent unknowns will exist that, nevertheless, should be assessed as they may be expected to interact with the rest of the system and impact overall system performance. Therefore, *Wizard-of-Oz* and similar types of simulation approaches that emulate operational principles instead of replicate actual architecture, or even instantiate actual autonomy, are often used during experimentation. With these types of studies, the goal is to make participants believe that a robotic vehicle is making the decisions. The potential problem with this approach is verifying that the perceived vehicle autonomy features are realistic and comparable to the future real-world system. One driving simulation use case was developed in support of the ARIBO driverless vehicle program. The purpose was to provide assessment of the human factors associated with the transition from manual to fully autonomous driving and help guide real-world vehicle design to better support trust and acceptance of a shuttle vehicle.

2.1 ARIBO

The US Army Tank Automotive Research, Development and Engineering Center (TARDEC) ARIBO program was a series of pilot programs using federal installations and universities as test beds for developing guidelines for operating autonomous vehicles in public, noncombat environments. The strategic objectives included socializing users and nonusers with autonomous systems, identifying operational issues and developing mitigation strategies to increase trust and use, and generating empirical data (e.g., performance, reliability, maintenance). A major goal of the ARIBO program was to produce technical and social-behavioral value through a cycle of data collection, reliability analysis, and technical and behavioral improvement. One specific research focus was on the prototype development of autonomy-enabled on-demand transportation vehicles, such as a shuttle vehicle for Soldier and civilian transport between the medical barracks and the medical center at Ft Bragg, North Carolina [2].

The driving simulation was developed through a computer-based simulation platform called Robotic Interactive Visualization Experimentation Technology (RIVET). RIVET was developed as a platform to develop and test autonomy algorithms using hardware-in-the-loop processes. In order to extend the capability of the simulated environment for HITL assessment, we developed the Control of Autonomous Robotic Vehicle Experiments (CARVE) interface to assess interaction, behaviors, and performance [3]. RIVET, with the integration of the CARVE interface, allowed for repeatability of missions, control over simulated robotic vehicles, integration of multiple levels of autonomy (from manual to fully autonomous), customization of feedback features (e.g., iconography, task information, etc.), customizable operational scenarios, redundancy in systems to account for failure, addition of secondary tasks to assess workload and stress, and customizable data logging.

The virtual environment mirrored the road structure and sidewalks of the real-world operational area. Customizations were made to the virtual environment to allow for interactions with pedestrians and other road traffic. Specific challenges with development of the ARIBO simulation included individual differences assessment [4], control allocation [5], performance measurement, user control integration, and display and feedback mechanism development [6].

2.2 Addressing the Technical Challenges

RIVET was chosen as the simulation platform for ARIBO because it can create a virtual terrain surface, static features such as buildings, and dynamic elements such as people and vehicles that mirrored the real-world environment. The system also provided the capability to integrate multiple types of control interfaces and develop graphic user interfaces to implement user displays, feedback mechanisms, and capabilities to assess human performance. Three challenges with understanding human interaction with systems that do not yet exist include mirroring real-world behaviors when those characteristics are yet to be completely defined, communicating environmental and vehicle constraints to the user without a full motion platform, and recording human performance.

Mirroring Real-World Vehicle Behaviors. Human factors assessments of driving simulations during early design stages often require participants to perceive various levels of vehicle autonomy. When these systems do not exist, *Wizard-of-Oz* type approaches that represent future capabilities are often employed. In driving simulation this requires the capability to mirror both throttle and steering autonomy. Since no vehicle autonomy software had been developed prior to the creation of the simulation, rather than having a person driving the vehicle (as is standard in Wizard-of-Oz approaches), CARVE was preprogrammed to mirror potential autonomous vehicle behaviors described in a requirements document for the real-world system (e.g., speed control, stopping behaviors, and distance to pedestrians and other road users). To accomplish this need, the vehicle autonomy was created through the integration of dynamic waypoints to control both throttle and steering. Traditional waypoint following provided the capability to set a path the vehicle would follow throughout a virtual environment. Dynamic waypoints provided a set number of actions that could occur at each point. These included a "stop" function to stop the vehicle for a set amount of time, and a "SetSpeed" option to mirror autonomous throttle control. The outcome of this capability was the appearance of real-time decision-making.

Availability of Vehicle Controls. With driverless vehicle autonomy, it is expected that there will be a human to be in-the-loop with the capability to intervene in the autonomous control modes of the vehicle, even if just to press an emergency stop button or call for assistance. Therefore, CARVE was designed to work with a number of controllers including: keyboard, mouse, touch screen, Logitech Gamepad, Logitech joystick, Logitech G27 wheel and pedals, Microsoft Kinect, and other user interfaces (e.g., autonomy engage and disengage buttons). This provided the functional capabilities to allow for changes in control allocation, such as switching between manual control and levels of autonomy [4, 5]. Figure 1 depicts the driving simulation setup for ARIBO.

Fig. 1. ARIBO simulation laboratory setup from the participant view with the three computer software experimenter control setup in the background.

Communicating to Participants. When working with computer-based simulation, it is essential to provide feedback to the user about the task, safety issues, and even environmental or system constraints. For ARIBO, graphical user interfaces (GUIs) were developed within RIVET. Feedback included notifications of arrival or planned departure from pick-up and drop-off locations, passenger and location information, and wrong way alerts during manual driving. The CARVE software was also adapted to provide transparency-related feedback (in line with the guidelines of the Situation Awareness agent-based Transparency Model [7]) to users in order to increase trust (for specific types of features see [5]). To provide reasoning information related to navigation-specific behaviors such as changes in route or speed, dynamic icons were added to the simulation through CARVE's dynamic waypoints feature. This technical feature provided the means to simulate autonomous obstacle detection and avoidance without the need for advanced software functionality (e.g., the vehicle slowed down in the middle of the road because a pedestrian walked across the roadway).

Recording Human Performance. When assessing human use and control allocation through simulation, human performance data collection is required but often difficult to accomplish due to the need for customized recording capabilities through the simulation. Through CARVE, we were able to integrate customized software to address this need. Data collection was customized to record vehicle behaviors as well as the frequency and duration of interventions in the vehicle's autonomy including event, time, and associated data (e.g., duration and associated times of button presses, type of button, as well as time and frequency between manual and autonomous modes). In addition, time stamps were recorded when the icons were displayed to compare events in the environment to user behaviors and psycho-physiological measures that could also be collected during trials. All data was reported in a customizable .csv output file.

3   Remote Operation of Unmanned Real-World Vehicles

Simulation can support development of real-world vehicles and assessment of new teams for remote operation of unmanned vehicles. Challenges in the development of a software-in-the-loop (SIL) simulation as a HITL test bed include integration of new human roles for remote operation, development of real-world virtual test courses, and software to capture performance and changes in human and vehicle behaviors. The US Army TARDEC Wingman program is included as the use case here because the goal of the program is to provide autonomous capabilities advances for manned and unmanned ground combat vehicles while decreasing the gap between autonomous vehicle control and the required level of human interaction.

3.1   Wingman

The US Army TARDEC Wingman JCTD program addresses a joint and combatant command warfighting need through the execution and demonstration of prototypes. This three-year effort couples Warfighters with an armed unmanned ground vehicle to improve the autonomous capabilities of mounted and unmanned combat support vehicles. At present, the Wingman team includes a single manned vehicle with a 5-man crew working together with a single unmanned robotic vehicle operating in a joint gunnery task. A major goal of this program is to advance manned-unmanned teaming (MUM-T) initiatives by iteratively defining and decreasing the gap between autonomous vehicle control and required level of human interaction while qualifying with a US Army gunnery certification on a test range.

3.2   Development of a Software-in-the-Loop Simulation System

Development of a Wingman SIL offers direct access to the real-world vehicle software within a simulated environment in order to provide a means to test and update software, advance integration, and assess teaming initiatives in a safe and cost-effective environment. Once complete, the SIL will also serve as a training platform for warfighters prior to using the real-world vehicles. Therefore, the SIL must provide an adequate representation of the real world and simultaneously work with users to meet qualification standards on the US Army gunnery team mission evaluations. These evaluations assess on-task human team communication, as well as performance (i.e., time and accuracy) associated with completing the gunnery exercises under a variety of environmental constraints.

3.3   Addressing the Technical Challenges

The SIL encompasses the entire software system including autonomous mobility of the vehicle platform (i.e., Robotic Technology Kernel or RTK), the weapon system autonomy (i.e., Autonomous Remote Engagement System or ARES), and the user displays (i.e., Warfighter Machine Interface or WMI) which allows the manned team the ability to interact with the autonomous vehicle systems. This software was integrated with two off-the-shelf simulation systems to support vehicle mobility (Autonomous Navigation and Virtual Environment Laboratory; ANVEL) and gunnery operations (Unity3D). Five challenges and possible solutions associated with supporting remote operation of unmanned real-world vehicles are discussed below. These include how to integrate proprietary software with off-the-shelf simulation systems, development of matching virtual environments across ANVEL and Unity3D, specific considerations for simulating human team member roles, controlling features in the environment, and human factors data collection.

Integrating Proprietary Software with Off-the-Shelf Simulation Environments.
There are multiple game-based off-the-shelf simulation environments available for development of driving simulation, each with their own advantages and limitations. When integrating with proprietary software as required with a SIL, it is necessary to understand the benefits and limitations. For the Wingman program, the driving simulation environment had to support integration with autonomous robotic mobility software, as well as an autonomous robotic engagement system for gunnery operations. Therefore, the two simulation systems were integrated into the SIL. The first and most critical technical challenge set to overcome designing a co-simulation system where ANVEL and Unity3D work in tandem, sharing the necessary data with each other, the

vehicle software, and the user displays, without interfering with the performance of any of these systems. The use of the Robot Operating System (ROS) as the basis for RTK and ARES was key in the integration of these systems. Since the engagement system is onboard the simulated vehicle, the vehicle orientation and localization data for robot mobility had to be passed from ANVEL to the RTK software, from RTK to ARES, and then to Unity3D. This was accomplished by using a ROS multimaster connection, which is the same implementation as the real-world Wingman vehicle for sharing localization data between RTK and ARES.

Generating Matching Environments. For the shared pose data to be useful, a procedure was needed for generating matching environments for both Unity and ANVEL based on real world gunnery test ranges. This is difficult to accomplish because the two simulation systems use two different base structures, including location of the origin, and left-handed versus right-handed coordinate systems, as well as two different file structures for developing terrain files. For Wingman, the final procedure takes USGS data that is converted to a raw format for Unity to utilize. This data was then output from Unity as a Wavefront (.obj) mesh, which was modified in Blender[1] to account for the differences in the orientation of the coordinate system, and then converted and used by ANVEL (for complete steps see [8]). As this procedure includes several manual steps, it remains a preliminary method until a more robust and automated procedure can be finalized.

Determining Which Human Roles to Simulate. When conducting HITL assessment, it is important to determine which roles and tasks will be integrated and which will be simulated. To increase consistency and repeatability between simulation runs, it may be pertinent to simulate certain human roles. For Wingman, the manned team includes a driver who is responsible for the command vehicle, team commander, LRAS3 operator to identify environmental targets, robotic vehicle operator to monitor and control the mobility operations of the unmanned vehicle, and robotic vehicle gunner to monitor and control gunnery operations. The current SIL has a simulated command vehicle driver and LRAS3 operator roles. The movement of the command vehicle was simulated through Unity by setting prescribed path segments where movement was based on either location triggers for the unmanned vehicle or through scripting, such as when the exposure time for all targets in an engagement expired. The target detection operations of the LRAS3 operator are simulated through data transfer when targets are raised and provide an icon on the commander's WMI display representing a possible target location. To simulate accurate time delays seen in the real-world, the experimenter can set a delay from when a target appears and when the LRAS3 target cue is sent to the WMI through a GUI in the Unity simulation.

Controlling the Environment. The test environments are gunnery ranges that are specific geographic locations with established vehicle routes and fixed targets, engagement areas, and firing positions. Still, there is great variability in designing test scenarios from the combinations of those fixed components. A GUI was designed to enable the

---

[1] Blender is a free and open source 3D creation suite that supports modeling, rigging, simulation, rendering, composition and motion tracking, and video editing and game design (https://www. blender.org/).

experimenter to cue targets' appearance times, exposure durations, and the delay before LRAS target cues are sent to the Wingman system (as described above). This queuing process can be done prior to an experiment run or modified during the run. An additional capability was added to enable preloading of the GUI form at run-time through use of an XML encoded configuration file. The ability to set or preload a script of target exposures enables any combination of targets in a scenario without having to modify the underlying simulation system and still enables a scenario to be reproduced.

Human Factors Data Collection. Another key element in active development is the capability to record human and system performance. Performance data collection during gunnery qualification exercises includes timing accuracy associated with target detection and firing, as well as the appropriateness of communications between the operators. If the SIL behavior or WMI does not accurately represent the real-world system or lacks some of the key roles, it reduces the value of the data collected. Therefore, Unity3D can timestamp and record each event of interest during the experiment run in a data file from a name and location set at run-time. In the autonomy software, data on the vehicle position and orientation, weapon orientation, and the weapon control inputs are collected in a ROS bag file and the relevant elements are then converted to a .csv file using a custom python script. The main challenges are determining which events are significant and a method to synchronize the times of the events between the three different logging systems (Unity, ROS, and WMIs).

4   Assessing Adaptive Teaming Capabilities

As autonomy and other intelligent technologies become increasingly pervasive both in civilian society as well as on the battlefield, increasingly heterogeneous groups of humans and intelligent agents with varied capabilities (e.g. smart sensors, AI software agents, or physical robots) will need to work efficiently and effectively to produce coordinated outcomes. An ever growing need is to conduct adequate test and evaluation of tools and technologies developed to support such mixed-initiative team operations. Herein, then, is another circumstance where driving simulation can be expected to provide important experimental methods for the early development and refinement of the required technologies to support an anticipated need for advanced teaming in mobile operations with heterogeneous groups [9]. In the following, we describe an aggressive program, funded by the Office of Secretary of Defense, aimed at advancing novel methods to address the problem of effectively assessing the human during teaming operations.

4.1   Privileged Sensing Framework Autonomy Research Pilot Initiative

The PSF research conducted under the ARPI program was aimed at developing a generalizable approach for human-autonomy teaming where, rather than being a presumed infallible supervisor, the human was treated as an intelligent sensor with special privileges that might be accorded based on known doctrine, tactics, or strategies not easily implemented through otherwise conventional control system means [10]. An important goal within this program was to assess HITL interaction with simulated mobile agents, which included driving autonomy, in order to characterize behavioral and physiological indicators of human operator states such as trust, which would thus be leveraged to predict behavioral outcomes like reliance on autonomy [11]. Early research successfully established a relationship between trust-based behaviors such as taking control *from* a driving aid or relinquishing control *to* a driving aid, automation reliability, and subjective workload as expressed through physiological state [12]. Together this data enabled initial proof-of-concepts that such variables may be integrated within a control system framework to facilitate a capability for predicting and mitigating inappropriate control authority mode switches. A modified leader-follower driving simulation paradigm was used to facilitate the measurement of behavioral and physiological indices of interaction (e.g., trust) [12], and to predict changes in control authority to later adapt vehicle autonomy based on these indicators [11, 13].

4.2 Addressing the Technical Challenges

There are a number of conceptual and technological challenges that need to be overcome when assessing teaming, including isolating psychophysiological indicators of trust, predicting trust-based behaviors, and managing those behaviors in real time. Challenges associated with assessing adaptive teaming capabilities include the level of platform fidelity, the process for integrating behavioral predictions - such as is needed to anticipate intended changes in control authority - in a real-time driving simulation, and determining which psychophysiological, behavioral, and task or environment features can be used for predicting and managing trust-based behaviors.

Simulation Platform Fidelity. The level of fidelity of a driving simulation platform is often a consideration for human factors assessment. The major components of the systems provide outputs through the visual driving screens and audio systems, and the vehicle dynamics must calculate at a high enough rate to refresh the physical displays. The TARDEC Ride Motion Simulator (RMS), depicted in Fig. 2a, is a high-performance, single and multiple occupant, six degrees-of-freedom motion based platform designed to recreate the 'ride' of military ground vehicles with high precision and accuracy [14]. SimCreator^{TM}, a commercial simulation package developed by Realtime Technologies, Inc. (Fig. 2b), was chosen for its flexibility with independent system integration and customizable creation of scenarios.

(a)   (b)

Fig. 2. The RMS (a) and simulated driving motion-synchronized visuals produced by SimCreator (b) supports real world vehicle dynamics

Integrating Processes for Trust-Based Decision-Making. Even with perfect *a priori* knowledge of expected operator behaviors, the ability to monitor and then integrate such information within a real-time simulation context is a considerable challenge. To address this challenge, one must determine a critical system behavior that can be reliably predicted far enough in advance of being enacted that, if necessary, mitigating actions may be taken. Leveraging tools from human neuroscience, impending decisions may be tracked and predicted by monitoring scalp-based electrical activity produced by frontal brain networks; such activity can be linked to decisions whereby risk and reward evaluation has been suggested to be supported by fundamental, and deep, brain structures (e.g., amygdala) [15]. Considering trust as an instance of such risky decision-making may thus enable a concrete method to address the challenge of behavioral prediction [16]. In the case of the simulated driving task described above, the A-B decision was classified as either a hand-off (automation is given control) or a take-over (regain manual control) based on perception of current environmental risk, task demands, and automation quality. To be able to elicit real-world type A-B decisions within a simulated environment, several technical challenges had to be addressed. Whether perfectly performing or imperfectly performing, it is reasonable to anticipate that there will be limited variability in the degree to which a person would engage such an autonomy (i.e. a perfectly "bad" autonomy would never be relied upon). To ensure the necessary variability to enable examination of the decision-making processes, an automation needed to be devised that people would sometimes allow to drive and would at other times take over. Therefore, with respect to both speed and heading control, an imperfect autonomy was created that emulated the driving performance of an average person (as determined by observation of manual driving of 10 individuals performing this same experimental task). This created regular instances where the human user would have to take over control to maintain lane position and possibly avoid a collision and likewise, other instances where the automation was driving such that the humans were willing to rely on it. Further, reward and risk were introduced through a scoring system in which participants gained or lost points, which later translated to payment, for a variety of successes and errors, which effectively monetized inadequate decision-making [11, 12].

Prediction and Management of Trust-Based Behaviors. There are both conceptual and technical challenges associated with assessing the human as a sensor that can advance human-autonomy teaming. From a conceptual standpoint, psychophysiological measurement and evaluation is traditionally developed for the laboratory environment and there is debate in the literature about what can be inferred from these measures. To address these issues, a data driven approach was used to select the relevant environmental (e.g., road curvature and time to contact with obstacles) and psychophysiological (i.e., electroencephalographic data and horizontal eye movement) input features that are correlated with changes in control authority [11]. Further technical challenges arose in deriving the predictive set of features in real-time and passing the data to the integrated system to influence changes in behavior. One of the primary challenges was that the environmental and psychophysiological features did not share the same sampling rate or scale. A unique architectural solution used a ROS-based architecture for managing integration and timing of heterogeneous signals, as well as asynchronous execution of the overall control algorithm informing the actuator interface. The architecture made use of native ROS-based management of signals in a publisher-subscriber interface, using the fastest sampling rate as the master system clock and buffers to manage windowed time slices of data for pre-trained classifiers. This data fusion framework successfully handled environmental and physiological signals at heterogeneous time scales with different architectural interfaces. It is amenable to modular implementation of newly integrated control algorithms [17].

5   General Discussion and Way Forward

Clearly there are a number of challenges to developing and assessing HITL driving simulation. While this account of the three examples does not provide a fully sufficient solution for the application of driving simulation to the conceptual and technical challenges involved in developing field-ready military systems, it provides potential methods of advancing driving simulations development to better support HITL integration and teaming. Future work will support the development of driving simulation for the US Army TARDEC Crew Augmentation and Optimization program. A recent focus on closed hatch operations and MUM-T has rejuvenated the interest and investment in advanced crew stations for manned platforms that interact with autonomy-enabled systems. This program brings together many of the simulation challenges presented in this paper to begin to understand in a manned vehicle crew station environment, human interaction with systems that do not yet exist, processes to support remote operation of unmanned real-world vehicles, and assessment and integration of adaptive teaming capabilities. Within this program, TARDEC will not only look at improvement to traditional crew station technologies such as vehicle displays and controls, but will address the problem of vehicle operation (both local and robotic) under hatch while possibly reducing the number of crew members without overburdening the team. Embedded intelligent agents can potentially manage and prioritize the overabundance of tasks, both physical and cognitive, so that crew members can perform at the crew's mission highest performance levels. Crucial to this is the development of an advanced simulation environment that can evaluate early crew station research and development work in virtual mission based scenarios by warfighters toward improving heterogeneous team interactions.

Acknowledgments. The views and conclusions contained in this document are those of the authors and should not be interpreted as representing the official policies, either expressed or implied, of the Army Research Laboratory or the U.S. Government. The U.S. Government is authorized to reproduce and distribute reprints for Government purposes notwithstanding any copyright notation herein.

References

1. Kaptein, N., Theeuwes, J., Van Der Horst, R.: Driving simulator validity: some considerations. Transp. Res. Rec.: J. Trans. Res. Board. **1550**, 30–36 (1996)

2. Mottern, E., Putney, J., Barghout, J., Straub, E.: Moving technology forward by putting robots to work on military installations: autonomous warrior transport on-base (AWTO). In: NDIA Ground Vehicle Systems Engineering and Technology Symposium, Novi, MI (2015)
3. Schaefer, K.E., Brewer, R.W.: A guide for developing human-robot interaction experiments in the robotic interactive visualization and experimentation technology (RIVET) simulation (ARL-TR-7683). US Army Research Laboratory, Aberdeen Proving Ground, MD (2016)
4. Schaefer, K.E., Scribner, D.N.: Individual differences, trust, and vehicle autonomy: a pilot study. Hum. Factors Ergon. Soc. **59**(1), 786–790 (2015). Los Angeles, CA
5. Schaefer, K.E., Straub, E.R.: Will passengers trust driverless vehicles? Removing the steering wheel and pedals. In: CogSIMA, pp. 159–165. IEEE, San Diego (2016)
6. Schaefer, K.E., Brewer, R.W., Putney, J., Mottern, E., Barghout, J., Straub, E.R.: Relinquishing manual control: collaboration requires the capability to understand robot intent. In: Collaboration Technologies and Systems, pp. 359–366. IEEE, Orlando (2016)
7. Chen, J.Y.C., Procci, K., Boyce, M., Wright, J., Garcia, A., Barnes, M.: Situation awareness based agent transparency (ARL-TR-6905). US Army Research Laboratory, Aberdeen Proving Ground, MD (2014)
8. Schaefer, K.E., Brewer, R.W., Pursel, E.R., Zimmerman, A., Cerame, E.: Advancements made to the Wingman software-in-the-loop (SIL) simulation: how to operate the SIL (ARL-TR-8254). US Army Research Laboratory, Aberdeen Proving Ground, MD (2017)
9. Piekarski, B., Sadler, B., Young, S., Nothwang, W., Rao, R.: Research and vision for intelligent systems for 2025 and beyond. Small Wars J. http://smallwarsjournal.com/jrnl/art/research-and-vision-for-intelligent-systems-for-2025-and-beyond
10. Marathe, A.R., Metcalfe, J.S., Lance, B.J., Lukos, J.R., Jangraw, D., Lai, K.-T., Touryan, J., Stump, E., Sadler, B.M., Nothwang, W., McDowell, K.: The privileged sensing framework: a principled approach to improved human-autonomy integration. Theor. Issues Ergon. Sci. **19**(3), 283–320 (2018)
11. Metcalfe, J.S., Marathe, A.R., Haynes, B., Paul, V.J., Gremillion, G.M., Drnec, K., Atwater, C., Estepp, J.R., Lukos, J.R., Carter, E.C., Nothwang, W.D.: Building a framework to manage trust in automation. In: SPIE 10194, Micro-and Nanotechnology Sensors, Systems, and Applications IX, Baltimore, MD (2017)
12. Drnec, K., Metcalfe, J.S.: Paradigm development for identifying and validating indicators of trust in automation in the operational environment of human automation integration. In: Augmented Cognition, pp. 157–167. Springer, Cham (2016)
13. Nothwang, W., Gremillion, G., Donavanik, D., Haynes, B., Atwater, C., Canady, J., Metcalfe, J., Marathe, A.: Multi-sensor fusion architecture for human-autonomy teaming. In: Resilience Week, pp. 166–171. IEEE, Chicago (2016)
14. Truong, N., Paul, V., Shvartsman, A.: Integration of the CAT crewstation with the ride motion simulator (RMS). SAE World Congress & Exhibition, SAE Technical Paper 200601-1171 (2006)
15. Basten, U., Biele, G., Heekeren, H.R., Fiebach, C.J.: How the brain integrates costs and benefits during decision making. Natl. Acad. Sci. **107**(50), 21767–21772 (2010)
16. Drnec, K., Marathe, A.R., Lukos, J.R., Metcalfe, J.S.: From trust in automation to decision neuroscience: applying cognitive neuroscience methods to understand and improve interaction decisions involved in human automation interaction. Front. Hum. Neurosci. **10** (290), 1–14 (2016)
17. Donavanik, D., Canady, K., Gremillion, G.M., Groves, P., Nothwang, W.D., Atwater, C., Fernandez, R., Haynes, B., Marathe, A.R., Metcalfe, J.: Data processing architectures for human machine fusion. In: Intelligent Virtual Agents, Physiologically Aware Virtual Agents Workshop, Los Angeles, CA (2016)

Trust in Automation Among Volunteers Participating in a Virtual World Telehealth Mindfulness Meditation Training Program

Valerie J. Rice[1]([⊠]), Rebekah Tree[2], Gary Boykin[1], Petra Alfred[3],
and Paul J. Schroeder[4]

[1] Army Research Laboratory, Fort Sam Houston, San Antonio, USA
{valerie.j.rice.civ,gary.l.boykin.civ}@mail.mil
[2] Inspired eLearning, San Antonio, USA
mariah.tree@inspiredelearning.com
[3] Pacific Science and Engineering Group, San Diego, USA
petraalfred@pacific-science.com
[4] DCS Corp, Alexandria, USA
pschroeder@dcscorp.com

Abstract. Trust is important in group interactions; however, little is known about trust in wellness-related telehealth training. This study examined self-reported trust in U.S. military active duty and veterans (n = 45) who participated in an 8-week mindfulness course offered in the Virtual World (VW) of Second Life. Participants completed a VW Trust Questionnaire (VWT, measuring relational trust such as communication, confidentiality, and self-representation) and a Trust in Automation Questionnaire (TIA, measuring confidence in system and perceived system security, integrity, dependability, and reliability) post training. Participants reported moderately high levels of TIA and high relational trust (VWT). Higher class attendance was associated with being comfortable speaking in the VW and belief in confidentiality (relational trust). Higher attendance was also associated with higher TIA. These results demonstrate that individuals are more likely to participate in virtual world telehealth interventions, and complete more of their training, when their trust is high.

Keywords: Trust · Automation · Telehealth · Virtual world

1 Introduction

Human factors issues in group-based teaching over a Virtual World (VW) include considerations of input devices, methods of communication, keeping attendees engaged, developing group cohesion, response timing, virtual environmental design, and technology competence and use [1]. Trust in automation is another important aspect for consideration. Trust in automation includes factors that relate specifically to technology, but not necessarily to trust among humans. These factors include the reliability, validity, utility, and robustness of the technology, as well as false alarms incurred while using the technology [2].

Trust is a key determinant of automation reliance and usage [3]. That is, individuals need to trust automation before they elect to use it. This may be especially true when

© Springer International Publishing AG, part of Springer Nature (outside the USA) 2019
D. N. Cassenti (Ed.): AHFE 2018, AISC 780, pp. 151–160, 2019.
https://doi.org/10.1007/978-3-319-94223-0_14

using automation for telehealth purposes, in which participants may express personal vulnerabilities. Trust in automation may impact their willingness to participate by virtually attending class or to communicate during training. In fact, health care staff acceptance of telehealth technologies for (and with) their patients are tied to their levels of trust in automation, including barriers such as difficulties with operating the system; apprehensions regarding information sharing, data security, technical and use issues; concerns about reliability and accuracy; and lack of confidence in technology [4].

This study focuses on trust in automation when participating in mindfulness meditation instruction in a VW. A VW is "A synchronous, persistent network of people, represented by avatars, facilitated by computers" [5]. In contrast to virtual reality, VW's allow users to interact with others and are persistent. Similarly, in a VW, users are represented by avatars, while in virtual reality users do not choose their representation or may not have a representation of themselves. VW's are highly social and interactive, and both users and developers create the environment, while virtual reality is not typically highly social and developers create the environment. Interactions are often more scripted in a virtual reality scenario, while they are free-flowing, person, and occur in real time in a VW. Second Life (SL) is a VW that is free to the public and provides a means for people to connect socially. It allows for extensive creativity, as users are able to script their own materials and have multiple avatars with various identities.

VW's are also used for communication and collaboration and trust has been found to be a key enabler of VW-based collaborations [6]. Some organizations initially moved forward in the use of VW such as IBM, Nissan, Toyota, Adidas, Reebok, Dell and Vodafone [7], possibly based on analysts' predictions that VWs will become the dominant method for accessing and sharing information over the internet [8]. Yet, some organizations have scaled back on the use of VWs, due to limited user response [7]. Chandra et al., 2011 postulated that perceived structural assurance and social presence were necessary for fostering initial trust in the VW, and that user trust is essential in influencing members of an organization to develop even the intention to use the VW for communications and collaborations. Gartner suggested that 90% of businesses fail in the VW, because the companies attend to the technology, instead of the needs, behaviors, and motivations of the users [9].

The purpose of this research was to examine research participants' self-reported relational trust (communication, confidentiality, and self-representation) and their trust in automation after attending a telehealth mindfulness meditation class via the VW of SL. It was expected that to freely and fully participate, participants would need to trust that the technology itself was safe and secure and their personal information was protected, as well as trusting in the functionality of the equipment and software.

2   Method

2.1   Participants

Forty-five (45) U.S. military active duty service members and veterans were recruited as part of a more extensive intervention-based study examining Mindfulness Meditation training offered in-person and via a VW.

2.2   Instruments

Demographic Survey. Demographic data included age, gender, education, ethnicity, military status, marital status, and deployment history. Participants also provided information on their hours of computer use per week (at home and at work) and self-ratings of their computer expertise.

Class Attendance. Each of the eight classes for the VW group were held for 1.5 h, and a single, silent, extended class was held for 3 h. Class attendance consisted of the participant logging in to the Virtual World of Second Life at the designated class time, maintaining avatar presence during the class, and participating in class exercises and discussions. Researchers took role for each class and tracked participation.

Virtual World Trust Questionnaire (VWT). Four questions were designed to capture how comfortable the participants were with communicating in the Virtual World (via voice and text), how they felt they were perceived while using an avatar, and how much they trusted their information to be kept confidential. These comprise aspects of relational trust. A Likert Scale of 1–5 was used, 1 = "strongly disagree", 3 = "neutral", 5 = "strongly agree". Reliability for the VWT was good ($\alpha = .77$).

Trust in Automation Questionnaire (TIA). The Trust in Automation Questionnaire consists of 12 items on a 7-point Likert scale with 1 being "not at all" and 7 being "extremely". Five questions address statements of distrust in the automation, six questions address statements of trust in the automation, and one question addresses familiarity with the automation. This scale was developed to examine trust in automated systems [10]. Reliability for the TIA was moderate ($\alpha = .62$).

2.3   Procedure

Following an Institutional Review Board approved protocol, participants completed an informed consent form and a demographic survey, as part of the pre-intervention data collection. Participants were assigned to one of three intervention groups: In Person (IP), Virtual World (VW), or Waitlist Control. Participants in the IP and VW groups completed nine mindfulness classes over eight weeks. After training, participants completed post-intervention assessments.

This paper addresses a subset of post-intervention questions administered *only* to the VW group, specifically asking about their experience using Virtual World technology during training and their trust in the automation. The in-person training group and the control group did not use the Virtual World technology and thus were not included in this assessment.

2.4   Data Analysis

Means and Standard Deviations were assessed for the Virtual World Questions and Trust in Automation Questionnaire. Pearson Product Moment correlations were used to determine relationships between demographic variables, trust, and class attendance. An ANOVA and the Student's t-test were conducted to detect differences in trust and class attendance based on demographics. IBM SPSS Statistics for Windows, version 22.0 (Armonk, NY) was used with all analyses having a .05 alpha level.

3   Results

Demographics. 45 individuals completed the study as part of the VW intervention group. Table 1 shows the demographic information. Participants ranged in age from 26 to 69 (M = 50.49, SD = 10.68).

Table 1. Participant demographics

	N (%)
Gender	
Male	25 (55.6)
Female	20 (44.4)
Military status	
Active duty	14 (31.1)
Reserve	1 (2.2)
Guard	1 (2.2)
Veteran	29 (64.4)
Education	
High school	3 (6.7)
Some college/associate's	10 (22.3)
Bachelors	13 (28.9
Masters/doctorate	16 (35.5)
Other professional	3 (6.7)
Ethnicity	
African American	9 (20.1)
Caucasian	29 (64.4)
Hispanic	4 (8.9)
Asian	1 (2.2)
Native American	1 (2.2)
Other	1 (2.2)
Marital status	
Married	28 (62.2)
Divorced/separated	9 (20.1)
Single	6 (13.3)
Widowed	1 (2.2)
Partnered	1 (2.2)
Computer experience	
Little to no	1 (2.2)
Basic	3 (6.7)
Intermediate	14 (31.1)
Expert	23 (51.1)
Advanced	4 (8.9)

Trust Measures. Tables 2 and 3 show the means, standard deviations, and number of responses for each item on the VWT and TIA. Although 45 participants completed the post-study questionnaire, not all participants answered all of the questions. Overall, participants felt comfortable communicating in SL, trusted their information would be kept confidential, and believed they were perceived as intended. For the TIA, mistrust in the system was low, while measures of trust in the system were relatively high.

Table 2. Means, standard deviations, and n's for the VWT

Item	Mean	SD	n
I felt comfortable speaking during my MBSR class over Second Life using my voice	4.00	.95	45
I felt comfortable speaking during my MBSR class over Second Life using my text	4.24	.83	45
I trust that my personal information will be kept confidential over Second Life	4.33	.67	45
I believe that my MBSR over Second Life classmates and instructors perceive me the way I intended using my avatar	3.95	.81	44

Table 3. Means, standard deviations, and n's for the TIA

Item	Mean	SD	n
The system is deceptive	1.86	1.34	42
The system behaves in an underhanded manner	1.67	1.22	42
I am suspicious of the system's intent, action or outputs	1.90	1.36	42
I am wary of the system	2.02	1.47	42
The system's action will have a harmful or injurious outcome	1.67	1.22	42
I am confident in the system	5.12	1.69	42
The system provides security	5.00	1.38	42
The system has integrity	5.00	1.33	42
The system is dependable	4.55	1.80	42
The system is reliable	4.48	1.85	40
I can trust the system	4.90	1.61	41
I am familiar with the system	4.59	1.50	39

Demographic and Trust Measures. No differences were found between trust measures for gender, military status (active duty and veterans) education level, ethnicity, marital status, or computer experience (intermediate, expert, advanced), p's > .05.

Hours of Computer Use per Week. Significant correlations were found between self-reported hours using a computer at work and responses on the VWT. Longer time spent using a computer at work was positively associated with feeling comfortable speaking, r (45) = .38, p = .009, and texting, r (45) = .36, p = .017, during class and trusting that personal information would be kept confidential, r (45) = .43, p = .004. Longer time spent using a computer at work was positively correlated with believing that classmates and instructors perceived the individual participant as they intended, r (44) = .53, p = .0001. No significant correlations were found between computer use at work and class attendance or hours of home computer use and class attendance on VWT questions (p's > .05).

Class Attendance. Table 4 shows the frequency of class attendance. The majority of participants who completed a post study assessment attended 5 or more classes (82.2%). No differences were found for class attendances among demographic groups, p's > .05.

Table 4. Class attendance

Classes attended	N	%
2	4	(8.9)
4	4	(8.9)
5	4	(8.9)
6	9	(20.0)
7	9	(20.0)
8	12	(26.7)
9	3	(6.7)

Class Attendance and Trust. Table 5 displays the bivariate correlations between class attendance and responses on the VWT and TIA. For the VWT Questionnaire, class attendance was significantly correlated with the participant's comfort in using their voice to communicate in the VW and with participants' trust that personal information would be kept confidential, p < .05. For the TIA Questionnaire, all six measures of trust were significantly correlated with attendance, p < .05. Familiarity with the system was not significantly correlated with attendance, p > .05. Being wary of the system was significantly associated with lower class attendance, p < .01.

4   Discussion

The purpose of this study was to examine active duty and veteran U.S. military volunteers self-reported relational trust and trust in automation after attending an eight-week mindfulness meditation training class delivered online via a VW.

     None of the demographic measures, except for hours of work-related computer use per day, were associated with trust. While America's seniors have been slower to adapt to technology, six of ten seniors now use technology [11]. The Pew report noted two groups of older Americans. The first group of younger, more highly educated, or more affluent seniors use technology and have a positive view of online systems. The second group of older and less affluent seniors often experience health challenges and are disconnected from digital technologies. Our results show that age did not influence responses to the trust questionnaires or class attendance, suggesting that this sample fell into the former category of seniors, and among the 60% who use technology. Indeed, our participants reported high levels of education and relatively high use of computers at home and work.

Table 5. Significant correlations between class attendance and responses to the VWT and TIA.

Measure	Item	r	p	n
VWT	I felt comfortable speaking during my MBSR class over Second Life using my voice	.53**	.00	45
	I trust that my personal information will be kept confidential over Second Life	.30*	.04	45
TIA	I am wary of the system	−.30*	.05	44
	I am confident in the system	.41**	.01	44
	The system provides security	.39**	.01	44
	The system has integrity	.37*	.01	44
	The system is dependable	.40**	.01	44
	The system is reliable	.41**	.01	42
	I can trust the system	.31*	.05	43

**p < .01, *p < .05

Our research showed an association between longer hours of work-related computer use and trust. These findings support those of Blank and Dutton [12] who found that between 2003 and 2009, age became less of a factor in trusting technology, while experience became more important. More investigation would be needed to identify if certain types of tasks at work lead to greater comfort in using VW technology.

In general, the participants in this study reported moderately high levels of trust and low levels of mistrust in the use of a VW technology system for attending a mindfulness meditation training class, and high positive levels of relational trust when attending a wellness intervention via a VW. This acceptance of using VW technology for attending a wellness-based training group was encouraging in terms of offering similar training to both military and civilian populations. Such training might include other forms of stress management, smoking cessation classes, assertiveness training, and perhaps support groups for those with serious illness, injury, or addiction.

Approximately 73% of the participants who returned for post-study assessments attended 6 or more classes during the study. While this does not account for the participants who left the study prior to the post-study assessment, it does indicate that the participants who completed the study were committed to the training. In a previous study looking at trust in telemedicine systems, researchers found the more interpersonal trust the participants had, the more trust they had in the system and vice versa [13]. Furthermore, the authors stated that increased experience fostered increased trust for both the interpersonal dynamic, as well as for the system, that is, greater technology (system) experience encouraged greater trust in both the instructor and other participants (and vice versa). It would also be expected that over the class duration (8 weeks), as more experience was gained, participant's trust would increase.

Greater trust was associated with higher attendance, while suspicion ("I am wary of the system" was associated with lower attendance. Correlations between class attendance and the questions about the use of the VW indicate that participants who felt comfortable speaking in the VW using their voice and who trusted their personal information would be kept confidential, attended more classes. This finding is consistent with reports that individuals are more likely to trust avatars when they are able to use voice communication [14]. Participants were encouraged to use their voice to communicate, but it was not required (texting was permitted). However, text was used by participants only when the technology failed and they could not be heard by other participants.

Interestingly, system familiarity was not significantly associated with class attendance. Twenty-two percent (22%) of participants had used a VW prior to participating in this study. However, none of the participants had ever used SL prior to the study. It was encouraging that familiarity was not required for participants to attend the classes, as this may indicate that those without familiarity can participate in, and gain from, participating in VW telehealth interventions.

As noted above, agreement with the statement "I am wary of the system" (indicating lower system trust) was associated with more absences from class. Since experience using the technology, and voice communication both increase trust in a technology system (VW), wariness may be mitigated by increasing practice time and using voice during training to use a VW prior to the start of an online group.

Another potential method for addressing lack of trust might include involving the users in the design considerations for the VW environment and for creating meaningful interactions within that environment. Most user involvement tends to stop at the technology interaction level [15]. Indeed, the value of co-creation could be considerable, especially in light of the type of services to be rendered and the experience levels of both healthcare professionals and the constituents they serve.

Trust in telehealth has also been shown to be connected to the perceived motives of the service providers [15]. That is, at a human systems integration level, all communications and interactions with patients should convey concern for the patient and his or her safety and privacy. Patients are likely to be concerned about the safety and reliability of the telehealth service and software, and their interest may be weighted around the perceived risks and benefits from the experience [15]. Finally, having a foundation of relational trust between the provider and the patient will likely impact patients' trust in the telehealth technologies used by their provider. Just as in in-person interactions, trust is progressive and molded by repeated interactions over time. Similarly, trust in the reliability, validity, usability, and integrity of the system build over time.

Limitations. The primary limitations of this study were the homogeneity of the sample population, which included U.S. military active duty and veteran volunteers, and the small sample size. Nevertheless, the findings provide valuable insight into the use of VW technology for providing health and wellness programs to U.S. military and veterans.

Conclusions. Overall, participants reported moderately high trust in automation and high relational trust. Relational trust that was high included feeling comfortable communicating, trusting in the confidentiality of the system and co-attendees, and believing their avatars (as representations of themselves) were perceived as they intended. Greater work-related computer use was related to feeling comfortable communicating with others during class and believing that classmates and instructors perceived participants as they intended. Finally, our research supports Lee and Moray's [3] proposal that trust is key to the use of automation, as an increase in trust was related to an increase in VW class attendance. Our research also supports the idea that trust is vital in VW group work, as it reduces uncertainty and equivocality and thereby helps establish shared understanding and virtual relationships [16–18].

Acknowledgments. Special thanks to the service members and veterans who participated in this study, as well as to Baoxia Liu, Jim Hewson, Angela Jeter, Cory Overby, and Jessica Villarreal. This research was supported by the Army Study Program Management Office. The views expressed in this article are those of the authors and do not reflect the official policy or position of the Department of the Army, Department of Defense, or the U.S. Government.

References

1. Rice, V.J., Alfred, P., Villarreal, J.L., Jeter, A., Boykin, G.: Human factors issues associated with teaching over a virtual world. Proc. Hum. Factors. Ergon. Soc. Annu. Meet. **56**, 1758–1762 (2012)
2. Hoffman, R.R., Johnson, M., Bradshaw, J.M.: Trust in Automation. IEEE Intell. Syst. **28**, 84–88 (2013)
3. Lee, J., Moray, N.: Trust, control strategies and allocation of function in human-machine functions. Ergonomics **35**, 1243–1270 (1992)
4. Brewster, L., Mountain, G., Wessels, B., Kelly, C., Hawley, M.: Factors affecting frontline staff acceptance of telehealth technologies: a mixed-method systematic review. J. Adv. Nurs. **70**, 21–33 (2014)
5. Bell, M.W.: Toward a definition of "virtual worlds". J. Virtual Worlds Res. **1**, 2–5 (2008)
6. Paul, D.L., McDaniel Jr., R.R.: A field study of the effect of interpersonal trust on virtual collaborative relationship performance. MIS Q. **28**, 183 227 (2004)
7. Chandra, S., Theng, Y.L., O'Lwin, M., Foo, S.: Exploring trust to reduce communication barriers in virtual world collaborations. In: ICA (2011). https://www.ntu.edu.sg/home/sfoo/publications/2011/2011-ICA_fmt.pdf
8. Gartner Research: Gartner says 80 percent of active internet users will have a "Second Life" in the virtual world by the end of 2011 (2007). https://www.gartner.com/newsroom/id/503861
9. Gartner Research: Gartner says 90 percent of corporate virtual world projects fail within 18 months (2008). https://www.gartner.com/newsroom/id/670507
10. Jian, J.Y., Bisantz, A.M., Drury, C.G.: Foundations for an empirically determined scale of trust in automated systems. Int. J. Cogn. Ergon. **4**, 53–71 (2000)
11. Pew Research Center: Older Adults and Technology Use (2014). http://www.pewinternet.org/2014/04/03/older-adults-and-technology-use/l
12. Blank, G., Dutton, W.: Age and trust in the internet: the centrality of experience and attitudes toward technology in Britain. Soc. Sci. Comput. Rev. **30**, 135–151 (2012)
13. Gogan, J., Garfield, M., Baxter, R.: Seeing a patient's eyes: system trust in telemedicine. In: BLED, p. 33 (2009)

14. Qui, L., Benbasat, I.: Online consumer trust and live help interfaces: the effects of text-to-speech voice and three-dimensional avatars. Int. J. Hum. Comput. Interact. **19**, 75–94 (2005)
15. Bhattacharya, S., Wainwright, D., Whalley, J.: Internet of Things (IoT) enabled assistive care services: designing for value and trust. Procedia Comput. Sci. **113**, 659–664 (2017)
16. McKnight, D.H., Cummings, L.L., Chervany, N.L.: Initial trust formation in new organizational relationships. Acad. Manag. Rev. **23**, 473–490 (1998)
17. Newell, S., Swan, J.: Trust and inter-organizational networking. Hum. Relat. **53**, 1287–1328 (2000)
18. McKnight, D.H., Choudhury, V., Kacmar, C.: Developing and validating trust measures for e-commerce: an integrative typology. Inf. Syst. Res. **13**, 334–359 (2002)

Perceived Workload and Performance in the Presence of a Malodor

William Y. Pike[1], Michael D. Proctor[2], Christina-Maile C. Pico[3], and Mark V. Mazzeo[1(✉)]

[1] United States Army Research Laboratory Human Research and Engineering Directorate Simulation and Training Technology Center, Orlando, FL, USA
{william.y.pike.civ, mark.v.mazzeo.civ}@mail.mil
[2] University of Central Florida Industrial Engineering and Management Systems, Orlando, FL, USA
michael.proctor@ucf.edu
[3] United States Army Corps of Engineers, Fort Bragg, NC, USA
christina.m.pico3.mil@mail.mil

Abstract. As part of an olfactory adaptation experiment, researchers collected NASA Task Load Index (TLX) data from United States Military Academy cadets who completed two complex tasks in either the presence or absence of a simulated malodor. Results showed that participants exposed to the odor twice tended to show a decrease in perceived mental demand during the second task. Furthermore, these participants also showed a higher correlation coefficient between decrease in mental demand and improvement in task performance. Taken together, these results indicate a possible link between olfactory adaptation and perceived mental demand, at least in the presence of a malodor.

Keywords: Perceived workload · Perceived mental demand
NASA Task Load Index · Malodor · Olfactory adaptation

1 Introduction

Extensive previous research has linked malodors to a variety of human performance issues, including stress, anger, and confusion [1], elevated flight or escape behavior, [2, 3] and reduced complex task performance [4].

Olfactory adaptation, or desensitization to a specific odor [5–7] has long been measured by two metrics: perceived intensity and detection threshold [2]. Perceived intensity, as the name implies, involves a subject's perception, on a numerical scale, of the intensity of the subject odor. Detection threshold is the concentration at which a subject first detects an odor. For olfactory adaptation to occur, perceived intensity should decrease and detection threshold should increase in a response period following an initial exposure to an odor in a treatment period. A follow-on to a pilot study [8] explored whether adaptation might be useful to ameliorate certain human performance issues. Results of that study are pending publication; however, this paper will review results of additional data collected, namely perceived workload as self-reported by the NASA Task-Load Index (TLX) [9].

© Springer International Publishing AG, part of Springer Nature (outside the USA) 2019
D. N. Cassenti (Ed.): AHFE 2018, AISC 780, pp. 161–169, 2019.
https://doi.org/10.1007/978-3-319-94223-0_15

Task performance can affect perceived workload [10]. Workload assessment techniques are required to be reliable, selective, sensitive, low disturbance, have sufficient diagnostic capabilities, and be easy to implement [11]. Hart and Staveland [12] developed the NASA TLX, which meets that criteria and measures perceived performance and workload [13]. There are both pen-and-paper and software versions of the TLX. The different media do not significantly affect results [10]. Due to its high usability, subjects also favor this assessment technique. Finally, the TLX has demonstrated a high correlation with performance [13].

NASA TLX assesses subjective mental workload through six subscales. Subjects complete the TLX immediately following performance of a complex task. Along with rating, the scales may be weighted from 0 to 5. This study used an unweighted or raw TLX, commonly used because researchers [14] demonstrated a high correlation between weighted and unweighted scores. The overall workload score is comprised of the ratings of the six subscales that are Mental Demand (MD), Physical Demand (PD), Temporal Demand (TD), Performance (Pe), Effort (Ef), and Frustration (Fr). The mental, physical, and temporal demand subscales measure demands imposed by the complex task, while the performance, effort, and frustration subscales measure the subject-task interaction. For this research, a 0-20 score per factor was used. An example of the pen-and-paper version of the NASA TLX is provided in the Appendix.

2   Methodology

2.1   Hypotheses

Relative to the TLX data collected in the experiment, researchers first hypothesized that participants initially exposed to a malodor while performing a relevant complex task would report a higher mental demand (MD) than participants not initially exposed. Secondly, participants initially exposed to the malodor during a treatment phase would show a reduced MD during a second exposure (the response phase). Thirdly, and conversely, participants not initially exposed to the malodor during a treatment phase would show an increased MD during a second phase (the response phase). Fourthly, participants not initially exposed to the malodor during a treatment phase would show a greater increase in MD than participants never exposed to the malodor. The fifth hypothesis is there will be an inverse correlation between the difference in MD and task performance across the two phases. Finally, the sixth hypothesis states the expected correlation will be more pronounced among participants initially exposed to the malodor.

2.2   Participants

A priori power analyses established a requirement of 156 participants among three groups (52 per group), with a desired significance level of .05 and a 0.2 type II error rate, with a medium effect size [15]. Freshmen, or "plebe" students at the United States Military Academy (USMA) at West Point, New York served as participants for the study. A total of 180 participants from this population participated, with slightly unequal numbers per group because of scheduling concerns. Prospective participants took part in the study only if they fulfilled specific criteria:

- No prior military service
- No prior olfaction of burnt human flesh
- No smoking within the previous four hours
- No current medications for cold or flu
- No current illnesses or conditions that could reduce olfaction capabilities (e.g. allergies, cold, flu)
- No one under 18 years of age.

2.3   Materials

Table 1 provides a summary of the items utilized for this study.

Table 1. Test materials. Item description, purpose, and provider.

Item description	Purpose	Provider
Checklist	Completed by observer to monitor experiment progress for each participant	PI
Sniffin' Sticks	Pre-screening participants for anosmia	US Neurologicals, LLC
Digital timer	Used by observer to monitor experiment progress and to report or control events	Unknown – simple commercial stopwatches
"Burnt human" scent	Simulated odor to serve as olfactory stimulus for perception and detection, manufactured by ScentAir, Inc.	ScentAir, Inc.
"ScentPop" odor delivery machine with custom housing	Used to disperse concentrated simulated odors within an enclosed area	ScentAir, Inc.
Quick set up dome tent	Enclosed area with room for participant, observer, and test materials	Pinnacle Tents
Table, chairs, lamp, pencils	Used to complete written portions of study	PI
AQS questionnaires	Used to assess perceived intensity of the simulated odor	Rotton, 1983 [4]
TLX questionnaires	Used to assess perceived workload	Hart & Staveland
Laminated guide sheet	Used to assist participant in completing the complex task	PI
SPSS version 24	Used for statistical analysis	IBM

2.4   Procedure

The experiment began with an overview of the study and a review of the informed consent form. Participant screening followed, based on the exclusion criteria previously described, with a demographics form afterwards. Test administrators then guided each participant through two successive phases (Treatment and Response) in two separate, identical tents. In each tent, the administrator and participant sat at a table with chairs. The administrator explained the activity and the time constraint for completion (12 minutes), and the participant would complete the activity within that time. During both phases, the administrator logged the time of tent entry, the time the simulated odor machine turned on (two minutes after entry), and the time the participant reported the odor's presence. If a participant did not report the presence of an odor, the administrator logged this time as 10 min (the remainder of the time in the tent for that activity). After each activity, the participant completed an Atmospheric Quality Scale (AQS) to describe the air inside the previous tent as well as a NASA TLX to describe their perceived workload during the just-completed phase.

2.5   Measurement

The objective for the overall research was to gauge the ability of olfactory adaptation to lessen certain human factor issues shown to degrade in the presence of malodors.

The AQS provided a means to quantify the perceived intensity of the simulated odor. The AQS is a standardized questionnaire designed to gather qualitative information from participants about air quality. It consists of eight parameters with Likert scale responses, which the participant filled in from a scale of 1 to 7. A "1" score indicated less intense or more pleasant ratings, while a "7" score indicated more intense or less pleasant ratings. For the purposes of this study, perceived intensity was defined as the sum of four key parameters from the AQS: bland vs. pungent; weak vs. strong; moderate vs. intense; and impotent vs. potent. Accordingly, the change in perceived intensity defined the difference between these calculated values as the participant moved from one phase to the next.

Second, administrator-reported event logs using timers provided a means to approximate detection threshold. The concentration at which an odor is first noticed traditionally defined detection threshold; Dalton [7] considers detection threshold as the primary metric in gauging adaptation. Since the dispersal rate of the odor machines is proprietary information, detection time was used instead. Note that concentration is equal to dispersal rate multiplied by time, divided by volume. With consistent dispersal rates (ensured by using odor canisters well within manufacturer specifications) and volume (ensured by using tents of the same dimensions), time was the only varying factor and thus a reasonable substitute for detection threshold.

2.6   Naming Convention

- Treatment Phase – the first phase of the experiment
- Response Phase – the second phase of the experiment
- Adaptation Condition – the group of participants ($n = 60$) who were subjected to the malodor during both phases

- No Adaptation Condition – the group of participants ($n = 59$) who were not subjected to the malodor during the Treatment phase, but were subjected to the malodor during the second phase
- Control Condition – the group of participants ($n = 61$) who were not subjected to the malodor during either phase.

3 Results

3.1 Tests for Normality

Kolmogorov-Smirnov analysis showed the vast majority of the data collected in this study displayed non-normal distributions. As a result, all analyses relied on non-parametric statistics.

3.2 Individual Statistical Analysis

Table 2 presents Wilcoxon Matched Pairs Signed Rank (WMPSR) p for related samples analysis of the mental demand (MD) TLX subscale. In addition, the table displays Mann-Whitney U results, rank-ordering significant differences between conditions.

Table 2. Median TLX values, Adaptation condition, Treatment and Response phases, with Wilcoxon Matching Pairs Signed Rank (WMPSR) test significance. Significant results (at $\alpha = .05$) highlighted in bold. Bonferroni corrections applied based on two sets of hypotheses (mental demand differences, correlation)

Condition	Phase			
	Mental demand: treatment phase	Mental demand: response phase	Mental demand: difference between phases (Within Subject)	WMPSR significance within condition $p = .05$
Adaptation (A)	12	11	−1	**.022**
No Adaptation (NA)	12	13	+1	.288
Control (C)	12	13	+1	.968
M-W U ($p = .025$*)			**A < NA, .017**	

Table 2 illustrates that only the Adaptation condition saw a significant change (decrease) from Treatment (first) phase to Response (second) phase (WMPSR, $p = .022$), and a significant difference exists between the Adaptation and No Adaptation conditions in change in Mental Demand (MD) (M-W U, $p = .017$), with the Adaptation condition decreasing and the No Adaptation condition increasing. However, M-W U did not show statistical differences between any pairs of conditions in the response phase. While the first finding is expected, the lack of significance between the pairs during the response phase is puzzling.

3.3   Correlation Analysis

Pursuant to the fifth and sixth hypotheses, correlation analysis (Spearman's Rho) compared the difference in task performance (percentage correct on the four-scenario medical evacuation quiz) to the change in perceived mental demand MD. Table 3 details Wilcoxon Matched Pairs Signed Rank (WMPSR) p for related samples analysis of the task performance. In addition, the table displays Mann-Whitney U results, rank-ordering significant differences between conditions.

Table 3. Significant WMPSR differences from treatment to response phase, listed by condition

Condition	Phase			
	Quiz score percent correct: treatment phase	Quiz score percent correct: response phase	Improvement (percentage points)	WMPSR significance within condition $p = .05$
Adaptation (A)	63.889	77.083	13.194	**.000**
No Adaptation (NA)	66.667	76.389	9.722	**.014**
Control (C)	65.278	72.222	6.944	**.000**
Between condition KW significance H	.247	.532	.064	
Order of difference in quiz score using between condition, Mann Whitney U $p = .025$	No significant between-condition differences			

While there were no significant between-condition differences, all three conditions improved quiz scores significantly from treatment phase to response phase. The Adaptation condition showed the highest increase.

Table 4 presents the condition, correlation significance, and correlation coefficient, comparing the change in MD to the change in quiz score, from treatment phase to response phase. Note that while all conditions significantly improved quiz scores, there was no significance between any condition pairs. The Adaptation condition's between-phase increase was the highest.

As shown by Table 4, there was a negative correlation overall, and in all conditions, between change in TLX perceived mental demand and change in test score. As test scores improved, perceived mental demand decreased. The correlation was significant overall (all three conditions examined together, $p = .035$, rho $= -.158$) though the majority of this negative correlation appears attributable to the Adaptation

Table 4. Correlation analysis by condition, comparing change in TLX MD to change in performance (quiz score percentage)

Condition	Significance	Correlation coefficient
Overall	**.035**	−.158
Adaptation	**.022**	−.298
No Adaptation	.289	−.14
Control	.898	−.017

condition, $p = .022$, rho $= -.298$. Interestingly, the other two conditions did not see a statistically significant correlation - both were negative. One might expect the No Adaptation condition to see a positive correlation if malodor had a meaningful impact on mental demand.

4 Discussion

The first hypothesis stated that participants initially subjected to a malodor would report a higher TLX Mental Demand than those not initially subjected to a malodor. This did not hold true, as all three conditions reported a median MD of 12. The second hypothesis stated that the Adaptation condition would see a decrease in MD from Treatment phase to Response phase. Indeed, the Adaptation condition saw a statistically significant decrease (12 to 11, $p = .022$). The third hypothesis did not hold true, as the No Adaptation condition, while increasing from a median MD of 12 to 13, failed to show statistical significance ($p = .288$). The fourth hypothesis which held that the No Adaptation condition would see a greater increase in MD than the Control condition, also did not hold true as both conditions increased from 12 to 13.

Related to correlation, the fifth hypothesis stated that there would be an inverse correlation between change in TLX MD and change in test scores, and, sixth, this correlation would prove stronger for the Adaptation condition. Overall, taking all 180 participants into account, there was a significant negative correlation ($p = .035$, Spearman's rho $= -.158$). Furthermore, the Adaptation condition was the only individual condition to show significance ($p = .022$), and it was a stronger correlation (Spearman's rho $= -.298$). Thus, this hypothesis also proved true in this test.

5 Conclusion

While not all hypothesized outcomes occurred, those that did hint at a potential links between exposure to malodors, mental demand, and the correlation between mental demand and task performance. While Mental Demand did not necessarily follow expectations and show higher scores in the presence of a malodor, it did see a reduction from participants afforded an adaptation experience.

The correlation analysis showed that, overall, as task performance (quiz scores) improved from Treatment to Response phase, the Mental Demand decreased, as expected. Furthermore, as expected, this correlation was strongest in the Adaptation condition. This finding hints to a potential link between mental workload and olfactory adaptation during the performance of a complex task relevant to the target population.

Additional self-report perceived workload tools, especially those compared in empirical studies to the TLX (e.g., [16]) should be considered, as well as alternate ways to rate the TLX (e.g., weighting factors [12]).

While this study used a single simulated malodor, odors do not exist singularly in the modern battlefield. Future studies should consider multiple odors. The odor was also presented with no visual context – a patient simulator (human or manikin) with burnt flesh moulage could provide more realism by providing more context. Furthermore, exposure time should be lengthened, both before beginning the complex task and overall.

Further analysis will explore potential correlation between the TLX subscales and the two olfactory adaptation metrics (perceived intensity and detection threshold).

Appendix: NASA TLX Form

Mental Demand   How mentally demanding was the task?

Very Low   Very High

Physical Demand   How physically demanding was the task?

Very Low   Very High

Temporal Demand   How hurried or rushed was the pace of the task?

Very Low   Very High

Performance   How successful were you in accomplishing what
                   you were asked to do?

Perfect   Failure

Effort   How hard did you have to work to accomplish
              your level of performance?

Very Low   Very High

Frustration   How insecure, discouraged, irritated, stressed,
                   and annoyed wereyou?

Very Low   Very High

References

1. Schiffman, S.S., Williams, C.M.: Science of odor as a potential health issue. J. Environ. Qual. **34**, 581–588 (2005)
2. Asmus, C.L., Bell, P.A.: Effects of environmental odor and coping style on negative affect, anger, arousal, and escape. J. Appl. Soc. Psych. **29**, 245–250 (2005)
3. Smeets, M.A.M., Dalton, P.H.: Evaluating the human response to chemicals: odor, irritation and non-sensory factors. Environ. Toxicol. Pharm. **19**, 129–138 (2005)
4. Rotton, J.: Affective and cognitive consequences of malodorous pollution. Basic Appl. Soc. Psych. **4**, 171–191 (1983)
5. Stuck, B.A., Fadel, V., Hummel, T., Sommer, J.U.: Subjective olfactory desensitization and recovery in humans. Chem. Senses **39**, 151–157 (2013)
6. Mainland, J.D., Lundström, J.N., Reisert, J., Lowe, G.: From molecule to mind: an integrative perspective on odor intensity. Trends Neurosci. **37**, 443–454 (2014)
7. Dalton, P.: Psychophysical and behavioral characteristics of olfactory adaptation. Chem. Senses **25**, 487–492 (2000)
8. Pike, W.Y., Proctor, M.D., Burgess, D.N.: Reliability and feasibility considerations in the assessment of a malodor adaptation technique: a pilot study. Mil. Med. **182**, e1521–e1527 (2017)
9. Hart, S.: NASA-task load index (NASA-TLX): 20 years later. In: Proceedings of the Human Factors and Ergonomics Society Annual Meeting, pp. 904–908. Sage Publishing, Los Angeles, CA (2006)
10. Cao, A., Chintamani, K.K., Pandya, A.K., Ellis, R.D.: NASA TLX: software for assessing subjective mental workload. Behav. Res. Methods **41**, 113–117 (2009)
11. Eggemeier, F.T.: Properties of workload assessment techniques. Adv. Psychol. **52**, 41–62 (1988)
12. Hart, S.G., Staveland, L.E.: Development of NASA-TLX (Task Load Index): results of empirical and theoretical research. Adv. Psychol. **52**, 139–183 (1988)
13. Rubio, S., Diaz, E., Martin, J., Puente, J.M.: Evaluation of subjective mental workload: a comparison of SWAT, NASA-TLX, and workload profile methods. Appl. Psychol. **53**, 61–86 (2004)
14. Byers, J.C., Bittner, A.C., Hill, S.G.: Traditional and raw task load index (TLX) correlations: are paired comparisons necessary? Advances in Industrial Ergonomics and Safety, pp. 481–485. Taylor & Francis, Abingdon, UK (1989)
15. Cohen, J.: A Power Primer. Psychol. Bull. **112**, 155–159 (1992)
16. Wiebe, E., Behrand, S.: An examination of two mental workload measurement approaches to understanding multimedia learning. Comput. Hum. Behav. **26**, 474–481 (2010)

Automatic Generation of Statistical Shape Models in Motion

Femke Danckaers[1]([✉]), Sofia Scataglini[2,3], Robby Haelterman[2], Damien Van Tiggelen[3], Toon Huysmans[1,4], and Jan Sijbers[1]

[1] imec – Vision Lab, Department of Physics, University of Antwerp, Universiteitsplein 1, 2610 Antwerp, Belgium
femke.danckaers@uantwerpen.be
[2] Department of Mathematics (MWMW), Royal Military Academy, Renaissancelaan 30, 1000 Brussels, Belgium
[3] Military Hospital Queen Astrid, Bruynstraat 1, 1120 Brussels, Belgium
[4] Applied Ergonomics and Design, Department of Industrial Design, TU Delft, Landbergstraat 15, 2628 CE Delft, The Netherlands

Abstract. Statistical body shape modeling (SBSM) is a well-known technique to map out the variability of body shapes and is commonly used in 3D anthropometric analyses. In this paper, a new approach to integrate movement acquired by a motion capture system with a body shape is proposed. This was done by selecting landmarks on a body shape model, and predicting a body shape based on features. Then, a virtual skeleton was generated relative to those landmarks. This skeleton was parented to a body shape, allowing to modify its pose and to add pre-recorded motion to different body shapes in a realistic way.

Keywords: Statistical body shape model · Motion capturing · Shape prediction

1 Introduction

Statistical body shape modeling (SBSM) is a well-known technique to map out the variability of body shapes and is commonly used in 3D anthropometric analyses. Statistical body shape models (SBSMs) can describe the variability of body shapes for a population of individuals. By adapting the parameters of the SBSM, a new realistic shape can be formed. Product developers may exploit SBSMs to design virtual design mannequins and explore the body shapes belonging to a specific percentile of a target group, allowing to visualize extreme shapes. Moreover, an SBSM allows to simulate a specific 3D body shape [1], which is useful for customization.

Nowadays, inertial motion tracking sensors (IMU) allow capturing human motion and acquiring the kinematic of the subject during a physical task. This information is translated as a skeletal animation as a Biovision Hierarchy (BVH) character animation file. In this study, we acquired the subject's motions with a real-time inertial motion tracking system (Yost Labs 3-Space Sensor).

This is especially relevant for people who have to perform physically demanding tasks in non-ideal circumstances. Their gear must have an optimal fit, to reduce the

© Springer International Publishing AG, part of Springer Nature 2019
D. N. Cassenti (Ed.): AHFE 2018, AISC 780, pp. 170–178, 2019.
https://doi.org/10.1007/978-3-319-94223-0_16

impact on their body. For example, reachability tests in vehicles or testing the freedom of movement when wearing their equipment or heavy backpacks [2, 3].

Unfortunately, to date, there is no framework available to generate body shapes in motion where both body shape as articulation is adaptable. Another possibility is to add motion to a specific body scan. We propose a new approach to integrate the movement acquired by an inertial motion capture system with the statistical body shape. This allows product developers to validate their designs for multiple poses and movements.

2 Methods

In this section, a framework to create moving SBSMs is described. First, a SBSM is built from a population of 3D human body shapes [4]. Next, the method to generate a body shape based on features is explained [1]. Finally, modification of a motion file and adding motion to a specific body shape is discussed.

2.1 Building a Statistical Shape Model

First, a reference surface, a digitally modeled body shape [5] with n uniformly distributed vertices, is registered in a marker-less way to N input surfaces to obtain a homologous point-to-point correspondence. All input surfaces were corrected for posture, in a way that every shape was standing in the average posture, determined from a population of 700 scans from the CAESAR database [4]. Then, a statistical shape model is built using principal component analysis of the population of N posture normalized corresponded surfaces. In an SBSM, the mean shape $\bar{x} \in \mathbb{R}^{3n}$ and the main shape modes, or the principal component (PC) modes of the SBSM $P \in \mathbb{R}^{3n \times (N-1)}$, are incorporated. This means that a new shape $y \in \mathbb{R}^{3n}$ can be formed by a linear combination of the PCs:

$$y = \bar{x} + Pb, \tag{1}$$

with b the vector containing the SBSM parameters.

A specific feature of a person's shape, such as height, can be adapted by adding a linear combination of principal components to the person's shape vector. The weights for this linear combination are computed via multiple linear regression of the PC weights on the body features $f = [f_1 f_2 f_3 \cdots f_f \, 1]^T \in \mathbb{R}^{f+1}$ (such as height, weight, gender,...) for the population of individuals. Every feature is defined by a scalar value. A mapping matrix $M \in \mathbb{R}^{N-1 \times (f+1)}$ describing the relationship between the biometric features $F = [f_1 f_2 f_3 \cdots f_N] \in \mathbb{R}^{(f+1) \times N}$ of every input shape and the principal component weights of every input shape $B \in \mathbb{R}^{(N-1) \times N}$ is calculated using multivariate regression, by

$$M = BF^+, \tag{2}$$

with F^+ the pseudoinverse of F.

By multiplying M with a given feature vector f, new principal component weights $b \in \mathbb{R}^{N-1}$ can be generated:

$$b = Mf. \tag{3}$$

From these principal component weights, a new body shape y can be built.

2.2 Skeleton Generation

The BHV file format is a way to provide skeleton hierarchy information in addition to the motion data. The skeleton is typically in T-pose. In such a BVH file, the skeleton is represented as a tree structure set of 18 joints, relative to each other. This is shown in Fig. 1. In most cases, the pelvis is the root of the skeleton. Every other joint is defined by an offset from the previous joint.

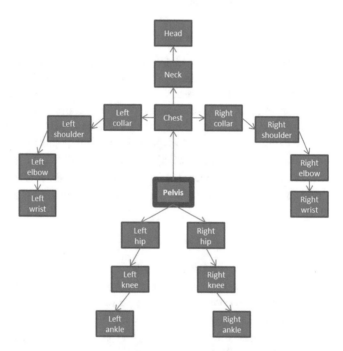

Fig. 1. Schematic visualization of the skeleton.

Forty-one landmarks available in the CAESAR database, such as olecranon, humeral epicondyle lateral, substernal are selected on the average body shape. Because of the correspondences, every vertex will remain at anatomically the same location, independent of shape.

A new skeleton of a Biovision Hierarchy (BVH) character animation file [6] was generated by calculating the optimal joint locations relative to these landmarks. Next, the skeleton is parented to the body shape by calculating skinning weights [7]. As a result, the pose of that body shape can be adapted. This means it can be adapted manually or a pre-defined movement can be executed.

2.3 Modification of Movement File

Motion is defined per frame, by a rotation offset per joint from the original skeleton. The body shapes available in the CAESAR database are standing in A-pose, as shown in Fig. 2. Therefore, the BVH files have to be adapted from T-pose to A-pose, as shown in Fig. 2. It is not sufficient to only convert the rest pose, as the motion is defined as rotation of the joints in rest pose. To solve this problem, we wrote Python code that can be run in Blender [8]. This code allows one to change the rest pose to the current adapted pose in Blender and to copy the original joint position to the new skeleton per frame.

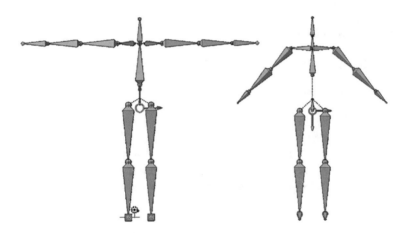

Fig. 2. The skeleton in rest pose. Left: T-pose, right: A-pose.

A skeleton $S \in \mathbb{R}^{3 \times j}$ is defined as a set of j joints $J \in \mathbb{R}^3$, whereas every joint per frame contains a rotation matrix $R \in \mathbb{R}^{3 \times 3}$ from the rest pose to the pose of the current frame.

$$S = \begin{bmatrix} J_1 \, J_2 \, J_3 \ldots J_j \end{bmatrix} \tag{4}$$

$$R = \begin{bmatrix} m_{00} & m_{01} & m_{02} \\ m_{10} & m_{11} & m_{12} \\ m_{20} & m_{21} & m_{22} \end{bmatrix} \tag{5}$$

The workflow is as follows: first, the original skeleton $S_O \in \mathbb{R}^{3 \times j}$ is manually put in A-pose by rotating the joints, by a transformation T, resulting in a transformed skeleton $S_T \in \mathbb{R}^{3 \times j}$. The new pose is applied as rest pose of the skeleton:

$$S_T = T \times S_o \qquad (6)$$

The following step is to update the actions per frame, as the movement is defined by rotation of the joints in rest pose. As a first step, the skeleton S_T is translated in a way so the root joints, in this case the pelvis, are on the same location. The copied joints are rotated to match the orientation of the target joints. This is done by inverting the rest pose matrix and multiplying this with the rotation matrix of the current inverse of the rotation matrix of the parent joint and the rotation matrix of the parent joint in resting position. The resulting rotation matrix has to be applied to the specific frame of the skeleton in A-pose as resting pose:

$$R'_T = R_T^{-1} \times \mathrm{parent}\left(R_{T,rest}\right) \times \mathrm{parent}(R_T)^{-1} \times R_O \qquad (7)$$

3 Experiments and Results

3.1 Building a Statistical Shape Model

A statistical shape model was built from the CAESAR database [9]. We selected 57 soldier-like (male, height 1m52–2m10, age 18y–35y, BMI 18.5–25) body shapes to build our model. The shapes were registered using the same template surface mesh, a digitally modeled body consisting of 100k uniformly distributed vertices. The average soldier, shown in Fig. 3, has a height of $1.84 \pm 0.07\,m$, a BMI of 22.4 ± 1.7 and is 27.9 ± 4.5 years old. From these meshes, posture variances were removed and a statistical shape model was built. In Fig. 4, the first three modes of variance of the SBSM are shown.

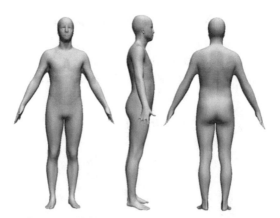

Fig. 3. Average soldier (male, height: 1.84 m, weight: 76.4 kg, age: 27.9 years, waist circumference: 846 mm, chest circumference: 951 mm, hip circumference: 990 mm, arm length: 654 mm, crotch height: 873 mm, knee height: 570 mm, shoulder breadth: 466 mm, sitting height: 953 mm, thigh circumference: 568 mm, BMI: 22.4).

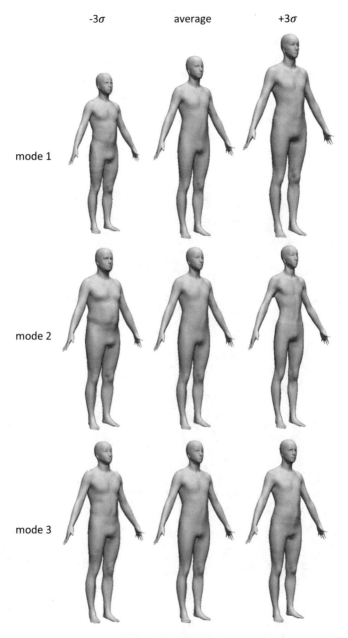

Fig. 4. The first three eigenmodes of the soldier SBSM plus and minus three standard deviations (σ) and the average body shape. The first mode mainly describes stature, the second mode mainly describes BMI, and the third mode mainly describes muscularity.

3.2   Shape Prediction and Skeleton Generation

A user interface was designed in which the following values can be specified: height, weight, age, waist circumference, chest circumference, hip circumference, arm length, crotch height, knee height, shoulder breadth, sitting height, and thigh circumference. We acquired the movement of a walking soldier using an inertial motion tracking system. After the shape had been generated, a new skeleton with associated movement was calculated. A screenshot of our implemented tool is shown in Fig. 5. This means that this feature will not be taken into account for shape prediction and the most plausible shape using the remaining values will be calculated.

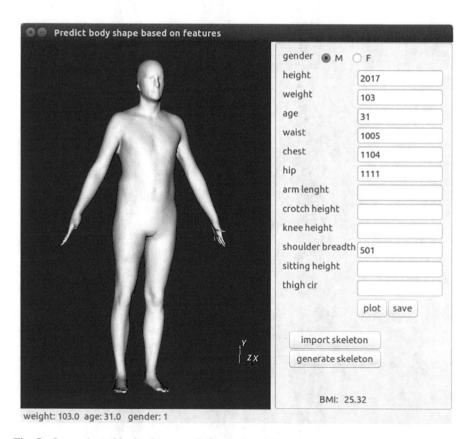

Fig. 5. Screenshot of body shape prediction and skeleton generation tool. The most plausible body shape of a male soldier with height 2.017 m, weight 103 kg, age 31, waist 100.5 cm, chest 110.4 cm, and shoulder breadth 50.1 cm is generated. The remaining values were unknown, so these values were not taken into account.

3.3 Adding Movement

The generated mesh and associated skeleton were imported in Blender, where the skeleton was parented to the mesh using automatic weights [7]. This approach resulted in a realistic body shape in motion, as can be seen in Fig. 6.

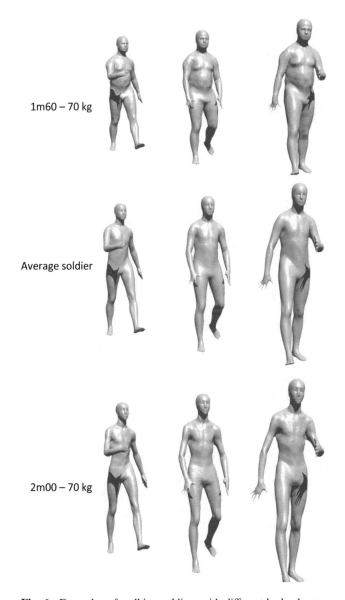

1m60 – 70 kg

Average soldier

2m00 – 70 kg

Fig. 6. Examples of walking soldiers with different body shapes.

4 Conclusion

We proposed an automatic technique to rig a statistical body shape model, allowing to simulate movement on a whole range of body shapes. Results show that our framework leads to detailed, realistic body shapes, moving in a natural way. This is especially useful for accessibility testing, e.g. when designing a vehicle, where the driver has to be able to perform specific movements to operate it in a correct way, while space is limited, or when optimizing comfort in wearing gear. Furthermore, static pose is adaptable by manipulating the armature, which is useful for designing near body products that require the body to be in a pose that is difficult to scan.

Acknowledgements. This work was supported by the Agency for Innovation by Science and Technology in Flanders (IWT-SB 141520).

References

1. Danckaers, F., Huysmans, T., Lacko, D., Sijbers, J.: Evaluation of 3D body shape predictions based on features. In: Proceedings of the 6th International Conference on 3D Body Scanning Technologies, Lugano, Switzerland, 27–28 October 2015, pp. 258–265. Hometrica Consulting - Dr. Nicola D'Apuzzo, Ascona, Switzerland (2015)
2. Knapik, J.J., Reynolds, K.L., Harman, E.: Soldier load carriage: historical, physiological, biomechanical, and medical aspects. Mil. Med. **169**, 45–56 (2004)
3. Haney, J.M., Wang, T., D'Souza, C., Jones, M.L.H., Reed, M.P.: Spatial and temporal patterns in sequential precision reach movements. Proc. Hum. Factors Ergon. Soc. Annu. Meet. **61**, 929–930 (2017)
4. Danckaers, F., Huysmans, T., Hallemans, A., De Bruyne, G., Truijen, S., Sijbers, J.: Full body statistical shape modeling with posture normalization. In: Cassenti, D.N. (ed.) AHFE 2017. AISC, vol. 591, pp. 437–448. Springer, Cham (2018). https://doi.org/10.1007/978-3-319-60591-3_39
5. 3D Scan Store: Male and Female Base Mesh Bundle. http://www.3dscanstore.com/index.php?route=product/product&product_id=679
6. Dai, H., Cai, B., Song, J., Zhang, D.: Skeletal animation based on BVH motion data. In: 2010 2nd International Conference on Information Engineering and Computer Science, pp. 1–4. IEEE (2010)
7. Baran, I., Popović, J.: Automatic rigging and animation of 3D characters. ACM Trans. Graph. **26**, 72 (2007)
8. Blender Online Community: Blender - a 3D modelling and rendering package (2015)
9. Robinette, K.M., Daanen, H.A.M., Paquet, E.: The CAESAR project: a 3-D surface anthropometry survey. In: Second International Conference on 3-D Digital Imaging and Modeling (Cat. No. PR00062), pp. 380–386. IEEE Computer Society (1999)

Multi-patch B-Spline Statistical Shape Models for CAD-Compatible Digital Human Modeling

Toon Huysmans[1,2]($\boxtimes$), Femke Danckaers[2], Jochen Vleugels[3],
Daniël Lacko[3], Guido De Bruyne[3], Stijn Verwulgen[3], and Jan Sijbers[2]

[1] Section on Applied Ergonomics and Design, Faculty of Industrial Design
Engineering, Delft University of Technology, Delft, The Netherlands
t.huysmans@tudelft.nl
[2] imec - Vision Lab, Department of Physics, Faculty of Science,
University of Antwerp, Antwerp, Belgium
[3] Department of Product Development, Faculty of Design Sciences,
University of Antwerp, Antwerp, Belgium

Abstract. Parametric 3D human body models are valuable tools for ergonomic product design and statistical shape modelling (SSM) is a powerful technique to build realistic body models from a database of 3D scans. Like the underlying 3D scans, body models built from SSMs are typically represented with triangle meshes. Unfortunately, triangle meshes are not well supported by CAD software where spline geometry dominates. Therefore, we propose a methodology to convert databases of pre-corresponded triangle meshes into multi-patch B-spline SSMs. An evaluation on four 3D scan databases shows that our method is able to generate accurate and water-tight models while preserving inter-subject correspondences by construction. In addition, we demonstrate that such SSMs can be used to generate design manikins which can be readily used in SolidWorks for designing well conforming product parts.

Keywords: Statistical shape modeling · B-splines · Computer-aided design
Digital human modeling

1 Introduction

In user-centered product design, ergonomics is pursued by putting the physical and mental characteristics of the human user center-stage [1]. For products that are worn on the body or closely interact with it, e.g. wearables, garments, seating furniture, etc., 'body shape' is a dominant physical characteristic with an impact on product affordance, comfort, and wearability [2]. Statistical shape models (SSM) are an elegant way to represent body shape variation observed in a population. SSMs typically represent shape variation by a shape space with the average shape at the origin and spanned by a small set of orthogonal shape modes. In this way, each point in the shape space represents a specific body shape and its coordinates define the contribution/weight of each shape mode. In this formulation, the weights are *body shape parameters* and such parametric body models are considered a valuable tool for user-centered product design [3].

A large part of the research on statistical shape modeling is devoted to methodologies for the construction of inter-subject correspondences and mathematical

© Springer International Publishing AG, part of Springer Nature 2019
D. N. Cassenti (Ed.): AHFE 2018, AISC 780, pp. 179–189, 2019.
https://doi.org/10.1007/978-3-319-94223-0_17

formulations of the statistical model that is derived from these correspondences. Many algorithms are available from the literature, both for organ modeling [4–6], mostly within the medical imaging field, as well as for full body modeling [7–12]. Most of these works construct SSMs from 3D triangle meshes, obtained from 3D scanning or medical imaging. Naturally, the derived SSM is also represented in triangular mesh form. Other geometric representations for SSMs are available, e.g. landmarks, medial models, level-sets, or basis-functions [4], but they are much less widespread.

For user-centered design, the predominance of triangular mesh SSMs is somewhat unfortunate since digital models in the form of triangular meshes are not well supported by popular CAD packages (e.g. SolidWorks). There, the majority of CAD operations is only available for B-spline geometry and while reverse engineering allows the conversion of triangular geometry to B-splines, for organic shapes it is often a slow and cumbersome process. In addition, it generally results in a varying number and distribution of patches, even for shapes generated by the same SSM. As a result, intersubject correspondences are lost and, with it, the ability to consistently link (parametric) designs to anatomical reference points on these shapes. SSMs with a spline-based representation could improve compatibility of SSMs with CAD-software.

Only limited work has been done on spline based SSMs. In [13], Quan et al. built a B-spline SSM of the face. They took as input triangular meshes with a pre-existing vertex correspondence, then parameterized the meshes using cylindrical coordinates and subsequently fitted the scans with a single B-spline patch. Unfortunately, the cylindrical mapping approach only handles simple geometries with disc-like topology. Hu et al. constructed SSMs of organs in the pelvic area using a NURBS representation and deformation [14]. The NURBS, however, was manually constructed using the loft tool in Rhinoceros and their NURBS deformation technique was only demonstrated on very coarse organ approximations. From the brief description of the method it is unclear whether the method can be applied to more detailed and/or complex shapes. In [15], Peng et al. employed conformal mapping to obtain a T-spline SSM of the face and its expressions. Similar to [13], their method is developed for shapes of disc topology and further modification would be required to apply the methodology to more complex shapes.

In this paper, an automated methodology is proposed that allows the construction of SSMs with a B-spline representation with less restrictions on topology or shape. Our method takes as input a set of pre-corresponded triangular meshes and generates as output a multi-patch B-spline SSM, where the B-spline patch topology is derived from a rough quadrilateral mesh approximation of the mean shape of the triangular SSM. Shape instances generated by our B-spline SSMs have consistent patch layout and seamlessly integrate with parametric CAD-software, where they can be utilized for ergonomic product design.

2   Methods

Starting from a set of triangular meshes (the population sample), with a one-to-one vertex correspondence, our method constructs an SSM with a B-spline representation in three phases. The process is visualized in Fig. 1.

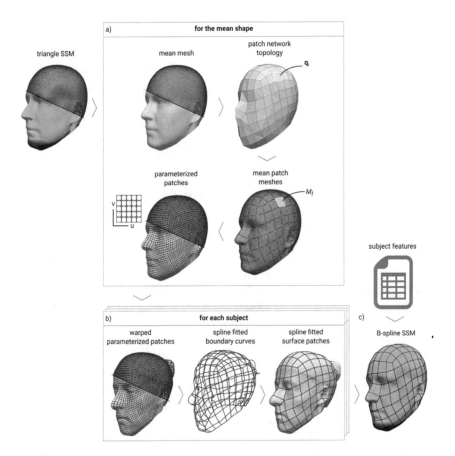

Fig. 1. A depiction of the steps of our method for the construction of a B-spline statistical shape model. See text for the explanation of the different steps.

In the first stage (Fig. 1a), the B-spline patch network topology for our SSM is defined, based on a rough quadrangulation of the mean shape. The mesh of the mean shape is then split into smaller patches, one patch per quad in the quadrangulation, and the vertices of the patches are equipped with uv-parameters, using a convex combination mapping, to facilitate B-spline fitting. In the second stage (Fig. 1b), the shape of each subject is approximated with a multi-patch B-spline surface. The approximation is achieved by first warping the modified mean mesh to the subject's shape via the vertex correspondence, followed by a B-spline curve approximation of the patch boundaries, and finally by a B-spline surface approximation for each of the patches where the approximation interpolates the B-spline patch boundary curves to ensure water-tightness of the surface. In the final stage (Fig. 1c), the SSM is calculated from the B-spline approximations by applying principal component analysis (PCA) to the control point coordinates and a further parametric body model can be derived via a regression analysis of the 3D shapes versus subject features.

In the following sections, a more elaborate description of the involved steps is provided.

2.1   Patch Layout

Our method starts from a set of n_s shapes $S = \{S_1, \ldots, S_{n_s}\}$ in triangular mesh form with a pre-existing vertex correspondence. Thus, each shape S_i is described by a set of n_v vertices $\left\{ v_1^i, \ldots, v_{n_v}^i \right\}$, with $v_j^i \in \mathbb{R}^3$ at the same anatomical location for each i. The surfaces S_i all share the same mesh connectivity consisting of n_t triangles. From these surfaces, the mean shape M can be derived with n_v vertices $\{m_1, \ldots, m_{n_v}\}$ and $m_i = \frac{1}{n_s} \sum_{j=1}^{n_s} v_i^j$. The mean shape M is fairly smooth but still contains the common characteristic features of the shapes under consideration. We therefore consider it as a good starting point to define the B-spline patch layout for our SSM. The patch layout is obtained by approximating M with a rough quadrangular (or quad) mesh Q using the mixed-integer quadrangulation method of Bommes et al. [17]. Any other method that generates a conforming quadrangulation, i.e. without T-junctions, could be used equally well [16]. The obtained quad mesh is described by a set of $n_p \ll n_v$ vertices in $\mathbb{R}^3$ and a set of $n_q \ll n_t$ quads. The connectivity of this quad mesh Q will be used to define the connectivity of the B-spline patches of the SSM in the next steps. There will be one patch per quad and the neighborhood relation of the quads will be adopted for the patches.

The mean mesh M is split into a set of smaller meshes, one mesh for each quad. Let us consider a quad q from Q with vertices (c_1, c_2, c_3, c_4). First, each of the four corners c_i is mapped to the closest vertex on M, denoted r_i. Here, the closest point search for points that are on the boundary of Q is limited to the boundary vertices of M. Then, for each edge $e = (c_k, c_l)$ of the quad, a geodesic curve g between the corresponding vertices (r_k, r_l) on M is traced using the method of Kimmel and Sethian [18]. Finally, by cutting the mesh M along the four obtained geodesic curves, the surface patch that corresponds to quad q can be extracted. This procedure is executed for each quad q_i in Q, resulting in the split of mean mesh M into a set of n_q patches $\{M_1, \ldots, M_{n_q}\}$. The i-th patch M_i is described with n_i vertices m_j^i, either directly inherited from M or introduced by the cutting of triangles along the geodesic paths.

Each of the meshes M_i will contribute one B-spline patch to the SSM. In order to facilitate B-spline approximations in the next phase of our methodology, the vertices m_j^i of mesh patches M_i are equipped with uv-coordinates $u_j^i \in [0, 1] \times [0, 1]$, defining their location in the parameter space of the B-spline. First, the perimeter of M_i is mapped to the perimeter of the parameter space. The four corner points of M_i, corresponding to the quad corners of q_i, are mapped to the four corners of the rectangular B-spline parameter space. The four boundary segments of M_i, connecting the patch corners, are mapped to the straight boundary segments of the parameter space using arc-length parameterization. The mapping of the boundary of M_i is extended to the interior vertices of M_i via the mean value surface parameterization method of Floater [19, 20]. This guarantees a one-to-one map between the surface of M_i and the B-spline parameter space, resulting in a complete and unambiguous parameterization of the patch for fitting. In addition, for neighboring patches, the parameterization of common boundary curves matches up, which is a requirement in our methodology in order to ensure water-tightness of the B-spline SSM.

2.2   B-Spline Approximation

For each shape S_i in the population, a multi-patch B-spline approximation is calculated based on the parameterized patches obtained in the previous step. First, the vertices of all the surface patches M_j are warped to match the shape of S_i and will be denoted M_j^i. For vertices on M_j inherited from M, this warping is trivially derived from the vertex correspondence between M and S_i. Vertices of M_j that were newly introduced during the splitting of M into patches, however, do not have an explicit correspondence with M or S_i. For these vertices, the warping is derived by barycentric interpolation with respect to the triangle of M in which the vertex was inserted.

The actual B-spline approximation of S_i, via the warped patches M_j^i, is obtained in two steps, in order to ensure water-tightness. First, the boundary segments of the warped patches are approximated with 3D uniform cubic B-spline curves, exactly interpolating the patch corners. This is followed by the approximation of each patch with a 3D uniform bi-cubic B-spline surface, exactly interpolating the previously calculated B-spline boundary curves. Robustness with respect to overfitting and undersampling is guaranteed by Tikhonov regularization promoting patch smoothness.

In this procedure, B-spline curves are specified by a set of n_k uniformly spaced knots $K = \{k_i\}$ in $\mathbb{R}$ and n_k control points $P = \{p_i\}$ in $\mathbb{R}^3$. The B-spline curve t, parameterized by a 1D parameter u, is then defined as

$$t(u|K,P) = \sum_{i=1}^{n_k} \beta_{i,4}(u)p_i, \tag{1}$$

where $\beta_{i,4}$ is the i-th 1D cubic B-spline kernel [21] corresponding to the knot sequence $\{k_1, k_1, k_1, k_1, k_2, \ldots, k_{n_k-1}, k_{n_k}, k_{n_k}, k_{n_k}, k_{n_k}\}$.

Similarly, B-spline surface patches are specified by a set of $n_k \times n_k$ knots $K = \{k_{ij}\}$ in $\mathbb{R}^2$ with uniform spacing and $n_k \times n_k$ control points $P = \{p_{ij}\}$ in $\mathbb{R}^3$. The B-spline surface s is parameterized by the 2D parameter $u = (u, v)$ and is defined as

$$s(u|K,P) = \sum_{i=1}^{n_k} \sum_{j=1}^{n_k} \beta_{i,4}(u)\beta_{j,4}(v)p_{ij}. \tag{2}$$

For a given patch boundary curve with n_c points x_i and parameters u_i, the approximation with a B-spline curve c entails finding the optimal control points $\hat{P}$ and is formulated as the following minimization problem:

$$\hat{P} = \arg\min_{P} \|BP - X\|^2 + \tau LP^2,$$
$$\text{subject to } CP = D. \tag{3}$$

The first term of the minimization objective measures the distance between the points x_i and the B-spline point $t(u_i)$, i.e. the fit of the curve. The $n_c \times 3$ matrix X contains the stacked boundary points x_i and the $n_k \times 3$ matrix P is formed by stacking the unknown control points. In the sparse $n_c \times n_k$ matrix B, the i-th row contains for each control point p_j the weight $\beta_{j,4}(u)$ it contributes to the B-spline point $t(u_i)$. The second term measures the smoothness of the B-spline curve, weighted by a factor τ, by means of summing the discrete Laplacian at each interior control point p_j. This Laplacian operator is formed by the rows of the sparse $(n_k - 2) \times n_k$ matrix L as

$$L_{ij} = \begin{cases} 2, & i = j \\ -1, & |i-j| = 1 \\ 0, & otherwise \end{cases}. \tag{4}$$

The equality constraint enforces the fixation of the two curve end-points at the patch boundary end-points. The $2 \times n_k$ matrix C is sparse and has a 1 on the first row at the index of the first control point and on the second row at the index of last control point. The 2×3 matrix D contains the stacked start and end point of the patch boundary. The minimum of Eq. (3) is obtained by solving the normal equations $(B^T B + \lambda L^T L)P = B^T X$ using a sparse LU factorization [22] with Karush-Kuhn-Tucker conditions to enforce the equality constraints.

For the fitting of the surface patches an analogous procedure as in Eq. (3) is followed. Matrix X is formed as before by stacking the patch points, P contains the grid of unknown control points in row-major order, and matrix B now contains the row-major ordered weights $\beta_{i,4}(u)\beta_{j,4}(v)$ for the grid of control points p_{ij}. The matrix L computes the 2D discrete Laplacian, with a central weight of 4 and a weight of -1 for the four incident control points on the grid. Finally, the equality constraints simply match the control points at the boundaries $u = 0$, $v = 1$, $u = 1$, and $v = 0$ with the control points of the previously calculated boundary B-spline curves. The resulting sparse system is again solved with LU-factorization.

The approach outlined in this section results in a water-tight, multi-patch approximation of each subject preserving the inter-subject correspondences.

2.3   SSM Construction

In order to calculate a B-spline SSM, each subject is represented by a n-dimensional vector obtained by concatenating the coordinates of all control points of all its B-spline patches calculated in the previous section. By applying a PCA on the resulting n_s vectors of dimension n a linear model of the variation is obtained:

$$\Theta = \bar{\Theta} + \sum_{i=1}^{n_s-1} w_i \ddot{\Theta}_i. \tag{5}$$

Here, Θ is the nD vector of control point coordinates of the modeled shape, $\bar{\Theta}$ is the n D vector of mean control point coordinates, $\ddot{\Theta}_i$ are the $n_s - 1$ principal component vectors of dimension n providing a basis for the space of shape variations, and the weights w_i control the contributions of the shape variations for a specific shape Θ. Typically, the weights w_i are allowed to vary in the range $|w_i| < 3\sqrt{\lambda_i}$, where λ_i is the i-th eigenvalue obtained in the PCA. In order to reconstruct the B-spline patches of the modeled shape, the control points of each patch in Θ are inserted in Eq. (2).

Following the approach of Allen et al. [7], a further body model is constructed that is parameterized by subject features, such as height, age, BMI, etc. The parameterization is captured by an $(n_s - 1) \times (n_f + 1)$ linear mapping matrix Π that maps subject features f_j onto the PCA weights $w_i = [w_1 \ldots w_{n_s-1}]^T$ for each subject:

$$\Pi[f_1 \ldots f_{n_f} 1]^T = w_i. \tag{6}$$

The mapping matrix is obtained by solving $\Pi = WF^+$, where W is the $(n_s - 1) \times n_s$ matrix of stacked subject PCA weights w_i, F is the $(n_f + 1) \times n_s$ matrix of stacked subject feature vectors, and F^+ denotes the pseudo-inverse of F. With the obtained mapping matrix Π, Eq. (5) can be rewritten to obtain a model parameterized by subject features:

$$\Theta(f_1, \ldots, f_{n_f}) = \bar{\Theta} + \ddot{\Theta}\Pi[f_1 \ldots f_{n_f} 1]^T, \tag{7}$$

where $\ddot{\Theta}_i$ is the $n \times (n_s - 1)$ matrix obtained by horizontally stacking the $n_s - 1$ principal components $\ddot{\Theta}_i$.

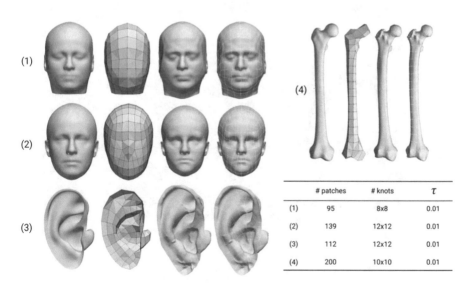

	# patches	# knots	τ
(1)	95	8x8	0.01
(2)	139	12x12	0.01
(3)	112	12x12	0.01
(4)	200	10x10	0.01

Fig. 2. The four datasets the methodology was evaluated on: (1) MRI heads, (2) CAESAR heads, (3) ear molds, and (4) femurs. For each dataset, from left to right, the mean shape, the quad mesh, a population subject, and the B-spline fit of that subject as calculated by our method. The table lists the chosen parameters that are required as inputs to our methodology.

2.4   SolidWorks AddIn

The authors developed a custom SolidWorks AddIn that allows the creation of instances of the parametric body models as native (B-spline) surfaces in SolidWorks. The AddIn with GUI was written in VB.net. The AddIn communicates with a custom written C#-library that encapsulates the model IO (HDF5 format) and shape generation process of Eq. (6).

3   Results

Our methodology was applied to four datasets: (1) 100 head surfaces, 50 male and 50 female subjects, extracted from MRI data in the ICBM database [23], (2) 1384 head surfaces from the CAESAR laser scan database [24], comprising 346 surfaces each of the groups male-Dutch, female-Dutch, male-Italian, and female-Italian, (3) 150 ear scans based on laser scans of silicone ear molds, and (4) 189 femur surfaces extracted from CT scanning. All surfaces were triangle meshes and inter-subject correspondences were calculated with previously published methods of the authors: the elastic surface registration method of [25] for datasets (1) and (2) and the cylindrical correspondence method of [6] for datasets (3) and (4). Intermediate results for each of the four datasets are shown in Fig. 2, along with the input parameters required by our methodology.

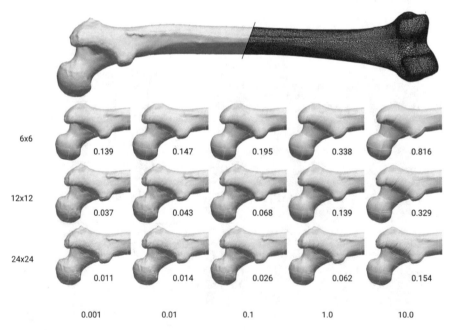

Fig. 3. Top: the triangle mesh of the first subject in the femur dataset. Bottom: resulting B-spline fits of this subject with varying number of control points per patch (rows) and Tikhonov regularization factors τ (columns). The fitting error is also provided for each case, measured as RMS surface distance from the mesh to the B-spline fit (in mm).

An experiment was performed on the femur dataset to investigate the influence of the number of control points per patch and the Tikhonov regularization factor τ (cfr. Eq. (3)) on the B-spline fitting. A multi-patch B-spline fit was calculated for the first subject of the database with combinations of, on the one hand, grids of 6×6, 12×12, and 24×24 control points per patch, and on the other hand, Tikhonov regularization factors of 0.001, 0.01, 0.1, 1.0, and 10.0. The resulting B-spline surfaces are shown in Fig. 3. There, also an estimate of the surface fit is provided by means of the root mean square (RMS) surface distance, measured from the triangle mesh vertices to the closest points on the B-spline surface.

The most computationally intensive part of our methodology is the fitting of the B-spline patches for all subjects in the population. For all datasets, the computation time per subject took up to 2s on a single core of an i7-5960X CPU.

In order to demonstrate our SolidWorks AddIn, a parametric model, with a variety of subject features such as 'head length' and 'head width', was calculated from the MRI head dataset and loaded into SolidWorks via our AddIn. Four head shapes (manikins) were generated: the average head and three heads with varying head width (percentile 5, 50 and 95). Finally, the average head manikin was employed to construct a part with perfect fit to the body using standard SolidWorks CAD-operations. The results are shown in Fig. 4.

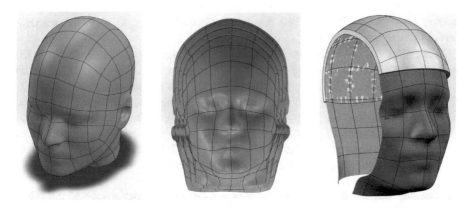

Fig. 4. Example body shapes (manikins) generated by our method and imported into SolidWork though our AddIn. Left: the average manikin of the MRI head model. Middle: cut-away of three manikins of this model with varying head/bi-tragion width (P5-50-95). Right: cutaway of the average manikin together with a perfectly fitting solid body obtained from the manikin by the SolidWorks features 'Intersect' and 'Shell'.

4 Discussion

The results in this paper show that our methodology is able to convert triangle mesh SSMs to B-spline parametric body models for a variety of shapes, without sacrificing on accuracy, water-tightness, or anatomical correspondence. Our methodology has three input parameters: number of patches, control point grid size, and the Tikhonov regularization factor. It was empirically determined that accurate models with a limited number of control points can be obtained by balancing the number of patches (here ranging from 90 to 200) with the number of control points per patch (ranging from 8×8 to 12×12). For the models considered in this paper, the Tikhonov regularization factor, used to regularize the B-spline fitting of the patches, could be kept constant at 0.01. A further analysis on the femur dataset demonstrated that the resulting models are not too sensitive to the regularization factor, i.e. the amount of smoothing and the fitting error remained acceptable for regularization factors differing several orders of magnitude: Fig. 4, τ ranging from 0.001 to 0.1 for the femur with a 12×12 grid size. Nevertheless, a too low value might lead to an ill conditioned system, especially when the number of control points far exceeds the number of vertices in the patch. Too high values can result in significant artifacts at the patch boundaries as can be seen from Fig. 4.

The resulting SSMs are also easily imported in CAD software, via our AddIn, and can be used to generate a variety of body shapes that are sufficiently detailed for the ergonomic design of products. Using standard CAD-operations, available in Solid-Works, products or product parts can be constructed directly from the manikins to nicely conform to the contours of the human body, as demonstrated in Fig. 4. It should be noted, however, that certain CAD operations, like surface offsetting, are notoriously complicated and are not always able to generate valid results for organic shapes. We anticipate that the robustness of these algorithms will keep on improving, but also acknowledge that it might also help if our models would be smooth at patch boundaries (on top of the water-tightness we currently provide).

For future work, we envisage two main directions. On the one hand, we would like to further improve the B-spline approximation by (a) integrating the splitting of the mean mesh in patches with the quad-remeshing step for improved stability and (b) by including the boundary curve normals in the patch fitting for improved smoothness at the patch boundaries. On the other hand, we would like to further improve the modeling step by investigating the use of weighted PCA to compensate for patch size variations in the SSM and to extensively evaluate the derived B-spline SSMs in terms of model compactness, generalization ability, and specificity.

Acknowledgments. This work was financially supported by VLAIO grants TETRA-130771 and SB-141520.

References

1. Pheasant, S., Haslegrave, C.M.: Bodyspace: Anthropometry, Ergonomics and the Design of Work. CRC Press, Boca Raton (2016)
2. Motti, V.G., Caine, K.: Human factors considerations in the design of wearable devices. In: Proceedings of the Human Factors and Ergonomics Society Annual Meeting, vol. 58, no. 1, pp. 1820–1824. SAGE Publications, Thousand Oaks (2014)
3. Baek, S.Y., Lee, K.: Parametric human body shape modeling framework for human-centered product design. Comput.-Aided Des. **44**(1), 56–67 (2012)

4. Heimann, T., Meinzer, H.P.: Statistical shape models for 3D medical image segmentation: a review. Med. Image Anal. **13**(4), 543–563 (2009)
5. Davies, R.H., Twining, C.J., Cootes, T.F., Taylor, C.J.: Building 3-D statistical shape models by direct optimization. IEEE Trans. Med. Imaging **29**(4), 961–981 (2010)
6. Huysmans, T., Sijbers, J., Verdonk, B.: Automatic construction of correspondences for tubular surfaces. IEEE Trans. Pattern Anal. Mach. Intell. **32**(4), 636–651 (2010)
7. Allen, B., Curless, B., Popović, Z.: The space of human body shapes: reconstruction and parameterization from range scans. ACM Trans. Graph. **22**(3), 587–594 (2003)
8. Anguelov, D., Srinivasan, P., Koller, D., Thrun, S., Rodgers, J., Davis, J.: SCAPE: shape completion and animation of people. ACM Trans. Graph. **24**(3), 408–416 (2005)
9. Hasler, N., Stoll, C., Sunkel, M., Rosenhahn, B., Seidel, H.P.: A statistical model of human pose and body shape. Comput. Graph. Forum **28**(2), 337–346 (2009)
10. Chen, Y., Liu, Z., Zhang, Z.: Tensor-based human body modeling. In: IEEE Conference on Computer Vision and Pattern Recognition (CVPR), pp. 105–112. IEEE (2013)
11. Park, B.K., Reed, M.P.: Parametric body shape model of standing children aged 3–11 years. Ergonomics **58**(10), 1714–1725 (2015)
12. Pishchulin, L., Wuhrer, S., Helten, T., Theobalt, C., Schiele, B.: Building statistical shape spaces for 3D human modeling. Pattern Recogn. **67**, 276–286 (2017)
13. Quan, W., Matuszewski, B., Shark, L., Ait-Boudaoud, D.: 3-D facial expression representation using B-spline statistical shape model. In: British Machine Vision Conference, Vision, Video and Graphics Workshop, 10–13 September 2007, Warwick (2007)
14. Hu, N., Cerviño, L., Segars, P., Lewis, J., Shan, J., Jiang, S., Wang, G.: A method for generating large datasets of organ geometries for radiotherapy treatment planning studies. Radiol. Oncol. **48**(4), 408–415 (2014)
15. Peng, W., Feng, Z., Xu, C., Su, Y.: Parametric T-spline face morphable model for detailed fitting in shape subspace. In: Proceedings of the IEEE Conference on Computer Vision and Pattern Recognition, pp. 6139–6147 (2017)
16. Campen, M.: Partitioning surfaces into quadrilateral patches: a survey. Comput. Graph. Forum **36**(8), 567–588 (2017)
17. Bommes, D., Zimmer, H., Kobbelt, L.: Mixed-integer quadrangulation. ACM Trans. Graph. **28**(3), 77:1–77:10 (2009)
18. Kimmel, R., Sethian, J.A.: Computing geodesic paths on manifolds. Proc. Natl. Acad. Sci. **95**(15), 8431–8435 (1998)
19. Floater, M.S.: Parametrization and smooth approximation of surface triangulations. Comput. Aided Geom. Des. **14**(3), 231–250 (1997)
20. Floater, M.S.: Mean value coordinates. Comput. Aided Geom. Des. **20**(1), 19–27 (2003)
21. Bartels, R.H., Beatty, J.C., Barsky, B.A.: An Introduction to Splines for Use in Computer Graphics and Geometric Modeling. Morgan Kaufmann, Burlington (1987)
22. Li, X.S.: An overview of SuperLU: algorithms, implementation, and user interface. ACM Trans. Math. Softw. **31**(3), 302–325 (2005)
23. Capetillo-Cunliffe, L.: Loni: Laboratory of Neuro Imaging (2007)
24. Robinette, K.M., Daanen, H., Paquet, E.: The CAESAR project: a 3-D surface anthropometry survey. In: Second IEEE International Conference on 3-D Digital Imaging and Modeling, pp. 380–386 (1999)
25. Danckaers, F., Huysmans, T., Lacko, D., Ledda, A., Verwulgen, S., Van Dongen, S., Sijbers, J.: Correspondence preserving elastic surface registration with shape model prior. In: 22nd International Conference on Pattern Recognition (ICPR), pp. 2143–2148. IEEE (2014)

Extreme Environments and Military Applications

Effects of Dynamic Automation on Situation Awareness and Workload in UAV Control Decision Tasks

Wenjuan Zhang, James Shirley, Yulin Deng, Na Young Kim, and David Kaber$^{(\boxtimes)}$

Department of Industrial and Systems Engineering,
North Carolina State University, Raleigh, NC 27695, USA
{wzhang28,jdshirle,ydeng2,nkim3,dbkaber}@ncsu.edu

Abstract. Complex unmanned aerial vehicle (UAV) operations place high information processing demands on operators. Dynamic automation or function allocation (DFA) has been proposed as a method to address operator "overload" situations without compromising system/situation awareness. This study made use of a high-fidelity UAV simulation to investigate any benefits of DFA when applied to decision making tasks. Three modes of UAV control automation, including DFA and static high and low level of automation, were compared. A preliminary analysis of operator responses to situation awareness queries showed no significant differences; however, response accuracy in an aircraft detection task indicated superior situation comprehension under DFA as compared to static high level of automation. Results of an embedded secondary task indicated that participants experienced lower cognitive workload when operating under DFA as compared to static low level of automation. Findings of this study could be used for guiding design of DFA approaches for UAV control.

Keywords: UAV · Dynamic automation · Situation awareness
Workload

1 Introduction

Unmanned aerial vehicle (UAV) ground control automation and interfaces have increased in complexity in order to provide access and information on all features of advanced UAV platforms. Consequently, UAV operators are under increasing information processing demands in using features while also ensuring vehicle safety and mission success. For this reason, many military UAVs require more operators than traditional manned aircrafts, leading to growing "manpower burden" for the military [1]. Automation has been applied to UAV functions to offload operator workload in terms of monitoring vehicle system status or target identification. However, highly automated systems not only pose technical challenges but can also inflate operator monitoring workload [2] and can cause decrements in operator ability to detect critical situations [3, 4].

Related to this, some research has demonstrated that intermediate levels of automation (LoAs) can reduce task workload while maintaining operator in critical

© Springer International Publishing AG, part of Springer Nature 2019
D. N. Cassenti (Ed.): AHFE 2018, AISC 780, pp. 193–203, 2019.
https://doi.org/10.1007/978-3-319-94223-0_18

aspects of a control loop [5]. Endsley and Kiris [3] found that intermediate LoAs resulted in fewer human out-of-the-loop performance decrements and greater situation awareness (SA) as compared with full automation in automobile navigation tasks. Endsley and Kaber [5] examined workload and SA effects of a wide range of LoAs with a multitasking simulation. Their results showed operator workload was reduced and SA was increased when intermediate level automation was applied to the decision making portion of the task (i.e., humans selecting options generated by computers).

Another line of research aimed at addressing workload and SA issues associated with automation has focused on adaptive automation or dynamic function allocations (DFAs). This approach dynamically allocates responsibilities between a human operator and computer in order to achieve optimal system performance [6]. Many studies have provided empirical evidence of benefits of adaptive automation. In a simulated UAV and unmanned ground vehicle (UGV) task, Parasuraman et al. [7] found dynamic activation of automated assistance to result in higher SA and lower workload as compared with manual system control. Calhoun et al. [8] also reported improved performance and reduced cognitive workload when adaptive automation was applied to an image analysis task as part of an unmanned vehicle simulation. Kaber et al. [9] investigated the effectiveness of adaptive automation for supporting different stages of information processing with a low-fidelity simulation of air traffic control (ATC) tasks. They observed superior task performance when automation was applied to lower-order information processing functions (i.e., information acquisition and action implementation), as compared to higher-order cognitive functions (i.e., information analysis and decision-making). More recently, Afergan et al. [10] assessed UAV operator workload and boredom in monitoring and simultaneously navigating multiple vehicles by collecting passive brain signals. They increased or decreased the number of UAVs based on real-time analysis of operator workload state. Results indicated that the adaptive task difficulty strategy resulted in lower failure rates and fewer vehicle collisions, as compared to the non-adaptive condition. The adaptive condition also resulted in quicker error detection by operators, possibly suggesting increased SA. Despite these well documented benefits of adaptive automation, most studies have focused on applying DFA to the information analysis portion of human information processing. Limited research has investigated effects of applying adaptive automation to decision making tasks.

In this study, we made use of a high-fidelity UAV simulation and realistic mission scenarios defined by a military professional having written hundreds actual UAV mission briefs. We developed a prototype vehicle control interface to apply DFA to a decision making task as part of the scenarios. The objective of the study was to determine whether DFA provides any benefits to operator workload and SA for decision making performance, as compared to static low and high automation, in UAV control. A designed experiment was conducted using the simulation with repeated trials by each participant with multiple subjective and objective measures of cognitive workload and SA.

2 Method

2.1 Participants

This preliminary study involved a small sample of six trained participants (all males; mean age = 19.8 yrs., S.D. = 1.7). The participants were recruited from the ROTC program at a local university. All participants were required to have 20/20 vision, no color vision impairment and general familiarity with computer usage. The study was approved by the University's Institutional Review Board.

2.2 Apparatus

The UAV control simulation was prototyped based on the ArduPilot Mission Planner software with slight modifications to allow for delivery of additional in-flight tasks (see Sect. 2.3). The simulation was developed using the Just-in-Mind prototyping tool (San Francisco, USA). The vehicle control interface (see Fig. 1) consisted of a primary flight display (upper-left section), a map display (upper-right section), a flight parameter display (lower-left section) and a mission planning display (middle-right section). The in-flight tasks were presented in the lower-right section of the interface, under the mission planning section, at various times during a mission.

During the experiment, the participant was seated in an office-like cubicle with a monitor in front of them, displaying the UAV control interface. The monitor was connected to a laptop computer located outside the cubicle. This configuration allowed an experimenter to observe participant task performance without distracting them. A mouse and keyboard was provided for participants to interact with the UAV simulation. Participants were allowed to adjust the position of the monitor, mouse and keyboard to ensure comfortable use of the setup.

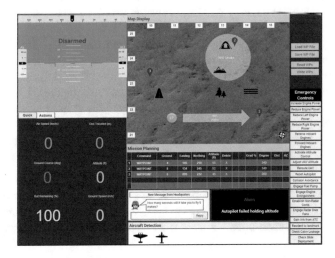

Fig. 1. UAV control interface

2.3   Scenario

The simulated UAV mission required participants to fly a basic reconnaissance mission while acting as a supervisory controller to the automation. They were also required to complete three types of embedded secondary tasks, including: (a) hostile aircraft detection; (b) flight time projections to waypoints and a launch pad; and (c) detecting and resolving system alarms with vehicle control functions. The following subsections provide details on each of these tasks, which also served as bases for some of the objective workload and SA measures.

Aircraft Detection. The aircraft detection task required participants to identify any hostile aircrafts among other friendly aircrafts based on radar display icons. Prior to the experiment, participants were presented with a total of 8 aircraft icons (4 friendly, 4 hostile) and given sufficient time to memorize them. During the experiment, random combinations of aircraft icons were presented in the aircraft detection section of the interface (bottom-right of Fig. 1). Participants were asked to identify all 'hostile' targets by clicking on the relevant icon(s). The task was presented with two levels of difficulty. For an 'easy' level, two icons (1 friendly and 1 hostile) were presented. For the 'hard' level, six icons (3 friendly and 3 hostile) were presented. A similar task was used by Roberts et al. [11] in a UAV simulation study. The aircraft detection task required operators to compare perceived information with working memory in order to interpret the meaning of the information (i.e., whether the detected vehicle was friendly or hostile). Therefore, task performance involved situation comprehension and decision making.

Flight Time Projection. The flight time projection task required participants to respond to messages from headquarters with a time estimation for the UAV to fly over a certain distance. To complete this task, participants had to divide the given distance by the vehicle's current ground speed, which ranged from 1 to 9 m/s. Like the aircraft detection task, there were two levels of difficulty for the calculation. For the 'easy' level, participants were given a two-digit division problem, where the divisor left no remainder (e.g., 36 divided by 6). For the 'hard' level, participants were given a three-digit division problem, where the divisor left a remainder (e.g., 555 divided by 4). When responding, participants were asked to round answers to the nearest single decimal place, if there was a remainder. Participants were allowed to use pen and paper for the calculation. The difficulty setting for this task was supported by studies that found division of larger numbers [12] and remainders [13] to increase the difficulty of division problems. As an embedded secondary task to the UAV mission, the flight time projection task provided a method of measuring reserve attentional capacity [14] and, therefore, was also considered as an objective indicator of overall workload for participants.

Alarm Handling. The alarm handling task was another decision making task that required participants to select the most appropriate emergency vehicle control action from a menu of nineteen potential actions in response to off-nominal events. Alarm content and appropriate solutions were constructed based on the Aviation Safety Reporting Systems (ASRS) database for manned aircraft incidents. The alarms were classified into six categories according to UAV flight subsystems, including: altitude control, route control, engine & fuel management, communication systems, and other. The emergency controls were grouped in the UAV control interface menu according to these subsystems. Participants were made aware of the subsystem categories during training.

During the experiment trials, two LoAs were applied to the alarm handling task. Under a 'high' LoA setting, the computer highlighted the three most relevant emergency control actions for addressing any detected off-nominal event. Participants were instructed to choose the alarm solution based on the computer suggestions. The high LoA condition represents a "rigid system", as defined by Endsley and Kaber [5], where the computer presents only a limited set of actions to the operator and human may not generate additional options. The 'low' LoA condition did not provide any decision making assistance for participants and they were required to evaluate all emergency control actions and make a selection to resolve an alarm.

Mission Structure. Each simulated UAV mission (i.e., experiment trial) consisted of three blocks of tasks with the first and third blocks containing easy tasks and the second block containing hard tasks (see Fig. 2). These difficulty blocks were constructed to represent task demand fluctuations in UAV control operations. Each block was divided into two phases by a temporary mission suspension (60 or 80 s). Each phase contained one aircraft detection task, one flight time projection task and one alarm handling task, presented in randomized order. Therefore, each block contained two instances of each task and each trial contained six occurrences of each task. Participants were allowed 20 s to complete each task. Participants were also instructed to maintain awareness of the status and changes with the UAV and all subsystems as well as the flight environment.

SA Queries and Workload Ratings. During the mission suspensions at the close of each phase, the control interface was blocked by a grey screen and three SA queries were presented. The queries were developed based on a goal-directed task analysis for the UAV mission and tested participant dynamic knowledge of vehicle flight parameters, flight trajectory, and geographical features presented on the map display. Participants were allowed 1 min to provide verbal responses to three queries, which were record and graded by an experimenter. The approach to the SA assessment was akin to application of the Situation Awareness Global Assessment Technique (SAGAT), as developed by Endsley [15]. During the suspensions at the end of each block, participants were also required to provide an overall (cognitive) workload (OW) rating by using a unidimensional scale from 0 to 100. The OW measure was adopted from Vidulich and Tsang [16].

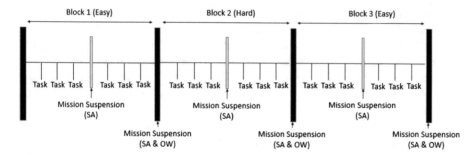

Fig. 2. A simulated UAV mission trial. (Note: "Task" represents an instance of aircraft detection, flight time projection or alarm handling).

2.4 Experiment Design and Variables

Independent Variables. Two factors were manipulated in this experiment, including the in-flight task difficulty and mode of task automation. The in-flight task difficulty had two levels. Under the easy level, both the aircraft detection and flight time projection tasks were set to easy. Under the hard level, both tasks were set to hard difficulty. The automation mode had three levels, including: the low LoA, high LoA and the DFA conditions. Under the high LoA, the interface assistance for the alarm handling task was active for all three trial blocks and inactive for all blocks under the low LoA. Under the DFA condition, the automation assistance dynamically activated during the hard block and deactivated during the easy blocks.

Experiment Design. The experiment followed a within-subject design. Each participant was exposed to all automation modes with one replication (i.e., 6 trials in total). The automation modes were presented in randomized order to mitigate potential carry-over effects of one automation mode on another.

Dependent Variables (DVs). The experiment captured four responses. The SA query accuracy (DV1) and aircraft detection accuracy (DV2) were objective measures of situation awareness. The OW rating (DV3) and flight time project accuracy (DV4) were subjective and objective measures of workload, respectively. DV1 was calculated as the ratio of correct answers to the total number of queries posed during each trial block.

2.5 Hypothesis

Based on the prior literature, we hypothesized that the DFA condition would result in better SA than the static modes of automation, as reflected by higher SA query accuracy (H1) and aircraft detection accuracy (H2). We also speculated that the DFA condition would result in lower workload than the low LoA, as reflected by lower OW ratings (H3) and higher flight time projection accuracy (H4).

2.6   Procedure

Consenting participants were asked to complete a brief demographic questionnaire on their age, gender, visual acuity, and any experience with UAV control interfaces. Subsequently, participants were provided with an introduction to the simulated UAV mission and the control interface features. Detailed explanation was provided for all components of the control interface. Participants also received instruction on how to complete in-flight tasks and were made aware of the experiment procedure (including random occurrences of the embedded secondary tasks). Participants were subsequently allowed sufficient time to memorize icons representing friendly and hostile aircraft. They were also presented with a list of possible alarms and corresponding emergency control actions. The list highlighted logical associations of off-nominal events and vehicle subsystems that could be used to address the events. The participants were instructed not to memorize the list and they did not have access to the solutions after the training session. Following the introduction, participants performed a training trial to develop proficiency with the UAV control interface and in-flight tasks. The training trial consisted of one easy block and one hard block, with high LoA applied only to the hard block. The training scenario was designed to ensure that participants were familiar with the control interfaces and capable of controlling the UAV under the different the LoAs.

After the training session, participants were provided with a final chance to review the friendly and hostile aircraft icon features. Before each test trial, participants were shown a mission information summary. (The summary of steps was not accessible during test trials.) The mission information included a relevant map and scheme of (vehicle) maneuvers, as well as descriptions of required activities during a mission. The participants were instructed to perform a total of seven experiment trials. The first trial was a pseudo trial with only two difficulty blocks, similar to the training trial and intended to reinforce participant familiarity with the control interface and task demands. The data collected from the pseudo trial was not included in the response data analysis. The training and pseudo trials lasted about 10 min each and each testing trial lasted around 15 min. A 3-minute break was given between trials to control fatigue.

2.7   Data Analysis

Repeated-measures Analysis of Variance (ANOVA) was applied to all DVs. Diagnostics on the aircraft detection (DV2) and flight time projection (DV4) accuracies revealed normality and homoscedasticity violations of the assumptions of the ANOVA procedure. Consequently, ranks of these two responses were submitted to the ANOVA to yield a non-parametric analysis. To account for individual differences in the subjective workload ratings, the OW ratings (DV3) for each participant were transformed to Z-scores using the mean and standard deviation of responses to each test condition. A total of 108 observations (6 participants × 6 trials × 3 blocks) were obtained for all DVs. However, OW ratings of one participants were excluded from data analysis because the participant failed to follow experiment instructions on workload rating. Consequently, the analysis on DV3 only included 90 observations.

The statistical model for analysis included the LoA as a main effect and participant as a blocking factor. The test trial number was also included in preliminary analysis of

200 W. Zhang et al.

the data and removed from the model due to lack of significance. For all analyses, the significance level was set to $\alpha = 0.05$. If there was a significant effect of LoA on a response measure, Duncan's multiple range test (MRT) was used for post-hoc analysis in order to compare the different modes. The significance groupings are represented by letters (e.g., "A", "B") on graphs in the results section. Conditions carrying the same letter are not significantly different from each other.

3 Results and Discussion

The ANOVA results revealed no effect of the mode of automation on SA query accuracy ($F(2,100) = 1.39, p = 0.2529$). These results were not in line with expectation (H1) or prior findings (e.g., [7]). However, the small sample size of this preliminary study might have limited the sensitivity of the analysis, resulting in the lack of statistical significance. Another reason for the finding could be that the DFA was only applied to one type of the task (i.e., alarm handling) as part of the simulated mission. The other tasks and the need to maintain awareness of UAV subsystems, as well as the environment, might have been more challenging for participants. As a result, the SA query accuracy was not significantly influenced by the decision making assistance provided by the DFA.

Alternatively, the ANOVA results on the ranked aircraft detection and classification accuracy showed a significant effect of automation mode ($F(2,100) = 67.68$, $p < 0.001$). Duncan's MRT results (see Fig. 3) indicated that both low LoA and DFA conditions resulted in better SA than the high LoA condition. These results supported our second hypothesis (H2). The findings were similar to those of Parasuraman et al. [7], where the adaptive automation was applied to an information analysis task (i.e., target recognition) as part of a UAV/UGV control simulation. They found that SA was better and that workload was lower for both adaptive and static automation as compared to manual performance. By applying DFA to decision making tasks, the present study further supported benefits of DFA in terms of situation awareness.

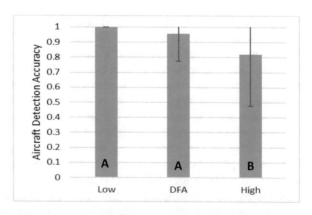

Fig. 3. Aircraft detection accuracy for different levels of automation

The ANOVA results on the z-transformed OW ratings revealed a significant effect of LoA ($F(2,83) = 3.43$, $p = 0.037$). Post-hoc analyses (see Fig. 4) showed that the high LoA resulted in lower OW ratings than both the LoA and DFA conditions. These results differed from the third hypothesis (H3). As mentioned earlier, the DFA was only applied to the alarm handling task. It is possible that the other tasks as part of the simulation posed greater cognitive demands on participants. It is possible that the more challenging embedded secondary tasks were major drivers of the perceived workload.

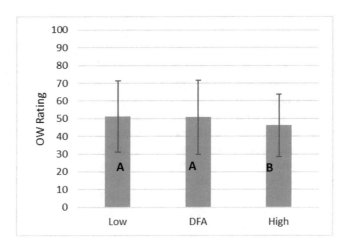

Fig. 4. Mean OW ratings for different automation modes

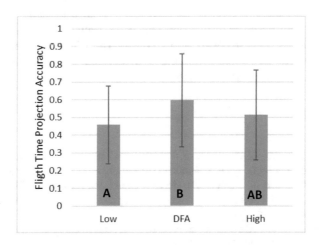

Fig. 5. Flight time projection accuracy for different levels of automation

Regarding flight time projection accuracy, the results showed a marginal effect of LoA ($F(2,100) = 2.93$, $p = 0.058$). Duncan's MRT results (see Fig. 5) showed that the

DFA produced higher accuracy than the low LoA condition, indicating lower workload for the adaptive automation approach. Opposite to the OW rating results, these findings were consistent with our expectation (H4) and other previous studies [7, 8]. A Greenhouse-Geisser statistic also indicated that the marginal effect was trending for significance with a larger sample size.

4   Conclusion

Prior studies have reported benefits of applying DFAs to information analysis tasks in complex automated systems. In this study, we focused on the application of DFAs to a decision making task in simulated UAV operations. We conducted an experiment to assess the effects of LoAs on operator SA and workload. We compared DFAs with static high and low level automation. Six ROTC students from a local university participated in the study.

The preliminary analysis showed no significant difference of the three LoAs on operator accuracy in responding to SA queries. However, results on aircraft detection task accuracy indicated superior situation comprehension with DFAs as compared to static high LoA. Results on flight time projection task accuracy indicated that participants experienced lower workload when operating with DFAs as compared to static low LoA. Ratings of overall cognitive workload revealed a different pattern of results. The findings of this study could be used for guiding UAV control system automation design with DFA capabilities.

One limitation of this study was investigation of the effects of LoAs on SA, workload and task performances across an entire UAV mission. It would be interesting to compare DFA effects with static automations during 'easy to hard' and 'hard to easy' task difficulty transitions between task blocks within each mission. Future analysis could also compare DFAs with low and high level static automation under critical workload conditions (e.g., overload with hard tasks) and SA-critical conditions (e.g., reduced engagement under easy tasks). Our next steps are to extend the present study with a larger sample size and address these limitations in future research and analysis.

References

1. Monfort, S.S., Sibley, C.M., Coyne, J.T.: Using machine learning and real-time workload assessment in a high-fidelity UAV simulation environment. In: Next-Generation Analyst IV, vol. 9851, p. 98510B. International Society for Optics and Photonics (2016)
2. Wiener, E.L., Curry, R.E.: Flight-deck automation: promises and problems. Ergonomics **23** (10), 995–1011 (1980)
3. Endsley, M.R., Kiris, E.O.: The out-of-the-loop performance problem and level of control in automation. Hum. Factors **37**(2), 381–394 (1995)
4. Scerbo, M.W., Freeman, F.G., Mikulka, P.J.: A brain-based system for adaptive automation. Theor. Issues Ergon. Sci. **4**(1–2), 200–219 (2003)
5. Endsley, M.R., Kaber, D.B.: Level of automation effects on performance, SA and workload in a dynamic control task. Ergonomics **42**(3), 462–492 (1999)

6. Kaber, D.B., Endsley, M.R.: The effects of level of automation and adaptive automation on human performance, situation awareness and workload in a dynamic control task. Theor. Issues Ergon. Sci. **5**(2), 113–153 (2004)
7. Parasuraman, R., Cosenzo, K.A., De Visser, E.: Adaptive automation for human supervision of multiple uninhabited vehicles: effects on change detection, situation awareness, and mental workload. Mil. Psychol. **21**(2), 270 (2009)
8. Calhoun, G.L., Ward, V.B., Ruff, H.A.: Performance-based adaptive automation for supervisory control. In: Proceedings of the Human Factors and Ergonomics Society Annual Meeting, vol. 55, no. 1, pp. 2059–2063. SAGE Publications, Los Angeles (2011)
9. Kaber, D.B., Perry, C.M., Segall, N., McClernon, C.K., Prinzel III, L.J.: SA implications of adaptive automation for information processing in an air traffic control-related task. Int. J. Ind. Ergon. **36**(5), 447–462 (2006)
10. Afergan, D., Peck, E.M., Solovey, E.T., Jenkins, A., Hincks, S.W., Brown, E.T., Jacob, R.J.: Dynamic difficulty using brain metrics of workload. In: Proceedings of the 32nd Annual ACM Conference on Human Factors in Computing Systems, pp. 3797–3806. ACM (2014)
11. Roberts, D.M., Taylor, B.A., Barrow, J.H., Robertson, G., Buzzell, G., Sibley, C., Baldwin, C.L.: EEG spectral analysis of workload for a part-task UAV simulation. In: Proceedings of the Human Factors and Ergonomics Society Annual Meeting, vol. 54, no. 3, pp. 200–204. SAGE Publications, Sage CA (2010)
12. Murata, A., Iwase, H.: Evaluation of mental workload by fluctuation analysis of pupil area. In: Engineering in Medicine and Biology Society Proceedings of the 20th Annual International Conference of the IEEE. Vol. 6, pp. 3094–3097. IEEE (1998)
13. Osburn, W.J.: Levels of difficulty in long division. Elem. Sch. J. **46**(8), 441–447 (1946)
14. Wickens, C.D., Gordon, S.E., Liu, Y., Lee, J.: An introduction to human factors engineering (1998)
15. Endsley, M.R.: Measurement of situation awareness in dynamic systems. Hum. Factors **37** (1), 65–84 (1995)
16. Vidulich, M.A., Tsang, P.S.: Absolute magnitude estimation and relative judgement approaches to subjective workload assessment. In: Proceedings of the Human Factors and Ergonomics Society Annual Meeting, vol. 31, no. 9, pp. 1057–1061. SAGE Publications, Los Angeles (1987)

Investigating the Large-Scale Effects of Human Driving Behavior on Vehicular Traffic Flow

Manuel Lindorfer[1]([✉]), Christian Backfrieder[1],
Christoph F. Mecklenbräuker[2], and Gerald Ostermayer[1]

[1] Research Group Networks and Mobility,
FH Upper Austria, Hagenberg, Austria
{manuel.lindorfer, christian.backfrieder,
gerald.ostermayer}@fh-ooe.at
[2] Institute of Telecommunications, TU Wien, Vienna, Austria
cfm@nt.tuwien.ac.at

Abstract. In recent years, understanding and modeling the human driver in the scope of traffic simulations has received considerable attention. With the advent and the ongoing development of new technologies in the field of Intelligent Transportation Systems, we are consequently moving towards an era where a majority of driving-related tasks will presumably be carried out by autonomous systems rather than humans. Notwithstanding, the transition from today's conventional traffic to tomorrow's highly automated traffic will not take place overnight. Up to that point, the available transportation infrastructure will most likely be shared among both human-driven and (partially) automated vehicles. Considering such scenarios of mixed traffic is therefore inevitable when developing new concepts and applications for the use in ITS, and requires a proper modeling of the human driver for simulation purposes. Although there have been diverse ways of integrating human factors with traffic simulation models, most existing studies focus on the impacts of human driving behavior in very constrained scenarios such as isolated platoons or bottleneck situations rather than on their large-scale effects. In this paper, we address this particular issue by performing large-scale simulations to investigate the impacts of human behavior on vehicular traffic flow under varying traffic conditions. We show how specific factors such as delayed reaction, distracted or anticipatory driving affect traffic efficiency and safety in terms of travel time, fuel consumption and accident frequency.

Keywords: Human factors · Driver behavior modelling · Traffic simulation

1 Introduction

Beyond doubt, the individual behavior of human drivers has a significant influence on vehicular traffic flow and affects both traffic efficiency and safety considerably. It is, therefore, no surprise that human factors are an extensively studied topic in the scientific community. This holds also true for the field of traffic simulation, which has gained increased attention in recent years, playing a crucial role in the development of technologies and applications designed for Intelligent Transportation Systems (ITS). In that regard, simulations are a widespread and frequently used method to model

© Springer International Publishing AG, part of Springer Nature 2019
D. N. Cassenti (Ed.): AHFE 2018, AISC 780, pp. 204–215, 2019.
https://doi.org/10.1007/978-3-319-94223-0_19

complex transportation networks and to investigate scenarios that cannot by studied in a real experiment or by other analytical methods. In the last decades, a vast number of models has been developed for that purpose to describe the physical propagation of traffic flows under various conditions [1–6]. Although these models have successfully been used to reproduce collective and self-organized traffic dynamics such as traffic breakdowns or the propagation of stop-and-go waves [7–9], many of these models lack the ability of describing human driving behavior adequately. A proper modeling of the human driver, however, is inevitable in order to allow for the realistic simulation of vehicular traffic, especially when considering scenarios of mixed traffic, where both human-driven and automated or even cooperative vehicles share the same infrastructure. For that reason, considerable efforts have been put into integrating human factors such as finite reaction times, perceptual limitations or driver distractions with – primarily microscopic – traffic models lately [10–15]. Although these studies provide valuable insights on how human factors affect both traffic flow and safety in either a positive or a negative way, a majority of the studies available in literature focuses on analyzing their impact in very constrained and simplified scenarios such as isolated platoons or bottleneck situations, while their large-scale effects remain disregarded.

In this paper, we address this particular issue. To be more specific, we study the large-scale impacts of human driving behavior on vehicular traffic flow by means of traffic simulations. We use the Extended Human Driver Model (EHDM) [15] and the microscopic traffic simulator TraffSim [16] to analyze how (i) the drivers' delayed reaction, (ii) anticipatory and (iii) distracted driving affect traffic flow and safety under consideration of different traffic densities. The results comprise a quantification of several key figures such as travel times, fuel consumption and accident frequency.

The remainder of this paper is organized as follows. In Sect. 2 related work in the area of driver behavior modeling and studies focusing on the impacts of human factors on vehicular traffic are elaborated. Section 3 gives a short overview of the simulation environment, the scenario used for investigation and applied models and parameters, followed by the presentation and discussion of simulation results in Sect. 4. Finally, Sect. 5 concludes the paper and provides an outlook on planned future work.

2   Related Work

The nature of human driving behavior is an extensively studied, yet controversial topic in traffic science. In the last decades, various aspects of human driving behavior and their impacts on traffic flow and safety have been investigated. In the following, we give a brief overview of what we consider to be the most prevalent factors characterizing the human driver and notable efforts to integrate these factors with (microscopic) traffic models.

An essential feature of human driving is a significant reaction time, resulting from the physiological aspects of perception, recognition and decision [17]. The resulting reaction time is naturally of the order of one second, but varies strongly between different drivers, situations and studies [18]. As reaction times were found to be a prime contributing factor to traffic instabilities, they are an essential element in many traffic models [2, 4, 5, 10, 19–21].

Another topic which has prompted considerable attention in recent years is distracted driving, not least since driver preoccupation with electronic devices is becoming increasingly common [22]. Distractions pose a severe safety risk on the road, contributing to large number of police-reported crashes [23], and generally result in a deterioration of driving performance, including an increase of the driver's reaction time or impacts on vision and steering behavior [24]. Despite their ascertained impact on road safety, distractions have barely been considered in existing traffic models so far, with the exception of [13, 15, 25].

Other characteristics of human driving behavior are driving errors and imperfect estimation capabilities, resulting from the perceptual limitation of human vision. A consequence thereof is that drivers are neither able to perceive small changes in stimuli nor to accurately estimate e.g. the spacing to or the velocity of neighboring vehicles [26]. Notable efforts to model such perceptual limitations and imperfect driving have been made by [3, 10, 14, 27].

Although being governed by a series of disadvantageous characteristics, human drivers are capable of driving safely in dense traffic or hazardous situations, even if the distance to the preceding vehicle is far below the average reaction time [12]. This suggests that human drivers achieve additional safety by considering not just the immediate vehicle in front, but also further vehicles ahead. Moreover, it was found that drivers are able to anticipate upcoming traffic situations adequately, especially while upon familiar situations [28]. These spatial and temporal anticipation capabilities have been considered in a number of traffic models, including [10, 15, 29, 30].

While all these contributions provide valuable insights on how particular characteristics of human driving behavior affect traffic flow and traffic safety in either a positive or a negative way, a majority of the presented studies has one limitation in common. In particular, they predominantly focus on investigating the effects of certain aspects of human driving in very constrained, but also idealized scenarios such as isolated platoons or bottleneck situations. Although these scenarios are quite useful to understand collective phenomena such as traffic stability or the propagation of stop-and-go-waves, they do not provide insights on how human factors affect traffic flow on a larger scale. Understanding these impacts, however, is vital in order to gain a better and more complete understanding of the nature of human driving behavior.

This paper tries to close this gap by investigating the large-scale effects of several human factors on both traffic efficiency and safety. To be more precise, it studies the impact of varying, situation-dependent reaction times, anticipatory and distracted driving on travel time, fuel consumption and accident frequency. All evaluations are proven for a large-scale scenario under consideration of different traffic densities with the aid of a microscopic traffic simulator [16].

3   Simulation Environment

For the simulations carried out in the scope of this work we use the microscopic traffic simulator TraffSim [16]. It allows for the time-discrete and state-continuous simulation of vehicular traffic and supports a wide range of configurable models and parameters, comprising different car-following and lane-change models as well as a physics-based fuel consumption model [31].

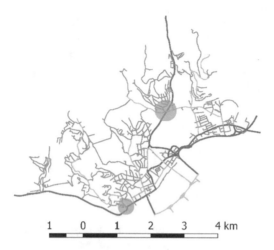

1   0   1   2   3   4 km

Fig. 1. Scenario used for simulation: the northern part of Linz, Austria. Differently colored segments represent different road types, e.g. freeways (*red*), expressways (*yellow*) and low-level roads (*gray*). *Highlighted areas* denote bottlenecks in the network causing a significant breakdown of traffic flow in case of a sufficiently high traffic density.

3.1   Scenario

The scenario used to investigate the effects of human driving behavior on vehicular traffic flow (see Fig. 1) comprises a real-world setup and represents an extract of the city of Linz, Austria. A similar scenario has also been used in previous studies [32, 33]. For our investigations we consider only unregulated intersections [34] (right-before-left rule) and single-lane traffic in order to mask out the influence of complex intersection regulation and lane-change logic. For all simulations we consider a setup with 4000 vehicles, which are distributed uniformly among all dangling road segments in the network.

In order to investigate the impacts of human driving behavior under different traffic conditions, we vary the vehicle density by changing the inter-arrival interval (IAI) between two consecutive vehicles arriving in the simulation. In particular, the arrival intervals are varied from 500 ms (dense traffic) to 2000 ms (free flow). Moreover, the start time of every single vehicle is changed randomly from one simulation run to the other in order to generate statistically reliable results.

3.2   Traffic Models

All simulations are carried out using a number of microscopic traffic models. For modeling car-following behavior we use the EHDM [15], an extension to the Human Driver Model (HDM) [10, 12] which allows to integrate numerous characteristics attributable to the human driver into time-continuous traffic models, namely (i) varying, situation-dependent reaction times, (ii) driver distractions and (iii) driving errors. The HDM, in turn, introduces additional features such as imperfect estimation capabilities and anticipatory driving. For further information on the EHDM and the HDM as well as recommended parameter ranges we refer to [10, 12, 15], respectively. In this work, we use the Intelligent Driver Model (IDM) [6] as underlying model for the EHDM and apply the same set of parameters for both models as in [15]. The fuel consumption of all vehicles is calculated using a physics-based fuel consumption model, as described in [31].

3.3   Modeling Driver Distractions

One aspect of this paper focuses on the large-scale effects of distracted driving on traffic flow. We use the Stochastic Distraction Model (SDM) [35] to integrate driver distractions with the simulations performed in the scope of this work. It uses stochastic processes to generate distraction profiles for all simulated vehicles, indicating whether the "driver" of a particular vehicle is distracted at a given point in time or not. The SDM is capable of simulating various distraction types simultaneously and thereby considers several parameters such as the drivers' exposure to specific types of distractive activities, their frequency of occurrence or their average duration. In this work, we use an identical parametrization as in [35], comprising 17 distinctive types of distraction. For a full list of distraction types and model parameters the authors refer to [23].

4   Results

Hereinafter, we present and discuss the findings obtained from simulating above-mentioned scenario under consideration of the simulation environment outlined in the previous section. In particular, we investigate the impact of (i) varying reaction times, (ii) anticipatory and (iii) distracted driving on vehicular traffic flow in terms of aggregated travel time, fuel consumption and accident frequency. Note that for all efficiency-related evaluations only simulations in which no collisions have taken place are considered. Moreover, all quantities outlined hereinafter represent mean values, whereby for each set of parameters at least 100 simulation runs were conducted. The corresponding standard deviation values are denoted by error bars.

4.1   Baseline

In order to quantify the impacts of human factors on both traffic efficiency and safety we have simulated the scenario shown in Fig. 1 using different vehicle inter-arrival intervals and without explicitly considering these factors, i.e. assuming an ideally attentive and instantaneously responding driver. Essentially, by neglecting the human factors incorporated by the EHDM, the behavior of all vehicles is solely determined by the dynamics of the underlying car-following model, i.e. by the IDM. The results obtained from these simulations serve as a reference for the forthcoming investigations. Note that the baseline scenario is entirely collision-free for all configurations.

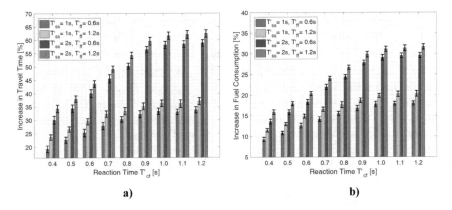

Fig. 2. Increase in travel time (*a*) and fuel consumption (*b*) as a function of varying reaction times compared to the baseline scenario under consideration of an inter-arrival interval of 600 ms, i.e. dense traffic.

4.2   Varying Reaction Times

In contrast to most existing traffic models, which presume reaction time to be an invariable parameter, the EHDM introduces the concept of varying, situation-dependent reaction times [15]. More precisely, the EHDM distinguishes between different driving regimes (car-following, free-flow, standstill), whereby the actual reaction time (denoted as T'_{cf}, T'_{ff} and T'_{ss}) depends on the regime the subject vehicle currently belongs to. In the following, we discuss how these varying reaction times affect traffic efficiency in terms of travel time and fuel consumption.

The results are depicted in Fig. 2, which outlines both aggregated travel time and fuel consumption as a function of reaction time and an inter-arrival interval of 600 ms, i.e. dense traffic. It can be seen that with an increase in reaction time, both quantities deteriorate likewise compared to the baseline scenario, even for moderate reaction times. Fundamentally, this degradation can be attributed to traffic instabilities caused by the drivers' delayed reaction [10, 12], which eventually lead to a breakdown of traffic flow and the emergence of stop-and-go waves, especially under dense traffic conditions. It should be noted that once the reaction time T'_{cf} exceeds a certain threshold (around 0.9 s), almost no significant degradation in both travel time and fuel consumption can be observed anymore. This complies with the findings in [12, 15], who showed that traffic instabilities arise or amplify in a platoon of vehicles only if the drivers' reaction time is sufficiently large. Obviously, once such instabilities or stop-and-go waves have emerged, travel time and fuel consumption depend primarily on how fast the resulting jam situation dissolves. This is to a large extent determined by the reaction time in the standstill regime T'_{ss}, which is applied when a vehicle has come to a complete halt, e.g. in front of an intersection or in stop-and-go traffic. Apparently, the lower T'_{ss}, the faster the dissipation of congestion, and vice versa.

As outlined by Fig. 2, travel time and fuel consumption are affected considerably under dense traffic conditions. The effects of a delayed reaction, however, become less influential with lower traffic densities. In such situations, less vehicles are subject to traffic congestion and stop-and-go traffic, thus the impact of T'_{ss} becomes negligible, i.e. the increase in both travel time and fuel consumption depends only on T'_{cf} and T'_{ff}. For a setup with an inter-arrival interval of 2000 ms, for example, the increase in travel time and fuel consumption ranges from 0.25% up to 0.8% and 1.5% up to 4.8% (depending on the individual reaction times), respectively. Thereby, the larger degradation in fuel consumption can be attributed to considerably larger fluctuations in the vehicles' acceleration caused by the drivers' delayed reaction.

Evidently, a delayed reaction also has an influence on traffic safety, as shown in previous studies [10, 12, 15, 25]. For the given scenario collisions can be observed when increasing T'_{cf} to values greater than 1.2 s. Please note that these configurations have not been considered for the results discussed in this section.

4.3 Anticipatory Driving

Remarkably, human drivers are able to navigate safely through dense traffic at relatively high speeds, even if the time gaps to the preceding vehicles are far below the drivers' reaction time [12]. One explanation therefor is the ability of drivers to achieve additional safety by anticipating on the surrounding traffic conditions. The HDM considers these stabilizing effects by integrating both a spatial and a temporal anticipation mechanism. The former enables drivers to consider more than just a single leading vehicle, i.e. to look also further vehicles ahead, while the latter takes the driver's ability to predict a future traffic situation accordingly over a short period of time into account [12]. Both stability mechanisms have been found to improve the collective stability of a vehicle platoon significantly [10].

Hereinafter, we present the findings obtained from simulating our reference scenario under consideration of varying reaction times and anticipative capabilities. Figure 3 shows the improvement in both travel time and fuel consumption compared to non-anticipatory driving as a function of varying traffic densities (Fig. 3a) and reaction times (Fig. 3b). It can easily be seen that considering anticipative capabilities leads to a significant reduction in both quantities, especially in situations with dense traffic and large reaction times. On the other hand, there is obviously less potential for improvement assuming lower reaction times, as it is the case for low traffic densities, where traffic can mostly flow freely. In fact, the influence of anticipatory driving is most striking in situations where the adverse effects of a delayed reaction are most significant. The differently shaped curves in Fig. 3b can be explained as follows. While in case of lower traffic densities vehicles are able to travel mostly freely through the network, a large number of vehicles ends up in congestion (e.g. in front of intersections) anyway for setups with high traffic density, simply because the network is overcrowded. The setups with 800 ms and 900 ms constitute an intermediate case. Here, traffic jams start to grow substantially only for larger reaction times, while heavy congestion can largely be avoided for smaller ones. Apparently, anticipatory driving can improve the former situation to a larger extent than the latter one, which results in a stronger dependency on the reaction time T'_{cf}.

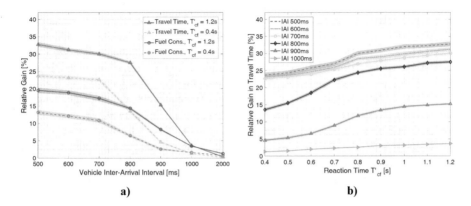

Fig. 3. Impact of anticipation on travel time and fuel consumption compared to non-anticipatory driving. In both cases we set the reaction times $T'_{ff} = 1.2$ s and $T'_{ss} = 2$ s, respectively. The transparent overlays in both figures denote standard deviation values.

Remarkably, one can obtain from Figs. 2 and 3 that anticipatory driving can compensate to a substantial extent for the negative effects of finite reaction times, which has also been hypothesized in [12, 25]. The reason therefore is that both temporal and spatial anticipation increase traffic stability significantly [10, 15], and thus avoid or at least reduce the number of instabilities and stop-and-go waves caused by the drivers' delayed reaction. Moreover, the consideration of more than just a single leading vehicle allows for a faster dissipation of jam situations, e.g. in front of an intersection, while at the same time providing smoother accelerations. Interestingly, by varying the spatial anticipation parameter n_a, i.e. the number of vehicles a driver looks ahead, we found that a substantial reduction in both travel time and fuel consumption can already be achieved when considering just one additional leader, while looking further vehicles ahead does not change the results qualitatively anymore.

4.4   Distracted Driving

It has been shown in various studies that distracted driving affects both traffic safety and efficiency considerably. In [15] it was found that distractions impair the collective stability of a vehicle platoon significantly, and eventually lead to rear-end collisions, even for relatively small reaction times. The EHDM distinguishes between two types of distraction, namely *minor* and *severe* distractions. The first type comprises cognitive and auditory distractions, which are modeled by a temporary increase of the driver's reaction time by a factor λ_r and a reduction of the vehicle's target speed by a factor λ_v to take into account a commonly observed compensatory behavior of distracted drivers

[36, 37]. The second type corresponds to situations where the driver looks at anything but the road ahead, e.g. when the driver's attention is attracted by outside objects or persons. In case of the latter, it is assumed that the driver is not able to react to changes in the traffic surrounding conditions anymore.

Using the SDM [35], we investigated the impacts of both types of distraction on vehicular traffic flow. In the first place, we explored how cognitive and auditory distractions affect traffic efficiency in terms of travel times and fuel consumption. Figure 4a shows the exemplary results of this evaluation for a setup with high traffic density, compared to an identical setup where distractions remain disregarded. It can be obtained that distracted driving negatively affects both travel time and fuel consumption, whereas this degradation compare to non-distracted driving can solely be attributed to the drivers' increased reaction time. Eventually, this increase in reaction time may also lead to the emergence of rear-end collisions, if only for large reaction times. In particular, we observed the (casual) occurrence of crashes only in scenarios with a baseline reaction time greater than 1.1 s. Considering the drivers' compensatory behavior, i.e. reducing the speed for the duration of the diverting task, does not change the results noticeable.

The second investigation focused on quantifying the impact of severe distractions on traffic safety. Therefore, we define the *accident frequency rate* as the number of collisions observed per 100 vehicle kilometers[1]. Our findings indicate that severe distractions are a prime contributing factor to rear-end collisions, which complies with empirical observations [23]. Note that collisions are caused either directly, i.e. by the distracted vehicle crashing into the preceding one, or indirectly as a result of the shock

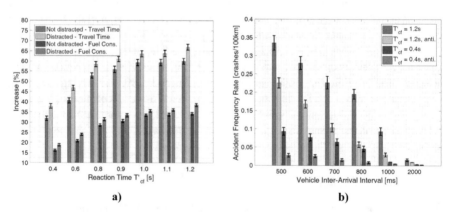

Fig. 4. Impact of distracted driving on travel time and fuel consumption (*a*) and accident frequency rate (*b*) compared to the baseline scenario. The response time increase and speed reduction factors λ_r and λ_v were set to 30% and 6%, respectively. (*a*) shows the results for an inter-arrival interval of 500 ms.

---

[1] For the reference scenario the total number of vehicle kilometers is 16789.48 km.

wave caused by the delayed response of subsequent vehicles. The number of collisions observed during a simulation thereby strongly depends on the drivers' reaction time, their anticipative capabilities and traffic density. Figure 4b shows these relations under consideration of different traffic densities and reaction times. As one might expect, anticipatory driving reduces the accident frequency rate markedly, whereby the greatest improvement can be achieved in situations where collisions are caused by the late reaction of vehicles following the distracted one. On the other hand, the stabilizing effect of anticipatory driving [10, 12] does not pay off in situations where drivers are not able to react quickly enough to changes in traffic conditions in order to avoid a rear-end collision after being distracted severely for a considerable amount of time, as shown in [15].

5   Conclusion and Future Work

In this paper, we studied, by means of simulation, the large-scale impacts of human driving behavior on vehicular traffic flow. In particular, we investigated how the drivers' delayed reaction, distracted and anticipatory driving contribute to both traffic efficiency and safety. The results comprise a precise quantification of these effects in terms of travel times, fuel consumption and accident frequency. All simulations have been carried with the aid of a large-scale scenario using the EHDM [15] and the microscopic simulator TraffSim [16].

Our results indicate that certain aspects of human driving behavior such as a delayed reaction and distracted driving do not only degrade the stability of a vehicle platoon [12, 15], but also impair traffic efficiency and safety considerably. The same holds true for stabilizing mechanisms such as the drivers' ability to anticipate future traffic conditions adequately and to consider more than a single leading vehicle. These anticipative capabilities were found to reduce both travel time and fuel consumption significantly, but also to have a positive impact on accident frequency. To what extent human driving behavior affects vehicular traffic flow – be it in a positive or a negative way –, however, strongly depends on the surrounding traffic conditions, i.e. the traffic density. Our findings provide evidence that human factors do not only play a crucial role when studying collective phenomena such as the stability of vehicle platoons or the propagation of stop-and-go waves, but also affect traffic efficiency and safety on a larger scale. Furthermore, our results indicate that anticipatory driving can compensate (to a substantial extent) for the negative effects of both a delayed reaction and distracted driving, which coincides with the hypotheses stated in [10, 12, 25]. It should be mentioned that, in this work, we focused on outlining the large-scale impacts of human driving behavior on vehicular traffic flow assuming identical driver-vehicle units. Therefore, future work could contain investigations considering both driver heterogeneity as well as mixed traffic scenarios consisting of both human-driven and (partially) automated vehicles in different compositions, i.e. assuming varying penetration rates of vehicles with a higher level of automation. Future work might also comprise to investigate the stochastic model components of the HDM and the EHDM, namely imperfect estimation capabilities and driving errors.

Acknowledgments. This project has been co-financed by the European Union using financial means of the European Regional Development Fund (EFRE). Further information to IWB/EFRE is available at www.efre.gv.at.

Europäische Union Investitionen in Wachstum & Beschäftigung. Österreich.

References

1. Pipes, L.A.: An operational analysis of traffic dynamics. J. Appl. Phys. **24**, 274–281 (1953)
2. Newell, G.F.: Nonlinear effects in the dynamics of car-following. Oper. Res. **9**, 209–229 (1961)
3. Wiedemann, R.: Simulation des Straßenverkehrsflusses. Institute for Traffic Engineering, University of Karlsruhe (1974)
4. Gipps, P.: A behavioural car-following model for computer simulation. Transp. Res. Part B: Methodol. **15**, 101–111 (1981)
5. Bando, M., Hasebe, K., Nakayama, A., Shibata, A., Sugiyama, Y.: Dynamical model of traffic congestion and numerical simulation. Phys. Rev. E **51**, 1035–1042 (1995)
6. Treiber, M., Hennecke, A., Helbing, D.: Congested traffic states in empirical observations and microscopic simulations. Phys. Rev. E **62**, 1805–1824 (2000)
7. Kerner, B.S., Rehborn, H.: Experimental features and characteristics of traffic jams. Phys. Rev. E **53**, R1297–R1300 (1996)
8. Cassidy, M.J., Bertini, R.L.: Some traffic features at freeway bottlenecks. Transp. Res. Part B: Methodol. **33**, 25–42 (1999)
9. Helbing, D.: Traffic and related self-driven many-particle systems. Rev. Mod. Phys. **73**, 1067 (2001)
10. Treiber, M., Kesting, A., Helbing, D.: Delay, inaccuracies and anticipation in microscopic traffic models. Phys. A **360**, 71–88 (2006)
11. Andersen, G.J., Sauer, C.W.: Optical information for car-following: the driving by visual angle (DVA) model. Hum. Factors **49**, 878–896 (2007)
12. Kesting, A.: Microscopic modeling of human and automated driving – towards adaptive cruise control. University of Technology Dresden (2008)
13. Yang, S., Peng, H.: Development of an errorable car-following driver model. Veh. Syst. Dyn. **48**, 751–773 (2009)
14. Jin, S., Wang, D.H., Huang, Z.Y., Tao, P.F.: Visual angle model for car-following theory. Phys. A **390**, 1931–1940 (2011)
15. Lindorfer, M., Mecklenbräuker, C.F., Ostermayer, G.: Modeling the imperfect driver: incorporating human factors in a microscopic traffic model. IEEE Trans. Intell. Transp. Syst. 1–15 (2017). https://doi.org/10.1109/TITS.2017.2765694. ISSN 1524-9050
16. Backfrieder, C., Ostermayer, G., Mecklenbräuker, C.F.: TraffSim – a traffic simulator for investigations of congestion minimization through dynamic vehicle rerouting. Int. J. Simul. Syst. Sci. Technol. **15**, 38–47 (2014)

17. Shiffrin, R.M., Schneider, W.: Controlled and automatic human information processing: II. Perceptual learning, automatic attending and a general theory. Psychol. Rev. **84**, 127–190 (1977)
18. Green, M.: How long does it take to stop? Methodological analysis of driver perception-brake times. Transp. Hum. Factors **2**, 195–216 (2000)
19. May, D.: Traffic Flow Fundamentals. Prentice Hall, Englewood Cliffs (1990)
20. Ahmed, K.: Modeling drivers acceleration and lane changing behavior. Massachusetts Institute of Technology (1999)
21. Davis, L.: Modifications of the optimal velocity traffic model to include delay due to driver reaction time. Phys. A **319**, 557–567 (2003)
22. Regan, M.A.: New technologies in cars: human factors and safety issues. Ergon. Aust. **8**, 6–15 (2004)
23. Stutts, J.: Distractions in everyday driving. Technical report, AAA Foundation for Traffic Safety (2003)
24. Dingus, T., Guo, F., Lee, S., Antin, J.F., Perez, M., Buchanan-King, M., Hankey, J.: Driver crash risk factors and prevalence evaluation using naturalistic driving data. Proc. Natl. Acad. Sci. **113**, 2636–2641 (2016)
25. van Lint, H., Calvert, S., Schakel, W., Wang, M., Verbraeck, A.: Exploring the effects of perception errors and anticipation strategies on traffic accidents - a simulation study. In: Cassenti, D.N. (ed.) Advances in Human Factors in Simulation and Modeling, pp. 249–261. Springer, Cham (2018)
26. Hamdar, S.: Driver behavior modeling. In: Handbook of Intelligent Vehicles, pp. 537–558. Springer, London (2012)
27. van Winsum, W.: The human element in car-following models. Transp. Res. Part F: Traffic Psychol. Behav. **2**, 207–211 (1999)
28. Tanida, K., Pöppel, E.: A hierarchical model of operational anticipation windows in driving an automobile. Cogn. Process. **7**, 275–287 (2006)
29. Lenz, H., Wagner, C., Sollacher, R.: Multi-anticipative car-following model. Eur. Phys. J. B – Condens. Matter Complex Syst. **7**, 331–335 (1998)
30. Eissfeldt, N., Wagner, P.: Effects of anticipatory driving in a traffic flow model. Eur. Phys. J. B – Condens. Matter Complex Syst. **33**, 121–129 (2003)
31. Treiber, M., Kesting, A.: Fuel consumption models. In: Traffic Flow Dynamics. Springer, Heidelberg (2013)
32. Backfrieder, C., Ostermayer, G., Mecklenbräuker, C.F.: Increased traffic flow through node-based bottleneck prediction and V2X communication. IEEE Trans. Intell. Transp. Syst. **18**, 349–363 (2017)
33. Backfrieder, C., Lindorfer, M., Mecklenbräuker, C.F., Ostermayer, G.: Impact of varying penetration rate of intelligent routing capabilities on vehicular traffic flow. In: 86th IEEE Vehicular Technology Conference (VTC-Fall) (2017, to be published)
34. Backfrieder, C., Ostermayer, G.: Modeling a continuous and accident-free intersection control for vehicular traffic in TraffSim. In: 8th European Modelling Symposium, pp. 333–337 (2014)
35. Lindorfer, M., Backfrieder, C., Mecklenbräuker, C.F., Ostermayer, G.: A stochastic driver distraction model for microscopic traffic simulations. In: 31st European Simulation and Modelling Conference (2017, to be published)
36. Liu, B.S., Lee, Y.H.: Effects of car-phone use and aggressive disposition during critical driving maneuvers. Transp. Res. Part F: Traffic Psychol. Behav. **8**, 369–382 (2005)
37. Cooper, J.M., Vladisavljevic, I., Medeiros-Ward, N., Martin, P.T., Strayer, D.L.: An investigation of driver distraction near the tipping point of traffic flow stability. Hum. Factors **51**, 261–268 (2009)

Agents in Space: Validating ABM-GIS Models

Kristoffer Wikstrom[2], Hal Nelson[1], and Zining Yang[2(✉)]

[1] Portland State University, Portland, OR, USA
hnelson@pdx.edu
[2] Claremont Graduate University, Claremont, CA, USA
{kristoffer.wikstrom, zining.yang}@cgu.edu

Abstract. The purpose of this paper is to spatially validate an agent-based predictive analytics model of energy siting policy in a techno-social space. This allows us to simulate the multitude of human factors at each level (e.g. individual, county, region, and so on). Energy infrastructure siting is a complex and contentious process that can have major impacts on citizens, communities, and society as a whole. Furthermore, the process is sensitive to varying degrees of human input, of differing complexity, at multiple levels. When it comes to validating ABMs, the virtual cornucopia of techniques can easily confuse the modeler. As useful as historical data validation is, it seems to be underutilized, most likely due to the fact that it is hard to find data suitable data for many models. For the purpose of In-Site, historical data availability is excellent due to Environmental Impact Assessments (EIA) providing us with citizen and community based organization (CBO) preferences, and regulatory decisions being public. For the model, citizen and CBO preferences were decided by coding comments on the EIA procedure so as to allow for quantitative analysis, and then geocoding the locations of the commenters. The end results of this is that, we can literally overlay our simulation results with the actual, real world, results of the historical project. This will allow for a high degree of confidence in the validation procedure, as well as the ability to deal with the complexity of the networks of human interactions.

Keywords: Agent-based model · GIS · Energy infrastructure siting
Community based organization · Validation

1 Introduction

Energy infrastructure siting is a complex and contentious process that can have major impacts on citizens, communities, and society as a whole. The International Energy Association (2015) estimates that $15 trillion in investments in renewable energy and transmission and distribution infrastructure is required in order to meet the Paris Accord agreements. Even without the requirements of this agreement, energy infrastructure is continuously built, updated and expanded to meet our need for energy. On a similar note, a 2010 World Health Organization report estimated that the world's urban

This research was supported by grants from the Haynes Foundation and the National Science Foundation (NSF award #1737191).

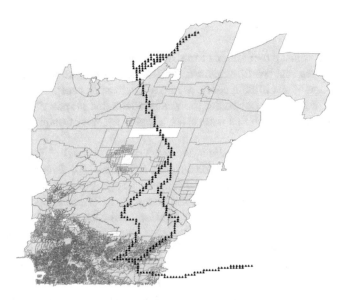

Fig. 1. Route of the power line.

population will increase by almost 3 billion people between 2009 and 2050 [1]. Such a migration would put heavy strain on energy infrastructure and create a need for expansion and consolidation of said infrastructure (Fig. 1).

These two trends illustrate the need for decision support tools to help facilitate the energy transition. Decisions support tools integrate Geographical Information Systems (GIS), a spatial modeling tool, with agent-based modelling, as humans and agents make decisions that are conditioned by their surrounding environment [2]. That is, both humans and agents move in space, make decisions dependent on the location of other agents in the space, as well as adapt to, and sometimes change, the resources and features in the landscape [3].

The purpose of this paper is to spatially validate an agent-based predictive analytics model of energy siting policy in a techno-social space. Unlike methods such as system dynamics and Bayesian networks, which are often of a more descriptive and correlational nature, agent-based modelling attempts to identify the cause of a particular phenomenon [4]. However, this also means that verification and validation becomes more difficult, as our prior uncertainty of what to expect makes it harder to establish whether what we are seeing is "[…] a legitimate result of the assumptions embedded in the model or, on the contrary, it is due to an error or an artefact created in the model design, its implementation, or its execution." [5]

The model results are validated against a historical project, the Tehachapi Renewable Transmission Project (TRTP), in the greater Los Angeles area. The project application was submitted to the California Public Utilities Commission (CPUC) in the Fall of 2004 and the project was finished in December of 2016 [6]. This project saw significant public opposition, especially in the city of Chino Hills, which even saw Southern California Edison have to tear down parts of already constructed power line towers and instead underground the power cables.

2   Literature Review

Validation, in its most parsimonious form, can be defined as assessing if the simulation outcomes matches real worth outcomes. The "model is considered valid for a set of experimental conditions if the model's accuracy is within its acceptable range, which is the amount of accuracy required for the model's intended purpose." [7] As such, we chose a validation approach that extensively employ the use of empirical data, as well as historical knowledge.

When it comes to spatial validation, validating the model differed from typical urban sprawl and land-use change models (such as Pontius [10]; Brown et al. [8]). Such models often focus on how the land changes through the simulation or modeling process and is thusly concerned with the accuracy of such changes. For instance, Brown et al. illustrates what they call "tension" between "two distinct notions of accuracy in land-use models [...] predictive accuracy and process accuracy." [8] To reconcile these two different notions they introduce the concept of "invariant vs variant regions", where the former is path independent and the latter is path-dependent [9].

Other efforts of tracing land changes, can be seen in Pontius, which traces land-use change in Massachusetts, with the outlined goal of comparing the similarity of two maps of land-use change [10]. Pontius which uses the example of deforestation in Costa Rica, further looks into these map comparisons by examining multiple resolution maps [11].

In the case of our model the land, or GIS layer, does not change over time (though as previously mentioned, the time period is usually not longer than the typical Environmental Impact Assessment (EIA) process of 20 months), but rather the focus is on the citizen agents and their actions. The similarities in spatial validation, however, are almost identical as the goal is to validate the spatial distribution of opponent citizens and community-based organizations (CBOs) across a particular landscape in Southern California (and, in so doing, the opposite, i.e. proponent citizens and CBOs, will naturally also be identified).

In terms of validating the output maps there are certain pitfalls to avoid. For instance, when it comes to statistically comparing how similar two maps are, statistical tests may not take into account patterns in the data. For instance, a normal chai-square test may be able to tell us how good the fit is between two maps, but it will not take into consideration the spatial pattern of the variables of interest [12]. This leads us to examine other measures that may be more robust in terms of validating them model, such as Moran's I and Geary scores. These measures also allow us to embrace a more non-binary view of validation, which is advocated by Crooks and Heppenstall, as well as Law and Kelton [25], where validity can be considered on a gradual scale, rather than as black and white [13].

An often overlooked issue with validation pertains to the changing of definitions of, for example, categories and objects. Using the example of cities, Batty and Torrens illustrates how validation becomes harder when the "system of interest" changes over time [14]. Fortunately, the short time-frame used in the model, as well as at the focus on easily defined structures, relationships, and agents (e.g. HVTL, CBOs, and citizens), means that this is less important in this validation exercise.

3 Verification

To test the coding, we ran parameter sweep simulations with individual parameter, i.e. changing one parameter per batch of runs, set to extreme values to see if the model returned expected values. For instance, by setting alternating parameters to 0, it is possible to confirm that the output indeed reflects this and returns null values where applicable.

We also performed unit-tests on the intermediate outputs of the ABM-GIS model. These were examined to ensure that the coding faithfully reflected the goals of each module.

4 Validation

When it comes to validating ABMs, there is a breadth of techniques available for the modeler [15] As such, in validating SEMPRO a clear preference was given to robust quantitative techniques, though proven qualitative methods, such as face validation, were also incorporated.

The most rigorous validation technique for an ABM-GIS model is that of using historical data to validate the model. As useful as historical data validation is it seems to be underutilized, most likely due to the fact that it is hard to find data suitable data for many models [4]. For the purpose of SEMPRO, historical data availability is excellent as the EIA documents provided us with citizen and community based organization (CBO) preferences, and regulatory decisions. For the model, citizen and CBO preferences were quantified by coding comments on the EIA procedure so as to allow for quantitative analysis, and then geocoding the locations of the commenters. The end results of this is that, we can literally overlay our simulation results with the actual, real world, results of the historical project.

These data resources lead us to choose a method of empirical ABM validation, often encountered in the economics literature, namely the History-Friendly approach [16, 17]. In transitioning this approach from an economics focus, we make slight changes where appropriate. For instance, in economics the focus is on time-series data, whereas we, by the nature of our topic, focus on a cross-section in a single period of time.

The History-Friendly models "are formal models that aim to capture, in stylized form, qualitative and 'appreciative' theories about the mechanisms and factors affecting industry evolution, technological advance and institutional change put forth by empirical scholars of industrial economics, technological change, business organization and strategy, and other social scientists." [18] In this approach empirical data and historical knowledge is used to inform the process of building, calibrating and validating a model of choice [19]. The approach also specifies that the model should be calibrated first, and then validated [20].

Finally, building on a critique from Windrum et al. [19], we augment the History-Friendly approach's lack of emphasis on sensitivity-analysis, by running a quasi-global sensitivity analysis on all input parameters in quintile steps.

In the case of SEMPRO, the model was calibrated to the TRTP project using U.S. Census data, for the area surrounding the proposed project. Historical knowledge regarding the project, such as CBO formation was also incorporated into model at this stage.

The GIS data containing census group data, park borders, as well as the TRTP power line data was loaded into NetLogo using various GIS shapefiles. Once into NetLogo, the coordinates were transformed, as well as manually reshaped to better fit the NetLogo display. This made it easier to verify and validate the model, as it enabled a visual understanding of the model's execution. This, however, created some problems with the model's output, as the coordinates were quite distorted. To ameliorate this situation, we transformed the NetLogo coordinates using MATLAB's Mapping Toolbox, a not entirely uncomplicated process that included resetting the projection, the geoid, and interpolating the citizen-agent locations. Having finalized the conversion, the new coordinates could be merged into the master data set, setting up the final step of the spatial validation.

In the penultimate step of spatially validating the model, we used ArcGIS to plot the citizen-agents, as well as the real citizen comments (EIA and scoping meetings). This involved conforming coordinate notations and creating shapefiles for use in GeoDa, our final step.

GeoDa was used to examine the spatial aspects of our output data, as well as our real world data. It allowed us to perform exploratory spatial analysis on our data, both simulated and historical, and create the choropleth maps in the results section.

5   Results

The choropleth maps below show and contrast our results with that of the real, historical comments. Using the count of citizen-agents and citizens, respectively, these maps display the percentiles for our area in question. There are 5960 blocks in the area of the project (seen below in the maps). Out of these 5960 we would expect a vast majority to be empty, as the literature suggests that proximity to the project is a significant factor in citizens' opposition to projects like the TRTP [21].

As such, we used the mean cutoff for citizen-agents' proximity to the project. The results can be seen in Map 2 and 3 that show the historical and simulated citizen comments respectively (Fig. 2).

Fig. 2. Historical Citizen Comment Addresses.

The addresses of the citizens that submitted comments on the TRTP show the main concentration along the powerline as it turned East into Chino Hills and Ontario California. There were a few comments along the north-south route, as well as engaged citizens along the northern coast.

The simulated citizen comments show a very similar pattern. The SEMPro model shows citizens activism in larger polygons in sparsely populated Kern County compared to historical data. In populated LA county, the predicted citizen locations are less concentrated directly adjacent to the power line, but exhibit similar locations otherwise (Fig. 3).

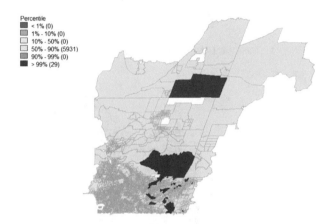

Fig. 3. Count of Citizen-Agent Locations, excluding agents below the mean proximity distance.

To further analyze the spatial relationships, we turn to the Local Moran's I maps below. The maps were created using queen contiguity rules for neighbors. The Local Moran's I measure was used rather than the Local G statistics, because the latter does not handle spatial outliers, as well [22]. The maps display spatial clusters (listed in the legend as "High-High" or "Low-Low") and spatial outliers (listed in the legend as "High-Low" and "Low-High"), with the former being positive spatial autocorrelation and the latter being spatial negative autocorrelation [22].

Figure 4 display a high degree of similarity in terms of spatial clustering, as evidenced by the High-High (red) areas for the historical citizen comments. The historical data also show significant spatial outliers where the jurisdiction has a high comment count with low comment counts in adjacent areas (pink, light blue). An example of this is the pink block group along the coast with high citizen comment count in this polygon with limited comments in adjacent polygons.

Turning the simulated citizen comments in Map 5, we see the High-High (red) in the same general areas in populated areas next to the power line. There are also similar counts of spatial outliers (Low-High-213, High-Low-8) as in the historical data (Fig. 5).

The differing sizes of the Census blocks can be deceiving to validation process. For instance, the upper half of the map has roughly 40 polygons, whereas the bottom half has over 5,000 polygons. This leads to the effect that the northernmost outliers look

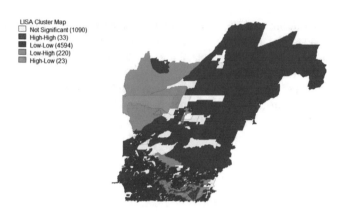

Fig. 4. Moran's I Map of Historical Citizen Comments

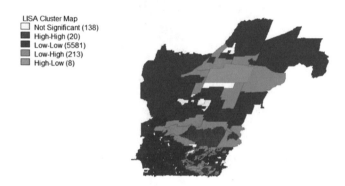

Fig. 5. Moran's I Map of Simulated Citizen Comments

disproportionately more influential than they are, for both maps. A similar issue is the sheer number of Census blocks (5960) to the number of Citizens (between 100–300, depending on the run).

6   Conclusion

The results were encouraging, but more work needs to be done. For one, further validation models should be used. The History-Friendly approach is not without its hazards, with some authors cautioning that the use of single-case historical data may not be applicable to wider industries, thus "[…] restricts the universality of the model." [23] An example of a further validation strategy to test would be the Werker-Brenner approach [24]. Furthermore, following the literature on proximity, reducing the target area of the ABM may be a worthwhile consideration.

Finally, the model needs to be tested against more historical cases. In California alone, there are at least a dozen different projects, spanning such areas as High-Voltage Power Lines, LNG terminals and windfarms. This would enable us to fully assess the generalizability of our results, and thus increase the confidence in the model. Again, due to the EIA process in California this data should be readily available, even if it at times may take quite a bit of work to get into analyzable shape.

The real world data shows the importance of proximity to a proposed infrastructure project, and the validation process confirmed that the SEMPRO model does indeed capture this relationship.

References

1. World Health Organization: Hidden Cities: unmasking and overcoming health inequities in urban settings. The WHO Centre for Health Development, Kobe, Japan, Chap. 1, p. 4 (2010)
2. Nelson, H., Cain, N., Yang, Z.: All politics are spatial: integrating an agent-based decision support model with spatially explicit landscape data. In: Campbell, H., et al. (eds.) Rethinking Environmental Justice in Sustainable Cities, pp. 168–189. Routledge Press, Abingdon (2015)
3. Johnston, K.M.: Agent Analyst. ESRI Press, Redlands (2013)
4. Duong, D.: Verification, validation, and accreditation (VV&A) of social simulations (2010)
5. Galán, J.M., Izquierdo, L.R., Izquierdo, S.S., Santos, J.I., del Olmo, R., López-Paredes, A., Edmonds, B.: Errors and artefacts in agent-based modelling. J. Artif. Soc. Soc. Simul. **12**(1), 1 (2009). http://jasss.soc.surrey.ac.uk/12/1/1.html
6. Southern California Edison: Project Timeline. https://www.sce.com/wps/portal/home/about-us/reliability/upgrading-transmission/TRTP-4-11. Accessed 28 Feb 2018
7. Sargent, R.G.: Validation and verification of simulation models. In: Proceedings of the 2004 Simulation Conference, Winter, vol. 1. IEEE (2004)
8. Brown, D.G., Page, S., Riolo, R., Zellner, M., Rand, W.: Path dependence and the validation of agent-based spatial models of land use. Int. J. Geogr. Inf. Sci. **19**(2), 153–174 (2005). https://doi.org/10.1080/13658810410001713399
9. Brown, D.G., Page, S., Riolo, R., Zellner, M., Rand, W.: Path dependence and the validation of agent-based spatial models of land use. Int. J. Geogr. Inf. Sci. **19**(2), 153 (2005). https://doi.org/10.1080/13658810410001713399
10. Pontius, R.G.: Quantification error versus location error in comparison of categorical maps. Photogram. Eng. Remote Sens. **66**, 1011–1016 (2000)
11. Pontius, R.G.: Statistical methods to partition effects of quantity and location during comparison of categorical maps at multiple resolutions. Photogram. Eng. Remote Sens. **68**, 1041–1049 (2002)
12. Costanza, R.: Model goodness of fit: a multiple resolution procedure. Ecol. Model. **47**, 199–215 (1989)
13. Crooks, A.T., Heppenstall, A.J.: Introduction to agent-based modelling. In: Heppenstall, A. J., Crooks, A.T., See, L.M., Batty, M. (eds.) Agent-Based Models of Geographical Systems, pp. 85–105 (2012). Chap. 5
14. Batty, M., Torrens, P.M.: Modelling and prediction in a complex world. Futures **37**(7), 745–766 (2005)
15. Ngo, T.A., See, L.M.: Calibration and validation of agent-based models of land cover change. In: Heppenstall, A.J., Crooks, A.T., See, L.M., Batty, M. (eds.) Agent-Based Models of Geographical Systems, pp. 181–196 (2012)
16. Malerba, F., Orsenigo, L.: Innovation and market structure in the dynamics of the pharmaceutical industry and biotechnology: towards a history-friendly model. Ind. Corp. Change **11**(4), 667–703 (2002)

17. Malerba, F., Nelson, R., Orsenigo, L., Winter, S.: History-friendly' models of industry evolution: the computer industry. Ind. Corp. Change **8**(1), 3–40 (1999)
18. Malerba, F., Nelson, R., Orsenigo, L., Winter, S.: History-friendly' models of industry evolution: the computer industry. Ind. Corp. Change **8**(1), 3 (1999)
19. Windrum, P., Fagiolo, G., Moneta, A.: Empirical validation of agent-based models: alternatives and prospects. J. Artif. Soc. Soc. Simul. **10**(2), 8 (2007)
20. Windrum, P., Fagiolo, G., Moneta, A.: Empirical validation of agent-based models: alternatives and prospects. J. Artif. Soc. Soc. Simul. **10**(2), 12 (2007)
21. Abdollahian, M., Yang, Z., Nelson, H.: Techno-social energy infrastructure siting: sustainable energy modeling programming (SEMPro). J. Artif. Soc. Soc. Simul. **16**(3), 6 (2013)
22. Anselin, L., Syabri, I., Kho, Y.: GeoDa: an introduction to spatial data analysis. Geogr. Anal. **38**(1), 5–22 (2006)
23. Windrum, P., Fagiolo, G., Moneta, A.: Empirical validation of agent-based models: alternatives and prospects. J. Artif. Soc. Soc. Simul. **10**(2), 11 (2007)
24. Werker, C., Brenner, T.: Empirical Calibration of Simulation Models, Papers on Economics and Evolution # 0410. Max Planck Institute for Research into Economic Systems, Jena (2004)
25. Law, A.M., Kelton, W.D.: Simulation Modeling and Analysis. McGraw-Hill, New York (1991)

Influence of Indirect Vision and Virtual Reality Training Under Varying Manned/Unmanned Interfaces in a Complex Search-and-Shoot Simulation

Akash K. Rao[1]([⊠]), B. S. Pramod[2], Sushil Chandra[3], and Varun Dutt[1]

[1] Applied Cognitive Science Laboratory, Indian Institute of Technology Mandi,
Mandi, India
akashrao.sse@gmail.com
[2] National Institute of Technology Surathkal, Surathkal, Karnataka, India
15co234.pramod@nitk.edu.in
[3] Institute of Nuclear Medicine and Allied Sciences, DRDO, Delhi, India
sushil.inmas@gmail.com

Abstract. In the real-world, manned and unmanned vehicles may be used for a number of applications. Visual technologies like indirect visual display (IVD) and virtual reality (VR) have been used to train operators in both manned and unmanned environments. The main objective of this research was to evaluate the effectiveness of manned and unmanned interfaces in IVD and VR display designs. Using an underwater search-and-shoot scenario, we developed two variations in display designs (IVD and VR) and two variations in type of interface-based training (manned and unmanned). A total of 60 subjects participated in the experiment, where 30 subjects were randomly assigned to simulations in IVD and the rest in VR. In both the simulations, 15 randomly selected participants executed the manned interface first and the remaining 15 executed the unmanned interface first. Results revealed that the subjects performed better in VR compared to IVD, and also performed better when they executed the unmanned interface first. We highlight the implications of our results for training personnel in scenarios involving manned and unmanned operations in IVD and VR interfaces.

Keywords: Virtual reality · Indirect visual displays · Human factors
Manned interface · Unmanned interface · NASA - TLX

1 Introduction

With the continuous evolution of technology, our capability to sense environments at a distance has grown exponentially over the last two or three decades [1]. Unmanned systems are one contemporary technical instantiation of this evolutionary vector in extending human perception action capabilities [2]. According to [2], the central problem of unmanned system development is not the feasibility and creation of the hardware, but the human factors issues in the assimilation of the relevant sensory

© Springer International Publishing AG, part of Springer Nature 2019
D. N. Cassenti (Ed.): AHFE 2018, AISC 780, pp. 225–235, 2019.
https://doi.org/10.1007/978-3-319-94223-0_21

inputs, the processing of information pertinent to user specified goals, and the translation of the user's subsequent decisions into effective action [2].

There has been a dramatic surge in the use of unmanned systems in military organizations around the world [3]. This surge has led to human factors research to understand human performance while monitoring and controlling unmanned vehicles [3]. Unmanned systems can be very difficult to control [4]. There are a number of documented human related problems when operating UUVs, including problems with perception, underwater navigation and orientation, interface design and situation awareness [4–6]. Perceptual issues have been long standing problem in unmanned systems, whereby the human operators are removed from the immediate environment. Operator is deprived of sensory cues, but must make navigational and control movements, and mission related decisions on sensor imagery that can lack resolution, color, field of view and depth perception [7]. Situation awareness issues for the human operator when supervising an unmanned underwater system can be far greater than surface and air unmanned systems due to constraints on perception and restricted communications available underwater [8]. Decision making in these situations, are hindered due to impoverished sensory information, attentional resource limits, task, and environmental stressors [2]. Most researchers agree that humans are needed to control or supervise the unmanned system, but what kind of knowledge, skills, and abilities they should possess has not been decided [2]. According to [2], for manned aircrafts, there are pre-selection criteria such as intelligence tests and medical examinations, but there are no pre-selection criteria for unmanned systems whatsoever.

Furthermore, since the past few years, indirect visual displays (IVDs) and virtual reality (VR) have been used as a means of training and assessing individuals in complex and dynamic tasks [10]. Indirect visual displays help in supporting full spectrum, 360-degree local area awareness operations, both remote and immersive [10]. Indirect vision systems were initially designed as tools to support mobility and means to assess and enhance situation awareness of the operator [20]. But due to the presence of three-dimensional objects in the environment, and the absence of localized acoustics detection [20], a technology shift to immersive VR has been recommended by researchers [10, 11], especially in training and assessment. According to [10], VR is the use of computer modeling and simulation to enable a person to interact with an artificial three-dimensional visual or other sensory environment. Virtual Reality allows the possibility for the individual to "dive" into the virtual world that allows the individual to build a better mental model of the scene, freely and seamlessly move around the virtual scene, and examine its descriptors from all possible perspectives [11]. Virtual environments are supposed to be more effective than other digital approaches with respect to the acquisition of several abilities [11]. Although these technologies are used to support human operations, an understanding of the issues related to cognitive and ergonomic challenges and how these interfaces enable better tactical thinking and decision-making is lacking in literature.

Instance-based learning theory (IBLT) [9], a theory of how individuals make decisions from experience, has elucidated decision-making in complex tasks very well. According to IBLT, decision-making is a five-step process: recognition of the situation, the judgment based in experience, choices among options based upon judgments, execution of chosen actions (decision-making), and feedback to those experiences that led to the chosen actions [9]. Hence, as per IBLT, when the complexity and the constraints in the task is higher, the decision-maker would be able to collect and store more experiences during training in the task. This collection of experiences would allow the decision-maker to get a better mental representation of the objective to be achieved and subsequently enhance decision-making [9]. Hence, owing to its enhanced telepresence and better immersivity, we expect that the VR simulation would help the individual to create a better mental model of the scenario and the objectives to be achieved. This in turn, would create more experiences in his/her's memory, which would lead to better performance due to optimal utilization of cognitive resources. We also hypothesize that unmanned interface-based training would lead to better performance. That is because the unmanned interface provides unconstraint visual conditions to human players.

In what follows, we investigate the implications of four simulations that differ in their display designs and type of interface-based training on human performance and cognitive workload. Then, we detail results and discuss the applications of our results on decision-making in the real world.

2 Materials and Methods

In this section, we describe an experiment to investigate the influence of an IVD/VR design involving unmanned/manned interfaces on one's decision-making performance.

2.1 Participants

A total of 60 participants (35 males and 25 females; mean age: 21.7 years, SD = 2.23 years) at the Indian Institute of Technology (IIT) Mandi, Himachal Pradesh, India took part in this study. The study was approved by an ethics committee at IIT Mandi. Participation was voluntary and a written consent was taken from all participants before they began the experiment. Fifty-four participants were right-handed. All participants had normal or corrected-to-normal vision, and no one reported any history of neurological disorders. All participants were from science, technology, engineering, or mathematics backgrounds. Seventy-two percent of participants reported that they had not experienced virtual reality before. All the participants received a flat payment of INR 50 for their participation in the study. They could also earn a performance-based incentive of INR 30 per successful simulation (a successful simulation is the one in which a participant is able to achieve all simulation objectives within a defined time period).

2.2 The Underwater Search-and-Shoot Simulation

An underwater based virtual search-and-shoot simulation was designed using Unity3D version 5.4.1 [12], and the avatars of the enemy submarines and the manned/unmanned interfaces were designed using Blender animation [13]. Figure 1 shows the overhead map of the terrain, with three headquarters located at different coordinates in the terrain. We followed the evolutionary prototyping SDLC model [14] in developing the

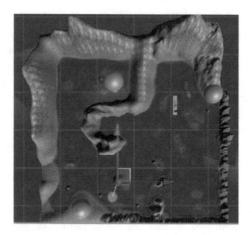

Fig. 1. The overhead map of the naval terrain designed in Unity3D. The blue spheres indicate the three headquarters in the simulation. The small red spheres indicated the enemies and the big red sphere indicated the position of the player.

simulation. As shown in Fig. 2(a), one of the display design was a typical first-person-shooter based IVD, providing a live-video feed with a horizontal field of view of 120° to the operator (which is comparable to the horizontal view of a healthy human being). The other was a VR design, catering to an artificial 3D environment, as shown in Fig. 2 (b). In addition to the variation in the display designs, variations in the computer interfaces were also introduced: a manned interface (as shown in Fig. 3(a)) and an unmanned interface (as shown in Fig. 3(b)). Table 1 shows the human factors issues introduced in the unmanned and the manned simulation. We introduced some physical and ergonomic changes in both the simulations in order to enforce realistic cognitive constraints in decision-making on the individual. The enemy submarines were positioned in such a way that all the headquarters contained approximately equal number of opponents guarding it. The participants' objective was to destroy all enemy submarines and protect all the naval headquarters within a specified time period (10 min). The participant's health was initialized to 100. Participants possessed a missile launcher for offense, consisting of 50 missiles.

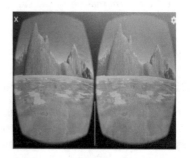

a

b

Fig. 2. The display design variations in the simulation (a) simulation in an indirect visual display (b) simulation in immersive mobile VR

Fig. 3. The different interfaces designed for the experiment (a) the manned interface (b) the unmanned interface

Table 1. Variation in the human factors issues in the computer interfaces in the simulation

Attribute	Manned simulation	Unmanned simulation
Embedded periscope view	Yes	No
Latency of response buttons	200 ms	800 ms
Delay in video transmission	15 frames	30 frames
Controls	Extreme™ 3D Scan Pro	Mouse and keyboard

One enemy could be killed in 5 missiles, and the player submarine could be destroyed with 12 missile hits. The total number of enemies in the simulation was kept to six. All enemies' health was initialized to 100. Two sub-types of enemy submarines were created: aggressive and defensive. The aggressive enemy submarine was programmed in a way so that they would pursue the player submarine relentlessly and try to shoot it down. The defensive enemy submarine would try to defend itself by dodging the player's attacks by moving in a random direction when under fire. These submarines were programmed to be offensive towards the three-naval headquarters, and they would start firing their missiles towards the headquarters as soon as they were in its vicinity. As shown in Fig. 2(a), we also implemented a SONAR sensor in both IVD and VR in the bottom right-side of the interface to enable the participants track the enemy submarines' suggestive positions. The SONAR sensor would detect the targets if it came under the radius of 50 units (in Unity3D calibration) from the center point (the black sphere in the Figure was suggestive of the player submarine's position). In addition, as shown in Fig. 2(a), the location of the three headquarters and the position of the player submarine relative to the headquarters were also shown in the bottom-left part of the interface. The IVD simulation was executed in a 21.5-in. HP desktop monitor, at a resolution of 1920 × 1080 pixel with noise-canceling headphones worn by the participant for receiving audio stimulus in a well-lit room. The participants used an Extreme™ 3D Pro joystick [15] for navigating and shooting in the manned simulation. Participants used a keyboard and a mouse for navigating and shooting respectively in the unmanned simulation. The VR simulation was executed using a mobile-based android system, through a 5.5-in. Xiaomi Redmi Note 3 smartphone [16] and My VR goggles [17], rendering a 120° horizontal field of view of the virtual environment. The participants used a DOMO MagicKey Bluetooth controller [18] to navigate and shoot in the VR simulation.

2.3 Experiment Design

In a lab-based setting, all the 60 subjects executed both the manned and unmanned interfaces (within-subjects) across both the IVD (N = 30) and VR (N = 30) designs (between-subjects). Within each display design (IVD or VR), half of the participants were given manned interface-based training first, and the other half were imparted unmanned interface-based training first. Behavioral measures like number of submarines destroyed and the total time taken to complete the simulation were recorded. In addition, various cognitive measures like the computerized version of the NASA-TLX [16] and simulator sickness questionnaire (SSQ) [17] were also recorded. The NASA-TLX questionnaire was recorded after every simulation executed by the participant. Owing to better telepresence and higher immersivity, we hypothesized that the performance would be better in the immersive VR design compared to the IVD design. Since unmanned interfaces produced an unconstraint view of surroundings, we expected an optimal transfer of cognitive skills in the unmanned interface-based training compared to the manned interface-based training.

3 Results

We carried out one-way ANOVAs to compare the effect of display design (IVD, VR) and type of interface-based training (manned, unmanned) on all the cognitive and behavioral descriptors mentioned above.

3.1 Performance Measures

As shown in Fig. 4(a), the percentage of enemy submarines destroyed were significantly higher in the VR design compared to IVD design (VR: 63% > IVD: 52%; F (1, 58) = 11.04, p = 0.002, r = 0.84). As shown in Fig. 4(b), time taken to complete the simulation was significantly higher in the VR design compared to the IVD design (VR: 267 s > IVD: 201 s; F (1, 58) = 7.41, p = 0.008, r = 0.88). Overall, these results show that the VR design led to superior participant performance compared to the IVD design.

Next, as shown in Fig. 5(a), the percentage of enemy submarines destroyed were significantly lower when the manned interface was presented first compared to when the manned interface was presented second (Manned first (MF): 43% < Manned second (MS): 61%; F (1, 58) = 49.45, p = 0.0001, r = 0.53). Second, as shown in Fig. 5 (b), the time taken to complete the task was significantly lower when the manned interface was presented first compared to when the manned interface was presented second (MF: 195 s < MS: 256 s, F (1, 58) = 12.02, p = 0.001, r = 0.83).

Furthermore, as shown in Fig. 5(c), the percentage of enemy submarines destroyed were significantly lower when the unmanned interface was presented first compared to when the unmanned interface was presented second (Unmanned first (UF): 44% < Unmanned second (US): 54%; F (1, 58) = 10.35, p = 0.002, r = 0.84).

Second, the time taken to complete the task was significantly lower when the unmanned interface was presented first compared to when the unmanned interface was presented second (UF: 190 s < US: 249 s, F (1, 58) = 13.29, p < 0.0001, r = 0.81; see Fig. 5(d)).

A two-way ANOVA was conducted that examined the effect of display design and the type of interface-based training on the differences in the percentage of enemy submarines destroyed (see Fig. 6). As shown in Fig. 6, a two-way ANOVA revealed a statistically significant interaction between the display design and the interface-based training order on the difference in the percentage of enemy submarines destroyed (F (1, 56) = 4.48, p = 0.039, η^2 = 0.074). Overall, as per our expectation, the unmanned training was much superior to the manned training in both IVD and VR design. However, the difference between unmanned and manned training was much less in the IVD design compared to the VR design.

3.2   Cognitive Measures

Next, we analyzed the NASA-TLX self-reported scores. The self-reported mental demand was significantly higher in the VR design compared to IVD design (VR: 6.63 > IVD: 5.00; F (1, 58) = 13.42, p = 0.001, r = 0.81) as shown in Fig. 7(a). However, as shown in Fig. 7(b), the participants also revealed significantly higher performance satisfaction in the VR design compared to IVD design (VR: 6.12 > IVD: 4.86; F (1, 58) = 11.2, p = 0.001, r = 0.84). As shown in Fig. 8(a) and (b), the mental demand was higher when the manned and unmanned interfaces were presented first compared to when they were presented second (MF: 6.73 > MS: 5.13, F (1, 58) = 16.71, p = 0.0001, r = 0.77; UF: 6.4 > US: 5.3, F (1, 58) = 7.13, p = 0.01, r = 0.59). In VR display and unmanned first training, the two-tailed Pearson correlation between the time taken and the nausea-related symptoms (in SSQ) was significant (r = 0.624, p < 0.05).

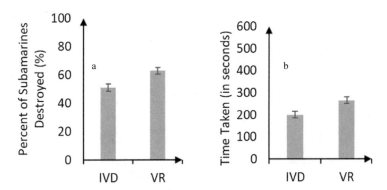

Fig. 4. Means and standard errors obtained in different display designs for time taken (a) and number of submarines destroyed (in percentage) (b).

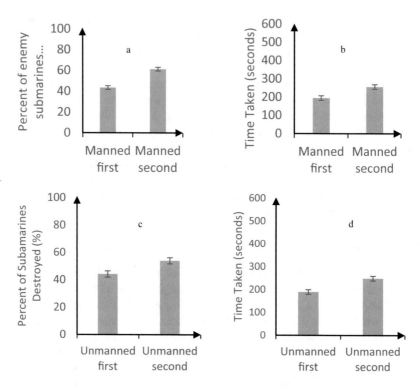

Fig. 5. Means and standard errors obtained in manned and unmanned interface-based training order (first or second). The number of enemy submarines destroyed (a) and the time taken (b) in manned interfaces. The number of enemy submarines destroyed (c) and time taken (d) in unmanned interfaces.

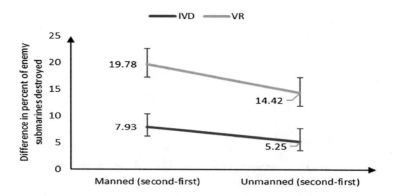

Fig. 6. Means and standard deviations of the difference in the percentage of enemy submarines destroyed due to interfaced-based training order (Manned second – Manned first, Unmanned second – Unmanned first) and display design (IVD, VR)

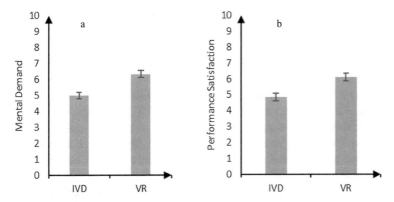

Fig. 7. Means and standard errors in IVD and VR display design for self-reported mental demand (a) and self-reported performance satisfaction (b).

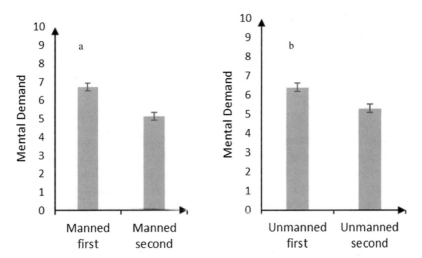

Fig. 8. Means and standard errors in interface-based training order for self-reported mental demand (a) and self-reported performance satisfaction (b).

4   Discussion

In this experiment, we evaluated the cognitive and behavioral implications of indirect vision and virtual reality display designs under varying manned/unmanned interfaces in a complex underwater search-and-shoot simulation. Results revealed that the participants performed better in the VR design compared to the IVD simulation and performed better in unmanned training compared to manned training. These results were found to be consistent with [11], where it was articulated that immersive VR enabled the participants to build a mental model of the simulation by seamlessly moving around the virtual world. Also, our results are in agreement with [19], where individuals were able to maximize their efficiency in comprehending the information conveyed in the virtual environments.

Our results on display design and interface training can also be explained by IBLT [9]. As per IBLT, situations that create more instances about the environment in decision-maker's memory would help to make good decisions. In agreement with this explanation, the unmanned training design provided an unconstraint training environment, which led to the creation of more instances in an individual's memory. These instances possibly lead to better performance. In fact, in agreement with IBLT, the VR design also led to an unconstraint perception of our environment causing it to produce enhanced performance. This explanation is supported by the self-reported mental demand score, which was higher in the VR design compared to the IVD design. Thus, more information availability and higher channelization of cognitive resources due to mental demand increased the information-processing speed from the instances created in memory.

Our results have important implications for decision-making in manned and unmanned operations as well as different display designs. Based upon our results, unconstraint training involving VR and unmanned scenarios seems to be the most potent in improving performance. Thus, it is advisable to train personnel in complex real-world applications using VR designs and unconstraint interfaces.

5   Conclusions

The results of our experiment indicate that VR design offers numerous advantages over the IVD design in complex decision-making scenarios. Also, our experiment proved that training individuals in unmanned (unconstraint) interfaces leads to efficient performance in a complex task compared to manned (constraint) interfaces. We expect to use these conclusions as a means of creating effective training environments for military personnel.

Acknowledgments. This research was supported by a grant from Defence Research and Development Organization (DRDO) titled "Development of a human performance modeling framework for visual cognitive enhancement in IVD, VR and AR paradigms" (IITM/DRDO-CARS/VD/110) to Varun Dutt.

References

1. Kurzweil, R.: The law of accelerating returns. In: Alan Turing: Life and Legacy of a Great Thinker, pp. 381–416. Springer, Berlin (2004)
2. Cooke, N.J.: Human factors of remotely operated vehicles. In: Proceedings of the Human Factors and Ergonomics Society Annual Meeting, vol. 50, no. 1, pp. 166–169. Sage Publications, Los Angeles, October 2006
3. Williams, K.W.: A summary of unmanned aircraft accident/incident data: human factors implications (No. DOT/FAA/AM-04/24). Federal Aviation Administration, Civil Aeromedical Institute, Oklahoma City, OK (2004)

4. Christ, D., Wernli, L.: The ROV Manual: A User Guide for Observation Class Remotely Operated Vehicles. Elsevier, Boston (2007)
5. Donovan, S.L., Triggs, T.: Investigating the effects of display design on unmanned underwater vehicle pilot performance (No. DSTO-TR-1931). Defense Science and Technology Organization, Maritime Platforms Division, Victoria, Australia (2006)
6. Donovan, S., Wharington, J., Gaylor, K., Henley, P.: Enhancing situation awareness for UUV operators. ADFA, Canberra (2004)
7. McCarley, J.S., Wickens, C.D.: Human factors implications of UAVs in the national airspace (2005)
8. Ho, G., Pavlovic, N.J., Arrabito, R., Abdalla, R.: Human factors issues when operating underwater remotely operated vehicles and autonomous underwater vehicles (No. DRDC TORONTO-TM-2011-100). Defence Research and Development, Toronto, Canada (2011)
9. Gonzalez, C., Dutt, V.: Instance-based learning: integrating sampling and repeated decisions from experience. Psychol. Rev. **118**(4), 523 (2011)
10. ter Haar, R.: Virtual reality in the military: present and future. In: 3rd Twente Student Conference IT (2005)
11. Freina, L., Canessa, A.: Immersive vs desktop virtual reality in game based learning. In: European Conference on Games Based Learning, p. 195. Academic Conferences International Limited, October 2015
12. Creighton, R.H.: Unity 3D Game Development by Example: A Seat-of-Your-Pants Manual For Building Fun, Groovy Little Games Quickly. Packt Publishing Ltd., Birmingham (2010)
13. Roosendaal, T., Selleri, S. (eds.): The Official Blender 2.3 Guide: Free 3D Creation Suite for Modeling, Animation, and Rendering, vol. 3. No Starch Press, San Francisco (2004)
14. Ragunath, P.K., Velmourougan, S., Davachelvan, P., Kayalvizhi, S., Ravimohan, R.: Evolving a new model (SDLC Model-2010) for software development life cycle (SDLC). Int. J. Comput. Sci. Netw. Secur. **10**(1), 112–119 (2010)
15. Behan, M., Wilson, M.: State anxiety and visual attention: the role of the quiet eye period in aiming to a far target. J. Sports Sci. **26**(2), 207–215 (2008)
16. Kim, Y.Y., Oh, M.A.: U.S. Patent Application No. 29/476,471 (2015)
17. Adhikarla, V.K., Wozniak, P., Barsi, A., Singhal, D., Kovács, P.T., Balogh, T.: Freehand interaction with large-scale 3D map data. In: 2014 3DTV-Conference: The True Vision-Capture, Transmission and Display of 3D Video (3DTV-CON), pp. 1–4. IEEE, July 2014
18. Wang, Y., Hasegawa, K., Terasaki, K.: U.S. Patent Application No. 13/013,072 (2011)
19. Hart, S.G., Staveland, L.E.: Development of NASA-TLX (Task Load Index): results of empirical and theoretical research. In: Advances in Psychology, vol. 52, pp. 139–183. North-Holland, Amsterdam (1988)
20. Metcalfe, J.S., Cosenzo, K.A., Johnson, T., Brumm, B., Manteuffel, C., Evans, A.W., Tierney, T.: Human dimension challenges to the maintenance of local area awareness using a 360 indirect-vision system. In: 2010 NDIA Ground Vehicle Systems Engineering and Technology Symposium: Modeling and Simulation, Testing and Validation Mini-Symposium (2010)

Cognitive Metrics Profiling: A Model-Driven Approach to Predicting and Classifying Workload

Christopher A. Stevens[1(✉)], Christopher R. Fisher[1],
Megan B. Morris[1,2], Christopher Myers[1], Sarah Spriggs[1],
and Allen Dukes[1]

[1] Air Force Research Laboratory,
Wright-Patterson Air Force Base, OH, USA
{christopher.stevens.28, christopher.fisher.27.ctr,
megan.morris.1.ctr, christopher.myers.29,
sarah.spriggs, allen.dukes.2}@us.af.mil
[2] Ball Aerospace and Technologies Corporation, Fairborn, OH, USA

Abstract. Workload management is integral to the success of human-machine teams, and involves measuring and predicting workload and implementing proactive interventions to mitigate the adverse effects of degraded performance. Common approaches to workload measurement rely on the use of subjective, behavioral, and physiological metrics. These approaches suffer from two important limitations. First, the mapping between workload, subjective ratings, behavior, and physiology is complex and noisy, resulting in high uncertainty. Second, metrics based on subjective ratings, behavior, and physiology often fail to explain *why* performance degrades, and consequentially does not inform the development of mitigation strategies. As an alternative, we propose using cognitive metrics profiling (CMP) to improve the measurement and prediction of workload. This approach uses computational cognitive models to simulate the activity within individual cognitive systems, such as vision, audition, memory, and motor, to measure and understand workload. We discuss how CMP can be used in an unmanned vehicle control task.

Keywords: Cognitive workload · Physiology · ACT-R
Computational cognitive models

1 Introduction

Sustained periods of high workload are associated with operator burnout, which can lead to negative outcomes such as health issues, decreased productivity, and increased turnover rates [1]. In addition, sudden changes in workload can lead to degraded performance [2]. Given these practical implications, workload assessment metrics have been the subject of intense research interest for decades [3].

Workload is commonly measured with subjective (e.g., retrospective self-reported workload), behavioral (e.g., primary task performance), and physiological (e.g., heart rate) metrics [3]. Although these metrics have demonstrated some utility [3, 4], they

D. N. Cassenti (Ed.): AHFE 2018, AISC 780, pp. 236–245, 2019.
https://doi.org/10.1007/978-3-319-94223-0_22

suffer from several limitations. One limitation is low sensitivity of the metric. For example, subjective metrics are often vulnerable to operator biases, and physiological metrics are often contaminated with reactions to other factors (e.g., emotional and environmental) resulting in high degrees of measurement error [3, 5]. Another limitation is low diagnosticity. These metrics have complex and noisy mappings with workload and provide limited insight into the underlying causes of high workload [5]. Without the ability to accurately measure workload and understand its causes, interventions designed to mitigate workload have a substantial risk of failure. What is needed is a formal theoretical framework for measuring workload and identifying the cognitive mechanisms that underlie workload. Here we discuss a method that we believe will help to make progress toward this goal: Cognitive Metrics Profiling.

2   Cognitive Metrics Profiling

Cognitive Metrics Profiling (CMP) is a technique for simulating and predicting cognitive load within the theoretical framework of a computational cognitive model [6]. This approach allows moment-to-moment *a priori* predictions of overall workload levels and classification of high workload intervals based on the type of capacity engaged by the current task (e.g., declarative memory, vision, motor, etc.). It is based on the ACT-R cognitive architecture, a computational theory of human cognition. CMP assumes a set of specialized, limited-capacity cognitive modules, including vision, motor control, declarative memory, and goal management. Each module can perform only one task at a time, a limitation that results in bottlenecks and performance degradation in high workload environments. With such a model it is possible to generate a projected time-series of workload levels based on a task protocol. Moreover, because CMP simulates behavioral outputs, it can provide performance predictions based on a person's current workload state and upcoming task requirements.

Variability in operator performance and strategy use presents another challenge for measuring and understanding workload. As a solution to this problem, the CMP approach can provide individualized workload assessments and predictions that capture differences in low-level cognitive function (e.g., speed of memory retrieval) and high-level strategies. By applying CMP at the level of the individual operator, it is possible to improve workload assessment and prediction. When individual differences are unknown, it is also possible to use a model of a "typical" individual as a starting point. As more information is acquired, CMP can become increasingly refined and individualized.

3   Overview

In the present work, we demonstrate the utility of cognitive metrics profiling using a simulated unmanned vehicle task. Throughout the 60-min session, workload varied on a moment-to-moment basis based on the number of subtasks performed by the individual. We used CMP to identify moments of high workload in each mission and moments when particular cognitive capacities were especially taxed. We then compared these predictions with the time course of physiological workload measures, noting points of overlap and departure. In the resulting analysis we show that combining CMP with traditional physiological analysis techniques provides meaningful insights that can inform real-time countermeasures for cognitive load.

4   Unmanned Vehicle Task

Human participants completed low and high workload versions of a 60-min simulated unmanned vehicle task (UVT) [7, 8] while wearing physiological sensors. During the UVT, participants monitored a military base at a computer station and dispatched UVs in response to events and queried the system for information. The interface for the computer station was realistic and fully functional, featuring a map of the base with landmarks and UV locations, an A.I. system for optimizing UV selection based on time of arrival and UV properties, a communication window for receiving and responding to messages, a headset and microphone for receiving and responding to auditory messages, a monitor displaying UV attributes and status, and a control system for allocating UVs to particular tasks. The low and high workload versions were similar, except the high workload version required participants to respond to more events than the low workload version.

Events in the UVT can be classified into two basic types. Base Defense Events required participants to manage resources and dispatch unmanned vehicles in response to planned and unplanned events. Examples of Base Defense Events include surveilling a specific building at a predetermined time, responding to a security breach, or reassigning unmanned vehicles due to mechanical failure. The other type of event was an information query. As an example of an information query, a participant might receive a message in a communication window asking for the altitude of a particular unmanned vehicle.

In order to complete the task, participants were required to continuously monitor the interface for new tasks and respond to them appropriately. New tasks could be found in three locations in the interface: the communication window, the base map, or the list of planned tasks. For some events, participants were notified via a headset.

5   Physiological Metrics

Research has identified several physiological metrics associated with mental workload. Three common physiological metrics that we examine in the current study are heart rate, EEG, and eye movement.

Research has commonly examined the relationship between heart rate (HR) and heart rate variability (HRV) metrics with mental workload. Increases in HR and decreases in HRV have been shown to relate to increases in mental workload, respectively [4]. HRV can be divided into two categories. The first includes time domain measures of HRV, which are calculated from normal-to-normal (NN) or R to R (RR) intervals. These include metrics such as mean NN, standard deviation of NN interval (SDNN), and root mean square of differences of successive NN intervals

(RMSSD) [9]. HRV can also be measured in terms of the frequency domain (i.e., spectral density analysis of NN interval data). Frequency domain metrics commonly include three ranges of power signals: very low frequency (VLF), which researchers have had difficulty interpreting; low frequency (LF), suggested to reflect sympathetic activity; and high frequency (HF), suggested to reflect parasympathetic activity. Research suggests that increases in mental workload are associated with increases in LF power and decreases in HF power [4, 9].

Several EEG metrics are associated with mental workload. Studies have found alpha band frequencies (8–13 Hz) to decrease with increased mental workload [4, 10]. Researchers have further suggested increased diagnosticity in dividing alpha into lower (8–10 Hz; reflects alertness) and upper (10–13 Hz; reflects information processing) band frequencies when measuring workload [11]. In addition, theta band frequencies have been found to have a positive association with mental workload [4, 10]. Studies have also shown ratios of frequency band power to be useful in measuring workload. Research has found that a ratio of theta and alpha, the Task Load Index (TLI), has a positive relationship with mental workload [12]. Another ratio metric, the Engagement Index (EI) - a ratio of beta to alpha and theta, has been shown to be diagnostic of alertness, and in some cases, might reflect increased workload when fatigued [12].

Research has also found associations between eye metrics and mental workload. Studies suggest that blink rate and duration decrease with increased mental workload [13]. Studies also suggest that pupil diameter has a positive relationship with mental workload [14–16].

6 Cognitive Model

We developed a model of the UVT based on the computational cognitive architecture Adaptive Control of Thought Rational (ACT-R) [17]. A computational cognitive architecture is a theoretical framework for developing and simulating unified theories of cognition. The primary goal of a computational cognitive architecture is to describe how cognitive systems—including memory, perception, reasoning, and motor movement—interact to produce cognition and task performance that emulates that of a human operator. Computational cognitive architectures are implemented as computer simulations and produce quantitative predictions of performance.

As illustrated in Fig. 1, the ACT-R architecture is organized as a centralized module with multiple peripheral modules. Each module is a specialized information-processing unit that operates independently of other modules. The procedural memory module is a centralized module that coordinates the information flow and behavior of the architecture. Procedural memory refers to memory for actions or procedures and is implemented as a production system in ACT-R. The procedural module has mechanisms for the evaluation, selection, and execution of production rules (e.g., IF-THEN rules). A simple example of a production rule is "IF the goal is to find a new message, and a message is in the visual system, then attend to the message". If that production rule is selected, the state of the model is updated and different production rules become eligible for selection. For example, an eligible production rule might attempt to retrieve the message from memory to determine if it is new (and requires action) or old (and can be ignored). The other modules include declarative memory for storing and retrieving factual information, a goal module for maintaining goal-relevant information, an auditory module for processing auditory stimuli, a visual module for locating and attending to visual stimuli, and a motor system for interacting with the environment. The information processing capabilities of modules are limited in two regards. First, modules interact with each other through buffers, and each buffer can only process one request at a time. Second, each buffer can maintain and share a maximum of one unit (or chunk) of information with the procedural memory module at any given time. These are the primary information processing bottlenecks in the ACT-R architecture, and are responsible for producing errors commonly found in human performance. A large and diverse body of research supports ACT-R as a viable account of human cognition and performance. Some successful applications of ACT-R include, memory [18], driving [19], problem solving [20], the effects of sleep loss [21], the effects of exposure to the solvent toluene [22], and meta-cognition in strategic interactions [23].

7 Goal Hierarchy

Our model of the UVT can be described in terms of a hierarchy of goals. Superordinate goals in the hierarchy involve searching the interface for new tasks and performing those tasks, whereas subgoals involve the intermediate steps necessary to accomplish a particular superordinate goal. Suppose, for example, that the current superordinate goal is to inspect the communication window for a new message containing instructions to perform a new task. If no message is found, the model will shift to a new superordinate goal, such as checking the map for a new event. However, if there is a new message instructing the model to inspect a suspicious vehicle, the new superordinate goal will be to perform the required task as instructed. In order to achieve this superordinate goal, the model must accomplish a series of subgoals that include, encoding the instructions into memory, searching the interface for additional information, selecting the correct UV, and dispatching it to the correct location.

7.1 Model Strategy

With a goal-structure in place, the model needs a strategy for achieving and prioritizing the goals. The space of potential strategies is quite large for a task as complex as the UVT. We adopted two criteria for narrowing down the strategy space. Our first criterion was simplicity, meaning that the strategy should be no more complex than is required to perform the UVT. Our second criterion was cognitive plausibility. The model should complete the task in a manner similar to human operators and commit errors that are also similar to that of human operators.

Based on these criteria, we adopted the following strategy as a starting point. The model employs a deterministic search strategy for identifying new tasks to complete. It searches the interface in a cyclical fashion, checking each section of the display for new information. This is in juxtaposition to other candidate search strategies, such as searching the interface according to a probability distribution based on the importance

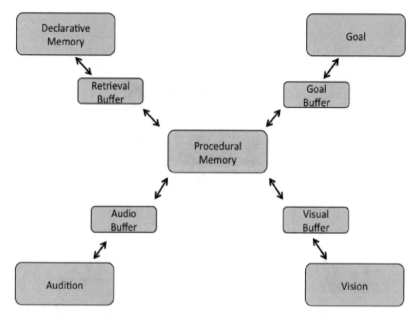

Fig. 1. A diagram of the ACT-R architecture.

of the events or the frequency with which they occur in those locations. In developing a strategy for the model, we must decide whether the introduction of a new task can interrupt an ongoing task. On one hand, the ability to interrupt an ongoing task could allow the model to prioritize more important tasks, thereby improving performance. On the other hand, interruptions impose additional workload demands, including, but not limited to, searching for competing tasks, deciding whether to switch a task, and remembering one's progress on the interrupted task. In keeping with our criteria, we decided to make the strategy ballistic. Once the model identifies and begins a new task, it cannot be interrupted (with one minor exception detailed below).

The model's strategy is represented as a flowchart in Fig. 2. The search process begins with checking the communication window for a message containing instructions for a new task. If no message is found, the model will search the next source for an event (in this case, the base map). However, if a task is found, the model will complete that task ballistically before checking for new events in the next source. The only exception to the rule occurs for messages sent via the headset. In such cases, the model will briefly suspend an ongoing task to encode the message and then resume the interrupted task. Once it completes the task, the model will respond to the instructions received through the headset (e.g., the "Interrupt?" node). Finally, once the task presented via the headset has been completed, the model goes to the node "Check Next Source" to resume the search cycle. Collectively, these policies form a meta-cognitive strategy for mitigating the model's cognitive limitations. The model completes tasks ballistically to alleviate cognitive load associated with task switching, and makes an exception for messages presented via the headset because, unlike information presented on the screen, auditory information has no enduring record, and can be easily forgotten.

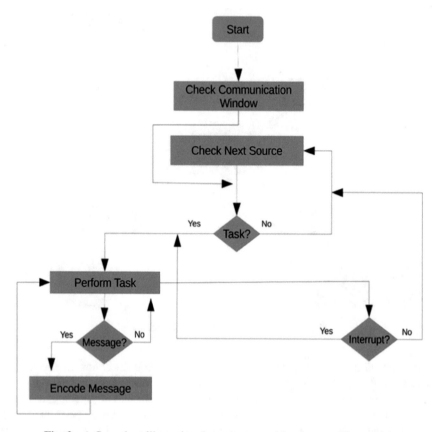

Fig. 2. A flow chart illustrating the strategy used by the cognitive model.

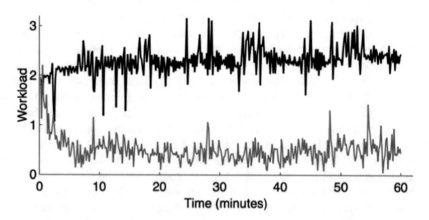

Fig. 3. The time course of composite workload (black) and Alpha band (gray) of a single subject scaled in arbitrary units for ease of presentation. Lower values of Alpha correspond to higher workload.

7.2   Applying Cognitive Metrics Profiling to the UVT

We used cognitive metrics profiling to analyze the workload demands of the UVT in two ways. First, we analyzed the composite workload across the entire 60-min UVT session. Second, we analyzed the specific contributions of each module to overall workload for a segment of the data in which composite workload showed the greatest variability. This detailed analysis provides insights into factors that lead to low and high workload in the UVT.

Figure 3 shows composite workload and Alpha band for a single operator across the 60-min UVT session. Composite workload is computed a weighted sum of activity in each module (1 = active, 0 = inactive) integrated over contiguous 10-s windows. Following previous research [24], we assigned a weight of 2 to the procedural memory module, a weight of 4 to the imaginal and declarative memory modules, and a weight of 1 to all other modules. As expected, the correlation between Alpha band and composite workload was negative although small ($r = -.21$). Both metrics indicate a quick build up in workload that is sustained throughout the task because the operator must continuously monitor the interface for new tasks. Deflections indicate momentary increases in workload in response to tasks.

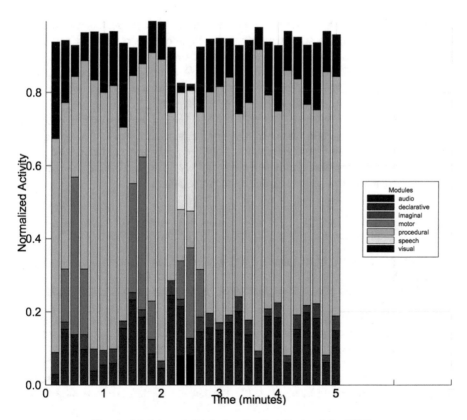

Fig. 4. Module activity during the first 5 min of the UVT.

Although the composite workload metric provides a theoretically grounded measure of workload, it cannot diagnose the cause of increased workload. To overcome this limitation, Fig. 4 decomposes workload into the relative processing times of each module. The lowest workload occurs around 2.5 min when the model communicates with the sensor operator. During this time, procedural module activity is low and speech comprises a large share of processing time. This suggests that procedural workload is low when communication is the primary task.

8   Conclusion and Outlook

Cognitive metrics profiling offers a theoretical framework that can potentially unify behavioral, physiological, and subjective workload assessments. With the same model, one can derive predictions for the time course of both physiological and behavioral workload indicators. In addition, CMP can identify which type of cognitive capacity (e.g., memory, motor, attention) is contributing to the workload. This information can refine the understanding of specific physiological signals. It is possible that these signals are more strongly related to some cognitive processes than others. If so, this approach could offer insights into the specific behavioral and physiological indices that should be used for workload monitoring in a given task.

Much work remains to be done applying this framework to workload assessment. First, the model output (both behavioral and physiological) needs to be validated against a variety of real-world tasks. Next, methods need to be developed for translating the model's predictions into algorithms capable of real-time inference. This will likely involve generating a set of user profiles from different model runs and performing a model-tracing analysis to determine the best fit [25].

Despite decades of research, we still do not have a comprehensive theory to describe and predict cognitive workload. Our present work is a step in that direction. If successful, this approach would improve not only our fundamental theoretical understanding of the factors that produce workload, but also our ability to recognize moments of high and low workload that occur in real-time.

References

1. Maslach, C., Schaufeli, W.B., Leiter, M.P.: Job burnout. Annu. Rev. Psychol. **52**(1), 397–422 (2001)
2. Cox-Fuenzalida, L.E.: Effect of workload history on task performance. Hum. Factors **49**(2), 277–291 (2007)
3. Miller, S.: Workload Measures. National Advanced Driving Simulator, Iowa City (2001)
4. Lean, Y., Shan, F.: Brief review on physiological and biochemical evaluations of human mental workload. Hum. Factors Ergon. Manuf. Serv. Ind. **22**(3), 177–187 (2012)
5. Matthews, G., Reinerman-Jones, E., Barber, D.J., Abich, J.I.: The psychometrics of mental workload: multiple measures are sensitive but divergent. Hum. Factors **57**, 125–143 (2015)
6. Gray, W.D., Schoelles, M.J., Myers, C.W.: Profile before optimizing: a cognitive metrics approach to workload analysis. In: CHI 2005 Extended Abstracts on Human Factors in Computing Systems, pp. 1411–1414. ACM (2005)

7. Rowe, A., Spriggs, S., Hooper, D.: Fusion: a framework for human interaction with flexible-adaptive automation across multiple unmanned systems. In: Proceedings of the 18th Symposium on Aviation Psychology, pp. 464–469 (2015)
8. Draper, M., Calhoun, G., Hansen, M., Douglass, S., Spriggs, S., Patzek, M., Rowe, A., Evans, D., Ruff, H., Behymer, K., et al.: Intelligent multi-unmanned vehicle planner with adaptive collaborative control technologies (impact). In: International Symposium of Aviation Psychology (2017)
9. Task Force of the European Society of Cardiology: Heart rate variability, standards of measurement, physiological interpretation, and clinical use. Circulation **93**, 1043–1065 (1996)
10. Borghini, G., Astolfi, L., Vecchiato, G., Mattia, D., Babiloni, F.: Measuring neuro-physiological signals in aircraft pilots and car drivers for the assessment of mental workload, fatigue and drowsiness. Neurosci. Biobehav. Rev. **44**, 58–75 (2014)
11. Klimesch, W.: EEG alpha and theta oscillations reflect cognitive and memory performance: a review and analysis. Brain Res. Rev. **29**(2–3), 169–195 (1999)
12. Kamzanova, A.T., Kustubayeva, A.M., Matthews, G.: Use of eeg workload indices for diagnostic monitoring of vigilance decrement. Hum. Factors **56**(6), 1136–1149 (2014)
13. Fogarty, C., Stern, J.A.: Eye movements and blinks: their relationship to higher cognitive processes. Int. J. Psychophysiol. **8**(1), 35–42 (1989)
14. Ahlstrom, U., Friedman-Berg, F.J.: Using eye movement activity as a correlate of cognitive workload. Int. J. Ind. Ergon. **36**(7), 623–636 (2006)
15. Beatty, J.: Task-evoked pupillary responses, processing load, and the structure of processing resources. Psychol. Bull. **91**(2), 276 (1982)
16. Porter, G., Troscianko, T., Gilchrist, I.D.: Effort during visual search and counting: insights from pupillometry. Q. J. Exp. Psychol. **60**(2), 211–229 (2007)
17. Anderson, J.R.: How Can the Human Mind Occur in the Physical Universe? Oxford University Press, Oxford (2007)
18. Anderson, J.R., Bothell, D., Lebiere, C., Matessa, M.: An integrated theory of list memory. J. Mem. Lang. **38**(4), 341–380 (1998)
19. Salvucci, D.D.: Modeling driver behavior in a cognitive architecture. Hum. Factors: J. Hum. Factors Ergon. Soc. **48**(2), 362–380 (2006)
20. Anderson, J.R.: Problem solving and learning. Am. Psychol. **48**(1), 35 (1993)
21. Gunzelmann, G., Gross, J.B., Gluck, K.A., Dinges, D.F.: Sleep deprivation and sustained attention performance: integrating mathematical and cognitive modeling. Cogn. Sci. **33**(5), 880–910 (2009)
22. Fisher, C.R., Myers, C., Reem, H.M., Stevens, C., Hack, C.E., Gearhart, J., Gunzelmann, G.: A cognitive-pharmacokinetic computational model of the effect of toluene on performance. In: Gunzelmann, G., Howes, A., Tenbrink, T., Davelaar, E. (eds.) Proceedings of the 39th Annual Conference of the Cognitive Science Society. Cognitive Science Society, Austin, TX (2017)
23. Stevens, C.A., Taatgen, N.A., Cnossen, F.: Instance-based models of metacognition in the prisoner's dilemma. Top. Cogn. Sci. **8**(1), 322–334 (2016)
24. Jo, S., Myung, R., Yoon, D.: Quantitative prediction of mental workload with the ACT-R cognitive architecture. Int. J. Ind. Ergon. **42**(4), 359–370 (2012)
25. Fisher, C.R., Walsh, M.M., Blaha, L.M., Glunzelmann, G., Veksler, B.: Efficient parameter estimation of cognitive models for real-time performance monitoring and adaptive interfaces (2016)

Simulation of Financial Systemic Risk and Contagion in the U.S. Housing Market

Faizan Khan[✉] and Zining Yang

Claremont Graduate University, Claremont, CA 91711, USA
{Faizan.Khan, Zining.Yang}@cgu.edu

Abstract. This paper presents an agent-based model (ABM) to model systemic risk in the housing market from 1986 to 2017. We provide a unique approach to simulating the financial market along with demonstrating the phenomenon of emergence resulting from the interconnected-behavior of consumers, banks and the Federal Reserve. Consumers can buy or rent properties, and these agents own characteristics such as income and may be employed or unemployed. Banks own balance sheets to monitor their assets and liabilities and participate in the interbank lending market with one another. This tool can assess the complexity of the United States' housing market, conduct stress tests as interest rates fluctuate, and explore the landmark financial crisis and epidemic of foreclosures. This is important because understanding the impact from increases in foreclosed properties and changes in rates can help policymakers and bankers have a better understanding of the complexity within the housing market.

Keywords: Agent-based modeling · Financial systemic risk · Contagion
Financial crisis · Banking

1 Introduction

For many Americans, the "American Dream" is an opportunity to own a home. Home ownership allows individuals to secure a potential appreciating asset. The real estate crash in 2007, which inevitably transitioned into a global financial crisis, is the closest event to a modern Great Depression to this day. It was initially a banking and securities crisis, which then evolved into an economic crisis. The diminishment of house prices influenced some individuals to strategically default on their mortgage payments because their mortgage was underwater (i.e., unpaid loan balance is greater than market value of property). Others defaulted without choice because of catastrophic events such as becoming unemployed. These defaults inevitably led to foreclosed properties. The foreclosure contagion effect mimics the concept of a virus spreading into the environment from carrying hosts; therefore, as the number of foreclosed properties in a neighborhood increases, the prices of homes surrounding the "infected host(s)" may be negatively impacted. The defaults and delinquent payments harmed the income of banks, and these financial institutions had difficulties in meeting debt obligations to other banks. As a result, housing prices depreciated and interbank lending tightened.

Researchers attempted to answer the housing market by analyzing various potential determinants, such as: monetary policy implemented by the Federal Reserve, the role of

© Springer International Publishing AG, part of Springer Nature 2019
D. N. Cassenti (Ed.): AHFE 2018, AISC 780, pp. 246–253, 2019.
https://doi.org/10.1007/978-3-319-94223-0_23

sub-prime mortgages and mortgage-backed securities (MBS), lending standards by the financial services sector, and the list continues; however, there is not a consensus as to what caused the U.S. to tumble into this detrimental position. What if it was a "perfect storm," which would include everything listed above plus more? Both, the housing market and financial crisis, are prime examples of complexity. Complex in the sense that there were several underlying factors, which contributed to the bubble and crisis. Although these factors may seem like dozens of mismatched puzzle pieces, they are quite interconnected.

Prior research provided a range of values to measure the foreclosure contagion effect. The purpose of this paper is to describe an agent-based tool, which may be helpful for facilitating historical, present, and future analyses of the foreclosure contagion effect. This model allows us to understand the behavior of consumers and banks; however, as previously mentioned, we will focus specifically on consumers in this demonstration. This agent-based model simulates the U.S. housing market—a complex adaptive system. Once the world initializes, we can perform stress testing and sensitivity analyses to assess a variety of outputs given specific inputs such as interest rates, contagion effect, etc. We find that clusters of foreclosed homes have a severe negative impact on neighboring homes. As homes enter foreclosure, this negatively affects a bank's balance sheet as well. In addition, monetary policy and control of interest rates by the Federal Reserve play a critical role in the probability of an interest rate foreclosure. Policymakers and banks should work together and develop programs to prevent massive amounts of foreclosures, as it is harmful for consumers, banks and a country's overall economy.

Moreover, this paper is organized as follows: Sect. 2 will be a review of prior literatures, Sect. 3 describes the methodology and research design, Sect. 4 analyzes results from the ABM, and Sect. 5 provides concluding remarks and improvements for future research.

2 Foreclosure Discount Literature

There is a variety of reasons as to why a borrower may involuntarily default on his/her mortgage such as unemployment, death, etc. Prior to a home legally entering foreclosure, it enters a stage called Real Estate Owned (REO), which allows the creditor (e.g., a bank) to retain full possession and then sell to another party. When a property enters REO and is no longer occupied, the property value may diminish due to neglecting maintenance, vandalism, squatting from drug dealers and gangs [1, 2]. Moreover, the property sells at an amount to cover the remaining balance of the mortgage.

Typically, it is in the interest of banks to work with borrowers on a plan for repayment because foreclosing a property inevitably leads to a bank not receiving the income it was anticipating over the life of the loan. Prior literature estimates the cost of foreclosure to range from $7,200 to $58,759 [2, 3]. As properties within a vicinity enter foreclosure, the neighboring property values diminish and lead to other properties foreclosing along with an increase in supply of homes available in the market.

This paper focuses on similar methodologies implemented by Gangel et al. [4] to represent the contagion effect and contributes to real estate, foreclosure, and financial literature. Similarly, we incorporate a range of values determined by prior literature to measure the contagion effect. Researchers have claimed the contagion effect to range from 0.9% to a high of 8.7% [2, 3, 5]. The literature has a consensus on the contagion effect being local among neighboring properties. We apply the range of 0.9% to 8.7% in our sensitivity analysis.

3   Research Design and Methodology

Agent-based modeling is a bottom-up computational approach that uses simple assumptions to simulate individual agent goals and interactive behavior over time whereas a method such as system dynamics is a top-down approach due to the focus on total system behavior [6]. Abdollahian et al. [7] created an ABM to model SemPro, which simulates how competing interests and barriers influence siting outcomes and policy for sustainable energy infrastructure. Abdollahian et al. [8] also used ABM to simulate the Human Development theory and analyze the effects from interactions between economics, culture, society, and politics. As Haldane [9] from Bank of England pointed out, more and more scholars start to use ABM to study economic, fiscal and monetary policies, leveraging the model's capability in studying crisis [10, 11]. According to Goldstein [12], "Housing market interactions are more complex than many other markets due to search costs coupled with nonzero costs for waiting, large product differentiation, and the frequent inclusion of mortgage financing."

The initial density parameter determines the density of houses in the world. The rental-house density parameter determines how many properties are rental at tick 0, and the percent-occupied parameter generates the initial number of people. Furthermore, the elements within each agent rely on a random uniform distribution (i.e., income). People choose to live in affordable homes based on their available income to spend on a mortgage or rent. The tool outputs a variety of results, such as average house price, average mortgage cost, balance sheets of banks, real and natural unemployment rates, and more. Additionally, the "world" is a 32 × 32 grid in Netlogo with agents as people, houses, banks and mortgages. Each tick represents one month, and one patch is equivalent to one mile. Income and housing prices are in thousands of U.S. dollars. People can choose to rent or own one or more houses. Each house can either have a mortgage affiliated with a bank, or no mortgage at all. Figure 1 demonstrates the interdependent relationships among the agents and a high-level overview of their attributes.

We assume that properties undergo formal and informal appraisals of property value. Informal in the sense that an owner can determine an approximate appraisal

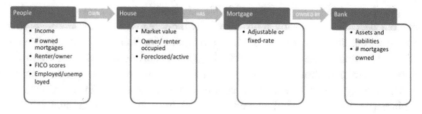

Fig. 1. Flowchart of the model, which lists entities, relationships and attributes at a high level.

value of his/her property through researching appraisal values of local and similar properties recently sold [13]. We also assume that neighboring properties have homogenous physical features, while the individuals who occupy each property may differ based on a distribution. The change in appraisal values of neighboring properties may influence an agent's property value through links. Every month, a random percentage of agents will assess the appraisal value of their property by observing the change in property values of neighboring homes with a maximum distance. The foreclosure discount is the negative percentage of price diminishment affiliated with foreclosed neighbors, and it is a function of change in price and distance from an appraised property. It is notated by μ as displayed in the contagion effect formulae. The formulae below outline important variables within the ABM, which are important for quantifying the foreclosure effect:

$$\Delta d_i = d_{max} - d_i, \quad \Delta p_i = p_{i_{t+1}} - p_{i_t}$$

$$Appraised\ Value_j = p_j - \sum_{i=1}^{n} \mu \cdot (p_j + \frac{\Delta p_i}{\Delta d_i}) \tag{1}$$

where:

Δd_i: difference between the ith property and maximum distance constant (d_{max})
d_i: distance from ith property and appraisal property (j)
μ: contagion effect severity for a single home
Δp_i: price change of ith property from t to t + 1
p_j: price of appraisal property

3.1 Types of Foreclosures

We embed the different foreclosure effects from Gangel et al. [4] within our model. Below is a summary description of each type of foreclosure:

Equity Foreclosure. We calculate an equity ratio using the current appraisal value and outstanding principal balance. If the equity ratio is below one, then the borrower acknowledges the presence of an underwater mortgage, which leads to an increased probability of the owner defaulting. Let C_{equity} be equal to a constant that reflects the effect of the equity ratio:

$$Equity\ Ratio = \frac{Appraised\ Value}{Unpaid\ Principal\ Balance}$$

If Equity < 1,

$$Equity\ Foreclosure\ Effect = \frac{(1 - Equity\ Ratio) * C_{equity}}{Unpaid\ Principal\ Balance} \tag{2}$$

If Equity ≥ 1,

$$Equity\ Foreclosure\ Effect = 0$$

Interest Rate Foreclosure. As previously mentioned, the ABM includes both fixed and adjustable rate mortgages. Fixed rate mortgages are not prone to interest rate fore-closures because the borrower has agreed to an affordable mortgage; however, ARMs are the opposite. After the fixed-rate period expires, the interest rate resets annually following LIBOR 12-month rates in this environment. An increase in interest rates leads to higher payments, which increases the probability of default whereas a decrease in rates reduces the probability of default. Let C equal a constant that reflects the effect of the probability of interest rate foreclosure, and let IC equal the percentage change between the prior and current monthly payments:

$$IC = \frac{Current\ Monthly\ Payment}{Fixed\ Period\ Monthly\ Payment}$$

$$Interest\ Rate\ Foreclosure\ Effect = \frac{(IC - 1) * C}{12}$$

(3)

Investor Foreclosure. An investor is likely to voluntarily ignore their mortgage obli-gation even when the investor can afford to make payments (i.e., exercise one's put option) if the renter's payment is below the monthly mortgage payment. While Gangel et al. [4] do not focus on the renter market, we chose to include these agents in our simulation; therefore, we can more accurately represent the investor foreclosure effect. Let $C_{investor}$ equal a constant that reflect the effect of the probability of an investor foreclosure:

$$Rent < Mortgage,$$

$$Investor\ Foreclosure\ Effect = \frac{C_{investor}}{12}$$

$$Rent > Mortgage,$$

$$Investor\ Foreclosure\ Effect = -\frac{C_{investor}}{12}$$

(4)

Catastrophic Foreclosure. A catastrophic event such as job loss may lead to an increase in the probability of foreclosure. Gangel et al. [4] simply include a constant to represent the probability of a foreclosure due to a catastrophic event; however, we only include this constant if the individual person is actually unemployed in the simulation. Let $C_{catastrophic}$ equal a constant that reflects the probability of foreclosure from becoming unemployed.

Total Probability of Foreclosure. We have described the different types of foreclosure effects, which ultimately calculate a final probability of a property entering foreclosure based on characteristics pertaining to individual mortgages, homes, and people.

4   Results

The baseline scenario initializes with the default parameters, and the goal is to replicate similar trends pertaining to the housing prices and foreclosure effects during the Financial Crisis. Figure 2 is a screenshot of the interface after it has initiated in NetLogo. In this scenario, we have set fixed-rate mortgages as 90% of the pool because most mortgages in the United States are in fact fixed-rate. We have also set mobility to 60 ticks (i.e., 5 years) and agents may move at any random tick under that threshold. Moreover, this will allow us to calibrate towards the behavior of 5/1 ARM home-owners, since individuals would typically move after the teaser rate has expired. People can place anywhere from a 10–25% down payment on their mortgage(s) representing the average range in the United States. We have set the foreclosure effect constant to be 35% for all of the types of foreclosures that can occur, which suggests that any of these effects have the same impact on the probability of foreclosure. The complete probability of a foreclosure must be equal to or greater than the threshold of 40% for this iteration, and/or 3 months delinquent on payments. Americans are typically 90 days delinquent on payments before a home enters REO. Additionally, the contagion effect for a neighboring home that enters a foreclosure is 8.7%, which is the upper estimation from prior literature. The neighborhoods are in a maximum radial distance of two patches to represent local communities with similar demographics.

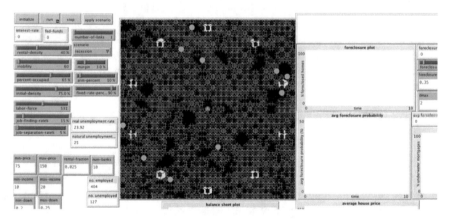

Fig. 2. Interface after initialization.

From January 1986 (time 0) to January 2000 (169 ticks), the average foreclosure property has remained quite flat compared to the initiation of the model, but the percentage of foreclosed homes has certainly fluctuated. Figure 3 displays three graphs, which include: average foreclosure probability, Federal Reserve rates, LIBOR, and the average house price. House prices drastically rose from 1986–2000. Just before the year 2000, we see a dip in the average house price with a slight recovery. Fed funds rate and LIBOR have diminished significantly since the initialization of the model.

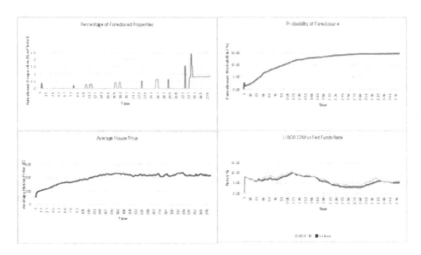

Fig. 3. Line graphs from the interface of the model.

By the time the model reaches September 2001, we apply two scenarios to the ABM: (1) a recession shock and (2) low welfare benefits. This is to account for a 9/11 political shock, which severely affected the economy. Although arbitrary, the impacts from the shock last until January 2004, which is after George W. Bush became President of the United States for a second term. We continue to run the model until Fall 2007, where we notice some effects from the financial crisis. At this period, the average house price has peaked but also taken a dip, which represents the burst of the housing bubble. In this run of the model, the average house price remains extremely high compared to historical prices. We also see a spike in the percentage of foreclosed properties.

5   Conclusion

This paper shows the power of simulation as an additional approach to scientific inquiry and builds on previous agent-based models of the real estate market. Currently, the results of the example simulation emphasize a tremendous influence from the Federal Reserve's monetary policy tool, banks' ability to lend to each other, and consumers' ability to make payments. The Fed's control over interest rates continuously demonstrated a real estate bubble bursting around 2007 regardless of how the parameters differentiated.

Our model incorporates many complex features representing real world characteristics of consumers and banks. After we set up the full model, we conducted a number of sensitivity analyses and focused on variables that influenced the probability of foreclosures. We find that neighboring properties entering foreclosure certainly have an impact on surrounding homes.

Most importantly, this agent-based model still has plenty of features that may be included to further model the real estate market in the United States. As we continue to build this model, we would like to place more emphasis on adding features to represent MBS trading, sub-prime mortgages, credibility and debt-to-income ratio of borrowers, credit default swaps, quantitative-easing (another Federal Reserve tool), Dodd-Frank Act, lending standards and deregulation, etc. Additional agents will include rating agencies such as Standard & Poor (rating agency), Fannie and Freddie (government-sponsored enterprises), and construction companies to control the supply side and pricing of homes. Another necessary component for the model is the ability for borrowers to prepay and refinance mortgages. In sum, agent-based modeling is a valuable tool and can help us monitor and prevent systemic risk in the housing market, and it allows us to gain a better understanding of emergence and interconnectedness in the "world."

References

1. Harding, J., Rosenblatt, E., Yao, V.: The contagion effect of foreclosed properties. J. Urban Econ. **66**(3), 164–178 (2009)
2. Rogers, W., Winter, W.: The impact of foreclosures on neighboring housing sales. J. Real Estate Res. **31**(4), 455–479 (2009)
3. Lin, Z., Rosenblatt, E., Yao, V.: Spillover effects of foreclosures on neighborhood property values. J. Real Estate Finance Econ. **38**(4), 387–407 (2009)
4. Gangel, M., Seiler, M.J., Collins, A.: Exploring the foreclosure contagion effect using agent-based modeling. J. Real Estate Finance Econ. **46**(2), 339–354 (2013)
5. Immergluck, D., Smith, G.: The external costs of foreclosures: the impact of single-family mortgage foreclosures on property values. Hous. Policy Debate **17**(1), 57–79 (2006)
6. Gilbert, N.: Agent-Based Models. Quantitative Applications in the Social Sciences. SAGE Publications, Beverly Hills (2008)
7. Abdollahian, M., Yang, Z., Nelson, H.: Techno-social energy infrastructure siting: sustainable energy modeling programming (SEMPro). J. Artif. Soc. Soc. Simul. **16**(3), 6 (2013)
8. Abdollahian, M., et al.: Human development dynamics: an agent based simulation of macro social systems and individual heterogeneous evolutionary games. Complex Adapt. Syst. Model. **1**(1), 18 (2013)
9. Haldane, A.G.: The dappled world. Speech (2016)
10. Dosi, G., et al.: Fiscal and monetary policies in complex evolving economies. J. Econ. Dyn. Control **52**, 166–189 (2015)
11. Geanakoplos, J., et al.: Getting at systemic risk via an agent-based model of the housing market. Am. Econ. Rev. **1023**, 53–58 (2012)
12. Goldstein, J.: Rethinking Housing with Agent-Based Models: Models of the Housing Bubble and Crash in the Washington DC Area 1997–2009. George Mason University (2017)
13. Ling, D., Archer, W.: Real Estate Principles: A Value Approach. McGraw-Hill Irwin, New York City (2009)

Micro-Simulation Model as a Tool for Evaluating the Reform of China's Personal Income Tax

Xiangyu Wan[1,2(✉)]

[1] Institute of Quantitative and Technical Economics of CASS, Beijing, China
45440549@qq.com, wanxy@cass.org.cn
[2] China Institute of Income Distribution of BNU, Beijing, China

Abstract. For evaluating the redistributive effects of the reform of China's personal income tax system, this research attempts to break through the limitations of traditional research methods and provide a micro, coherent and structural framework for experimental research on policy reforms. Based on the microcosmic survey data of CHIPs and the system of personal income tax in China, a micro-simulation model was established to comprehensively analyze and quantify the redistributive effect of the reform of the personal income tax system and to clarify the final destination of the effect of the policy. This research provides the modeling tools and the quantitative basis for the design and evaluation of the reform of redistribution system.

Keywords: Personal income tax · Micro simulation · Income distribution
Policy effect attribution

1 Research Background

In recent years, the objective seriousness and subjective urgency of the inequality in income of Chinese people have been greatly highlighted due to the comprehensive factors such as the continuous fermentation caused by the follow-up of the international economic crisis and the slowdown of China's economic growth. According to public economics, the government can effectively adjust the income gap by formulating and implementing the corresponding redistribution policy, and the fiscal and taxation policies are its main tools. The personal income tax ("personal income tax"), which is called "Robin Hood Tax", is not only an important component of fiscal revenue, but also the main means of adjusting the income gap across the country because of its strong redistributive ability and it has long been the focus of attention in the society on redistribution issues.

From the point of view of micro-incidence analysis, Micro-Simulation is undoubtedly the most effective research approach [1]. Micro-Simulation has rapidly developed in the field of evaluating redistribution of public policy, also has become a powerful tool for policy analysis in government departments and even the legislative basis for the reform of relevant policies in some countries [2]. Applying Micro-Simulation method to analyzing the economic policies is also the frontier of

© Springer International Publishing AG, part of Springer Nature 2019
D. N. Cassenti (Ed.): AHFE 2018, AISC 780, pp. 254–261, 2019.
https://doi.org/10.1007/978-3-319-94223-0_24

economic research [3]. Among them, tax system evaluation has been the most important research field of Micro-Simulation. Its application focuses on the following aspects: (1) Combining Micro-Simulation with economic measurement or input-output. For example, Eason combined micro-simulation with econometric methods to analyze the distributional effect and fiscal effect of tax policy in the UK under the framework of input-output [4]; (2) Comparative application of transnational system, for example, Bourguignon et al. comparatively analyzed the distribution effect of tax systems of France, Italy and England [5]; (3) Comprehensive analysis of the tax system and transfer payment system, for example, Sutherland expanded the empirical study of EUROMOD in the EU tax and transfer payment model [6]; (4) The impact of tax system on micro-individual behavior, for example, Labeaga et al. took the labor supply behavior as the simulation object to analyze the fiscal effect and the distribution effect of the tax system reform [7].

However, in China, although Micro-Simulation model has been introduced successfully [8] and has long been used for the evaluation of personal income tax [9], due to the limitations of data and modeling techniques it still in the initial stage, and there is an urgent need to put Micro-Simulation model into reality to form a systematic research system. Based on this, the study focuses on establishing an effective Micro-Simulation model to quantify the multidimensional economic effects of various specific policy parameters of the current personal income tax system and finally provide a quantitative basis for the evaluation and design of the reform of the individual income tax system.

2 Modeling Frameworks

In order to build a bridge between macroeconomic analysis and microeconomic analysis, the micro-simulation method no longer adopts the traditional aggregate analysis model and typical individual analysis model, but based on the actual simulation of the relevant characteristic quantity for each micro-individual. The basic idea can be summarized as follows: (1) Composing the data files according to the sample of micro-individuals in the economic system, that is, establishing the simulation environment of the model; (2) Constructing the simulation model according to the individual behavior in the real economic system, that is, structuring micro-individual behavior patterns of model; (3) Applying computer simulation method to simulate the change of eigenvalues related to micro-individuals due to the changes of micro-individuals' characteristics and related policy variables (such as price, tax, etc.), that is, obtaining the simulation results through the operation of the model; (4) Through the statistics, analysis, inference and synthesis of the characteristic variables to obtain the impact of policy changes on the micro-individuals, and to further obtain the macro-level and the effect of the policy implementation at all levels, that is, accumulating and analyzing the simulation results.

The modeling ideas shown in Fig. 1.

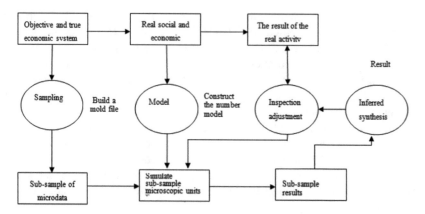

Fig. 1. Based on the database of micro-simulation model of the basic ideas

Based on the data of Chinese resident income survey project (CHIPs) and Micro-Simulation technology, we have realized the process of implementation from the individual income tax system to the corresponding micro-levy object, specifically, constructing the model of personal income tax can be divided into six steps: (1) problem identification; (2) data collection and processing; (3) to establish a model framework; (4) simulation experiment design; (5) computer implementation; (6) results analysis.

3 Empirical Results and Evaluation

3.1 Micro-data Description

Firstly, Table 1 presents the micro individual level information of CHIPs (2013) data. This study collected a total of 18379 samples of urban residents in China. Among them, the actual employed population with operating employment was 10740, representing 59.31% of the total sample by weighted representation, this indicator reflects China's current high labor force participation rate and actual employment level. Among all the working-employed population, the total number of taxable individuals applying to the wage income tax is 9845, while the weighted occupies 91.27% of the employed population. The taxable individuals applying for the service income tax are 118, weighted by the employed population 0.37% of the taxable individuals. The taxable individuals applying for operating income tax 1512, accounting for 14.35% of the employed population. Among the working population, the gross individual income tax is the absolute majority of taxable individuals. According to relevant statistics, the total amount of tax levied on labor income in 2013 accounted for over 84% of the total personal income tax in China. Among them, the wage income tax a proportion is as high as 62.3%. In addition, based on the results of the micro-data measurement, the Gini index of the pre-tax income of wage-earning individuals is about 0.4540, indicating that the data of China's pre-tax wage income have great inequality.

According to the economic point of view, the family is the basic unit of income distribution. The sample of urban residents collected in this study comes from 6,262 households, and the population weighted averagely by the representative weighted average is 2.96. Each household annual average disposable income is 81,092 yuan and each person of the household possess disposable average income of 29,510 yuan. According to the micro-data, the Gini index of pre-tax income of residents is 0.3424, while the Gini index of per capita income of residents before tax is 0.3420. The data shows that without adjusting income through personal income tax, the comprehensive income gap of urban households in China is far less than that of wage earners. In fact, this is mainly due to the characteristics of micro-family structure in China The structure of the labor supply structure formed on the basis of the comprehensive effect of the existing social security systems.

Table 1. Micro-sample household statistics

Household sample size (unweighted)	Average household size	Pre-tax household disposable income	Per capita disposable income before taxes	Gini index	
				Household income before tax	Household income before tax per capita income
6262	2.9569	81092.20	29509.94	0.3424	0.3420

3.2   The Micro-fate of the Current Personal Income Tax System

Based on the above micro-data foundation, we applied the micro-simulation model to apply the current personal income tax system to the micro-individuals in strict accordance with its levy rules and calculation methods.

Table 2. The current personal income tax system on micro-individuals

	Wages and income tax	Service income tax	Operating income tax
The actual individual taxable individuals taxable proportion	0.2660	0.7700	0.1721
Overall tax rate	0.0202	0.1393	0.0432
The amount of tax paid	0.7768	0.0308	0.1924
The overall tax rate	0.0231	Gini index for after-tax wage	0.4442

Table 2 first gives the current personal income tax system for micro-individuals, especially the influence of the labor people's income distribution status. As China has different deduction projects and deduction standard in salaries income, remuneration for labor services and other operating income, so the number of individuals actually taxed on personal income tax differs from the number of taxable individuals, and the taxable income actually paid to individuals of different subspecies is not the same in the proportion of tax individuals. In the table, we can see that due to all kinds of relief, especially the exemption amount, only 26.6% of the individuals actually paid the wage income, while 73.4% of the wage-earning individuals are totally covered by the exempted items and the deduction standard. This confirms that the administrative objective of previous personal income tax reform is to reduce the individual income tax burden on wage-earners, however maximize tax exemption does not mean tax fair. At first, it can be confirmed from the situation of the individual income tax that due to the differentiation of the levy rules and the exemption criteria, so the Remuneration which is mostly similar with wage and salary income in income attributes, accounting for as much as 77% of the actual taxpayers' income. At the same time, the operating income which should be mainly focused on income adjustment is actually only 17.21% of the actual taxpayers due to some specific implementation measures of the local taxation departments in the actual levy. Similarly, for individual income earners, the overall current tax rate for working-class income tax is 2.02%, the overall operating income tax rate is 4.32%, but the overall labor tax rate is as high as 13.93%. At present, there are serious drawbacks in the individual income tax system, and it is impossible to realize the tax fairness at the micro-individual level only through the continuous adjustment of the wage and income tax. The third row of Table 2 shows the proportion of Wages Income Tax, Service Income Tax and Operating Income Tax in the labor income tax. We can see that, due to the large number of working-class people, the tax accounted for 77.68%. Income tax on services Taxable income is limited, the tax accounted for 3.08%, and operating income tax accounted for 19.24%. If other income tax items, such as property income tax and contingent income tax, are included in the calculation, the proportion of wage income tax, service income tax and operating income tax in the total personal income tax will be approximately 64%, 2.5% and 16%., this data is consistent with relevant macro statistics. In addition, irrespective of property income and contingent income, the overall tax rate on personal income tax in China is about 2.31%, this is to say, for the labor income in our country, 2.3% of the total amount is taxed as personal income tax. Obviously, as a revenue which main target is adjusting income, it's proportion and scope is not matched with its propose, so increasing the direct tax and controlling the transition of the tax burden will be the key point in the future tax system reform. Finally, the Gini index of after-tax wage income was 0.4442, Compared with the pre-tax fell by 0.0098 percentage points. The data can be used as a measure of Redistribution capacity of the wage and income tax. We can see from it that the personal income tax system has a limited effect on decreasing the wage gap. It is not very objective to overemphasize the current redistribution effect of Wage Income Tax.

Although the subject of personal income tax is individual in china, but we still select family as our subject when we took into account the objective of taxation of Robin Hood tax revenue. The "All" row in Table 3 depicts the overall redistributive effect of the current personal income tax system at the household level. Data show that at present our country residents average personal income tax per household is 1391.79 yuan, the per capita level is 480.94 yuan, accounting for residents' pre-tax income of 1.6%. For several households, the average expected rate of personal income tax is 0.69%. Compared with individual level of labor income, the tax rate had reduced to less than 2% at the household level. In addition, after calculation, the household after-tax income Gini index of 0.3341, lower than the pre-tax income Gini index of

0.0083, while the household per capita income Gini index of 0.3346, lower than the pre-tax income per capita income Gini index of 0.0074. The above indicators reflect the current personal income tax system does not effectively improve the residents' income redistribution ability.

Table 3. The current personal income tax system at the household level

Group	Average household size	Household average disposable income	Household per capita disposable income	Household tax amount	Households per capita tax	Group overall tax rate	Household mean tax rate	Tax burden
1	3.66	27713.01	7426.23	34.18	10.56	0.0011	0.0007	0.0032
2	3.46	45572.54	13203.24	77.91	21.80	0.0016	0.0014	0.0058
3	3.23	54611.85	16939.90	142.62	41.71	0.0025	0.0022	0.0108
4	3.01	61085.05	20261.21	166.56	49.17	0.0026	0.0022	0.0132
5	2.92	69030.43	23630.35	186.58	79.79	0.0026	0.0019	0.0150
6	2.80	76696.31	27381.06	350.24	103.38	0.0043	0.0035	0.0273
7	2.80	88181.65	31568.82	573.32	172.13	0.0062	0.0050	0.0422
8	2.69	100332.50	37370.50	1056.46	329.84	0.0099	0.0081	0.0837
9	2.53	119209.02	47281.58	2193.04	693.83	0.0172	0.0133	0.1525
10	2.39	184750.60	77174.13	10203.92	3691.08	0.0504	0.0341	0.6463
All	2.96	81356.38	29588.14	1391.79	480.94	0.0160	0.0069	1.0000
Gini index for households after-tax	0.3341			Gini index for household per capital after-tax		0.3346		

Further, we conducted a declassified group study of micro-household data based on the per-capita disposable income level of households, and we can see the corresponding results in line 1 to 10 of Table 3. Due to the mismatch between the subject of personal income tax and the target of income adjustment, even the lowest 10% of the income group that should be exempted from tax obligations still need undertake the personal income tax of 34.18 yuan per household. The overall personal income tax of this income group is about 0.11%, and the expected rate is 0.07% per capital at the household level, so this group borne 0.32% of the total personal income tax. This phenomenon of "poor families failing to exempt from personal income tax" is mainly due to "there may be individuals with higher incomes from poor families with relative higher dependency ratio" and "some of the poor families can't be partially exempt from income tax relief" and so on reality reasons. The results further show that the current personal income tax system deviates from family economic attributes, which has resulted in the mismatch between the actual target and expected target, which may damage the policy accuracy and the effect of redistributes of personal income tax. From the perspective of protecting humanism of low-income groups or from the objective of ensuring scientific and modern state-governing capacity, it is of great realistic significance to carry out the "comprehensive income reform" and the "reform according to the family" system. Sub-low-income groups face a similar situation, average household pay a tax of 77.91 yuan, the overall tax rate of about 0.16% of the group, and this group undertook 0.58% of total tax. The highest income group is the main subject of personal income tax, the average household bear a tax of 3691.08 yuan, the overall tax rate of 5.04% and the expected average tax rate of 3.41% per household, the tax paid up to 64.63%. The second highest income groups pay the tax of 693.83 yuan each household and the tax rate is 1.72%, which slightly higher than the overall sample of 1.60%. This group of taxes accounted for 15.25%, with the highest income group accounting for 79.88%.

4   Conclusion

Based on the results of the above studies, we conclude the following main conclusions:

First, due to the structural characteristics of China's overall tax system and the measures for the calculation of individual income tax, the scope and amount of income actually levied on personal income tax in China are limited, and the overall tax rate is low and accounted for a small share in the total tax, which limited the advantage characteristics of facilitating the subject and transferring tax burden.

Second, the existing personal income tax system in china has a serious mismatch between the tax object and the policy target, so we should adopt a measure that adapts to the family structure and matches the means of income adjustment and units of income distribution. This will be the most important links of personal income tax reform in the future.

Third, the policy design of China's personal income tax system lacks the test of micro-foundation and scientific tools, and the policy design lacks rationality. Therefore, it is necessary to establish a scientific evaluation tool and systematically select to optimize the policy design.

Fourth, china's personal income tax system has too many categories and its classification basis lacks practical significance. However, the different taxation of different income seriously undermines the income adjustment ability, so the comprehensive income reform has great significance.

Finally, it should be pointed out that aside from the above empirical results, this study systematically introduces micro-simulation technology into the field of public policy evaluation and provides the basic module for future integrated modeling, which is of great theoretical and practical significance to China's income distribution, public management and application Economics and other related research.

Acknowledgements. This research is supported by Innovation Project of Chinese Academy of Social Sciences.

References

1. Orcutt, G.: A new type of social-economic system. Rev. Econ. Stat. **58**, 773–797 (1957)
2. Gupta, A., Kapur, V.: Microsimulation in Government Policy and Forecasting. Elsevier Science, Amsterdam (2000)

3. Bourguignon, F., Spadaro, A.: Microsimulation as a tool for evaluating redistribution policies. J. Econ. Inequal. **4**(1), 77–106 (2006)
4. Eason, R.: Microsimulation of direct taxes and fiscal policy in the united kingdom. In: Harding, A. (ed.) Microsimulation and Public Policy, pp. 23–46. Elsevier Science, New York City (1996)
5. Bourguignon, F., O'Donoghue, C., Sastre-Descals, J., Spadaro, A., Utili, F.: Eur3: a prototype european tax-benefit model. In: Gupta, A., Kapur, V. (eds.) Microsimulation in Government Policy and Forecasting, pp. 173–202. Elsevier Science, New York (2000)
6. Sutherland, H.: EUROMOD: the European Union tax-benefit microsimulation model. Int. J. Microsimulation **6**(1), 4–26 (2013)
7. Labeaga, J.M., Oliver, X., Spadaro, A.: Discrete choice models of labour supply, behavioural microsimulation and the Spanish tax reforms. J. Econ. Inequal. **6**(9), 247–273 (2008)
8. Gao, J.: Modeling of micro analysis and simulation and application of. Comput. Simul. Reform Pension Insur. Syst. **16**(2) (1999). (in Chinese)
9. Zhang, S., Wan, X.: A micro-simulation model and public policy analysis. J. Quant. Tech. Econ. **26**(8) (2009). (in Chinese)

Design of a New Setup for the Dynamic Analysis of the Recoil-Shoulder Interaction

Elie Truyen[1]([⊠]), Patrik Hosek[1], Niels Maddens[2], and Johan Gallant[1]

[1] Department of Weapon Systems and Ballistics, Royal Military Academy, Renaissance Avenue 30, 1000 Brussels, Belgium
elie.truyen@rma.ac.be
[2] Industrial Engineering Faculty, KU Leuven, Campus Oostende, Ostend, Belgium

Abstract. We addressed the measurement and modeling of contact forces during intense, short impact loading of the shoulder. Experimental recoil force measurements show significant intra- and inter subject dispersion, advocating a an approach without real shooters but based on a biomechanical model for the dynamic analysis of the recoil-shoulder interaction. This mechanical prototype would replace actual shooters and is being developed and optimized for the standard, military, standing shooting position, but the framework could easily be extended to other positions and applications like clay shooting and hunting. The obtained recoil parameters are more realistic than current measures. The latter are not registered in realistic conditions as they are not based on direct measurements of the recoil forces on the shoulder but evaluated in a free-to-move or blocked setup. We illustrate the effects of these conditions on the recorded recoil force.

Keywords: Recoil · Shoulder · Dynamic analysis · Contact force

1 Introduction

Firing a weapon generates a short intense force acting on the shoulder. High repetitive exposure can lead to injuries for hunters, sport shooters and soldiers. The maximum allowable level of recoil is imposed by the U.S. Army Test Operations Procedure TOP 3-2-504 [1]. This norm imposes the maximum number of shots that can be fired per day, based on the free recoil energy of the weapon-ammunition system. This global measure is evaluated with a free moving pendulum and is not based on direct measurements of the recoil forces on the shoulder. However, individual body characteristics, posture and the presence of personal protective equipment on the shoulder will influence the intensity and rate at which the recoil force is transmitted to the shoulder and these parameters should therefore be included in an adequate regulation.

Recoil can be expressed in measurable units as recoil force, recoil impulse or recoil energy. A clear link between the measurement of these quantities and perceived pain levels and injury criteria has not been reported yet in scientific literature. Burns [2] noted that pain and bruising may even occur at lower levels than the maximum allowable recoil. Blankenship et al. [3] attempted to quantify injury from recoil. Different techniques to identify injury risks, when firing a weapon with high recoil were assessed. Their focus was on the effects of recoil and not on the interaction between the shooter and the weapon or how the recoil force was transmitted to the shoulder. Cho et al. [4] reported a case of recurrent posterior shoulder instability after practicing rifle shooting. Bruno et al. [5] considered the transmission of impulse and vibration energy from the shoulder to the cervical spine and ipsilateral maxilla responsible for otoconia detachment. This resulted in observed dizziness within the first 3 days after shooting.

Most current impact- and crash models, related to the shoulder- and neck region are either passive and inertia driven or limited by the inherent difficulty of modeling and registering the shoulder joint kinematics due to lack of sufficient body landmarks to register all degrees of freedom experimentally. Early modelling efforts to describe the man-weapon interaction date back to the 1970's [6] but haven't developed drastically since [7]. These models are based on a Langangian formulation of the conservation laws, resulting in second order differential equations describing the movement of the body segments. Zakharenkov et al. [8] tried to model the interaction between the shooter and his weapon for different shooting positions. Lee et al. [9] modelled the human impulse characteristics through MSC.ADAMS and BRG.LifeMOD. They identified the impulse transfer paths in the horizontal direction in standing posture, proving that the effect of recoil is not restricted to the shoulder but propagated throughout the body. More recently Suchocki and Ewertowski [10] used a combined experimental and FEA approach to evaluate the contact forces during recoil. They showed that the impact force is greater for the prone position than for the standing position. Ewertowski [11] had already identified two distinct phases in the recoil loading. An initial impulsive action of the gun followed by a second pushing action. Although the maximum recorded forces were lower, the impulse during the second phase proved to be higher than for the first phase.

2 Recoil Parameters

2.1 Free- and Blocked Recoil

Recoil can be measured in two equivalent ways. In the case of free recoil, the rearward translation of the recoiling mass can be measured or the impulse can be estimated by recording the sway of a weapon mounted in a pendulum. The movement of the weapon can be recorded by high speed imaging or tracking of markers on the weapon body. Blocked recoil is evaluated by integrating a piezoelectric force sensor between the stock butt of the weapon and a rigid support. It is common to have only one degree of freedom in these setups. We evaluated the most common recoil parameters of six different weapon systems in this configuration as show on Fig. 1. The weapon is clamped on a movable sledge that can be blocked or move freely. The recoil force is determined in a blocked setup while recoil energy is derived from velocity, acceleration and displacement during free recoil. The movement during free recoil is stopped by a spring that is mounted in the longitudinal direction. The evaluated weapons were the following: FNC and SCAR-L (5.56 mm), SCAR_H and FAL (7.62 mm) and two riot guns (short barrel and long barrel). It is important to note that the recoil is always the result of a weapon-ammunition combination and does not solely depend on the used weapon.

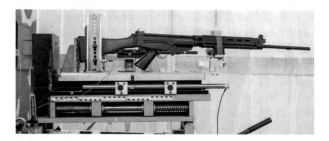

Fig. 1. Classic test setup for recoil evaluation. The weapon is clamped on a sledge that can be blocked or slide freely, depending on the targeted output parameters.

Every considered weapon was fired 10 times in a blocked setup and 5 times in a free-to-move setup. The average results are show in Table 1. We used a PCB M221B05 force sensor, a Kistler 8002 accelerometer and a LVDT (Linear Variable Differential Transformer) with a sensitivity of −0.3 V/cm for the displacements. The time history of the force signal was recorded for 0.15 s. The internal mechanics of the weapon are clearly reflected in the shape of this signal and vary greatly from one weapon system to the other. Hence one should be careful in the comparison solely based on F_{max}.

Table 1. Overview of recoil parameters for different weapons.

Weapon system	F_{max} (N)	Impulse (Ns)	v_{max} (m/s)	a_{max} (m/s^2)
FNC	790 ± 28	6.4 ± 0.1	−0.51 ± 0.00	−48 ± 3
SCAR-L	727 ± 15	7.4 ± 0.2	−0.55 ± 0.00	−51 ± 4
SCAR-H	1606 ± 39	13.5 ± 0.2	−1.00 ± 0.02	−158 ± 20
FAL	1714 ± 23	12.3 ± 0.2	−0.84 ± 0.04	−188 ± 5
RIOT long	2667 ± 139	14.4 ± 0.4	−1.71 ± 0.06	−375 ± 23
RIOT short	2692 ± 153	15.5 ± 0.6	−1.66 ± 0.01	−354 ± 18

2.2   Influence of the Shooter

The presence of the shooters' shoulder behind the weapon has a large influence on the amplitude and temporal characteristics of the recorded force signal. This can be illustrated by comparing the output from a blocked setup with the output from a trial with a shooter and an instrumented butt stock. The same PCB M221B05 force sensor was used as can be seen in Fig. 2. The force sensor is mounted in the middle of the shoulder piece. The bars on either side are sliding freely in the contact plate with the shoulder. This ensures that they do not cause additional resistance in the longitudinal direction. They are merely there to prevent buckling.

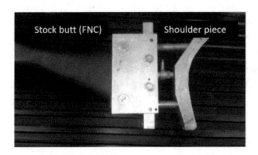

Fig. 2. Instrumented stock butt for recoil evaluation.

Figure 3 shows the time history of the recoil forces generated by the FNC in both setups. The higher frequency content of the blue curve is apparent. This clearly shows how the shoulder behind the weapon acts as a low-pass filter/damper. The blue curve, shows typical characteristics that can be linked to the mechanics of the weapon. The first peak (400 N) is due to the action of the gases on the breech during the shot. The second and most important peak (800 N) is due to the internal pieces moving backwards and hitting the weapon body on their rearmost position. This peak would not be visible for weapons with a different functioning principle. The last negative peak (−500 N) indicates the return of the internal pieces in the forward position. Figure 3 clearly illustrates the need to take the presence of the shoulder into account. The amplitude of the contact force between weapon and shoulder is much smaller than in a blocked setup. The blue curve, however, is still useful. The blocked recoil method allows to see the contribution of different weapon mechanics on the total recoil. Therefore this method is still useful in relative comparisons.

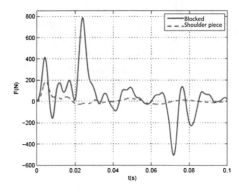

Fig. 3. Comparison between recorded force signals for a blocked recoil setup and with a shouldered instrumented stock butt.

3   Modeling the Shoulder-Weapon Interaction

3.1   Experimental Observations

The movement of the shoulder was registered with a Krypton camera at a sampling rate of 1000 Hz and an accuracy of 50 μm. Two markers were tracked in the sagittal plane, as can be seen on Fig. 4. The marker for the shoulder was attached to the right acromion and the second marker was attached to the stock butt. Their initial positions are marked by the vertical lines in the images. Two distinct phases can be identified. In Fig. 4b it is clear how the weapon initially moves backwards without a movement of the shoulder marker. It our hypothesis that this phases corresponds to the indention of the soft tissues. During the second phase (Fig. 4c), the shoulder and weapon move conjointly backwards.

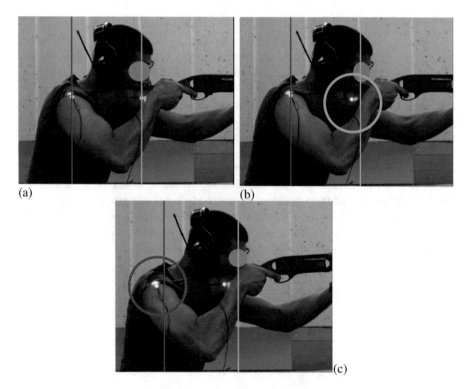

Fig. 4. Different stages of the recoil. (a) Neutral position. (b) Shoulder indentation (0.02 s). (c) Conjoint movement of weapon and shoulder (0.2 s).

The two phases are also clearly visible on Fig. 5, showing the horizontal displacement of a Riot gun for three different shots, fired by the same person. The first phase takes around 0.02 s. In this period the weapon moves back 4 cm. This phase is not as much affected by changes in posture. The second phase however clearly differs. The shooter adjusted his posture after the first shot. However, considering the fact that we use passive models without active muscle contractions, the validity of this region for our analysis doesn't extend more than a few tens of a second. This is related to the human reaction time of around 0.15 s. After this point the shooter will consciously react to the postural disturbance caused by the recoil. Hence, passive models will no longer be valid.

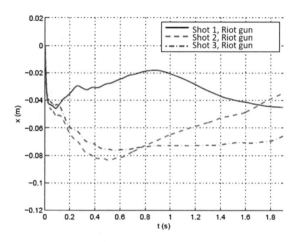

Fig. 5. Horizontal displacement of the RIOT gun

3.2   Mechanical Model Development

Our mechanical concept has only two degrees of freedom: a translation in the longitudinal shooting direction and a pivoting rotation around the hip in the sagittal plane. This is less than other models who also propose a secondary rotation around the shoulder. However, implementations of such models all show that the rotation around the shoulder was an order of magnitude smaller than the rotation around the hip ($< 10^{-2}$ rad). The effect of fixing the angle of the shoulder on the recoil parameters was minimal. The weapon is mounted on a sledge whose rearward movement speed is controlled by a dual spring damper system, i.e. similar to Sect. 3.1 the spring constants are changed after a certain threshold value of the gun displacement is reached. This threshold corresponds with the experimentally observed indentation of the soft tissues. Figure 6 depicts the mechanical concept for the recoil measurement. The framework is realized with aluminum profiles. This allows the flexibility to adapt dimensions and add accessories to the model. Analytical modeling has shown that the length and moment of inertia of the "trunk" have a significant influence on the movement of the weapon after the shot. The transition of the spring constants, from the first phase of the recoil to the second phase, is realized with only one spring. This becomes possible due to the principle, shown in Fig. 7. For every value of the angle α there is a change in resistance in the x direction. The spring, modeling the shoulder is mounted vertically, but creates

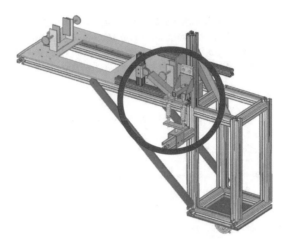

Fig. 6. Mechanical concept for the recoil measurement.

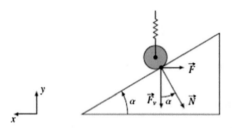

Fig. 7. The spring constant, modeling the resistance of the shoulder is adapted by changing the angle α.

a "virtual" spring in the longitudinal direction due to the inclination of the ramp. The relationship between the angle α and the virtual spring constant k' is derived in Formulas 1–6.

$$\tan(\alpha) = \Delta y \,/\, \Delta x \tag{1}$$

$$F_y = k \,\Delta y \tag{2}$$

$$\tan(\alpha) = \Delta y / \Delta x \tag{3}$$

$$F = \tan(\alpha)\, F_y = k \, \tan^2(\alpha) \, \Delta x \tag{4}$$

$$k' = k \, \tan^2(\alpha) \tag{5}$$

4 Conclusions

We discussed the traditional methods to determine recoil parameters and showed their shortcomings. The force recorded during blocked recoil differs considerably from the method with an instrumented stock butt, both in amplitude and in dynamic characteristics. It is shown that the movement of the weapon against the shoulder can be divided in two distinct phases. We proposed a simple mechanical concept to register more realistic values of the recoil force against the shoulder. We also showed how a change in mechanical resistance can be obtained with only one spring and a wheel running up a ramp.

References

1. TOP 3-2-504. Safety Evaluation of Hand and Shoulder Weapons; U.S. Army Test and Evaluation Command Test Operations Procedure 3-2-504, DRSTE-RP-702-104, Aberdeen Proving Ground, MD (1977)
2. Burns, B.P.: Recoil Considerations for Shoulder-Fired Weapons U.S. Army Research Laboratory, Aberdeen Proving Ground, MD (2012)
3. Blankenship, K., Evans, R., Allison, S.C., Murphy, M., Isome, H.: Shoulder-Fired Weapons With High Recoil Energy: Quantifying Injury and Shooting Performance. In: U.S. Army Research Institute of Environmental Medicine Report no. T04-04. Military Performance Division, Natick, MA (2004)
4. Cho, J.H., Chung, N.S., Song, H.K., Lee, D.H.: Recurrent posterior shoulder instability after rifle shooting. Orthopedics 35(11), e1677–e1679 (2012)
5. Bruno, E., Napolitano, B., Girolamo, S.D., Padova, A.D., Alessandrini, M.: Paroxysmal positional vertigo in skeet shooters and hunters. Eur. Arch. Oto-Rhino-Laryngol. 264(4), 381–383 (2007)
6. Hutchings, T.D., Rahe, A.E.: Study of Man-Weapon Reaction Forces Applicable to the Fabrication of a Standard Rifle Firing Fixture. No. RIA-R-TR-75-035, Rock Island Arsenal IL General Thomas J Rodman Lab (1975)
7. Taraszewski, M., Ewertowski, J.: Complex experimental analysis of rifle-shooter interaction. Def. Technol. 13(5), 346–352 (2017)
8. Zakharenkov, V.F., Arseniev, S.I., Belov, A.V., Agoshkov, O.G., Lee, Y.S., Kim, I.W., Chae, J.W.: Modeling and numerical investigation of the stochastic biomechanical interaction human-rifle system. In: 20th ISB – Ballistics International Symposium on Ballistics, pp. 216–224 (2002)
9. Lee, Y.S., Choi, Y.J., Han, K.H., Chae, J.W., Choi, E.J., Kim, I.W.: A study on the human impulse characteristics with standing shooting posture. Key Eng. Mater. 297, 2314–2319 (2005)
10. Suchocki, C., Ewertowski, J.: Experimental and finite element studies on man-Rifle reaction force. Probl. Mechatron. Armament Aviat. Saf. Eng. 1, 7–22 (2016)
11. Ewertowski, J., Piekarski, R.: The impact of selected sorts of rifles on shooter-comparative analysis. Issues Armament Technol. Sci. Bull. Mil. Inst. Armament Technol. 138, 73–86 (2016)

Cognitive Modeling

From Cognitive Modeling to Robotics: How Research on Human Cognition and Computational Cognitive Architectures can be Applied to Robotics Problems

Troy Dale Kelley[1](✉) and Christian Lebiere[2]

[1] U.S. Army Research Laboratory Aberdeen Proving Ground,
Aberdeen, MD, USA
troy.d.kelley6.civ@mail.mil
[2] Carnegie Mellon University Pittsburgh, Pittsburgh, PA, USA

Abstract. Cognitive psychology and Artificial Intelligence (AI) have long been intertwined in the study of problem solving, learning, and perception. The early pioneers of AI, Herbert Simon and Allen Newell, drew as inspiration chess masters and from their study developed computer programs to mimic the problem solving abilities identified in chess masters. The understanding of chess strategies relied heavily upon characterizing the problem space as a combination of symbolic inference and statistical pattern matching, which allowed for a quick understanding of the environment by computer systems. Recently, robotics has emerged as an AI domain, and the problem space has proven a difficult one due to the sub-symbolic nature of the knowledge. As robotics has emerged as a field in AI, cognitive architecture researchers have continued to refine their understanding of cognition in new ways that allow for the duplication of human problem solving with limited resources. The goal of this manuscript is to inform the AI world of the successes cognitive architectures have produced with the hope that this knowledge can be transferred to AI, and more specifically, robotics.

Keywords: Cognitive architectures · Artificial Intelligence · Robotics

1 Introduction

Allan Newell's classic 1973 paper, "You can't play twenty questions with nature and win" was a seminal treatise concerning science and the direction of scientific research. In that paper, Newell noted that much of psychology had fallen into a trap where - to use the old saying - the forest could not be seen through the trees. Psychology researchers at the time were dividing the task of understanding the complexity of human cognition by repeatedly dividing it into studies in narrow, specialized areas, updating individual algorithms to support existing theory and model piecewise data, while not being aware of how each theory fit into an overall theoretical framework. As Newell noted, while "the policy appears optimal – one never risks much, there is feedback from nature at every step, and progress is inevitable. Unfortunately,

© Springer International Publishing AG, part of Springer Nature (outside the USA) 2019
D. N. Cassenti (Ed.): AHFE 2018, AISC 780, pp. 273–279, 2019.
https://doi.org/10.1007/978-3-319-94223-0_26

the questions never seem to be really answered, the strategy does not seem to work". [1] And while Newell at the time was lamenting the lack of cumulative progress in the field of psychology, this reasoning could be applied to other areas of scientific inquiry, not just psychology, but also to recent research in the area of Artificial Intelligence (AI) and Robotics. The research has concentrated on expanding existing theory (neural networks) by the use of brute force methodology (convolutional neural nets and deep learning) and dimensionality reduction (Principle Components Analysis) which are limited approaches and do not address underlying theoretical issues of adaptability and generalizability which are hallmarks of human cognition. Newell formulated a dozen such criteria for unified theories of cognition [2] that were later refined and used to evaluate competing approaches to human cognition and AI [3].

Newell's answer to the problems addressed in his 20 questions paper was to develop a Cognitive Architecture (CA), which is a computational instantiation of a *unified* theory of cognition. A CA would allow for all aspects of cognition to be tested and ultimately integrated in a coherent framework. Studies of attention, perception, and problem solving could be done within the context of a unified theory. CAs allowed for computational implementations of human cognition that mimicked aspects of human intelligence, but within a unified architecture rather than in isolation, bringing converging constraints to bear on each aspect of cognition.

The goal of this manuscript is contrast the AI approach to problem solving (neural nets, PCA/vector space representations, POMDP) with the cognitive approach to problem solving within CAs (hybrid architectures, memory separation, adaptivity) with an emphasis on the successes CAs have produced. It is hoped that AI and robotics researchers will take the CA research more seriously, not be destined for the 20 questions trap, and transfer lessons learned from CAs to robotics.

2   Artificial Intelligence Approaches

Neural networks, were developed and refined as a sub-symbolic distributed processing systems, much like collections of neurons in the brain [4]. However, neural networks are lacking in the critical component of adaptivity. Once a neural network is trained, it cannot be updated in a continuous fashion, or this leads to catastrophic forgetting [5]. The neural network research community has not aggressively attempted to solve the problem of neural network adaptivity, instead, they have focused on a more brute force methodology of deep learning [6]. This problem of adaptivity is also apparent in vector space approaches (vector support machines, singular value decomposition) where the space is not easily manipulated in real time based on new information. Typically, the space is well defined and the distances between vectors in that space are functions of calculations that represent the entire space, so changes to the representation effect the entire space.

Some algorithms used in AI have been co-opted from other disciplines without regard for the assumptions that were used to develop the algorithms in the first place. For example, Principle Component Analysis (PCA) has been a popular technique for the extraction of the most important vectors in a problem space. However, PCA was developed as a statistical analysis tool [7] and certain statistical assumptions must be met before the technique can be applied. The major assumptions of PCA are that (1) the data is linear, (2) the vector with the highest variance is the most important, and that (3) the relationship amongst the vectors is orthogonal. PCA has been applied to image recognition where these assumptions are violated [7]. For example, lighting can cause the most variance in images but is not a principle component for object identification. Newer analysis techniques such as Independent Components Analysis address the problem of linear relationships, but still violate the variance problem.

Markov decision making has been used in robotics for several decades [8]. Like PCA, Markov models use a certain set of assumptions in order to calculate optimal behavior given a state. One of the basic assumptions is that a Markov model is *independent* at each time step of all other previous states and actions. This is the Markov property [9]. However, much of the information received by a robot, in a real world situation, is related to the previous information. This is known as *context* in psychology and is used heavily by humans for object recognition and decision making. Additionally, another basic assumption of Markov models is that there are a certain number of identifiable states from which to base decisions upon. However, many robotics researchers have found that there are actually and infinite number of states a robot can be in, and even in what appears to be the same state (perceiving the same wall from the same location) can be different due to sensor error and lighting changes. This has led to the development of Belief Networks [10] but these are computationally intractable.

Navigation and localization is one of the most important algorithms for a robot. One of the most popular algorithms for navigation and localization is SLAM (Simultaneous Localization and Mapping). The SLAM algorithm creates an exact representation or map of the environment. While this seems useful as a mapping algorithm and for navigation, it has proven to be an inflexible representation of the environment. We know from the psychological research that humans create a landmark-based representation of the environment, not an exact map. For example, a human cognitive map represents locations as landmarks (a church) and the connections between the landmarks (the church is north of the school) are used to indicate general distances between landmarks (the church is about a mile north of the school) [11]. In contrast, using the AI approach means there are no cognitive landmarks on a SLAM map. This means a blade of grass on a SLAM map is just as important as a church! Additionally, a human landmark-based map allows for dynamic representations, since changes between the landmarks are not important (snow, rain, lighting), if the landmarks are still recognizable. This generalized sparse representation is typical for human knowledge organization and it contrasts greatly with AI knowledge organization of exact and fully complete representations.

3   Lessons Learned from Cognitive Architectures

One of the most important discoveries in the development of CAs is the organization of memory. Early research on subjects with amnesia [12] found that memory was separated into distinct units even at the neurochemical level [13]. The computational advantage of this is not yet known, but in the broadest sense we know that humans separate memories into declarative (facts), procedural (how to accomplish tasks) and episodic memories (events), and these are well reflected in CAs. However, this idea is not well represented in AI. Typical robotics systems use a "World Model" which is a collection of all types of memory, irrespective of origin or the eventual use of the memory [14]. These memories are typically organized by the developer based on convenient programming conventions and not intended usage.

Not only is memory separation (declarative, procedural, and episodic) important within a CA, but the dynamic properties of memory, such as decay, are also important. While decay might seem to be a dis-advantageous aspect of human cognition, and perhaps to a robot as well, memory decay serves a function of removing irrelevant unimportant or outdated information [15]. Conversely, the strengthening of memories enables the most important information to become more accessible. Moreover, these processes of decay and strengthening follow pervasive regularities such as the power law of practice and forgetting [16]. Not only does memory decay or strengthening serve a purpose of removing or emphasizing information, but memory strength distributions allow the brain to be *adaptive*. Adaptivity is one of the most important aspects of human cognition. By allowing the strength of memories to vary as part of the inter-action with the environment, continuously adaptive behavior emerges as a byproduct of functional task selection using efficient algorithms.

The use of memory decay and strength has been used in robotics applications, and there have been three insights from this unpublished research by the authors. The first insight was that the most powerful way of endowing the robotic architecture with cognitive capabilities is not by using CAs to implement specific modules but rather by embedding cognitive mechanisms as a central component of the robotic architecture. A second insight was that increasing robot autonomy results in ever growing world model size and complexity, which combined with limitations in CPU, gives rise to the same problem that human memory evolved to solve: how to determine the most relevant piece(s) of information in memory, given a specific history and processing context? The third insight was that in order to achieve the full spectrum of human cognitive capabilities, a tight combination of symbolic representations, for properties like generality, and statistical mechanisms, for properties like adaptivity, are required [17]. Combining those insights, the CA approach can augment a robotics world model approach by endowing every object with the same activation computations as those in a CA's long-term memory [18]. Thus, activation of world model objects reflect factors such as recency and frequency, contextual relevance, and semantic similarity to information requests. Potential uses of this process include the disambiguation of human statements, execution monitoring, detection of unexpected events, and rele-vance judgments.

Another CA approach is to combine cognitive architectures and neural networks to develop a hybrid system where each component contributes what it does best. SAL (Synthesis of ACT-R and Leabra) [19] is a hybrid CA resulting from the integration of the symbolic Atomic Components of Thought-Rational (ACT-R) architecture [20] and the neural Leabra architecture [21]. The basic principle is to leverage each framework for what it does best: the neural architecture for its perceptual abilities, and the sym-bolic architecture for its complex memory and control abilities. One possible appli-cation of this hybrid methodology is to leverage the adaptivity of the symbolic architecture to provide metacognitive oversight of neural perception. For example, Vinokurov [22] used quick-learning symbolic declarative memory to introspect into the neural perception layers and store associations between those distributed representations and corresponding object categories in order to detect previously unseen objects and start training neural perception to recognize those new categories. Similarly, Vinokurov [23] used memories of previous objects and their locations to provide top-down expectations to bottom-up neural perception in order to improve object recognition and localization. In both cases, the integration of neural processing with a symbolic CA provided unique benefits by combining fast adaptive learning of symbolic memories with slower learning in neural perception.

Additionally, CAs can inform the structure and content of machine learning models. Fields [24] have shown that there are broad parallels between modeling patterns in CAs such as instance-based learning and machine learning techniques such as k-nearest neighbors. Those parallels can be exploited further by using the constraints in CAs to develop cognitive models that can be used to train machine learning models such as Markov networks. Sycara [25] developed a cognitive model of human control of robotic swarms that provided a high-fidelity account of human learning and decision-making. The cognitive model generated large amounts of training data over a broad range of conditions that was used to train a Markov model. The state representation in the Markov model was a discretization of the continuous representation of the control instances in the cognitive model's memory. The resulting methodology can be applied directly to any instance-based cognitive model to generate a machine learning model such as Markov or Bayesian network.

Because they are based on symbolic representations, CAs can be used to integrate knowledge bases with statistical mechanisms such as those providing adaptivity and generalization in cognitive models. Oltramari and Lebiere [26] describe a general methodology for integrating CAs and knowledge bases that unifies lexical (e.g., WordNet), structural (e.g., FrameNet) and statistical (e.g., latent semantic analysis) ontologies in a memory representation that can leverage activation-based knowledge retrieval mechanisms. The resulting integrated cognitive knowledge representation and mechanisms can be used to recognize human actions in video [27], behavioral patterns in a physical environment [28], and physical structures such as buildings from partial observation [29]. Further developments to address scaling and generality limitations of this approach are discussed by Lieto [30].

4 Conclusions

This paper is meant to inform the robotics community about principles implemented in cognitive architectures and their applicability to robotic problems. We argue that CAs can provide a coherent framework for testing the interactions amongst algorithms, and the use of CAs can guard against incrementally changing algorithms to produce small advancements without regard to the overall theory (the 20 questions trap). We highlight recent advances in CAs by using memory decay/strength to control information access for large memory stores, and hybrid architectures to leverage the strength of knowledge representation techniques (symbolic and sub-symbolic). We argue that robotics and

CAs are not incompatible, but rather that they can be combined to endow robots with human characteristics such as adaptivity, generalizability, and robustness thus enabling robots to work naturally with humans in the future.

References

1. Newell, A.: You can't play 20 questions with nature and win: Projective comments on the papers of this symposium (1973)
2. Newell, A.: Unified Theories of Cognition. Harvard University Press, Boston (1990)
3. Anderson, J.R., Lebiere, C.: The Newell test for a theory of cognition. Behav. Brain Sci. **26**, 587–637 (2003)
4. Rumelhart, D.E., McClelland, J.L.: PDP Research Group: Parallel distributed processing, vol. 1, p. 184. MIT press, Cambridge (1987)
5. Fahlman, S.E., Lebiere, C.: The cascade-correlation learning architecture. In: Advances in neural information processing systems, pp. 524–532 (1990)
6. Arel, I., Rose, D.C., Karnowski, T.P.: Deep machine learning-a new frontier in artificial intelligence research [research frontier]. IEEE Comput. Intell. Mag. **5**(4), 13–18 (2010)
7. Shlens, J.: A tutorial on principal component analysis. arXiv preprint arXiv:1404.1100 (2014)
8. Koenig, S., Xavier, R.: A robot navigation architecture based on partially observable markov decision process models. In: Artificial Intelligence Based Mobile Robotics: Case Studies of Successful Robot Systems, pp. 91–122 (1998)
9. Kallianpur, G., Mandrekar, V.: The Markov property for generalized Gaussian random fields. Ann. Inst. Fourier **24**(2), 143–167 (1974)
10. Cooper, G.F.: The computational complexity of probabilistic inference using Bayesian belief networks. Artif. Intell. **42**(2–3), 393–405 (1990)
11. Hanford, S.D., Janrathitikarn, O., Long, L.N.: Control of mobile robots using the soar cognitive architecture. J. Aerosp. Comput. Inf. Commun. **6**(2), 69–91 (2009)
12. Milner, B., Corkin, S., Teuber, H.L.: Further analysis of the hippocampal amnesic syndrome: 14-year follow-up study of HM. Neuropsychologia **6**(3), 215–234 (1968)
13. Nissen, M.J., Knopman, D.S., Schacter, D.L.: Neurochemical dissociation of memory systems. Neurology **37**(5), 789–794 (1987)
14. Dean, R.M.S.: Common world model for unmanned systems. In: Unmanned Systems Technology XV, vol. 8741, p. 87410O. International Society for Optics and Photonics, May 2013
15. Newell, A., Rosenbloom, P.S.: Mechanisms of skill acquisition and the law of practice. In: Anderson, J.R. (ed.) Cognitive Skills and Their Acquisition, pp. 1–55. Lawrence Erlbaum Associates, Hillsdale (1981)
16. Anderson, J.R., Schooler, L.J.: The adaptive nature of memory. In: Tulving, E., Craik, F.I.M. (eds.) Handbook of Memory, pp. 557–570. Oxford University Press, New York (2000)
17. Laird, J.E., Lebiere, C., Rosenbloom, P.S.: A standard model of the mind: toward a common computational framework across artificial intelligence, cognitive science, neuroscience, and robotics. AI Mag. **38**(4) (2017). https://doi.org/10.1609/aimag.v38i4.2744
18. Anderson, J.R., Lebiere, C.: The Atomic Components of Thought. Lawrence Erlbaum Associates, Mahwah (1998)
19. Jilk, D.J., Lebiere, C., O'Reilly, R.C., Anderson, J.R.: SAL: an explicitly pluralistic cognitive architecture. J. Exp. Theor. Artif. Intell. **20**(3), 197–218 (2008)

20. Anderson, J.R., Bothell, D., Byrne, M.D., Douglass, S., Lebiere, C., Qin, Y.: An integrated theory of the mind. Psychol. Rev. **111**(4), 1036 (2004)
21. O'Reilly, R.C., Munakata, Y.: Computational Explorations in Cognitive Neuroscience: Understanding the Mind by Simulating the Brain. MIT Press, Cambridge (2000)
22. Vinokurov, Y., Lebiere, C., Wyatte, D., Herd, S., O'Reilly, R.: Unsupervised Learning in Hybrid Cognitive Architectures. In: Proceedings of AAAI-12 Workshop on Neural-Symbolic Learning and Reasoning (2012)
23. Vinokurov, Y., Lebiere, C., Szabados, A., Herd, S., O'Reilly, R.: Integrating top-down expectations with bottom-up perceptual processing in a hybrid neural-symbolic architecture. In: Proceedings of the Fourth Annual Meeting of the BICA Society (BICA-2013) (2013)
24. Fields, M., Lennon, C., Lebiere, C., Martin, M.K.: Recognizing scenes by simulating implied social interaction networks. In: Proceedings of the 8th International Conference on Intelligent Robotics and its Applications, Portsmouth, UK, 24–27 August 2015 (2015)
25. Sycara, K., Lebiere, C., Pei, Y., Morrison, D., Tang, Y., Lewis, M.: Abstraction of analytical models from cognitive models of human control of robotic swarms. In: Proceedings of the 13th International Conference on Cognitive Modeling (ICCM-2015), Groningen, NL (2015)
26. Oltramari, A., Lebiere, C.: Knowledge in action: Integrating cognitive architectures and ontologies. In: Oltramari, A., Vossen, P., Qin, L., Hovy, E. (Eds.) New Trends of Research in Ontologies and Lexical Resources: Ideas, Projects, Systems. Springer, Germany (2013)
27. Oltramari, A., Lebiere, C.: Using ontologies in a cognitive-grounded system: automatic action recognition in video-surveillance. In: Proceedings of The 7th International Conference on Semantic Technologies for Intelligence, Defense, and Security (STIDS 2012). Fairfax, VA (2012)
28. Kurup, U., Lebiere, C., Stentz, A., Hebert, M.: Predicting and classifying pedestrian behavior using an integrated cognitive architecture. In: Proceedings of the Behavior Representation in Modeling and Simulation (BRIMS-12) Conference, Amelia Island, FL (2012)
29. Oltramari, A., Vinokurov, Y., Lebiere, C., Oh, J., Stentz, A.: Ontology-based Cognitive System for Contextual Reasoning in Robot Architectures. Presented at the AAAI Spring Symposium on Knowledge Representation and Reasoning in Robotics. AAAI Spring Symposium Technical Report SS-14-04. Menlo Park, CA: AAAI Press (2014)
30. Lieto, A., Lebiere, C., Oltramari, A.: The knowledge level in cognitive architectures: Current limitations and possible developments. J. Cogn. Syst. Res. (2017). http://dx.doi.org/10.1016/j.cogsys.2017.05.001

Adaptive Automation in Cyber Security

Daniel N. Cassenti$^{(\boxtimes)}$, Vladislav D. Veksler, and Blaine Hoffman

Human Research and Engineering Directorate,
United States Army Research Laboratory, 519 Mulberry Point Rd.,
Aberdeen Proving Ground, MD 21005, USA
{daniel.n.cassenti.civ, vladislav.d.veksler.ctr,
blaine.e.hoffman.civ}@mail.mil

Abstract. Cyber analysts must work long hours and under intense pressure all while performing complex mental tasks. Research is needed to improve the execution of their tasks. We describe a research plan that can be used to provide cyber analysts with the assistance they need using adaptive aids. The Multi-Level Cognitive Cybernetics (MLCC) [1] approach provides a methodological approach to studying adaptive automation and advancing its development across multiple levels of analysis. We follow up on a previous paper [1] by focusing on how MLCC can be used to improve cyber analyst tasks. We use the breakdown of cyber situation awareness by D'Amico and Whitley [2] to analyze the different cyber tasks and provide general examples of aids that may assist with each of the three stages. A more in-depth example of a specific type of cyber adaptive aid is then provided. We conclude with a call for action to expand adaptive automaton into the cyber realm.

Keywords: Cyber security · Cybernetics · Cognitive psychology
Human factors · Research methods

1 The Role and Tasks of a Cyber Analyst

Cyber analysts face difficult and complex tasks in their roles. Compounding the problem is the long hours, stress, and vigilance that the job entails. With such high stakes for any error the analyst commits (e.g., a breach in the network leading to the theft or manipulation of critical data), it's incumbent on human factors research to design software solutions that will simplify the analyst's tasks, reduce stress, and bring attention to problematic cyber events so they are not missed. In Subsects. 1.1–1.2, we discuss the cyber analyst role in more depth. In Sect. 2, we outline the Multi-Level Cognitive Cybernetics perspective, which can be used to help cyber analysts by providing adaptive aids that adjust to the cyber analyst's needs. Sections 3 and 4 reviews general adaptive automation that may help with cyber analyst tasks, then two specific examples.

1.1 Difficulty of Cyber Work

Cyber analysts work under fairly taxing conditions. These often include twelve-hour shifts, half of which are night-shifts, comprised of staring at a glowing screen with the purpose of sifting out rare records that may signify danger among thousands of similar-looking benign records of network activity. Delays in processing potential threats can lead to catastrophic problems (e.g., a Chinese government cyber attack on the Office of Personnel Management that led to the collection of the private data of millions of United State federal workers [3]), which puts a great deal of stress on a cyber analyst suffering from the effects of fatigue. Given these influences on performance, it makes sense to offer cyber analysts solutions via human factors on the software they use [4]. Improved software means aids that positively impact imperfect work performance and decreases odds that the potential catastrophic failures will occur.

1.2 Work Activities

Cyber analyst work performance relies on an analyst's use of perceptual, information processing, and decision making processes to complete their tasks. Therefore, cyber analysis is a cognitive problem, rather than a physical or motivational one [5]. Although cyber analysts use a wide variety of specific tools and application to perform their jobs, in general there work activity involves visually scanning strings of code representing cyber traffic into and out of the employing organization's network.

The data that a cyber analyst must process does not represent all of the traffic that a network receives or transmits. Doing so would put impossible burdens on the analysts. Instead, Intrusion Detection Software (IDS) filters the traffic so that only potentially malicious activity is presented to the analyst. However, IDS is designed to be cautious so that even traffic with the slightest deviation from normal patterns reaches the analyst (see [6]). This design choice is made to ensure that malicious activity does not go unnoticed since IDS is the first line of defense, however this puts analysts under greater workload demands than they would face with stricter filters [7].

The code that a cyber analyst looks for to determine whether activity is a threat or benign covers a wide range of potential warning signs. The code of interest could be from a certain IP address or class of IP addresses (indicating a specific attacker), a certain type of activity (e.g., attempt to access a secret file), or outbound traffic from an insider threat. Needing to track a host of different warning signs only serves to increase the cognitive workload demands on a cyber analyst.

Alcan and Başar [8] equate cyber analyst work to game theoretic puzzles, which means that analyst's must also weight complex schemas of attacker-defender dynamics, which make cyber analyst work complex and difficult to process. This is especially true when the analyst must process historical activity to anticipate future attacks. Underlying this type of analysis is the types of psycho-social considerations to discern motivation and goal states, so that similar pattern in the future will be met with successful counter-measures.

Given all of these cognitive demands on analysts, it makes sense to devise ways to provide the analyst with relief by offsetting some of the cognitive burdens through the use of human factors techniques to designing software aids (see [9]).

2   Multi-Level Cognitive Cybernetics

Cyber analysts are highly trained in the tasks they must perform, however the stressors and complexities discussed above would make predicting performance difficult. Aids to help cyber analysts should therefore be delivered only when the analyst needs them. If a cyber analyst is performing well, the introduction of an aid would not serve to increase performance much, but could decrease performance by serving as a distraction. However, if stressors overwhelm the analyst's cognitive processing, an activated aid will minimize the decline in performance. The solution is adaptive automation, software aids that only engage when the user needs them.

We proposed using the Multi-Level Cognitive Cybernetics (MLCC; [1]) approach to developing adaptive automation for cyber security. It helps to break down the parts of this term to understand what goes into the approach.

First, cybernetics [1, 10] is the study of communication and control between animals and their environments, operating within closed-feedback loops. In the present work however, we can express this as communication and control in a closed-feedback loop between a user and software. The communication and control typically extends from the user to the software in a unidirectional way in which the software only responds to deliberate keyboard, mouse, and touch-screen activity. In the MLCC approach, the software could perceive other types of communication from the user with the intent of deciding when the user is in a state that has or will lead to degraded performance. When this happens, an automated aid will engage. The software will continue to monitor these signals and turn off the aid when a state supporting good performance re-commences.

The cognitive portion of the MLCC approach emphasizes that users need help with mental processing. Since software is typically designed to facilitate perceptual incorporation of information and actions are made using well-practiced mouse, keyboard, and touch screen movements, users tend to need help with cognitive processing only. In MLCC, cognition is broken down into stages of processing as represented in Fig. 1. We will discuss this in great detail in Sect. 2.3. Figure 1 below outlines the stages of cognition along with corresponding stages in the technology portion in MLCC.

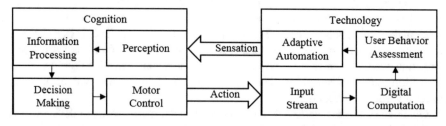

Fig. 1. The MLCC framework including stages of cognition and adaptive automation [1].

Sensory output from the technology (e.g., the visual interface, warning sounds) enter through the user's sensory organs and initiate cognitive processing, beginning with the conversion of raw sensations to perceptual representations that can be understood by the user. These representations are filtered in the information processing stage to facilitate focus on the important elements from the perception stage. Once these elements are isolated, a decision-making stage works to process a course of action and feed that desired action into motor control processing. Motor behavior produces the action and this is in turn processed by the technology.

The multi-level term applies to the ways that adaptive automation can interpret the user's state to provide the timing as well as the type of adaptive automation that the user needs to boost performance. This will be discussed in greater detail in Sect. 2.2.

2.1   Closed-Feedback Loops

The innovation of cybernetics is that it expresses interactions within closed feedback loops, thus framing communication and control in a way that can account for how all the activity of the intentional entity is supported or detracted by the unintentional (i.e., the environment) or other intentional entity. As it applies to a user interacting with software, cybernetics would seem to suggest that the along with the user, the software closes the feedback loop as a supportive, unintentional entity.

Though MLCC does not include philosophical argument for the intentionality of software, it takes the position that the software at least may have some degree of agency, by changing how it supports the user based on the user's behaviors. In traditional human factors work, software is more static that under MLCC as designers create software that blindly follows whatever deliberate behavior the user offers. Traditional human factors still results in a closed-feedback loop, but the software doesn't adapt except to deliberate user action. Adaptive automation expands to interpret other kinds of behavior.

2.2   Levels of Analysis

MLCC includes four levels of analysis that may be used to develop adaptive automation algorithms for when adaptive aids are necessary. It may seem intuitive to decide to have computer aids on all the time, but when the user is not having difficulties, the aids cannot improve performance much and therefore would be more likely to serve as a distraction than a help. Additionally, an active aid introduces the problem of automation bias [11], which is a heuristic to accept the results of an automated aid, even though the aids are fallible. If aids were not fallible, then the user would not be necessary (i.e., the process could be automated), so the risk of automation bias should only be taken when the user is in a low-performance state. Dutt et al. [12] found that automation bias is particularly prevalent for cyber analysts using decision support tools, which are generally more complex than adaptive automation, but the warning should still apply.

The first level is metacognitive. With this, the user evaluates her or his ability to perform the task at hand. This could be done a priori and thus adaptive automation triggers whenever the task requires a skill that the user is not confident can be performed well. It can also be done online such that the user make an explicit call for adaptive aids to engage.

The second level is performance-based, where an adaptive aid will engage if some continuous measure of performance dips below a threshold. If there is no continuous measure of performance, then the designers could devise an aid to trigger based on some physiological indicator or indicators of performance such as heart-rate variability [13] or pupil diameter [14]. These may not be based on a threshold model like performance-based measures, but could require a more algorithmic complicated approach.

The final level is computational modeling (see [15]), which is different from the others in that it is employed based on data derived from past use of the tool. The computational modeling level allows the designer to figure out what thresholds or algorithms to set for the engagement of adaptive aids. Without this level, it would be guesswork as to what level of performance or physiological response is appropriate for adaptive automation. The computational modeling level also is used to determine what types of aid should be deployed, which is discussed in the next section.

2.3   Stages of Cognitive Processing

Although the timing of when to engage adaptive automation is critical, not all adaptive aids should be considered equal. Some may help more than others. To figure out what aids will help the most, a human factors analyst will needs to divide the cognitive processes involved in a task into meaningful units. The left-hand portion of Fig. 1 divides cognition into general stages. Any of the four may be at issue, however well-designed software should present less difficulty to the perception and motor control stages. MLCC focuses mostly on the information-processing (i.e., filtering and sorting information to focus on important elements) and decision-making (i.e., arriving at a single course of action) stages.

Researchers can use a cognitive task analysis (see [16]) to figure out which stage of processing needs more help, but MLCC recommends running studies to derive empirical data, which may then lead to cognitive models. The cognitive models have a host of information that can be used to derive measures of complexity and a narrowing down of what aids may work best in reference to the stages of processing. We discuss this more in [15].

3   Stages of Cyber Situation Awareness

Just as MLCC breaks down the closed-feedback loops of human-computer interaction into stages, D'Amico and Whitley ([2]; see also [17]) break down the situation awareness (see [18]) process of a cyber analyst into stages. These stages form a type of closed-feedback loop themselves as the results of stage 1 feed into stage 2, then stage 2 into stage 3. Stage 3 results effect how stage 1 data is processed. In the next three sub-sections, we will discuss these three stages and how adaptive automation may help each.

3.1   Level 1: Perception/Detection

The first stage of cyber security situation awareness is also one of two stages that D'Amico and Whitley [2] describe as the tactical level. D'Amico and Whitley [2] frame the tactical level as a process of applying filters to select out increasingly dangerous activity. Phase 1 is the perception and detection stage and moves from raw data to suspicious activity. The first layer of filtering is from raw data to interesting activity and is controlled by the automated IDS system. The cyber analyst sees the interesting activity, but since the IDS errs on the side of false alarms, much of the interesting activity is actually benign, but must be checked regardless (Fig. 2 display what post-IDS data may look like). The analyst flags the activity that may have mal-intent as suspicious activity and this set moves to Level 2 processing.

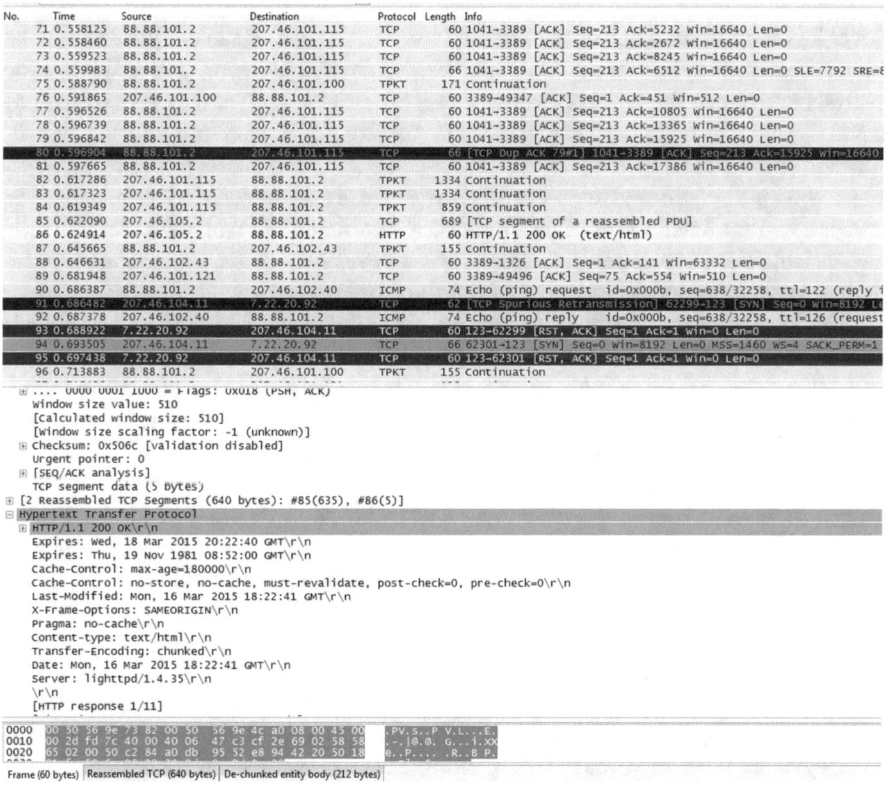

No.	Time	Source	Destination	Protocol	Length	Info
71	0.558125	88.88.101.2	207.46.101.115	TCP	60	1041-3389 [ACK] Seq=213 Ack=5232 Win=16640 Len=0
72	0.558460	88.88.101.2	207.46.101.115	TCP	60	1041-3389 [ACK] Seq=213 Ack=2672 Win=16640 Len=0
73	0.559523	88.88.101.2	207.46.101.115	TCP	60	1041-3389 [ACK] Seq=213 Ack=8245 Win=16640 Len=0
74	0.559983	88.88.101.2	207.46.101.115	TCP	66	1041-3389 [ACK] Seq=213 Ack=6512 Win=16640 Len=0 SLE=7792 SRE=8
75	0.588790	88.88.101.2	207.46.101.100	TPKT	171	Continuation
76	0.591865	207.46.101.100	88.88.101.2	TCP	60	3389-49347 [ACK] Seq=1 Ack=451 Win=512 Len=0
77	0.596526	88.88.101.2	207.46.101.115	TCP	60	1041-3389 [ACK] Seq=213 Ack=10805 Win=16640 Len=0
78	0.596739	88.88.101.2	207.46.101.115	TCP	60	1041-3389 [ACK] Seq=213 Ack=13365 Win=16640 Len=0
79	0.596842	88.88.101.2	207.46.101.115	TCP	60	1041-3389 [ACK] Seq=213 Ack=15925 Win=16640 Len=0
80	0.596904	88.88.101.2	207.46.101.115	TCP	66	[TCP Dup ACK 79#1] 1041-3389 [ACK] Seq=213 Ack=15925 Win=16640
81	0.597665	88.88.101.2	207.46.101.115	TCP	60	1041-3389 [ACK] Seq=213 Ack=17386 Win=16640 Len=0
82	0.617286	207.46.101.115	88.88.101.2	TPKT	1334	Continuation
83	0.617323	207.46.101.115	88.88.101.2	TPKT	1334	Continuation
84	0.619349	207.46.101.115	88.88.101.2	TPKT	859	Continuation
85	0.622090	207.46.105.2	88.88.101.2	TCP	689	[TCP segment of a reassembled PDU]
86	0.624914	207.46.105.2	88.88.101.2	HTTP	60	HTTP/1.1 200 OK (text/html)
87	0.645665	88.88.101.2	207.46.102.43	TPKT	155	Continuation
88	0.646631	207.46.102.43	88.88.101.2	TCP	60	3389-1326 [ACK] Seq=1 Ack=141 Win=63332 Len=0
89	0.681948	207.46.101.121	88.88.101.2	TCP	60	3389-49496 [ACK] Seq=75 Ack=554 Win=510 Len=0
90	0.686387	88.88.101.2	207.46.102.40	ICMP	74	Echo (ping) request id=0x000b, seq=638/32258, ttl=122 (reply i
91	0.686482	207.46.104.11	7.22.20.92	TCP	62	[TCP Spurious Retransmission] 62299-123 [SYN] Seq=0 Win=8192 Le
92	0.687378	207.46.102.40	88.88.101.2	ICMP	74	Echo (ping) reply id=0x000b, seq=638/32258, ttl=126 (request
93	0.689523	7.22.20.92	207.46.104.11	TCP	60	123-62299 [RST, ACK] Seq=1 Ack=1 Win=0 Len=0
94	0.693505	207.46.104.11	7.22.20.92	TCP	66	62301-123 [SYN] Seq=0 Win=8192 Len=0 MSS=1460 WS=4 SACK_PERM=1
95	0.697438	7.22.20.92	207.46.104.11	TCP	60	123-62301 [RST, ACK] Seq=1 Ack=1 Win=0 Len=0
96	0.713883	88.88.101.2	207.46.101.100	TPKT	155	Continuation

```
⊞ .... 0000 0001 1000 = Flags: 0x018 (PSH, ACK)
    window size value: 510
    [Calculated window size: 510]
    [window size scaling factor: -1 (unknown)]
  ⊞ Checksum: 0x506c [validation disabled]
    Urgent pointer: 0
  ⊞ [SEQ/ACK analysis]
    TCP segment data (5 bytes)
  ⊞ [2 Reassembled TCP Segments (640 bytes): #85(635), #86(5)]
  ⊟ Hypertext Transfer Protocol
    ⊞ HTTP/1.1 200 OK\r\n
      Expires: wed, 18 Mar 2015 20:22:40 GMT\r\n
      Expires: Thu, 19 Nov 1981 08:52:00 GMT\r\n
      Cache-control: max-age=180000\r\n
      Cache-control: no-store, no-cache, must-revalidate, post-check=0, pre-check=0\r\n
      Last-Modified: Mon, 16 Mar 2015 18:22:41 GMT\r\n
      X-Frame-Options: SAMEORIGIN\r\n
      Pragma: no-cache\r\n
      Content-type: text/html\r\n
      Transfer-Encoding: chunked\r\n
      Date: Mon, 16 Mar 2015 18:22:41 GMT\r\n
      Server: lighttpd/1.4.35\r\n
      \r\n
      [HTTP response 1/11]
```

```
0000  00 50 56 9c 73 82 00 50  56 9c 4c a0 08 00 45 00   .PV.s..P V.L...E.
0010  00 2d fd 7c 40 00 40 06  47 c3 cf 2e 69 02 58 58   .-.|@.@. G...i.XX
0020  65 02 00 50 c2 84 a0 db  95 52 e8 94 42 20 50 18   e..P.... .R..B P.
```

Frame (60 bytes) | Reassembled TCP (640 bytes) | De-chunked entity body (212 bytes)

Fig. 2. A screenshot of cyber traffic that a cyber analyst might see in stage 1 processing.

The software that cyber analysts use is diverse across organizations, however the Stage 1 activity is typically displayed on the screen with a row for each incident and columns dividing up the elements of the activity such as IP address of the internet user and file location the user attempted to access. To classify the activity as suspicious, the analyst must scan the code to figure out if an element of the activity matches a warning sign of mal-intent. For example, the IP address may be from a known cyber attacker or the file location is known to have limited-access, sensitive content.

Adaptive automation at Level 1 is relatively easy to visualize. The use of an information-processing aid here would comprise information-filtering. Each item presented to the analyst has some type of risk associated with it that the IDS used to determine that the cyber activity was a threat. The information-processing aid would sort the list of activity by the type of threat and highlight the column of code that is most indicative of that risk type (e.g., IP addresses that have common elements from a previous attack). For the decision-making task, the aid would add a flag to records judged to be the most probable malicious activity (without further indication as to which part of the record should be examined).

3.2   Level 2: Comprehension/Situation Awareness

Level 2 is the second stage of the tactical level. At the end of Level 1, the analyst has isolated cases of cyber traffic that could be events of mal-intent. The original cyber analysts may guide the process through Level 2 or a new analyst may take over. If the Level 2 analyst determines that there was mal-intent, it does not suggest that the attacker was successful, so more processing needs to take place to determine if the attack was successful and thus classify the event as an incident. The incidents are grouped into the smallest set of data because everything up to that point was a filtering mechanism. The Level 2 analyst concludes the Level 2 process by initiating responses to the incidents to optimally recover from the attack.

The Level 2 analyst's role is categorically different from the Level 1 analyst's role. Instead of just looking at the screen showing cyber activity, the role requires looking into other data sources within the network. For example, did the interaction include downloading a secret file or did the user discontinue the download before it completed? In the latter case, the intent was likely accidental and not malicious. An information-processing aid may help by populating a list of possible questions that the analyst should consider in response to a set of incidents. A decision-making aid would be more specific by tying one event or incident to the place in the network where the check needs to occur, then repeated for each of the events or incidents.

3.3   Level 3: Projection/Threat Assessment

The final stage of situation awareness is Level 3 for projection into future cyber defense actions and assessment of threats (see [20]). This level is likely to be performed by other cyber analysts than those who perform Level 1 and 2 tasks. As opposed to the tactical level from the first two stages, Level 3 is strategic in that it updates the signs of cyber-attack by updating what sorts of warning signs will be watched for in future cyber action at Level 1 and 2 based on the instances of events [21]. Interestingly, the use of deception against cyber attackers may help deter future attacks [22].

Adaptive automation at this level could also be divided into aiding the information-processing and decision-making stages. The analysis at Level 3 is meant to result in prediction. So an information-processing aid could determine categories of threat, such as common segments of IP addresses of attackers, to help the Level 3 analyst filter a class of incidents down to a common predictive assessment (without making the decision for them). A decision-making aid would make a more specific recommendation than the information-processing aid for a select few instances rather than across all activity.

4 Example of Adaptive Aids in a Cyber Task

An example of adaptive automation is included in this section to help illustrate how MLCC can be applied to cyber security tasks. We chose the Level 1 task because it is the most susceptible to information overload (see [19]) with the highest volume of data. We will discuss Level 1 tasks in the context of the difficulties of the task and how adaptive automaton can help overcome these problems.

4.1 Classifying Activity as Potential Threats or Non-threats

To classify activity that makes it past the IDS as a threat or a non-threat, the analyst must match up the type of task with the elements of the activity that would illustrate whether the activity is potentially compromising or not. With a list of activity, the analyst needs to visually scan the information and know when to stop scanning to pick up on the information that will indicate the threat level. The analyst may also scan the timestamps of the activity and try to understand if common elements, such as a user or target within the host network, appear in multiple instances of activity. These types of trends may also indicate suspicious activity and a potential threat.

The analyst must also pinpoint what incidents should be flagged as potential threats. Too many false alarms will be a distraction to the Level 2 analyst and too many misses could compromise the network, so the analyst must be careful to not select either too many or too few records of activity.

4.2 An Information Processing Aid

An information processing aid should help with the sorting and filtering function in the Level 1 process. If analysts are demonstrating performance difficulties, it would make sense for a software aid to intervene. Intervention would essentially comprise an intelligent sort of network activity records by type of potential threat (e.g., attempt to access a forbidden file). As an additional help, highlighting of the column for each grouping of activity will help draw the analyst's attention to the element or most common interest for that type of potential threat. Note that this type of aid would help only in the information-processing stages of cognition.

4.3 Decision-Making Aid

A decision-making stage should help with the decision about whether each instance of activity is a potential threat or non-threat. The aid would therefore place a flag at the start of the line of the most suspect activity according to the IDS software. The aid would not indicate why the activity is a potential threat, but it might help the analyst increase performance, if the software detects that performance is declining or about to decline.

4.4   Empirical Test

The MLCC approach calls for empirical testing of the effectiveness of the aids and how performance-decline state should be detected (i.e., metacognitive, performance-based, physiological, or some combination thereof). The aids themselves can be tested by having study participants perform the task and setting up conditions in which the aid is never engaged, always engaged, or engaged when the participant's behavior suggests a decreased-performance state. The timing of aid activation could be crossed with type of aid (i.e., information processing or decision making) to see how the two types of aids affect performance relative to one another. Other independent variables of aid-type, timing of aid engagement, and effectiveness of each aid may be tested, as well. The data may then be used to construct computational models of the experiment and further refinement of the aids.

5   Conclusions

Cyber analysis is a complicated and stressful activity that can result in catastrophic problem if done incorrectly. Given the state of the field, it is incumbent upon human-factors designers to help provide relief to the cognitive burdens put on an analyst. We recommend using the MLCC approach for designing and testing adaptive automation, which would allow cyber analysts to have help when they need it, and avoid the distraction or overreliance that the aid may produce when they do not. We outlined the MLCC approach, and described the stages of cyber analyst situation awareness with specific examples of how adaptive automation can help at each level. We suggest two possible aids that could help Level 1 analysis. The hope is that the arguments presented in this paper can help engage more research in adaptive automation for cyber analysts. We believe this approach can increase the performance of cyber analysts in protecting the networks of their organizations and call for other researchers to join in investigating this important topic.

References

1. Cassenti, D.N., Gamble, K.R., Bakdash, J.Z.: Multi-level cognitive cybernetics in human factors. In: Hale, K.S., Stanney, K.M. (eds.) Advances in Neuroergonomics and Cognitive Computing, pp. 315–326. Springer, New York (2016)
2. D'Amico A., Whitley K.: The real work of computer network defense analysts. In: Goodall, J.R., Conti, G., Ma, K.L. (eds.) VizSEC 2007, pp. 19–37. Springer, Berlin (2008)
3. Gootman, S.: OPM hack: The most dangerous threat to the federal government today. J. App. Sec. Res. 11, 517–525 (2016)

4. Gutzwiller, R.S., Fugate, S., Sawyer, B.D., Hancock, P.A.: The human factors of cyber network defense. In: Proceedings of the Human Factors and Ergonomics Society, pp. 322–326. Sage, Los Angeles, September 2015

5. Maqbool, Z., Makhijani, N., Pammi, C., Dutt, V.: Effects of motivation: rewarding analysts for good cyber-attack detection may not be the best strategy. Hum. Fac. **59**, 420–431 (2017)

6. Gonzalez, C., Ben-Asher, N., Oltramari, A., Lebiere, C.: Cognition and technology. In: Kott, A., Wang, A., Erbacher, R. (eds.) Cyber Defense and Situational Awareness, pp. 93–117. Springer, Switzerland (2014)

7. Boyce, M.W., Duma, K.M., Hettinger, L.J., Malone, T.B., Wilson, D.P., Lockett-Reynolds, J.: Human performance in cybersecurity: a research agenda. In: Proceedings of the Human Factors and Ergonomics Society, pp. 1115–1119. Sage, Los Angeles (2011)

8. Alpcan, T., Başar, T.: Network security: A decision and Game-Theoretic Approach. Cambridge University Press, Cambridge (2010)

9. Knott, B.A., Mancuso, V.F., Bennett, K., Finomore, V., McNeese, M., McKneely, J.A., Beecher, M.: Human factors in cyber warfare: alternative perspectives. In: Proceedings of the Human Factors and Ergonomics Society, pp. 399–403. Sage, Los Angeles (2013)

10. Wiener, N.: Cybernetics or Control and Communication in the Animal and the Machine. John Wiley & Sons, New York (1948)

11. Kaber, D.B., Endsley, M.R.: The effects of level of automation and adaptive automation on human performance, situation awareness and workload in a dynamic control task. Theo. Iss. Ergon. Sci. **5**, 113–153 (2004)

12. Dutt, V., Moisan, F., Gonzalez, C.: Role of intrusion-detection systems in cyber-attack detection. In: Nicholson, D. (ed.) Advances in Human Factors in Cybersecurity, pp. 97–110. Springer, Cham (2016)

13. Segerstrom, S.C., Nes, L.S.: Heart rate variability reflects self-regulatory strength, effort, and fatigue. Psychol. Sci. **18**, 275–281 (2007)

14. Recarte, M.A., Nunes, L.M.: Effects of verbal and spatial-imagery tasks on eye fixations while driving. J. Exp. Psy. App. **6**, 31–43 (2000)

15. Cassenti, D.N., Veksler, V.D.: Using cognitive modeling for adaptive automation triggering. In: Cassenti, D.N. (ed.) Advances in Human Factors in Modeling and Simulation, pp. 378–390). Springer, New York (2017)

16. Schraagen, J.M., Chipman, S.F., Shalin, V.L. (eds.): Cognitive Task Analysis. Lawrence Erlbaum, Mahwah (2000)

17. D'Amico, A., Whitley, K., Tesone, D., O'Brien, B., Roth, E.: Achieving cyber defense situational awareness: a cognitive task analysis of information assurance analysts. In: Proceedings of Human Factors and Ergonomics Society, pp. 229–233. Sage, Los Angeles (2005)

18. Endsley, M.R.: Toward a theory of situation awareness in dynamic systems. Hum. Factors **37**, 32–64 (1995)

19. Eppler, M.J., Mengis, J.: The concept of information overload: a review of literature from organization science, accounting, marketing, MIS, and related disciplines. Inf. Soc. **20**, 325–344 (2004)

20. Arora, A., Dutt, V.: Cyber security: evaluating the effects of attack strategy and base rate through instance based learning. In: Proceedings of International Conference on Cognitive Modeling, pp. 336–341. ICCM, Ottawa (2013)

21. Dutt, V., Ahn, Y.S., Gonzalez, C.: Cyber situation awareness modeling detection of Okaycyber-attacks with instance-based learning theory. Hum. Factors **55**, 605–618 (2013)

22. Aggarwal, P., Gonzalez, C., Dutt, V.: Cyber-security: role of deception in cyber-attack detection. In: Nicholson, D. (ed.) Advances in Human Factors in Cybersecurity, pp. 85–96. Springer, New York (2016)

An Integrated Model of Human Cyber Behavior

Walter Warwick[1](✉), Norbou Buchler[2], and Laura Marusich[2]

[1] TiER1 Performance Solutions, LLC, 100 E. RiverCenter Blvd. Suite 100,
Tower 2 Covington, Kentucky 41011, USA
w.warwick@tier1performance.com
[2] Army Research Laboratory – Human Research and Engineering Directorate,
2800 Powder Mill Road, Adelphi, MD 20783, USA
{Norbou.buchler.civ, laura.marusich.ctr}@mail.mil

Abstract. Agent-based models are commonplace in the simulation-based analysis of cyber security. But as useful as it is to model, for example, adversarial tactics in a simulated cyber attack or realistic traffic in a study of network vulnerability, it is increasingly clear that human error is one of the greatest threats to cyber security. From this perspective, the salient features of behavior are those of an agent making decisions about how to use a system, rather than an agent acting as an adversary or as a "chat bot" which functions merely as a statistical message generator. In this paper, we describe work to model a human dimension of the cyber operator, a user subject to different motivations that lead directly to differences in cyber behavior which, ultimately, lead to differences in the risk of suffering a "drive-by" malware infection.

Keywords: Human behavior representation · Cyber behavior
Model integration

1 Introduction

Despite the digital and virtual nature of the cyber domain, the dynamics of cyberspace are fundamentally human and adversarial. Broadly defined, the human dimension of cybersecurity involves the dynamic interaction of *attackers*, *defenders*, and *users*. Users pursue their defined goals (work and personal) that often require interacting with others and online systems to socialize, craft work products, and communicate using networked technology. Attackers seek to exploit both networked system vulnerabilities and increasingly the user community with social engineering attacks, whereas defenders monitor systems and attempt to thwart and mitigate any actions taken to compromise them. Many studies have exposed the vulnerabilities posed by the user, for instance in maintaining compliance with security policies [1–3] or in identifying insider threats [4, 5]. Working in that spirit, we examine how a computational models of human behavior can provide insights into the vulnerabilities posed by various types of users.

While a number of organizations are beginning to recognize the need for a human-centered perspective for the development of effective cyber security policies

© Springer International Publishing AG, part of Springer Nature 2019
D. N. Cassenti (Ed.): AHFE 2018, AISC 780, pp. 290–302, 2019.
https://doi.org/10.1007/978-3-319-94223-0_28

[6, 7], research into the human factors of cyber security is still in its infancy. This is especially evident in the way that the human is typically modeled in simulation-based analyses of cyber behavior. At best, the human might be represented as a set of rules in agent-based simulations used to understand network-level behavioral effects; at worst, the role of the human is often reduced to that of a "traffic generator" that ensures the appropriate statistical characteristics of communications within a network. Human cyber behavior is heterogeneous and complex, depending on many aspects of the user, such as risk tolerance, productivity, and capacity for multi-tasking. We suggest user modeling as a way to take these aspects into account and to provide a more sophisticated and informative representation of the human operator. User modeling is a domain of human-computer interaction (HCI) focused on developing and modifying conceptual models of the user. Typically, the main goal of user modeling is to customize system adaptations to the specific needs of the user. However, there are a few instances where behavioral user models have been derived from cyber data, typically to detect insider attacks.

Our approach seeks to develop simulated user profiles by integrating two formalisms commonly used to model human behavior—task network models and Markov decision processes. Task network modeling tools are commonplace in HCI research and provide an intuitive graphical user interface built around hierarchical task analysis as a general method for representing human behavior. One of the chief benefits of task network modeling tools is their ability to support different levels of abstraction. In the context of a cyber domain, at the most detailed level, task network models have been implemented to represent a user clicking on individual messages [8]; at a much higher level of abstraction, they have been implemented to represent the steps a user takes to meet a tactical goals [9]. For this work, we chose an intermediate level of abstraction, focusing on the various tasks a cyber operator performs during the course of day (check email, surf the web, use any one of the Microsoft Office products) while abstracting away both the low-level button clicks needed to perform those tasks as well as the high-level goals that the work must accomplish throughout the day. As we describe in more detail later, this choice allows us to represent the objective features of the cyber operator's task environment while also allowing us to introduce some subjective features of the operator in a straightforward manner.

Markov decision processes (MDP) represent a class of stochastic processes that have a wide spectrum of practical applications for simulating human behavior. MDPs provide a natural representation of decision making under uncertainty. And like task network models, MDPs can be applied at different levels of abstraction. For instance, Markov models have been used to simulate network-oriented human behaviors for the purpose of network traffic generation in large-scale simulations [10]. At the other end of the spectrum, MDPs have been used to represent fine-grained interactions with desktop applications. For example, in the case of Outlook, this would include (a) read mail, (b) send mail, (c) reply to mail, and (d) quit Outlook.

Again, in our use of Markove models, we have focused on an intermediate level of abstraction, aligning the decision making of the MDP with the three-state topology of the task network. More specifically, we used the reward function of the MDP as a proxy for operator motivation and used the policy generated by the MDP to provide transition probabilities for the task network model. By modeling the interaction between a cyber operator's motivation and his activities, we gain a unique, human-centered perspective on the simulation-based analysis of cyber security. For example, we can start to explore how changes in cyber policies and procedures, to the extent that they affect task-level behavior, impact risk. In addition, the integrated model we describe below provides a principled method for linking descriptive statistical data and qualitative data to produce quantitative predictions about performance. Finally, while computational models of human behavior are usually developed using existing empirical data, such models often reveal the need for new kinds of data. Indeed, as we describe later, in developing our own model of cyber behavior, we found gaps in the existing empirical data. And while we were able to employ a sort of bootstrapping to patch over these gaps, this workaround highlights the need for new empirical study. In this way, human centered modeling promotes a virtuous cycle of theorizing, modeling and experimenting.

2   The Model

Our model of the cyber operator comprises three major components: first, there is a task network model that represents objective features of the work environment (e.g., a high-level representation of work day activities and the risks associated with them); second, there is a Markov decision process model (MDP) that represents subjective features of the worker (e.g., the reward the worker associates with each task); third, there is a "middleware" layer that effects communication between the task network and Markov models.

2.1   The Task Network Model

Task network modeling tools provide a framework for representing human behavior as a decomposition of operator goals or functions into their component tasks, which themselves can be further decomposed. This framework is visualized by way of an intuitive graphical representation of the process being modeled in which nodes represent tasks and directed edges represent the flow of control among the tasks. While the modeling framework is quite general, task network models are generally used to simulate the time course of a series of discrete events. We used C3Trace, a government-owned task network modeling tool, to implement a model of the cyber work environment.

There are three noteworthy features of the task network model. First, unlike more traditional task network models, the representation of the task environment is very sparse. While task network modeling environments are designed specifically to help the analyst represent complex work environments by way of hierarchical task decomposition, with each level containing potentially several interdependent threads of task execution, the current model has only a single level of task decomposition with a complete subgraph of just three tasks. Conceptually, this decomposition represents a worker that continually cycles between three basic (i.e., unanalyzed) activities while doing computer work. Our cyber operator is either managing email, using an application (e.g., Microsoft Word, Excel etc.) or browsing the internet. Second, the branching among the three tasks is determined by a *set* of probabilities which is conditioned on the output of the MDP model (described next). So, for example, the likelihood of transitioning from the browser task to the email task is given by selecting from a set of two probabilities that correspond to the two-action policy output from the MDP. In this way, the task network diagram recapitulates the structure of the MDP graph without explicitly representing every state-action edge. Third, despite the simplicity of the task network, a variety of different data have been aggregated within the model and we have even used the model itself to help "bootstrap" some data that were not readily available. We return to this issue when we describe how we initially validated the model's predictions.

2.2 The Markov Decision Process Model

Markov decision processes are mathematical models of decision making. Formally, an MDP is a five-tuple that consists of a set of states, a set of actions, a transition function that maps state action pairs to the probability of arriving in a subsequent state given that state-action pair, a reward function that, for each action, maps state-to-state transitions to real values, and a discount factor used to weight the difference between immediate and future rewards. Intuitively, an MDP reflects the reasoning of a decision maker who much must balance short- and long-term gains in an environment in which the relationship between actions and outcomes is probabilistically defined.

As we alluded to previously, the structure of the task network model has a direct representation as the three-state MDP. There are just two actions for our notional cyber operator, working and loafing. The tasks (email, Office, browser) can be represented as states while the probabilistic branches among the tasks realize the transition function. The probabilities here represent the "stickiness" of each task while taking a certain action. So, it is more likely that the cyber operator, once browsing, will continue to browse while loafing and, likewise will continue to engage an Office task while working. But the cyberworld is uncertain and it is always possible that work on an Office task will be interrupted by an email or that after enough loafing on the internet that work on an Office task will resume. We treat these probabilities as objective features of the cyber operator's work environment. By contrast, we can define both the reward function and discount factor in the task network model and use them to represent subjective features of the cyber operator. We use the reward function to represent different operator profiles. For example, a "slacker" who chooses to loaf while browsing will associate a larger reward with continued browsing than with a transition to an Office task. Conversely, a diligent worker (what we subsequently call a "John Henry") busy working on an Office task will associate a higher reward with continued work rather than an email interruption (Fig. 1).

There are well-known algorithms for finding the optimal action to take in any state for a given MDP. In this context, optimality is couched in terms of greatest expected reward given a sequence of actions to take from each state. Rather than implement those algorithms directly in the task network model, we instead chose to use existing libraries in R, a statistical programming environment. This division of labor between the two applications is intended to support inter-operability between R and C3Trace at run time. Specifically, the analyst initializes the task network model by choosing a

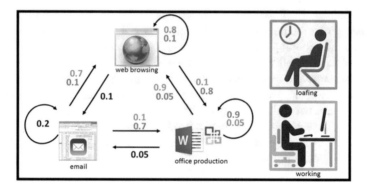

Fig. 1. The MDP model in detail, showing transition probabilities for the two different actions

Action	State	browser	Transitions email	office
LOAF	browser	0.8	0.1	0.1
	email	0.7	0.2	0.1
	office	0.9	0.05	0.05
WORK				
	browser	0.1	0.1	0.8
	email	0.1	0.2	0.7
	office	0.05	0.05	0.9

Action	State	Slacker Rewards browser	email	office
LOAF	browser	10	5	1
	email	10	10	1
	office	20	5	1
WORK				
	browser	2	2	2
	email	1	1	1
	office	1	1	1

Action	State	John Henry Rewards browser	email	office
LOAF	browser	1	1	1
	email	1	1	1
	office	1	1	1
WORK				
	browser	5	5	10
	email	5	7	10
	office	3	7	10

Fig. 2. The transition function that defines the probabilistic environment and two worker profiles as defined by different reward functions

transition function, a reward function and a discount factor. Then, once the task network model starts running, these functions are passed to R and a policy is computed and returned to the task network model and used to determine which of the two available probabilities should be used in selecting the next task to perform. The result is a pseudo-random sequence of task execution along with any measures associated with the performance of the different tasks over time. Although we have treated the transition function, reward function and discount factor as fixed initial conditions for each run of the task network model, this is not a hard requirement; a new policy could be computed should those functions change during run time which, in turn, could affect the time-ordered sequence of tasks the simulated cyber operator performs.

2.3   The Middleware

Working under a separate Phase II Air Force STTR, we developed a middleware that allows existing models to be combined to produce agent behavior. We have used that middleware here to effect the run-time interoperability between the task network and MDP models. The middleware uses standard TCP/IP sockets to connect component models. TCP/IP socket connections are well-supported in a wide variety of programming languages, and they impose a minimal burden on the developer of the component model. The middleware itself implements a multi-threaded socket server, with the middleware serving as the server and the component models as clients. Communications among the middleware and component models follow a request-reply pattern, with the component models making requests to and receiving replies from the workspace.

There are two important features of the multi-threaded architecture. The first and more obvious is that the communication between the middleware and each client model is executed as a separate process. Consequently, the interactions between the middleware and the client models are handled asynchronously and there is no requirement (or safeguard) that the model interactions are brokered in any specific order. The second feature, related to the first, is that the client models interact with the workspace and not each other. Although it is not a pressing concern for the present application of the middleware, we note that by brokering model interaction through the middleware, rather than allowing direct model-to-model connections, we mitigate the complexity when multiple models are integrated. Specifically, the number of connections managed in the middleware grows linearly rather than quadratically in the number of models.

We are using JavaScript Object Notation (JSON) as the standard for sharing information among between C3Trace and R. Like our choice to use TCP/IP sockets, we chose JSON primarily because it is lightweight and well-supported across a variety of programming languages. While the JSON standard provides the syntax, the semantic content of the object consist of: (1) a client attribute-value pair that consists of the attribute label "client" as a string and another string that contains the name given to the client model (e.g., "C3Trace" or "R"); (2) a mode attribute-value pair that consists of the attribute label "mode" as a string and another string that is either "set" or "get" (e.g., R gets the transition probabilities set by C3Trace; R sets the policy for C3trace to get); (3) a type attribute-value pair that consists of the attribute label "type" as a string and another string that contains the a name given to this kind of object (e.g., "TransitionProbs"); (4) an argument attribute-value pair that consists of the attribute label "values" and another JSON object as the value (e.g., the row-major enumerations of the probability values from the transition function expressed as an array).

3   The Data

The goal of the cyber operator model is to predict the prevalence of malware infection given different user profiles. As with any model-based prediction, the output is only as good as the input. But, again, like any usefully complex model, there are many free parameters to be set in the model and finding the data to anchor those settings is not always straightforward.

We begin with the one of the key parameters: the risk of a malware infection associated with browsing activity. (We have limited our attention to browsing as a malware vector.) As we describe more fully later, this task-related risk interacts with several other independent parameters in the model to produce a dependent measure of the total number of malware infections over a known period of time for a given user profile. The risk parameter is expressed as a real-valued number between zero and one and represents the probability that the execution of the browser task will result in a malware infection. The browser task in unanalyzed in the sense that it doesn't represent individual websites visited or the operator's interactions with the web page; it is just a composite representation of all the activities and their possible malware risks that occur in a browsing session. Not surprisingly, we did not find a numeric value to express risk in exactly those terms in the literature. Instead, we turned to the work of Ovelgonne, Dumitras [11] who provide real-world incidence data on malware attacks grouped by user type (e.g., gamers, software developers, office workers etc.). These data were collected from 1.6 million machines over an eight-month period and provide concrete counts of the malware present on each machine. Ovelgonne et al. then grouped those data according to user profiles based on the software applications on the machines. While these data are very detailed, we still had to extrapolate from them to arrive at a numeric risk value we could associate with browsing activity. Specifically, given a worst-case count of 8.1 malware infections per machine per 200 day period from Ovelgonne et al. and using other data sources (described below), we "bootstrapped" an estimate for the number of browser sessions per day to arrive at value for the probability of malware infection for each browsing session.

Our sense of bootstrapping is informal but apt insofar as we used a simulation-based estimate of one model parameter to underpin another simulation-based prediction from the model. First, using data from Mark et al. [12], we bounded the length of time the cyber worker spends each day at the computer at 274 min. Second, using the same data source, we took the average frequency for email checks during the day together with the report of the total time spent on email to derive an average time for individual email tasks. Third, following the work of Liu et al. [13] we used a Weibull distribution with a scale parameter set to 1 and the shape parameter set to.7 to model the duration of browser sessions. Finally, we used a simple guess to set the time the cyber worker spends using Office tools without interruption at 3 min. Having fixed the time (and distribution) for each of the three tasks in our model along with the total length of the "cyber day", we then ran the model to determine estimates for how often the browser task was executed during a day. In the interest of arriving at a conservative estimate of the risk (i.e., not underestimating the risk), we ran the model using the John Henry profile. Finally, using the estimate for the number of browser sessions per day we derived the risk of malware infection per browser by simply dividing the total number of browser sessions in a simulated 200 day period into the 8.1 malware infections reported by Ovelgonne et al.

Clearly, there is a direct mathematical dependency between the model-based prediction of the total number of malware infections due to browsing and the model-derived risk of malware infection per browsing session; changing the denominator in the calculation will change the numeric value of the risk, but not the model-based prediction of the total number of malware infection (modulo stochastic variability between runs). We maintain that this dependency is not a vicious circle but, rather, a virtuous one. Indeed, having fixed the other task times based on empirically grounded and independent estimates, we arrive at a model-based estimate for baseline level of malware risk that, in turn, puts us in position to address the more salient question of how risk varies by user profile.

4   Model Mechanics

As the previous discussion suggests, the dependencies in our model are more complex than they might seem given the simple three task structure of the task network and the corresponding three state topology of the MDP. Moreover, there are several intermediate processes in the model that complicate the relationships between input and output. Figure 3 depicts the "mechanics" of the input-output relationships of the combined task network and MDP models.

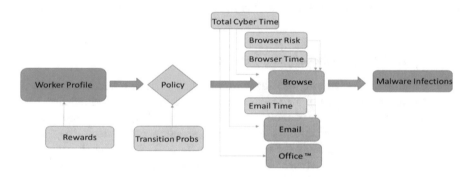

Fig. 3. Input-output mechanics in the combined task network MDP model

The chunky blue arrows denote the main input-output relationship implemented in the model: from a worker profile, specified by way of a reward function, as input to count of total malware infections as output. The relationship is moderated by two intermediate processes. The first, denoted by the yellow diamond, is the determination of a policy by the MDP given both the reward and transition function. The second, denoted by the green squares, is the execution of a sequence of tasks given the policy and the timing information described previously.

The skinny blue arrows and the attached pink squares denote independent parameters that condition the model processes. Again, the worker profile is given by a reward function that summarizes the subjective goodness the cyber operator associates with the transition to a new state given a previous state-action pair. Referring to the lower two panels of Fig. 2, we note that the absolute values for a given profile are of less consequence than the relative values between two profiles, and in both cases values were chosen by hand manner to show off differences between two extreme worker profiles. Similarly, we chose transition probabilities by hand to reflect gross features of the cyber environment. Taken as inputs to the MDP, the rewards and transition probabilities (along with a discount factor) produce a policy as output which, in turn, is used as one of several inputs to the task network model. Specifically, the policy dictates which action to take, loaf or work, from each state, browse, Office, or email, which ultimately dictates which set of probabilities to use in the branching conditions in the task network model.

Although these branching probabilities dictate the relative likelihood that a task will execute, these likelihoods by themselves do not determine how many times a task will execute during a given run. That depends on a more complex interaction between the branching probabilities and the timing information for both the individual tasks and the duration of the entire simulation run. For example, given a simulation of a fixed duration, even the relatively infrequent execution of a task with a very long duration could simply use up time that might otherwise be taken up by the more frequent execution of another task. And just as the branching probabilities for the three exclusive and exhaustive tasks sum to one, the sum of the individual task times must sum to the total duration of the run. Given this zero-sum nature of both the task times and branching probabilities, the number of times the browser task will execute depends on the policy output of the MDP and the independently specified task times and run time of the simulation.

4.1 Interpolation, Prediction, and Validation

By leveraging the varied empirical data we obtained from the literature against the dependencies implemented in the model, we were able to generate a simulation-based estimate for the total number of browser sessions per day. This, in turn, allowed us to estimate the risk of malware infection per browsing session and, ultimately, to generate a simulation-based estimate for total malware infections. To the extent that we "boot-strapped" our way to this estimate, we do not consider this to be a prediction of the model. Instead, we view this baseline estimate as one of several pieces of evidence that the model is well calibrated. In addition, we looked at the distribution of tasks under the different worker profiles and compared them to some of the data reported in Mark, et al. [14]. Although worker activity was grouped into different categories in those data, we were still able to see some similar patterns between the human activity and the task distributions in the model. In particular, Mark et al. report that a worker in a "focused state" (a perpetual state for our notional John Henry worker profile), will use "produc-tivity apps" (analogous to our Office task) roughly twice as often as they frequent email or Facebook and internet surfing (two activities that we do not distinguish under our more generic "Browser" task). As shown in Table 1, the model output for a John Henry profile shows that Office tasks occurred an average of 14,382 times, with the browser task averaging 6092 occurrences, a ratio of 2.36. We also looked at the total time spent loafing for our simulated slacker (using the browser task as a proxy for loafing) and found reasonable agreement with the lower (and shocking) bound reported by Ugrin and Pearson [15] that 60% of information worker's day is spent "cyber loafing" (see Table 2). While none of these qualitative comparisons constitute a formal validation of the model, they do suggest at least some degree of plausibility and, moreover, demon-strate how a deliberately abstract representation of the cyber worker's task environment can be anchored in useful ways by varied and disparate sources of empirical data.

Table 1. Results from ten runs of the model using the "bootstrapped" risk for browsing over 200 simulated days. The results constitute our baseline behavior for our John Henry worker profile.

Run	Browser hits	Malware infections	Email	Office task
1	6138	7	6756	14400
2	6049	7	6743	14385
3	6129	10	6574	14420
4	5980	6	6814	14378
5	6174	2	6674	14403
6	6081	10	6730	14395
7	6025	15	6800	14357
8	6042	5	6749	14386
9	6119	7	6812	14326
10	6180	7	6674	14367
Mean	6091.7	7.6	6732.6	14381.7
Std. Dev.	66.81990721	3.470510689	74.85274	26.67520697
Mean/200	30.4585	0.038	33.663	71.9085

The more interesting predictions come from comparing numbers of malware infections given different worker profiles. For this, we ran the model under the same initial conditions, for the same 200 simulated days and changed only the reward function (see the lower two panels of Fig. 2).

Table 2. Results from ten runs of the model under running for 200 simulated days under the slacker profile. Note the total time spent browsing which is, on average, roughly 60% of the 274 min a cyber worker spends at the computer.

Run	Browser hits	Malware infections	Email	Office task	Time browsing
1	45155	65	6041	5450	31790.05678
2	45775	55	5999	5439	31893.23064
3	45319	54	6058	5442	31807.37617
4	45991	57	5928	5408	32080.11721
5	45493	67	5950	5435	31954.02981
6	45394	75	6092	5404	31862.19834
7	45464	69	6063	5441	31810.08756
8	45279	55	6085	5427	31833.66988
9	45817	62	6208	5393	31797.53165
10	45421	51	6008	5459	31797.04336
Mean	45510.8	61	6043.2	5429.8	31859.53414
Std. Dev.	252.0868104	7.416198487	75.53383348	20.370567	90.42448721
Mean/200	227.554	0.305	30.216	27.149	159.2976707

Using the "John Henry" profile of as our baseline, our simulated "Slacker" had roughly eight times the number of malware infections. While this factor is much larger than the worst-to-best case factor of 2.5 reported by Ovelgonne et al., we note that we calibrated our baseline model using a worst-case estimate of the malware risk and applied this single value of risk across the two user profiles. That is, rather than identify a set of features that supports a post hoc mapping of various user profiles to different risks, we have a model in which the changes in a user profile are intended to predict risk. For example, by changing the reward function to represent a worker who is less inclined to slack all the time, we get the following results: (Table 3)

Table 3. Behavior that results from modifying the slacker's reward function to represent a worker less inclined to slack all day.

Run	Browser hits	Malware infections	Email	Office task
1	10255	13	7044	13336
2	10373	19	6821	13369
3	10312	13	6682	13466
4	10220	11	6832	13362
5	10246	16	6974	13315
6	10363	11	6934	10363
7	10199	12	6940	13327
8	10123	21	6864	13391
9	10251	15	6697	13384
10	10257	11	6863	13378
Mean	10259.9	14.2	6865.1	13069.1
Std. Dev.	74.89466677	3.521363372	115.012511	951.761694
Mean/200	51.2995	0.071	34.3255	65.3455

We note that this profile leads to an increased risk of malware infections over the baseline profile and that the difference between profiles (an average of 7.6 malware infection for John Henry versus an average of 14.2 infections for this modified slacker profile) is closer to the factor of 2.5 that Ovelgonne reports. We also note that we did not attempt to fit this reward function to produce a desired total number of malware infections. Instead, our goal here is to demonstrate that intuitively plausible changes to a user profile result in plausible changes to the model's predictions.

5   Discussion

Like any other intelligent behavior, human cyber behavior is complex, dynamic and purposeful. The model just described is one attempt to manage and understand the manifest complexity of cyber behavior. We accomplish this in two ways. First, the relatively sparse structure of the task network model leads to succinct data requirements (e.g., time spent browsing during the workday). That is, there are relatively few "free parameters" in the task network model and most of those can be anchored directly to empirical estimates. And in those cases where empirical data is lacking, the task network provides a principled mechanism for "boostrapping" estimates that, ultimately, serve as placeholders until future research provides better estimates. Second, despite the sparse structure of the task network, the behavior of the model is rich and dynamic owing to the integration of the MDP. Like the task network model, the data requirements for the MDP are succinct (e.g., estimates of the subjective rewards associated with each action and outcome). It is the combination of the objective features of the task environment with the subjective and uncertain nature of the decision-making process that allow us to explore the richness of the cyber operator's behavior.

References

1. Besnard, D., Arief, B.: Computer security impaired by legitimate users. Comput. Secur. **23** (3), 253–264 (2004)
2. Fulford, H., Doherty, N.F.: The application of information security policies in large UK-based organizations: an exploratory investigation. Inf. Manage. Comput. Secur. **11**(3), 106–114 (2003)
3. Werlinger, R., et al.: Towards understanding diagnostic work during the detection and investigation of security incidents. In: Steven, N.L.C., Furnell, M. (eds.) Third International Symposium on Human Aspects of Information Security & Assurance, Athens, Greece, pp. 119–132. University of Plymouth (2009)
4. Bishop, M., et al.: Insider threat identification by process analysis. In: IEEE Security and Privacy Workshops, San Jose, CA, pp. 251–264. IEEE (2014)
5. Costa, D.L., et al.: An Insider Threat Indicator Ontology (2016)
6. Cranor, L.F., Garfinkel, S. (eds.): Security and Usability: Designing Secure Systems That People Can Use. O'Reilly & Associates Inc, Sebastopol (2005)
7. Winnefeld, J.A., Kirchhoff, C., Upton, D.M.: Cybersecurity's Human Factor: Lessons from the Pentagon, in Harvard Business Review. Harvard Business Publishing, Cambridge (2015)
8. Kilduff, P.W., Swoboda, J.C., Katz, J.: A platoon-level model of communication flow and the effects on operator performance. U.A.R. Laboratory, Editor. Human Research and Engineering Directorate, Aberdeen Proving Ground, MD (2006)
9. Brett, B.E., et al.: The Combat Automation Requirements Testbed (CART) Task 5 Interim Report: Modeling a Strike Fighter Pilot Conducting a Time Critical Target Mission. A.F.R. Laboratory, Editor, Dayton, OH (2002)
10. Heegaard, P.E.: GenSyn - a Java based generator of synthetic Internet traffic linking user behaviour models to real network protocols. In: ITC Specialist Seminar on IP Traffic Measurement, Modeling and Management, Monterey, CA (2000)
11. Ovelgonne, M., et al.: Understanding the relationship between human behavior and susceptibility to cyber-attacks: a data-driven approach. In: ACM Transactions on Intelligent Systems and Technology (TIST) - Special Issue: Cyber Security and Regular Papers, vol. 8, no. 4 (2017)
12. Mark, G., et al.: Email duration, batching and self-interruption: patterns of email use on productivity and stress. In: 2016 CHI Conference on Human Factors in Computing Systems, San Jose, California, USA, pp. 1717–1728. ACM (2016)

13. Liu, C., White, R.W., Dumais, S.: Understanding web browsing behaviors through Weibull analysis of dwell time. In: 33rd International ACM SIGIR Conference on Research and Development in Information Retrieval, Geneva, Switzerland, pp. 379–386. Association for Computing Machinery, Inc. (2010)
14. Mark, G., et al.: Bored mondays and focused afternoons: the rhythm of attention and online activity in the workplace, In: SIGCHI Conference on Human Factors in Computing Systems, Toronto, ON, Canada, pp. 3025–3034. ACM (2014)
15. Ugrin, J.C., Pearson, J.M.: The effects of sanctions and stigmas on cyberloafing. Comput. Hum. Behav. 29(3), 812–820 (2013)

Conditional Deterrence: An Agent-Based Framework of Escalation Dynamics in an Era of WMD Proliferation

Zining Yang[✉], Jacek Kugler, and Mark Abdollahian

Claremont Graduate University, Claremont, CA 91711, USA
{zining.yang, jacek.kugler, mark.abdollahian}@cgu.edu

Abstract. We offer a revised conditional deterrence agent based model applied to global and regional nuclear proliferation issues. Further extending the dyadic logic already established in the deterrence literature helps anticipate more recent 21st century challenges generated by the proliferation of nuclear capabilities and their acquisitions by dissatisfied non-state actors. Key elements include relative capabilities, risk propensity associated with the status quo, and physical exposure to preemptive-attack or retaliation. This work continues to extend our previous complex adaptive system framework to generalize insights to deterrence environments with multiple competing actors. Our preliminary analysis suggests that deterrence is stable when the capabilities of a dissatisfied challenger are inferior to that of a dominant and satisfied defender. Conversely, deterrence is tenuous when a dissatisfied challenger approaches parity in capability with a more dominant and satisfied defender, or when a violent non-state actor obtains nuclear capability or other WMDs.

Keywords: Agent-based model · Game theory · Security · Deterrence
Proliferation

1   Introduction

Under a currently fractured, multipolar security environment, the ability of agents to deter others by threats to use weapons of mass destruction (WMD) warrants serious reexamination. The specter of nuclear war continues to be a policy priority as weapon proliferation to smaller nations continues and potential expansion to non-state agents creates new security challenges. Specifically, Mutual Assured Destruction has preserved stability among the great powers - but a nuclear balance is resisted vigorously many countries, clearly questioning the stability of a balance of nuclear capabilities At the same time, the persistent pursuit of nuclear capabilities by dissatisfied states such as North Korea raise doubts about the ability of far larger nuclear powers to prevent proliferation or even consider a pre-emptive attack. Moreover, pre-emptive strikes against nascent WMD installations in Iraq and Syria and the threat of such capabilities by Iran challenge the notion of that non-nuclear nations have been persuaded to abandon nuclear proliferation. While nuclear weapons have only been used once and bi-lateral nuclear deterrence has been stable thus far, one must question the long-term

© Springer International Publishing AG, part of Springer Nature 2019
D. N. Cassenti (Ed.): AHFE 2018, AISC 780, pp. 303–312, 2019.
https://doi.org/10.1007/978-3-319-94223-0_29

stability of nuclear deterrence because WMD devices have been employed after a long successful deterrence after World War I. Retaliatory threats by far superior nations consistently fail when only one side has such weapons. Above all, nuclear proliferation raises the specter of dirty bomb use by dis-satisfied agents that include terrorist groups like ISIS and Al-Qaeda were they to acquire such capabilities.

We propose an integrated simulation model that identifies and addresses security conditions for the success and failure of WMD deterrence in the twenty-first century. Particularly, our agent-based models respond to the need to reassess the relevance of deterrence as both theory and practice in light of current foreign policy challenges. Decision makers have a need to rapidly evaluate, assess, and reason about deterrence stability when they encounter new political and security crises in an era of expanding proliferation. However, it is still difficult to find an integrated predictive model for deterrence that takes account of both global and regional interactions in an increasingly globalized and concurrently fractured security environment.

In the real world, decision to initiate or escalate a conflict is made by a number of agents within each nation or group. To capture this reality, we employ a complex adaptive systems framework with agent based modeling. The micro behavioral foundations stem from game theoretical advances in conditional deterrence. The macro structural environments, which constrain and/or incentivize micro individual behavior, can help provide the necessary but not sufficient conditions for a deterrence failure and can act as an early warning diagnostic to guide to more stable outcomes. Gone are the bipolar days of US and USSR nuclear rivalry. Today we are seeing increased multipolarity emerge from the meso level of international politics. Networks of dyadic behavior, with cross-cutting cleavages and interests, demand that we take a much more nuanced, and sophisticated approach to deterrence. While the micro motivational game theoretic component above provides the behavioral motivation for deterrence or proliferation. Our agent-based formulation incorporates policy option assessments to defuse a conflict or anticipate the long-term effect of decisions made today. Detailed information about actors, their positions on critical issues related to WMD, the importance they attach to such outcomes are modeled in the assessment of capability, exposure, risk and trust.

This research is a theoretical response. Here we raise questions over both the plausibility of disarmament as well as the validity of classical deterrence. Past nuclear stability does not guarantee that all future wars will be waged with conventional weapons. As Zagare and Kilgour [1] observe, "nuclear war has been avoided not because of nuclear weapons, but in spite of them." Much of previous literature does not address anticipated challenges that may be generated by the acquisition of WMD capabilities through new mechanisms, such as rising regional competitors, rogue regimes and violent non-state actors. By contrast, this work argues that deterrence remains relevant as a primary strategic goal, yet we should not be condemned to the hindsight of our deterrence experience of the twentieth century. Our research seeks new guidance on what kinds of threats can and cannot be deterred under which conditions. Then we look to move deterrence strategy frameworks beyond its Cold War roots and construct a more modern, valid policy guide that would best promote stability and peace in the contemporary world.

2 Classical Deterrence

To this day, classical deterrence theory remains the basis of nuclear weapons deployment and targeting policy. The logic can be traced to Brodie's [2, 3] seminal works, with extensive formalization by realist theorists [4]. In the classical deterrence setting, stability depends on increasing the absolute cost of war to its parties. Conflict is least likely under a nuclear"] balance of mutually-assured-destruction (MAD), where the unacceptable costs of war exceed any possible gains. In formalizing classical deterrence, Powell [5] assume that actors are risk-averse, thus an uncertain preemptive challenge is inferior to a certain *status quo ex ante*.[1] From this perspective, all members of the "nuclear club" are assumed to prefer the maintenance of the status quo. By implication, no rational actor can consider war as a desirable outcome eliminating the possibility of initiating an attack. The key finding is that deterrence is optimized under MAD. The key implication tested here is that regardless of the number of nuclear-armed actors present in the system, deterrence is always stable.

3 Conditional Deterrence

Our analysis differs fundamentally from classical deterrence theory. Conditional deterrence shows that it is unrealistic to assume that war is *always* inferior to the status quo. Initiating a nuclear war is not an irrational decision as extensive evidence shows conventional conflicts that were started at parity or by inferior actors [6–9]. We do not delay that nuclear war is indeed a rare event. However, the fact that previous crises have not escalated to nuclear conflict since 1945 does not provide a guarantee that future crises and wars will remain conventional. Conditional deterrence considers uncertain payoffs of war along a continuum where hostile action may involve a dirty bomb, a limited retaliatory nuclear response to terminate a conflict or an all-out nuclear exchange concurrent with MAD when victory and defeat are possible outcomes. To achieve such results, we fundamentally challenge classical deterrence's rigid assumption that decision-makers have zero risk tolerance, assuming away events such as the World Wars and 9/11 as outliers or episodes of pathological "craziness" [5, 10]. In our view, the historical record indicates that countries are at times willing to bear the risk of war, even when costs are great. Extreme risk-takers – like Hitler, Mao Zedong or Bin Laden –viewed conflict as a reasonable chance to change the status quo under certain circumstances. Rogue challengers dissatisfied with the status can strategically choose to take the risk associated with a nuclear pre-emption when they are not exposed to

---

[1] A similar argument is made by Intriligator (1975) and Intriligator and Brito (1984), who achieve stable deterrence in their model by excluding risk, but advocate non-proliferation by including risk (Intriligator and Brito 1981). If risk were included in both models, then proliferation is dangerous and deterrence is unstable. Fearon (1995) argues that risk-acceptance is equivalent to irrationality and implies that a risk-acceptant leader such as Hitler is "a possible exception" that cannot be rationally explained (p. 388). I concur with Zagare and Kilgour (2000) and Zagare (2004), who show that such exceptions are self-serving.

retaliation. A viable deterrence framework must reflect these variations in risk tolerance particularly as the number of nuclear nations expands.

Furthermore, we disagree that irrationality is the cause of war. Our signaling model seeks to filter out the noise inherent in diplomatic demands by discerning revealed, sincere claims from obfuscating bluffs. Consequently, we explore the conditions associated with variations in capability, risk propensity and exposure. Successfully accounting for these factors would greatly aid policymakers in discerning genuine threats from those that are simply bluffing.

4   Modeling Proliferation

To examine the conditions of deterrence, we model deterrence as a strategic risk-taking process. Since decision-makers hope to attain the best ex post result, they consider the ex-ante position as a trade-off between the anticipated benefit of choosing war, and the risk-adjusted likelihood of the actual outcome. In the context of deterrence, we postulate two interacting parties of a Challenger and Defender, where Challenger has some level of nuclear capabilities that generates risk. Challenger can deliberately threaten the Defender, indicating the foreseeable risks of harm. Accordingly, Defender can decide whether to avoid the risks by making a concession, or choosing to bear the risks through confrontation. Throughout the course of the interaction, withdrawal by either side is permitted. Figure 1 depicts the conditional deterrence framework.

This game-theoretic examination of the conditional deterrence approach highlights key shortcomings of the classical deterrence framework. In the classical model, WMD initiation is systematically prevented and can happen only by accident or irrational action. Along with perfect deterrence [1] conditional deterrence implies that WMD initiation can be intentional and rational.[2] Conditional deterrence shows that initiation can take place both under perfect credibility and under extreme conventional asymmetry.

Following Yang [11, 12], Yang et al. [13, 14] and Abdollahian et al. [15], We incorporate game theory and system dynamics in an agent-based framework to understand the interactions between different of actors and simulate various scenarios. The agent based formulation incorporates assessments of policy options that can defuse a conflict or anticipate the long-term effect of decisions made today for the future. To do so, we capture detailed information about stakeholders, their positions on critical issues related to WMD, the importance they attach to such outcomes must be added to the assessment of capability, exposure, risk and trust.

In our model, agents include both state actors and non-state actors. Agent attributes include capability, satisfaction, trust, exposure, and risk. With the game theoretical framework, strategy decisions and outcome histories determine agent's preferences for proliferation, cooperation, deterrence and conflict. Decisions are constrained or incentivized by the changing global structural environment via conditional deterrence

---

[2] Zagare and Kilgour's (2000) perfect deterrence framework derives instability from the lack of credibility. This contrasts with the continuous variations in capability and risk used in the conditional deterrence model.

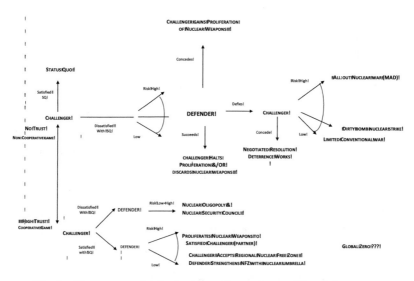

Fig. 1. Challenger and defender interaction under conditional deterrence.

theory, subject to individual agent attributes at any particular time. At micro level, proliferation and deterrence results from agents' interaction that creates current feasible choice set, conditioned upon past behavior and future expectation. At the macro level, regional or global proliferation and deterrence of WMD usage emerge from agents' behavior interactions.

The model is initialized with the creation of actors, which includes global and regional powers as well as nuclear non-state actors, measured by destructive capabilities. All state actors are exposed while non-state actors are not exposed in terms of their nuclear capability. Another critical agent attribute is level of satisfaction with the current order of the international system. The lower the level of satisfaction is, the more inclined that actor wants to challenge the dominant powers. Risk attribute contributes as a multiplier when actors evaluate their expected utility of challenging or maintaining the status quo. Different from satisfaction, which reflects the actor's current status, the trust attribute reflects the level of trustworthy measured by previous behavior. Repeated cooperative behavior in the past generates high level of trust, while non-cooperative behavior decreases the level. This attribute is altered by proliferation of WMD in two ways. Trust can reinforce and cooperation enhanced by the addition of WMD within a cooperative arrangement, while proliferation can reduce trust and have a negative impact on disarmament agreements or cooperation.

We allow individual agent memory and behavioral learning from dyadic outcomes as many others do [16–18]. Trust and risk components allows sum of all prior agents behavioral histories and evolutionary through iterations to contribute to each actor and current macro states of proliferation or deterrence. Agents calculate their expected payoff for different options, and the perception of a pair of agents determines the behavior and outcome of their interaction.

Stability in conditional deterrence is non-monotonic and depends on the distribution of conventional capabilities, nuclear arsenals, challenger's evaluation of risk, and exposure to retaliation. In our model, deterrence becomes tenuous as nuclear arsenals proliferate under two extreme conditions. First, risk-acceptant terrorists facing extreme conventional asymmetry, but low exposure to retaliation, may use low-yield nuclear devices as well as limited dirty bombs when available.[3] Second, dissatisfied and risk-acceptant nations that reach conventional parity and possess large nuclear or WMD arsenals may initiate all-out war.[4] Regardless of risk propensity, conditional deterrence anticipates that deterrence will remain stable between large and relatively small powers where conventional asymmetry is in place. Nuclear deterrence can deter war among dissatisfied and small nations fully exposed to retaliation from the overwhelming nuclear arsenals held by major powers.[5] Thus, conditional deterrence shows that the relationship between the number of nuclear weapons and stability depends on the distribution of capabilities, a challenger's evaluation of the status quo, and their exposure to retaliation. Deterrence becomes tenuous with the proliferation of nuclear or WMD arsenals to dissatisfied nations or entities, thereby increasing the risk of war. The most direct opportunity emerges after a conventional conflict is initiated and a challenger chooses to escalate while a defender chooses to retaliate. Such actions have direct effects on relative capabilities that directly connected to the costs inflicted by participants on each other.

5   Preliminary Results and Policy Implications

Our preliminary result indicates that without further restrictions on proliferation, deterrence will become increasingly fragile. With the costs of acquiring WMDs declining, there is an increased risk of WMD proliferation to dissatisfied countries that reach parity within a region, as well as non-national entities.

Although a general solution still being researched, several policy implications follow. Deterrence is strengthened by proposals that reduce the size of nuclear arsenals among the established nuclear powers and prevent proliferation, bolster regional and global nuclear asymmetry, field anti-nuclear devices among satisfied powers, and deny WMD capabilities to violent non-state actors that seek weapons of mass destruction (Fig. 2).

Reducing the size of nuclear arsenals among the established nuclear powers can increase the stability of deterrence. Reducing the size of arsenals below MAD can only be accepted if the two sides agree to follow common principles and support the status quo. Although major nuclear powers are exposed to the weapons of other major powers, the United States does not fear the development of such capabilities by Britain and France, who are considering joint deployment of submarine-based nuclear

---

[3] These are conditions associated with a weak, but risk-acceptant al-Qaeda.

[4] This would have been the case had the USSR's (Warsaw Pact) conventional capabilities matched those of the USA and NATO. By 2050, if China is risk-acceptant and dissatisfied, such conditions would be met.

[5] Situations where a nuclear umbrella is extended, as in Northeast Asia, are expected to remain stable.

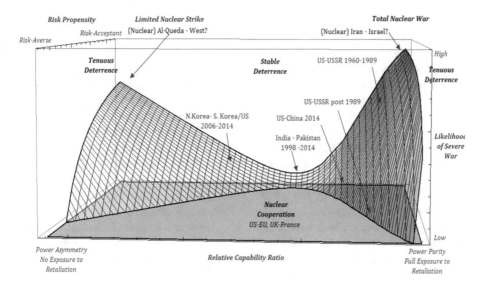

Fig. 2. Analytical implications of conditional deterrence.

deterrents.[6] Furthermore, although Israel has the capacity to destroy most of Europe, no one considers this capability to be a major danger. Satisfied countries "trust" other satisfied countries and willingly accept the risk that nuclear weapons pose. A very different perspective is adopted for weapon development by China, Russia or even India. In these instances, the United States sees such developments with caution, reserve or fear, and these potentially dissatisfied nations are seen as potential aggressors who must be deterred. Limiting the size of nuclear arsenals below the mutual assured destruction threshold can only happen as the perception of risk diminishes. For this reason, the reduction of arsenals is one of the most stabilizing policies that nations can adopt in the nuclear era.

Beyond the horizontal proliferation of weapons within an arsenal, the proliferation of nuclear weapons to other states is dangerous. The Nuclear Non-Proliferation Treaty (NPT), the Proliferation Security Initiative and the Global Initiative to Combat Nuclear Terrorism (GICNT) all aim, through different means, to reduce the proliferation of nuclear technology and weapons proliferation. These programs should be expanded. The NPT's formalized institutional structure creates a set of international norms against proliferation, while the latter two vehicles facilitate interstate cooperation through vehicles that do not impose burdens on their political sensitivities. The initiatives complement non-proliferation norms by taking an active role in preventing WMD proliferation. To enhance the success of the non-proliferation endeavor, all major powers – the United States, Russia, China, France and Britain, and India – must

---

[6] For an example of proposed nuclear weapons cooperation between France and Britain, see Marcus (2012).

cooperate. They still do not today. Conditional deterrence indicates that preventing WMD proliferation is the second most important policy initiative at hand.

With respect to mitigating the risk of regional and global WMD war, a strategy that preserves the power preponderance of countries satisfied international order preserves deterrence. Our analysis indicates that the threat posed by North Korea is limited given the large conventional asymmetry with South Korea and Japan, and the presence of the US nuclear umbrella that assures extended deterrence commitments in Northeast Asia. Conversely, nuclear proliferation to Iran represents a considerable regional danger. As Iran increases its conventional capabilities vis-à-vis Israel, this nuclear proliferation can lead to conflict; a nuclear confrontation is a possibility because Iran can achieve nuclear parity with Israel with only a limited nuclear arsenal.

In addition to addressing the offensive use of weapons, the role of defensive systems must also be considered. As the United States expands the development of regional and global ballistic missile defenses, deterrence is strengthened in the short term but not in the long term. Nuclear defenses offer two principal technical advantages. First, anti-ballistic devices aim at ballistic missiles that are the most effective means of delivery of nuclear weapons. In the pressing case of Israel and Iran, theses states lack a shared border. Consequently, Israel and Iran's primary means of delivering nuclear warheads against one other lie in their medium-range-ballistic-missiles (MRBM). It is worth noting that in this dyad, Israel has fielded and continues to deploy anti-ballistic missile defenses. Iran currently maintains no such capability, and its defense industry currently lags behind. Iran's relatively isolated diplomatic position makes obtaining and fielding defensive systems difficult. Effective defenses in the hands of the satisfied power, in this case Israel, bolsters the regional security architecture and favors the status quo. The same holds for the global system.

Denying nuclear or other WMDs to violent non-state actors is critical. Conditional deterrence shows that only coordinated policies to stem nuclear weapons away from dissatisfied non-state actors can avert the initiation of risky attacks. Given the past attempts of violent non-state actors to acquire nuclear materials, interdiction measures designed to prevent their acquisition of weapons is an essential means of preventing their use. Conditional deterrence is persuasive because it connects the full range of unstable conditions within a single framework while concurrently accounting for successful deterrence. Dissatisfied non-state actors are a source of instability because they are not exposed to retaliation. With the knowledge of a terrorist group's inability to be deterred, only preventing WMD proliferation can ensure stability.

6   Conclusion and Discussion

Conditional deterrence implies that WMD can prevent conflict under narrow conditions, but can lead to WMD imitation, retaliation or exchange under many other structural conditions. Understanding the dynamics of transition between outcomes is critical to minimize the likelihood of WMD use in war. Global powers cannot abandon WMD because the technology is in place and small rough nation or non-state actors acquiring such capabilities can effectively blackmail the great powers. While a monopoly is no longer feasible, a security oligopoly in which the position of WMD is dominated by the hierarchy with great powers holding WMD with global reach can minimize the likelihood of their use. This voluntary or institutionalized collusion of WMD oligopolies would reduce access to WMD and prevents first use by the threat of immediate retaliation, which would provide safety for participants that lack such weapons. The great power could only maintain within the WMD oligarchy if they develop mutual trust.

The policy options we suggest are not a panacea as the specter of nuclear conflict is unfortunately alive and well. No single policy or set of actions by a single state is likely to ensure stability; rather a host of actions from many actors in concert is required. Conditional deterrence implies that doing nothing will eventually lead to conflict. This work may provide a starting point to more creative recommendations to maintain peace in the coming decades.

We strongly believe that integrating macro, meso and micro theories and approaches in a complex adaptive systems framework gives us increased understanding of nuclear actors that differ in their assessments of risk, exposure and willingness to accept costs. We believe such approaches could provide practitioners with a more accurate and updated guide to proliferation and the initiation and escalation conflict.

References

1. Zagare, F.C., Kilgour, D.C.: Perfect Deterrence. Cambridge University Press, Cambridge (2000)
2. Brodie, B.: The Absolute Weapon: Atomic Power and World Order. Ayer Co Pub, Manchester (1946)
3. Brodie, B.: Strategy in the Missile Age. Princeton University Press, Princeton (1959)
4. Intriligator, M.D., Brito, D.L.: Can arms races lead to the outbreak of war? J. Conflict Resolut. **28**, 63–84 (1984)
5. Powell, R.: Crisis bargaining, escalation, and MAD. Am. Polit. Sci. Rev. **81**(03), 717–735 (1987)
6. Kang, K., Kugler, J.: Assessment of deterrence and missile defense in East Asia: a power transition perspective. Int. Area Stud. Rev. **18**(3), 280–296 (2015)
7. Kugler, J.: Terror without deterrence: reassessing the role of nuclear weapons. J. Conflict Resolut. **28**(3), 470–506 (1984)
8. Kugler, J., Zagare, F.C.: Exploring the Stability of Deterrence. Lynne Rienner Publishers, Boulder (1987)
9. Organski, A.F.K., Kugler, J.: War Ledger. University of Chicago Press, Chicago (1980)
10. Blair, B., Brown, M., Burt, R.: Can disarmament work? Foreign Affairs **90**(4), 173–178 (2011)
11. Yang, Z.: The freedom of constraint: a multilevel simulation model of politics, fertility and economic development. J. Policy Complex Syst. **2**(2), 3–21 (2015)
12. Yang, Z.: An agent-based dynamic model of politics, fertility and economic development. In: Proceedings of the 20th World Multi-Conference on Systemics, Cybernetics and Informatics (2016)
13. Yang, Z., Abdollahian, M., de Neal, P: Social spatial heterogeneity and system entrainment in modeling human and nature dynamics. In: Proceeding of Asian Simulation Conference, pp. 311–318. Springer, Singapore (2016)

14. Yang, Z., de Neal, P., Abdollahian, M.: When feedback loops collide: a complex adaptive systems approach to modeling human and nature dynamics. In: Advances in Applied Digital Human Modeling and Simulation, pp. 317–327. Springer International Publishing (2017)
15. Abdollahian, M., Yang, Z., Coan, T., Yesilada, B.: Human development dynamics: an agent based simulation of macro social systems and individual heterogeneous evolutionary games. Complex Adapt. Syst. Model. **1**(1), 1–17 (2013)
16. Axelrod, R.: The evolution of strategies in the iterated Prisoner's Dilemma. In: The Dynamics of Norms, pp. 199–220 (1987)
17. Nowak, M., Sigmund, K.: A strategy of win-stay, lose-shift that outperforms tit-for-tat in the Prisoner's Dilemma game. Nature **364**(6432), 56–58 (1993)
18. Nowak, M.A., Sigmund, K.: Evolution of indirect reciprocity by image scoring. Nature **393**(6685), 573–577 (1998)

Human Behavior Under Emergency and Its Simulation Modeling: A Review

Yixuan Cheng[1](✉), Dahai Liu[1], Jie Chen[1], Sirish Namilae[1],
Jennifer Thropp[1], and Younho Seong[2]

[1] Embry-Riddle Aeronautical University, 600 S Clyde Morris Blvd,
Daytona Beach, FL, USA
{chengy5, chenj5}@my.erau.edu,
{liu89b, namilaes, throppj}@erau.edu
[2] North Carolina Agricultural and Technical State University,
1601 E Market St, Greensboro, NC, USA
yseong@ncat.edu

Abstract. An emergency is a serious, unexpected, and potentially life-threatening situation requiring immediate action. Emergency evacuation is the most critical step to save people's lives. The purpose of this paper is to provide a review of various factors to investigate human behavior under emergency situations. Computational modeling and simulation as a practical way to replicate human behavior change requires quantifying psychological and physical parameters. Previous studies on humans and animals, as well as common simulation approaches were reviewed. According to the results of this literature review, future experiments or simulations can consider not only physical parameters such as human dynamics, but also quantifying psychological parameters such as interpersonal relationship, goal-seeking behavior, decision-making differences, and many more.

Keywords: Human factors · Simulation modeling · Evacuation
Emergency

1 Introduction

An emergency can be defined as an unpredictable situation that we are not familiar with but forced to deal with immediately [1]. The scale and severity of emergencies vary in different situations. Examples of emergencies include natural disasters such as floods, hurricanes, tornados, and fire [2, 3]; technological disasters such as airplane crashes, boat collisions, and bus accidents [4]; hazardous material releases such as toxic gas releases, chemical spills, radiological accidents, and explosions [5]; terrorist attacks [6]; pandemics [2]; and civil disturbances and workplace violence resulting in bodily harm and/or trauma [2, 5, 7]. The cognition, behavior, emotion, and previous experience of individuals can also directly impact the outcomes of emergency situations. Human beings encountering emergency situations commonly develop negative psychological feelings such as stress, panic and show altered decision-making patterns and exhibit non-adaptive behaviors such as panic behavior and crowd behavior [2, 8, 9]. Apart from the psychological impacts, human behavior as pedestrians in emergency situations can lead to serious physical consequence such as injuries or deaths caused by stampeding, pushing, knocking and trampling [3, 10]. Understanding the characteristics of such events is of utmost importance for decision-makers to effectively deal with these situations. There are some unique factors associated with emergency that need to be considered when planning for emergency evacuations, including environmental factors, human factors and policy factors.

Emergency evacuation planning has become an indispensable tool to protect the general public in such unforeseen situations [1]. Many previous modeling studies have simulated the environmental and policy factors [1]; however, human factors are very difficult to replicate or quantify for use as inputs for simulation models. Some prior simulation models have simulated pedestrian moving velocity, pedestrian density, required space to move, and step length. However, human factors such as relationships with other evacuees, purpose of the trip or event, and other psychological differences among people are difficult to quantify. Techniques such as interviews [11], evacuation video review, and questionnaire analyses have been used to investigate occupancy types, human behaviors, and psychological changes during emergency, especially during evacuations [12, 13]. Research integrating psychological parameters in emergency human behavior and pedestrian movement models is extremely limited. The purpose of this review is to provide an overview of the efforts of emergency planning and responses from previous studies and identify the factors that relate to emergency response, in order to provide insights into the improvement of the emergency response in the future.

2 Human Behavior Change Under Emergencies

Human behavior changes in response to emergency situations are complex. Psychological changes are influenced by different environmental, policy, and human factors. These different factors can potentially be used as parameters to predict human behavior change and can be useful in understanding and formulating responses to rapidly changing emergency situations.

2.1 Human Behaviors and Psychological Parameters

Human factors have been studied extensively in the field of disaster psychology, a domain that focuses on the psychological changes people experience before or during the disaster and the resulting behavior [14]. Psychological parameters can be measured based on Leach's Dynamic Disaster Model [14, 15]. This model describes a disaster in three phases and five stages. In each phase and stage, people have different psychological reactions. The three phases are the pre-impact phase, which contains the threat stage and warning stage; and the impact phase and post-impact phase, which contain the recoil stage, rescue stage, and post-traumatic stage [14]. This model also provides a method of grouping psychological variables by which the psychological behavior parameters can be implemented in evacuation models [14].

Both the pre-impact phase and impact phase can be viewed as pre-movement process. Human behavior such as ignoring or denying the emergency situations and being apathetic to the upcoming danger is common in the first pre-impact phase in an emergency [14]. The pre-movement process refers to occupants receive warnings and respond, before they begin to evacuate. Movement process refers to occupants start to evacuate towards exists with the intention to leave or reach a safe place [16, 17]. Researchers have studied human behavior under emergency during the pre-movement process and the movement process [1, 17]. Procedures used to estimate the psycho-logical behavior parameters were recommended to be collecting empirical data from national research centers, referring to previous literature, and making smart guesses when developing models [14].

Inefficient behavior is commonly seen in the impact phase. Some commonly observed human behavior in the pre-movement process involves collecting information about the situation, collecting important belongings, and choosing an optimal exit to escape [16]. The response to warnings is sometimes slow because the event may not yet be life-threating, and thus evacuees are not urged to evacuate immediately [16]. For example, fires, hurricanes and flooding may not threaten people initially; however, once the risk or danger become visible, it can grow rapidly thus leave a very short period of time to evacuate. In this case, people whose responses are delayed for a long period of time during pre-movement process may become trapped in dangerous situations. Incident analyses have shown that there is a connection between a delayed evacuation and a high number of fire deaths or injuries, particularly in residential and hotel buildings [16]. Hence the process in the pre-movement phase is believed to be more decisive to survival than the actual movement process [18].

During the movement process, evacuees exhibit behaviors such as wayfinding, choosing an escape route and alternatives if necessary, and movement towards the selected exit [16]. For evacuees who are not familiar with the environment, wayfinding or choosing an escape route usually depends upon the way the evacuees entered the building, and other escape routes or emergency exists may be easily overlooked [19].

2.2   The Decision-Making Process

During an emergency, before humans act, they first undergo the decision-making process [20, 21], which is typically conducted under time pressure [22]. Effective decision-making depends on the available information and the time available to process the information [23]. Risk perception and decision-making are the two of the most important determinants of human behavior [24]. People first perceive, recall, and think about the emergency situations, which can be viewed as their perceptions of risks. Next, people process the information and make decisions before behaving. The output of this process consists of the actions that they execute during an emergency.

During evacuation, evacuees use environmental cues to process information to select an escape route. Limited time pressure and stress from physical threats (e.g. fire, smoke, flood) are some of the environmental cues that affect evacuees' information processing. Different environmental cues provide different information for decision-making, ultimately resulting in variations in evacuees' behaviors. The information may be incomplete, unacceptable and invalid [25]. As human beings have limited information-processing capacity, they tend to seek the most beneficial information to optimize decisions [26]. Other than environmental cues, people also rely on warning messages from governmental authorities, information from others, and previous experience with similar scenarios to correctly identify and assess risks, make optimal decisions, and take appropriate actions [27]. In addition, evacuees continuously review their decisions during an evacuation by assessing the surroundings and processing additional information to determine if they need to change their decisions [21, 24].

Stress, anxiety, and arousal are the immediate consequences of threat, and individuals undergoing such states tend to make decisions based on internal hypotheses and dominant cues [28]. Inappropriate dominant response, however, can impair decision-making because it results in the neglect of processing unusual information or by reinterpreting the unusual information to fit previous experience or their expectations [29]. In the worst case, when evacuees start to feel hopeless about escaping from the danger, they may enter the highest level of stress for decision-making phase, which is hypervigilance and show panic emotion and panic behavior as a consequence [8].

Prior experiences also shape evacuee decision-making. In dangerous situations, decision-making is influenced by previous experiences with similar cases and psychological state and physiological abilities determine behavior [30]. Factors such as familiarity with the environment, availability of external cues (i.e. warnings, signage, presence of other evacuees or staff), and personal experience affect evacuees' decisions [20]. Noticeably, familiarity with the environment and personal experience may not promote optimal decision-making. It may promote a selective knowledge of the environment and lead to ignorance of alternative exit routes. Evacuees may try to escape from the further exits with which they have had previous experience, instead of making rational decisions based on environmental cues and feasible methods, which may involve moving towards the closest exit about which they may have no prior knowledge [31–33].

2.3   Crowd Behavior

When people are stuck in a crowded environment, the negative emotions and resulting behaviors can impact a large group of people and therefore lead to serious consequences. Negative psychological reactions such as insecurity, anxiety, worry or fear experienced in a group during an emergency can lead to distress and worsen a threatening or harmful situation [22]. This phenomenon has been described as "herding behavior", a type of irrational behavior that often leads to dangerous overcrowding and impaired escape [9]. Herding behavior is the result of social contagion, which is the transition of experience from individual psychology to mass psychology [9]. Conformity in behavior can be observed, in which individuals tend to follow others' actions. People dissolve their individual identities, motivations and rationalities into a collective mind when being in crowds; therefore, behavior including decision-making differs in crowds in comparison to being alone or in a small group [34]. Further, behaviors like stampeding, pushing, knocking, and trampling on others are commonly seen in crowds. These destructive actions are described as non-adaptive crowd behaviors and may result from an individual's high stress level, inability to make decisions, social identity within a group, loss of personal space, high crowd density, severe external crises or emergencies, and high emotional arousal [34].

Evacuees sometimes experience the phenomenon called 'going with the crowd', when they abandon their own thinking and adopt actions by following others [34]. For instance, when the visibility is low due to smoke in a fire, evacuees who are prone to this phenomenon may be easily affected by others and follow the crowd movement. Although following the crowd is not always harmful, doing so irrationally can reduce the efficiency of using exits, lead to mistaken route choices, and result in jamming.

On a macroscopic level, social structures of interaction also affect human behaviors in crowds. Pre-existing structures (e.g. family or friends) and structures formed at the time of emergency (e.g. queues) are the two social structures of interaction to be considered when studying crowd behavior [35]. Previous studies have demonstrated that pre-existing social structures of interaction play a significant role in human behavior in crowds [36–38], specifically, people who come together to a location as a group also tend to move together, orient toward each other, and leave the location together [14, 39]. The closer the relationship is among individuals within a group, the more likely they will behave as single entity [39]. Family, friends, or colleagues tend to influence evacuees more than groups formed by strangers [40]. This effect could potentially slow down the flow of the crowd if a large group with strong relationship between people tries to move together or move slowly to wait or look for other group members [12, 13, 16, 34]. The study of the social attachment model in crowd behavior has indicated that in times of emergency, people normally display affiliative behaviors [11]. Affiliative behaviors include moving from unfamiliar situations towards familiar people and/or places [11]. When people need to escape from an urgent situation immediately, the time they spend to seek for familiar people or move towards familiar place could slow down the evacuation process. After the reunion with the familiar group, the chance of individual escape is decreased. The larger the group one is in; the longer people take to evacuate. In other words, social attachment often delays egress.

2.4   Panic Emotion and Panic Behavior

Panic-related emotions and panic behavior also influence decision-making and consequent human behavior changes. The risk level of disasters affects the severity and distribution of panic [41]. Environmental and situational cues that may generate and facilitate panic behavior are: (1) perception of an urgent and immediate threat to him/herself and/or loved ones; (2) belief that escape from the emergency situations is possible, however, the escape routes are becoming inaccessible and time to escape is rapidly decreasing; and (3) feelings of helplessness, especially when others are not willing or not able to help [27]. Three common tiers of distress experienced include anxiety, which is the least intense in terms of emotional response, followed by fear, and panic, which is a more intense extension of fear [42].

In addition to fear, the bodily sensation of physical pressure caused by contact with other bodies confined in a space also has the potential to cause panic [43]. Under emergency situations, aggressive human behavior may happen due to the competition for resources like space or escape opportunity. These behaviors can lead to a faster-is-slower movement trend, in which individuals attempt to move faster but cause slower flow through a bottleneck [44, 45]. The harder the evacuees push towards an exit, the more pressure and interpersonal friction forces will occur in the crowd. In a crowded environment, physical pressure, fear and anxiety, as well as panic develop quickly. Unavoidably, moving or passing a bottleneck frequently becomes uncoordinated [44].

In contrast to the literature reviewed above, previous interview studies of survivors from different emergencies [11] including the 9/11 terrorist attack [46] showed that egocentric, adverse or non-adaptive panic behavior is not necessarily common during emergency situations [11, 46–49].

3   Factors Impacting Human Behavior Change

We classify the different factors affecting the human behavior in emergencies as (1) human factors, (2) environmental factors, and (3) policy factors. Each of these factors are discussed in detail below.

3.1   Human Factors

Disaster and evacuation models require incorporation of human characteristic variables to achieve realistic predictions and realistic problem solving. These human characteristic variables can be grouped according to attributes such as age, gender, mobility (e.g., healthy or disabled), prior evacuation experience, knowledge about the surrounding environment, and intelligence (e.g. consistent evacuation strategies versus changing evacuation strategies) [14]. In non-rushed situations, individuals prefer to walk at a comfortable walking speed which differs as a function of gender, age, disability, scenario, surrounding environment, the time of the day, and the purpose of the trip [9]. During an emergency evacuation, individuals have different pre-movement times in which they recognize and respond to the situation initially [50, 51]. Evacuees' original status (e.g. age, body size, gender, mental and physical ability to detect and respond to warnings, mobility) [14, 16], occupant temporary status (e.g. disease, state of sleep or wakefulness, alcohol or drug influence, tasks or activities underway at the time) [15, 16], familiarity with the building environment [14, 16, 19, 20, 31–33, 50], moving speed [50, 51], preferred personal space [50, 51], step length [50, 51], educational level, profession, and relevant experience will affect evacuation efficiency [16, 24]. Other influential factors include interactions between evacuees and the density of evacuees in the vicinity and distribution of evacuees within the building [16]. In addition, human effectiveness (e.g., problem-solving style and achievement motivation) was found to have a positive impact on the effectiveness of evacuation in technological disasters (e.g. airplane crash, boat collision, and bus accident) [4]. Income has a negative relationship with stress during the emergency [52]. Personality traits, social features, and evacuees' situational features affect their choice of egress route, final exit, and whether to stay with the crowd and follow their familiar paths or follow their own paths and make diverse choices [53].

3.2   Environmental Factors

Physical threats are environmental factors that lead to time pressure and stress and consequently cause impairment of evacuees' information processing [21]. Effective environmental information can provide evacuees clear and accurate spatial cues to assess the emergency situations during evacuation [21].

Experiments have shown that the usability of building properties, including the architectonical constructions (layout, number and widths of exits, working sprinklers in case of fire, flammability of a structure, e.g., the materials with which it is constructed and its contents) [16, 18, 24, 51, 54]; design of escape routes; escape route indications; presence of escape maps in building corridors; presence of escape route signage (recommended to be placed closer to the ground to prevent loss of sight in smoke) [16, 18, 51, 55]; facility types (e.g. airport terminal, classroom, hospital, hotel, restaurant, shopping malls, stadium, theatre, underground subway, or well-managed offices) [18, 51]; effective communication system [55]; type of alarm system (broadcasting or pre-recorded warning sound) [51]; volume of warnings [16, 53]; and presence of a focal point that can attract evacuees' attention (e.g., a stage or screen to indicate warnings or an evacuation message) [51] can have significant effects on evacuation time.

3.3   Policy Factors

There are numerous policies, regulations, and laws existing to deal with the egress and evacuation of individuals [30]. For instance, the Robert T. Stafford Disaster Relief and Emergency Assistance Act (Stafford Act) is a national level regulation for emergency management [56]. The Occupational Safety and Health Administration (OSHA) sets standards and provides guidance for evacuation policies, evacuation procedures, emergency escape procedures, route assignments, as well as rescue and medical duties for designated workers in workplaces [5]. OSHA has Emergency Action Plans, Fire Prevention Plans, and standards for fire detection systems, employee alarm systems, fixed extinguishing systems and portable fire extinguishers [5]. The safety management system, emergency response plan and procedures, fire alarm evacuation policy, and emergency management strategy enforced by company, public venue, residential building and so forth; as well as rescue response from governmental organizations can also impact human behavior during an evacuation [17, 30, 50].

4   Experimental and Simulation Studies

Experimental studies have been widely used in emergency evacuation studies. Due to the nature of the emergency studies, it is challenging to include human participants, however, there are a few studies that have used human subjects. For example, in the aviation field, human participants have been used to investigate the influence of spatial factors such as narrow configurations of an airplane cabin. The McDonnell Douglas MD-11 evacuation experiment in 1991, the Airbus A380-800 evacuation experiment in 2006, and other aircraft evacuation experiments as required by the Federal Aviation

Administration (FAA) were carried out successfully, however, injuries to participants have occurred [57, 58]. To ensure participants' safety, any simulation studies that involve human participants must be initially approved by the organization's institutional review board (IRB), though this process may preclude the use of hazardous situations, which are the nature of emergency scenarios, in the protocol [59, 60].

Because of the ethical difficulties associated with measuring higher risk factors such as panic, the consequent human behaviors are largely unpredictable [30, 45]. As a result, researchers studied responses in animals to predict human panic behavior. In medical simulation studies of clinical trials, researchers [61, 62] regularly use mice to instead of human beings. This is due to the high similarity of both biology and sequence between human and mouse. The dynamics of escape panic in mice has been used as an analog for the escape panic experienced by human evacuees [63]. This study concluded that panic behavior in mice (and thus to an extent in humans), was influenced by the architecture of the space to which they are confined [63]; exits with larger and wider doors resulted in a higher escape rate. However, an increased number of exits did not result in a higher escape rate. This study was based on the fact that mice are allelomimetic (i.e. follow or copy others' behaviors), therefore they all tried to escape from one exit door. While, human beings are not allelomimetic [63], but may be influenced by 'going with the crowd' phenomenon [34]. Though mice have become a primary animal species in various experimental studies [61] and successfully demonstrated theories such as the spatial influence during evacuation [63] and the faster-is-slow effect [64, 65], there are also difference(s) between animals and humans that animal studies cannot avoid.

Other animal species have also been used for evacuation studies. For instance, Argentine ants' movement patterns were observed and studied in a series of experiments [66, 67]. The ants' movement patterns were affected by the layout or the geometrical structures at the escaping areas [66]. However, these studies failed to observe the faster-is-slow effect because the ants did not display a selfish evacuation behavior during the experiment [67]. Further analysis of the failed study found that the ants had specific characteristics that yielded them inadvisable for representing human behaviors, especially under emergency situations [68].

4.1 Computer Emergency Simulations

Computer-based simulation is a method for studying and researching different scenarios in a real-world system, which can overcome the restriction of using human participants and the deficiencies of using animals. The advantages of using computer-based simulation are: safer, cheaper, better performance, easier manipulation, randomization in scenario assignments, capability of modeling sophisticated conditions (e.g. physical systems dynamical models, spatial environment models, and agent decision models) [69]. Commonly used simulation systems include ARENA, Any-Logic, airEXODUS, AvatarSim, AIEVA, and GPSS. The primary drawback in computational simulations is the inability to incorporate complex behavior of individual responses to emergency. Only behaviors such as pedestrian dynamics [20] and simple imitation of pedestrian stress, anger, and panic behavior in aircraft [70] can be incorporated. However, the simulation studies provide an effective tool for evaluating behavior of large crowds, where individual decision making may not have significant impact.

4.2   Modeling and Simulation of Pedestrian Movement

Analyzing pedestrian motion helps planning for facilities and predicts evacuation strategies to suppress the risk of loss of life. Social force models and queuing theories are approaches that have been suggested to study human dynamics (e.g. pedestrian density versus speed change). Social force models have been applied to crowd simulations situations in panic situations [44], traffic dynamics [71], evacuation [72] and animal herding [73]. Algorithmic developments have included generation of force fields using visual analysis of crowd flows [74], explicit collision prediction [75], and collision avoidance [76]. Pedestrian behaviors in normal situations are different from in evacuation situations and a comparison study was established [9]. The social force model alters between these analyzed cases since the nervousness factor is implemented. In non-emergency situations, the self-organization of pedestrians is emphasized through line formation along hallways and oscillations at bottlenecks. On the other hand, panic circumstances are more chaotic. The tendency of herding, lane breakdown and clogging are observed which in return reduces the chance of survival. Pedestrian density is one of the primary factors affecting the movement of pedestrians [77]. This effect is expected to be more important during emergency and high crowd density situations. There is significant experimental evidence for reduction of pedestrian speed with increase in pedestrian density [78].

5   Conclusion

The most significant contribution of this review was the summary of human factors, environmental factors and policy factors that affect human behavior change. Previous studies established mathematical models to measure environmental factors and physical parameters of human factors such as human dynamics. However, there are insufficient studies that have measured human psychological parameters due to the difficulty of quantifying them and ethical issues of exposing human participants to hazards in experimental protocols. In order to understand, predict, and manage human behavior, this literature review has summarized possible parameters to be modeled and measured.

References

1. Alexander, D.E.: Definition of emergency. In: Penuel, K.B., Statler, M., Hagen, R. (eds.) Encyclopedia of Crisis Management, pp. 324–325. SAGE Publications, Thousand Oaks (2013)
2. Fagel, M.J., Krill, S., Lawrence, M.: Policy and laws relating to emergency management planning. In: Fagel, M.J. (ed.) Crisis Management and Emergency Planning: Preparing for Today's Challenges, pp. 3–17. CRC Press, Boca Raton (2014)

3. Guha-Sapir, D., Hoyois, P., Below, R.: Annual Disaster Statistical Review 2015: The Numbers and Trends. CRED, Brussels (2016)
4. Cassidy, T.: Problem-solving style, achievement motivation, psychological distress and response to a simulated emergency. Couns. Psychol. Q. **15**, 325–332 (2002)
5. States Department of Labor. https://www.osha.gov/SLTC/etools/evacuation/evaluate.html
6. Van de Walle, B., Turoff, M.: Decision support for emergency situations. Inf. Syst. e-Bus. Manag. **6**, 295–316 (2008)
7. Neria, Y., Nandi, A., Galea, S.: Post-traumatic stress disorder following disasters: a systematic review. Psychol. Med. **38**, 467–480 (2008)
8. Aldag, R.J.: Decision making: a psychological analysis of conflict. Acad. Manag. Rev. **5**, 141–143 (1980)
9. Helbing, D., Farkas, I.J., Molnar, P., Vicsek, T.: Simulation of pedestrian crowds in normal and evacuation situations. In: Schreckenberg, M., Sharma, S.D. (eds.) Pedestrian and Evacuation Dynamics, pp. 21–58. Springer, Berlin (2002)
10. Pine, J.C.: Natural Hazards Analysis: Reducing the Impact of Disasters. CRC Press, Boca Raton (2009)
11. Cocking, C., Drury, J., Reicher, S.: The psychology of crowd behaviour in emergency evacuations: results from two interview studies and implications for the fire and rescue services. Irish J. Psychol. **30**, 59–73 (2009)
12. Purser, D.A., Raggio, A.J.T.: Behaviour of crowds when subjected to fire intelligence. Building Research Establishing Report CR 143/95. Building Research Establishment Ltd., Watford (1995)
13. Purser, D.A., Bensilum, M.: Quantification of escape behavior during experimental evacuations. Building Research Establishment Report CR 30/99. Building Research Establishment Ltd., Watford (1999)
14. Vorst, H.C.M.: Evacuation models and disaster psychology. Procedia Eng. **3**, 15–21 (2010)
15. Leach, J.: Survival Psychology. Palgrave Macmillan, London (1994)
16. Purser, D.A., Bensilum, M.: Quantification of behaviour for engineering design standards and escape time calculations. Saf. Sci. **38**, 157–182 (2001)
17. International Organization for Standardization. https://www.iso.org/obp/ui/#iso:std:iso:tr:13387:-8:ed-1:v1:en
18. Kobes, M., Helsloot, I., de Vries, B., Post, J.G., Oberijé, N., Groenewegen, K.: Way finding during fire evacuation; an analysis of unannounced fire drills in a hotel at night. Build. Environ. **45**, 537–548 (2010)
19. Sime, J., Breaux, J., Canter, D.: Human Behaviour Patterns in Domestic and Hospital Fires. BRE Report, UK (1994)
20. Gwynne, S., Galea, E.R., Lawrence, P.J., Filippidis, L.: Modelling occupant interaction with fire conditions using the buildingEXODUS evacuation model. Fire Saf. J. **36**, 327–357 (2001)
21. Ozel, F.: Time pressure and stress as a factor during emergency egress. Saf. Sci. **38**, 95–107 (2001)
22. Knuth, D., Kehl, D., Hulse, L., Schmidt, S.: Perievent distress during fires-the impact of perceived emergency knowledge. J. Environ. Psychol. **34**, 10–17 (2013)
23. Heliovaara, S., Kuusinen, J., Rinne, T., Korhonen, T., Ehtamo, H.: Pedestrian behavior and exit selection in evacuation of a corridor-an experimental study. Saf. Sci. **50**, 221–227 (2012)
24. Mu, H.L., Wang, J.H., Mao, Z.L., Sun, J.H., Lo, S.M., Wang, Q.S.: Pre-evacuation human reactions in fires: an attribution analysis considering psychological process. Procedia Eng. **52**, 290–296 (2013)

25. Yoon, S.W., Velasquez, J.D., Partridge, B.K., Nof, S.Y.: Transportation security decision support system for emergency response: a training prototype. Decis. Support Syst. **46**, 139–148 (2008)
26. Ben Zur, H., Breznitz, J.S.: The effect of time pressure on risky choice behavior. Acta Physiol. **47**, 89–104 (1981)
27. Gantt, P., Gantt, R.: Disaster psychology: dispelling the myths of panic. Prof. Saf. **57**(8), 42–49 (2012)
28. Staw, B.M., Sandelands, L.E., Dutton, J.E.: Threat-rigidity effects in organizational behavior: a multilevel analysis. Adm. Sci. Q. **26**, 501–524 (1981)
29. Rice, R.E.: From adversity to diversity: applications of communication technology to crisis management. Adv. Telecommun. Manag. **3**, 91–112 (1990)
30. Abolghasemzadeh, P.: A comprehensive method for environmentally sensitive and behavioral microscopic egress analysis in case of fire in buildings. Saf. Sci. **59**, 1–9 (2013)
31. Bode, N.W.F., Codling, E.A.: Human exit route choice in virtual crowd evacuations. Anim. Behav. **86**, 347–358 (2013)
32. Turner, R.H., Killian, L.M.: Collective Behaviour. Prentice-Hall, Upper Saddle River (1957)
33. Kahnemen, D., Tversky, A.: Prospect theory: an analysis of decision under risk. Econometrica **47**, 263–292 (1979)
34. Pan, X., Han, C.S., Dauber, K., Law, K.H.: Human and social behavior in computational modeling and analysis of egress. Autom. Constr. **15**, 448–461 (2006)
35. Tucker, C.W., Schweingruber, D., McPhail, C.: Simulating arcs and rings in gatherings. Int. J. Hum. Comput. Stud. **50**, 581–588 (1999)
36. Aveni, A.F.: The not-so-lonely crowd: friendship groups in collective behavior. Sociometry **48**, 96–99 (1977)
37. McPhail, C.: The Myth of the Madding Crowd. Aldine de Gruyter, Hawthorne (1991)
38. McPhail, C., Wohlstein, R.T.: Collective locomotion as collective behavior. Am. Sociol. Rev. **51**, 447–463 (1986)
39. Sime, J.D.: Affiliate behaviour during escape to building exits. J. Environ. Psychol. **3**, 21–41 (1983)
40. Nilsson, D., Johansson, A.: Social influence during the initial phase of a fire evacuation - analysis of evacuation experiments in a cinema theatre. Fire Saf. J. **44**, 71–79 (2009)
41. Armfield, J.M.: Cognitive vulnerability: a model of the etiology of fear. Clin. Psychol. Rev. **26**, 746–768 (2006)
42. Zakaria, W., Yusof, U.K.: Modelling crowd behaviour during emergency evacuation: a proposed framework. In: 2016 International Conference on Advanced Informatics: Concepts, Theory And Application (ICAICTA) (2016)
43. Schneider, B.: The reference model SimPan - agent-based modelling of human behaviour in panic situations. In: Tenth International Conference on Computer Modeling and Simulation (2008)
44. Helbing, D., Farkas, I., Vicsek, T.: Simulating dynamical features of escape panic. Nature **407**, 487–490 (2000)
45. Hu, Z., Sheu, J., Xiao, L.: Post-disaster evacuation and temporary resettlement considering panic and panic spread. Transp. Res. Part B **69**, 112–132 (2014)
46. Blake, S.J., Galea, E.R., Westeng, H., Dixon, A.J.P.: An analysis of human behavior during the WTC disaster of 11 September 2001 based on published survivor accounts. In: 3rd International Symposium on Human Behavior in Fire, pp. 181–192. InterScience Communications, Belfast (2004)
47. Aguirre, B.E.: Commentary on "understanding mass panic and other collective responses to threat and disaster": emergency evacuations, panic, and social psychology. Psychiatry **68**, 121–129 (2005)

48. Bohannon, J.: Directing the herd: crowds and the science of evacuation. Science **310**, 219–221 (2005)
49. Mawson, A.R.: Understanding mass panic and other collective responses to threat and disaster. Psychiatry **68**, 95–113 (2005)
50. Oswald, M., Lebeda, C., Schneider, U., Kirchberger, H.: Full-scale evacuation experiments in a smoke filled rail carriage-a detailed study of passenger behavior under reduced visibility. In: Waldau, N., Gattermann, P., Knoflacher, H., Schreckenberg, M. (eds.) Pedestrian and Evacuation Dynamics. Springer, Heidelberg (2005)
51. Proulx, G.: Evacuation time. In: SFPE Handbook of Fire Protection Engineering, 4th edn, pp. 3–355. National Fire Protection Association Quincy, MA (2008)
52. Benight, C.C., Harper, M.L.: Coping self-efficacy perceptions as a mediator between acute stress response and long-term distress following natural disasters. J. Trauma. Stress **15**, 177–186 (2002)
53. Zhan, X., Yang, L., Zhu, K., Kong, X., Rao, P., Zhang, T.: Experimental study of the impact of personality traits on occupant exit choice during building evacuation. Procedia Eng. **62**, 548–553 (2013)
54. Purser, D.A.: Behavioural impairment in smoke environments. Toxicology **115**, 25–40 (1996)
55. Proulx, G., Sime, J.D.: To prevent panic in an underground emergency, why not tell people the truth? Fire Saf. Sci. **3**, 843–852 (1991)
56. Association of State and Territorial Health Officials. http://www.astho.org/Programs/Preparedness/Public-Health-Emergency-Law/Emergency-Authority-and-Immunity-Toolkit/Key-Federal-Laws-and-Policies-Regarding-Emergency-Authority-and-Immunity/
57. The Guardian. https://www.theguardian.com/business/2006/mar/27/theairlineindustry.travelnews
58. A380 successful evacuation trial. Aircr. Eng. Aerosp. Technol. **78**(4) (2006)
59. Office for Human Research Protections. http://www.hhs.gov/ohrp/regulations-and-policy/regulations/45-cfr-46/index.html
60. Penslar, R.: Protecting Human Research Subjects. National Institutes of Health, Washington DC (1933)
61. Battey, J., Jordan, E., Cox, D., Dove, W.: An action plan for mouse genomics. Nat. Genet. **21**, 73–75 (1999)
62. Wasserman, W.W., Palumbo, M., Thompson, W., Fickett, J.W., Lawrence, C.E.: Human-mouse genome comparisons to locate regulatory sites. Nat. Genet. **26**, 225 (2000)
63. Saloma, C., Perez, G.J., Tapang, G., Lim, M., Palmes-Saloma, C.: Self-organized queuing and scale-free behavior in real escape panic. Natl. Acad. Sci. U.S.A. **100**, 11947–11952 (2003)
64. Lin, P., Ma, J., Liu, T., Ran, T., Si, Y., Li, T.: An experimental study of the "faster-is-slower" effect using mice under panic. Phys. A: Stat. Mech. Appl. **452**, 157 (2016)
65. Shiwakoti, N., Sarvi, M.: Enhancing the panic escape of crowd through architectural design. Transp. Res. Part C **37**, 260–267 (2013)
66. Shiwakoti, N., Sarvi, M., Rose, G., Burd, M.: Animal dynamics based approach for modeling pedestrian crowd egress under panic conditions. Procedia – Soc. Behav. Sci. **17**, 438–461 (2011)
67. Soria, S., Josens, R., Parisi, D.: Experimental evidence of the "faster is slower" effect in the evacuation of ants. Saf. Sci. **50**, 1584–1588 (2012)
68. Parisi, D., Soria, S., Josens, R.: Faster-is-slower effect in escaping ants revisited: ants do not behave like humans. Saf. Sci. **72**, 274–282 (2015)
69. Shi, J., Ren, A., Chen, C.: Agent-based evacuation model of large public buildings under fire conditions. Autom. Constr. **18**, 338–347 (2009)

70. Sharma, S., Singh, H., Prakash, A.: Multi-agent modeling and simulation of human behavior in aircraft evacuation. IEEE Trans. Aerosp. Electron. Syst. **44**, 1477–1499 (2008)
71. Treiber, M., Hennecke, A., Helbing, D.: Derivation, properties, and simulation of a gas-kinetic-based, nonlocal traffic model. Phys. Rev. E **59**(1), 239 (1999)
72. Wei-Guo, S., Yan-Fei, Y., Bing-Hong, W., Wei-Cheng, F.: Evacuation behaviors at exit in ca model with force essentials: a comparison with social force model. Phys. A **371**(2), 658–666 (2006)
73. Li, F., Chen, S., Wang, X., Feng, F.: Pedestrian evacuation modeling and simulation on metro platforms considering panic impacts. Procedia-Soc. Behav. Sci. **138**, 314–322 (2014)
74. Mehran, R., Oyama, A., Shah, M.: Abnormal crowd behavior detection using social force model. In: Computer Vision and Pattern Recognition IEEE Conference, pp. 935–942. IEEE Press, New York (2009)
75. Zanlungo, F., Ikeda, T., Kanda, T.: Social force model with explicit collision prediction. EPL (Europhy. Lett.) **93**(6), 68005 (2011)
76. Flötteröd, G., Lämmel, G.: Bidirectional pedestrian fundamental diagram. Transp. Res. Part B: Methodol. **71**, 194–212 (2015)
77. Lakoba, T.I., Kaup, D.J., Finkelstein, N.M.: Modifications of the helbing-molnar-farkas-vicsek social force model for pedestrian evolution. Simulation **81**, 339–352 (2005)
78. Bruno, L., Venuti, F.: The pedestrian speed–density relation: modelling and application. In: 3rd Footbridge International Conference (2008)

ACT-R Modeling to Simulate Information Amalgamation Strategies

John T. Richardson, Justine P. Caylor[✉], Eric G. Heilman,
and Timothy P. Hanratty

Computational Information Science Directorate, United States Army Research
Laboratory, Aberdeen Proving Ground, MD, USA
{john.t.richardson7.civ, justine.p.caylor.ctr,
eric.g.heilman.civ, timothy.p.hanratty.civ}@mail.mil

Abstract. Today, military decision-making is dependent on the ability to amalgamate information across sources of varying degrees of agreement. Given the increasing volume of information, automated methods to assist in the identification and prioritization of the most valuable or relevant information has become paramount. Relevant information is not only critical to situational awareness and the military decision-making process, but vital to mission success. Towards this end, the US Army Research Laboratory (ARL) has undertaken a research initiative to model and test how analysts perceive the Value of Information (VoI) in varying military context. The goal of this effort is to develop methodologies useful in the development of automated information agents. As a part of the VoI initiative, ARL conducted an experiment with Subject Matter Experts (SMEs) at the US Army Intelligence Center of Excellence (ICOE), where data was collected on how intelligence analysts' amalgamate information given information content and source reliability within complementary and contradictory conditional associations. The resulting experimental data was incorporated into an Adaptive Control of Thought-Rational (ACT-R) model. Exercising the ACT-R cognitive model resulted in some interesting response behaviors not observed in the initial experiment. In an effort to better understand the perceptions (cognitive underpinnings) of a military intelligence analyst, this paper extends the previous effort and utilizes a crowdsourced experiment within Amazon Mechanical Turk (Mturk). The experiment captures many of the same conditional ratings encountered by the military analysts. Data gained from the Mturk experiment will be examined using the ACT-R model as a simulation to determine whether the same data distributions exist within a wider audience and as a direct comparison to the analyst's responses. This paper will examine the Mturk experimental design, discuss the experimental apparatus implementation and provide an overview of the ACT-R model utilized to replicate the amalgamation strategies.

Keywords: Value of Information · Adaptive Control of Thought-Rationale
Information amalgamation · Cognitive modeling

© Springer International Publishing AG, part of Springer Nature (outside the USA) 2019
D. N. Cassenti (Ed.): AHFE 2018, AISC 780, pp. 326–335, 2019.
https://doi.org/10.1007/978-3-319-94223-0_31

1   Introduction

In order to make an informed decision it is necessary to gather relevant information. In military operations, especially as tempo increases, time constraints make traditional methods of information review unattractive. Consequently, commanders have battle staffs to execute timely information review to support decision-making. At the foundation of this hierarchy is the individual analyst. At the individual level, cognitive overload is a rising concern as the volume of information continues to increase with the number and variety of sources [1].

To unload this cognitive burden the US Army Research Laboratory (ARL) envisions assistive methods for gathering and ranking information at the analyst level. Establishing a foundation for these methods requires understanding how an analyst determines value. To investigate the process ARL has previously conducted experiments with analysts to measure their evaluation of a single piece of information and how it changes when related information is provided side-by-side [2].

This paper presents a re-creation of ARLs previous experiment, replacing analysts with the general population. The purpose is to identify similarities and differences in the cognitive process of determining value between the expert population and the general population. The first section provides background on the current abstraction that ARL uses for valuing information and its Value of Information (VoI) experiment. The next section details our comparison of the performance of the expert and general population in the experiment. A proposed ACT-R [3] model to examine the cognitive processes in information ranking follows. The paper concludes with a discussion of the results and future directions.

2   Background

2.1   Quantifying the Value of Information

Determination of the value of a piece of information is a complex cognitive task performed by the analyst. ARLs initial goal is to capture this process at an abstract level. In other words, is it possible to reduce information to a set of parameters that can be judged for value by an analyst in an experimental setting? To accomplish this goal ARL used existing US Army doctrine [4] which defined ordinal parameters for the reliability and content of a piece of information. Reliability is a measure of the information source and has rankings of A to E (reliable, usually reliable, fairly reliable, not usually reliable, and unreliable). Content ranks the verisimilitude of the information from 1 to 5 (confirmed, probably true, possibly true, doubtfully true, improbable). Combining both parameter provides twenty-five distinct ratings, such as A1 (Tables 1, 2).

While it is possible to assign twenty-five ratings, doctrine does not provide a ranking for them. Previous work by ARL has established rankings of the ratings by using empirical data from analysts who ordered the ratings in an abstract setting [1]. No details of the information were provided beyond the content and reliability parameters. ARL created a fuzzy associative memory (FAM) model, based on the collected analyst data that quantifies the ratings allowing for ranking [5].

Table 1. Source reliability

A	Reliable	**No doubt** of authenticity, trustworthiness, or competency; has a history of complete reliability
B	Usually reliable	**Minor doubt** about authenticity, trustworthiness, or competency; has a history of valid information most of the time
C	Fairly reliable	**Doubt** of authenticity, trustworthiness, or competency but has provided valid information in the past
D	Not usually reliable	**Significant doubt** about authenticity, trustworthiness, or competency but has provided valid information in the past
E	Unreliable	**Lacking** in authenticity, trustworthiness, and competency; history of invalid information

Table 2. Information content

1	Confirmed	**Confirmed** by other independent sources; **logical** in itself; **Consistent** with other information on the subject
2	Probably true	Not confirmed; **logical** in itself; **consistent** with other information on the subject
3	Possibly true	Not confirmed; **reasonably logical** in itself; **agrees with some** other information on the subject
4	Doubtfully true	Not confirmed; possible but **not logical; no other information** on the subject
5	Improbable	Not confirmed; **not logical** in itself; **contradicted** by other information on the subject

The quantitative score is termed the applicability score of the piece of information. Devoid of context, this is the value of the piece of information. ARL has shown how the combination of additional context parameters, such as timeliness and operational tempo can alter the applicability score and change the value of the information.

2.2 Quantifying the Value of Information Interaction

Extending their previous research ARL used empirical data to quantify the value of information interaction and amalgamation [6]. In this experiment, analysts considered two pieces of information linked by four different qualitative relationships.

- *Totally Complementary* - The information supports the same conclusion.
- *Somewhat Complementary* - The information supports a related conclusion.
- *Somewhat Contradictory* - The information does not support the same conclusion.
- *Totally Contradictory* - The information supports a conflicting conclusion.

In the experiment, analysts were given an initial piece of information, a relationship, and an additional piece of information (Fig. 1). The combination of information and relationship is designated as a 'couplet' by ARL. Analysts were asked to provide two rankings on a sliding applicability scale (extremely applicable to not applicable). The first ranking was their applicability score for the initial piece of information. Data from this question directly relates to earlier experiments where analysts were tasked with ranking individual pieces of data. The second ranking is the adjusted applicability score of the initial piece of information given its relationship to the additional piece of information. The slider only displays an ordinal scale of applicability, but there are underlying numerical rankings that are recorded. From this numerical data, ARL was able to expand their FAM model to account for value adjustments caused by related information. Furthermore, ARL was able to use the empirical data to estimate ranking and interaction distributions. These distributions were used to build an ACT-R model to simulate the analyst ranking process [6].

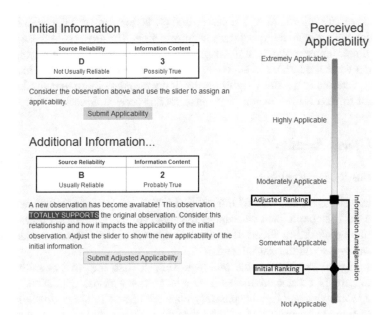

Fig. 1. VoI experiment interface and example

2.3   Expanding the Participant Pool

The original experiment to quantify information interaction had twenty US Army analysts as participants. ARL repeated this experiment with an expanded participant pool. This group included 829 Amazon Mechanical Turk (Mturk) participants. Mturk is an online crowd-source platform where users are compensated for their participation.

Differences in this version of the experiment include training material, a pretest, and reduced task load. In the original experiment, ARL was available in person to train and discuss the experiment purpose with the analysts. This is not possible in the Mturk environment, thus several pages of training and instructions were provided to the participant to familiarize them with the objective of the experiment. Following the training the participants were given a pretest consisting of three trials of the VoI task described in the previous section. The purpose of the pretest was to decide if the participant understood the provided training material. Couplets for the pretest tasks were selected such that the expected result should be obtained. For example, D3 totally complemented by a B2 should yield a positive adjustment to original applicability of D3. If the participant answered at least two out of three tasks as expected, they were allowed to take the experiment and earn compensation. Finally, each analyst completed 125 unique tasks. This required too much time for the Mturk environment, thus each Mturk participant was limited to ten tasks. The couplets used for the ten tasks were randomly selected from the original 125, with the goal of getting at least sixty user responses per couplet.

The goal of expanding the participant pool was to answer two questions. (1) Is there any difference between a trained analyst and the general population when assigning value based on our information parameters, and (2) Is there any difference in how these two populations process multiple pieces of information. The answers to these questions will help direct future research. If the cognitive processes used by both populations are similar then we can continue to augment our experiments with crowd source data and have an expectation to get meaningful results. If the populations are not similar then we can attempt to identify the factors that make each process unique.

3   VOI Task Results

3.1   Initial Rating Ranking

The ranking of the initial piece of information is the first part of the VoI task. In this part of the task, the participant determines the applicability of the initial information by adjusting the applicability slider. This determination is made without consideration of any relationships to additional information.

In the original experiment, the rankings were divided into five cognitive groups (CG). The groups connected the ordinal slider values (most applicable, probably applicable, somewhat applicable, possibly applicable, and least applicable) with the underlying numerical values produced by the slider. Comparing the median ranking of the initial information shows that 21 out of 25 ratings were in the same cognitive group for both populations (Fig. 2). The ratings B5, D1, D2, and D4 were all ranked one cognitive group lower by the Mturk population. For the majority of the ratings the two populations qualitatively perceive the same applicability for the initial information, as categorized by cognitive group. This similarity indicates a common starting point before processing the perceived effects of information amalgamation.

3.2   Adjusted Rating Ranking

The Mturk adjusted rankings diverged greatly from the expected results. ARL expected complementing relationship adjustments to increase the initial ranking and conflicting relationship adjustments to decrease the initial ranking. Limited deviations from the expectation were observed in the previous experiment. These observations were attributed to a 'dilution effect' [6]. Examining the rankings in this experiment does not indicate the presence of a dilution effect. Rather it indicates the users employed a

	1	2	3	4	5
A	9	8	7	6	6
B	8	7	6	6	5
C	6	6	5	4	4
D	6	5	4	3	2
E	5	4	3	2	1

Analyst

	1	2	3	4	5
A	9	7	7	5	5
B	7	7	6	5	4
C	6	6	5	4	3
D	4	4	4	2	2
E	5	4	4	2	2

MTurk

CG1:	CG2:	CG3	CG4	CG5
9	8,7	6,5	4,3	2,1

Fig. 2. Rating rankings by cognitive group

strategy that disregarded the relationship linking the two pieces of information. The hypothesis is the users adjusted the ranking to the average ranking of the two pieces of information.

Evidence of the averaging strategy can be drawn from the data. The initial information rankings exist for all twenty-five ratings. If these values are substituted into the couplets three cases are possible: (1) the initial information is ranked higher than the additional information, (2) the initial information is ranked lower than the additional information, and (3) the initial information is ranked the same as the additional information. If the averaging strategy is assumed then case one would generate negative adjustments, case two would generate positive adjustments, and case three would generate no adjustment. To illustrate, using the Mturk values in Fig. 2, a B1 (seven) supported by a B4 (five) would give an adjusted value six which is lower than the initial ranking and counter to expectation.. Examining the entire data set reveals that the majority of couplets exhibited the averaging behavior (Table 3). The averaging strategy is not considered a viable strategy for this experiment because it does not take into consideration the connecting relationship.

The averaging behavior was not detected in the original experiment. As a result, this behavior was not anticipated, and unfortunately two of the three pretest couplets that were used produced expected results if the averaging strategy was employed (D3 totally complements B2 and C4 somewhat complements A3). Consequently, the pretest did not eliminate participants that did not understand the purpose of the relationships linking the two pieces of information.

Table 3. Evidence of averaging strategy (I: initial, X: additional)

	Couplets	Averaging
I > X	53	47
I < X	52	47
I = X	23	18
Total	128	112

Due to the averaging strategy dominating the adjusted rankings, a comparison to the values from the original experiment was non-informative. An ACT-R model to simulate this approach is addressed in the next section.

4 ACT-R Method

4.1 Architecture Background

Within the VoI experimental paradigm, we observed a vast array of applicability values and trends that emerged when valuing information. The online experiment required the novice user to use different cognitive mechanisms when performing each task, and may have a role in the way the user made decisions. An ACT-R model created from a previous ARL experiment was used to gain a better understanding of the cognitive mindset that the Mturk participants had during the experimental process, and test whether the model can simulate equivalently how the experiment was performed.

ACT-R is an architecture that is structured on the theory of human knowledge and human cognition [3]. Widely accepted among research communities, ACT-R has been used to simulate human cognitive tasks that capture how humans perceive and perform these various tasks [7]. The three main properties of the ACT-R framework are the environmental stimulus of visual and motor modules, procedural, and declarative memory (Fig. 3). The ACT-R model will use visual and motor functionality to buffer the information given from the experimental environment, and then use both procedural and declarative memory to understand and perform the task assigned.

The ACT-R cognitive framework model is an important tool utilized by the Department of Defense (DoD), because the model recreates human cognition during the performance of tactical military tasks. Some of the military tasks an ACT-R model can incorporate are the processes of situational awareness, prediction, and multitasking

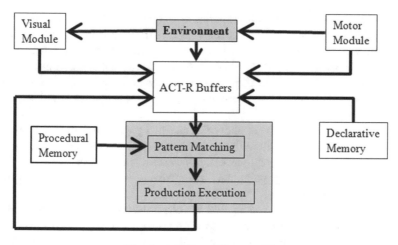

Fig. 3. ACT-R architecture [8]

[9, 10]. The Mturk experiment enabled novices to value data ratings similar to a subject matter expert within the situational awareness process. To investigate cognitive properties and observed trends, ARL will modify an existing ACT-R model that simulates novice amalgamation strategies exhibited in the Mturk experiment.

4.2 Simulation of Amalgamation Strategies

ARL is modifying the existing ACT-R 5.0 model, created for a previous VoI experimental effort involving military analysts, to represent the novice user [11]. The modified ACT-R model will take the role of a novice user and complete the same experimental tasks of valuing information after combining complementary and contradictory information. The model follows production rules that enable the simulated novice user to complete specific tasks within a certain period. The tasks will include the perception of displayed data, a determination of value for the initial piece of information, perception of the conditional and the additional card, and adjustment of the applicability value.

To observe trends within Mturk experimental data, the ACT-R model will simulate the cognitive productions that novice users perform in valuing information. The model uses newly created value distributions derived from the Mturk data set to produce a larger data set from experimental couplets. To discover emerging trends, the simulated data set will be compared to the Mturk data set using statistical analysis methods.

5 Conclusion and Path Forward

As information sources proliferate, information overload is a growing concern for the modern warfighter. ARL is modeling the decision making process in order to develop methods to assist the warfighter in information prioritization.

In this paper, ARL repeated its experiment to quantify information amalgamation. Originally, the experiment was conducted with US Army analysts. In this iteration, the experiment was performed using crowd-sourcing. The comparison between trained analysts and the wisdom of the crowd addresses two fundamental questions: (1) Is there any difference between the two population when assigning value to information, and (2) Is there any difference in how these two populations amalgamate multiple pieces of information.

The results of the experiment show that the two populations qualitatively rank a single piece of information within the same cognitive group. This important similarity means the fundamental value of a piece of information is consistent across populations. With this common starting point, any comparisons in information amalgamation strategies should be direct.

In this experiment, ARL was unable to collect useful information amalgamation data. The participants resorted to an information averaging strategy that did not account for the linking relationship. This failure is likely due to a weakness in the training material. The path forward includes identifying participants that did not use the averaging strategy and repeating the analysis with this subset of data.

Previously, ARL created an ACT-R model to simulate the cognitive processes undertaken when performing the experiment. This model will be updated with the crowd-sourcing results. ARL will use the ACT-R participant to simulate the experiment using ranking and amalgamation distributions calculated from the Mturk data. Future uses of the simulation will be to test 'couplets' that were not part of the experiment.

ARL will continue to investigate VoI and its role in decision-making. Quantifying this process, will lead to assistive battlefield technologies that will reduce decision time and increase reaction time on the battlefield.

Acknowledgments. This research was supported in part by an appointment to the Student Research Participation Program at the U.S. Army Research Laboratory administered by the Oak Ridge Institute for Science and Education through an interagency agreement between U.S. Department of Energy and USARL.

References

1. Hanratty, T., Heilman, E., Dumer, J., Hammell II, R.J.: Knowledge elicitation to prototype the value of information. In: Proceedings of the 23rd Midwest Artificial Intelligence and Cognitive Sciences Conference (MAICS 2012) (2012)
2. Hanratty, T., Heilman, E., Richardson, J., Mittrick, M., Caylor, J.: Determining the perceived value of information when combining supporting and conflicting data. In: Proceedings of the SPIE 10207, Next-Generation Analyst, vol. 102070 N, 3 May 2017. https://doi.org/10.1117/12.2264820
3. Anderson, J.R., Lebiere, C.: The Atomic Components of Thought. Erlbaum, Mahwah (1998)
4. Anonymous: US Army Field Manual (FM) 2-22.3, Human Intelligence Collection Operations, US Army, September 2006

5. Hammell II, R.J., Hanratty, T., Heilman, E.: Capturing the value of information in complex military environments: a fuzzy-based approach. In: Proceedings of the IEEE International Conference on Fuzzy Systems 2012 (FUZZ-IEEE 2012), Brisbane, Australia, 10–15 June 2012
6. Hanratty, T., Heilman, E., Richardson, J., Caylor, J.: A fuzzy-logic approach to information amalgamation. In: FUZZ-IEEE, Naples, Italy, 9–12 July 2017
7. Cassenti, D.N., Reifers, A.L.: Counting ACT-R to represent time. In: Proceedings of Human Factors and Ergonomics Society Conference (2005)
8. The University of North Carolina at Chapel Hill, P.: ACT-R Components [Digital image] (2011). https://sakai.unc.edu/access/content/group/e0fe0057-42ab-4f4d-abde-a150a98 989c4/public/classnotes/chap10.html
9. Buwalda, T.A., Borst, J.P., Taatgen, N.A., Van Rijn, H.: Evading a multitasking bottleneck: presenting intermediate representations in the environment. In: Proceedings of the 33st Annual Conference of the Cognitive Science Society (2011)
10. Kelley, T.D., Patton, D.J., Allender, L.: Predicting situation awareness errors using cognitive modeling. In: Smith, M.J., Salvendy, G., Harris, D., Koubek, R.J. (eds.) Proceedings of Human-Computer Interaction International 2001 Conference, Usability Evaluation and Interface Design: Cognitive Engineering, Intelligent Agents and Virtual Reality, vol. 1, pp. 1455–1459. Erlbaum, Mahwah (2001)
11. Caylor, J., Hanratty, T., Heilman, E., Cassenti, D., Richardson, J., Asher, A.: Evaluation of information amalgamation strategies using ACT-R. In: Proceedings of the 8th International Conference on Applied Human Factors and Ergonomics, Los Angeles, California, 17–21 July 2017

No Representation Without Integration! Better Cognitive Modeling Through Interoperability

Walter Warwick[1]([⊠]), Christian Lebiere[2], and Stuart Rodgers[1]

[1] TiER1 Performance Solutions, LLC, 100 E. RiverCenter Blvd. Suite 100,
Tower 2, Covington, KY 41011, USA
{w.warwick,s.rodgers}@tier1performance.com
[2] Carnegie Mellon University, 5000 Forbes Avenue, Pittsburgh, PA 15213, USA
cl@cmu.edu

Abstract. Historically, cognitive modeling has been an exercise in theory confirmation. "Cognitive architectures" were advanced as computational instantiations of theories that could be used to model various aspects of cognition and then be put to empirical test by comparing the simulation-based predictions of the model against the actual performance of human subjects. More recently, cognitive architectures have been recognized as potentially valuable tools in the development of software agents—intelligent routines that can either mimic or support human performance in complex domains. While the introduction of cognitive architectures to what has been regarded as the exclusive province of artificial intelligence is a welcome turn, the history of cognitive modeling casts a long shadow. In particular, there is a tendency to apply cognitive architectures as monolithic, one-off solutions. This runs counter to many of the best practices of modern software engineering, which puts a premium on developing modular and reusable solutions. This paper describes the development of a novel software infrastructure that supports interoperability among cognitive architectures.

Keywords: Human behavior representation · Cognitive architectures
Model integration

1 Introduction

Historically, cognitive modeling has been an exercise in theory confirmation [2, 3]. Cognitive architectures were advanced as computational instantiations of theories that could be used to model various aspects of cognition and then be put to empirical test by comparing the simulation-based predictions of the model against the actual performance of human subjects. While a unified approach to cognition has proved to be a productive avenue of research, the very features that make cognitive architectures theoretically appealing also make them often hard to use in more applied settings. First, we note that cognitive architectures are themselves complex software systems that come with significant theoretical and computational overhead, requiring expertise in both cognitive and computer science to develop models. Gluck [4] estimates the time and cost of developing a model of a "moderately complex task" at 3.4 years and

© Springer International Publishing AG, part of Springer Nature 2019
D. N. Cassenti (Ed.): AHFE 2018, AISC 780, pp. 336–345, 2019.
https://doi.org/10.1007/978-3-319-94223-0_32

$400,000. This does not include the follow-up costs of upgrading the model as the systems it interacts with change. The problem is exacerbated by the fact that there are so few experts capable of building such models [5], and that the original designer is often the only one with the knowledge to change the model.

Second, in addition to these costs there is a deeper challenge to overcome. Even when they work in an applied context, cognitive modelers are still cognitive scientists whose interests and biases are shaped by their training. At the very least, this leads modelers to use the tools with which they are familiar. In more extreme cases, agent development can be mired in what is essentially a proxy war for long standing academic debates as design decisions are second-guessed more for their theoretical implications than their practical benefits. Either way, the result is that architectures are often applied monolithically, even when they are not especially well suited to the problem at hand.

We maintain that these two challenges—steep development costs and the monolithic application of architectures—are inextricably related and they become even more acute when cognitive models are applied to the development of agent-based behaviors. Accordingly, we think both challenges can be addressed by recognizing that each modeling formalism has distinct strengths and weaknesses [6] and that there are efficiency gains when design choices are driven by the requirements of the desired behavior rather than by the architecture used to model the behavior. This paper describes the development of a novel software infrastructure that allows the analyst to choose the 'right tool for the job' by promoting the decomposition of agent behaviors so that cognitive architectures can be applied in a piecemeal manner with the requirements of the behavior driving the choice of architecture.

2 The DREAMIT Workspace: Design Goals

Working under a Phase II Air Force STTR, we are developing software that allows existing models to be combined to produce agent behaviors. This software development has been driven by two design goals:

The first goal is to allow the development of component models using different modeling formalisms (Fig. 1). Component models can represent any one of the wide variety of human behaviors that underpin intelligence—from perception, through cognition, to action. The models are basic only in that they represent isolated processes; they potentially map to all levels of behavior. Examples include component models of vision, auditory processing, recognition, decision-making, and motor planning. Component models can be constructed using different modeling formalisms (e.g., ACT-R, SOAR, and C3TRACE) and programming languages (e.g., MATLAB and R).

Component models provide the basic blocks for assembling more sophisticated behaviors (Fig. 2). This enables re-use; the same component model can be used for new behaviors. This also enables composability; complex agents can be assembled from simpler component models. Behaviors can be built using components from one modeling formalism (Fig. 2, Panel 1), two modeling formalisms (Panel 2), or many modeling formalisms (Panel 3). Finally, this enables specialization. This pluralistic approach extends the capabilities of an intelligent agent beyond those supported by any

Fig. 1. Libraries of basic component building blocks created using different modeling formalisms.

single formalism. Additionally, it enables trade-offs between representational fidelity and computational cost. When multiple modeling formalisms can be used to represent the same process, the developer can choose among component models with fidelity and computational cost in mind. For example, high-fidelity cognitive architectures such as ACT-R and Soar provide detailed accounts of human behavior at a significant complexity cost, while task network or mathematical models may provide an adequate low cost alternative.

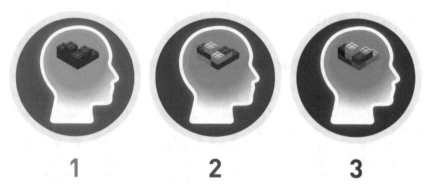

Fig. 2. Composition of increasingly sophisticated agents from basic component building blocks.

We note that the workspace is not a development environment for the models themselves but, rather, a software framework for connecting them at run time, as we describe next.

The second key idea is to design a "workspace" that enables interactions among component models. This is the "glue" that holds the component models together. The workspace supports the definition of persistent local objects (i.e., local to the workspace) that define the input-output relationships among the component models at run time. In this way, the workspace ensures system level modularity; a well-specified interface can allow different human behavior representations (HBRs) to "plug and play" with a given simulation. On the other hand, the workspace ensures modularity internal to the HBRs as well. Although it is commonplace to decompose tasks during the analysis of complex behavior, modelers often use a single computational architecture to develop the corresponding HBR. Even if the architecture provides some degree of modularity (e.g., separate modules to represent perceptual and cognitive function), the modularity of the architecture rarely recapitulates the decomposition of the task. The workspace aligns the decomposition of complex behaviors with the modularity of the HBRs used to model them. This allows existing HBRs to be re-used and rearranged to construct novel complex behaviors.

The chief benefits of the workspace are scalable agent development via the combination of multiple model components, efficient agent development via model re-use, and enhanced agent capability via model specialization. Aside from these benefits, the workspace creates greater transparency in the agent's internal processes. Each of the component models that make up the agent receives inputs from and returns outputs to the workspace. The meaning of these inputs and outputs are related to the component tasks, and are human-interpretable. Finally, by brokering model interaction through the workspace, rather than allowing direct model-to-model connections, we mitigate the complexity when multiple models are integrated within the workspace. Specifically, the needed number of connections among component models grows linearly rather than quadratically in the number of models.

3   The DREAMIT Workspace: Implementation Details

We are using standard TCP/IP sockets to connect component models within the workspace. TCP/IP socket connections are well-supported in a wide variety of programming languages and they impose a minimal burden on the developer of the component model. The workspace itself is implemented as a multi-threaded socket, with the workspace serving as the server and the component models as clients. Communications among the workspace and component models currently follow a request-reply pattern, with the component models making requests to and receiving replies from the workspace.

There are two important features of the multi-threaded architecture. The first and more obvious is that the communication between the workspace and each client is executed as a separate process. Consequently, the interactions between the workspace and the client models are handled asynchronously and there is no requirement (or safeguard) that the model interactions are brokered in any specific order. The second feature, related to the first, is that the component models interact with the workspace and not each other. We previously mentioned this feature and its implication for managing complexity. We mention it here to underscore a more prosaic benefit, namely, that there is no hard requirement to synchronize the execution of the component models. Instead, the only requirement is that the request-reply cycles of the connected models eventually overlap. Until then, if a component model is expecting output from another model that has yet to connect, the workspace supports a soft failure mode in which the expecting model receives empty replies to its request, effectively corresponding to a missing piece of skill or knowledge. Of course, the modeler must be aware of this possibility, and engineer his model accordingly to exhibit the required robustness, but the connection between any one model and the workspace does not depend on the presence or absence of another model to complete a handshake or respond to a request.

We are using JavaScript Object Notation (JSON) as the standard for sharing information among component models within the workspace. Like our choice to use TCP/IP sockets, we chose JSON primarily because it is lightweight and well-supported across a variety of programming languages. While the JSON standard provides the syntax, the semantic content of the object consist of: (1) a timestamp attribute-value pair that consists of the attribute label "timestamp" as a string and a base-10 numeral that represents the (server) clock time at which the JSON object was passed to or from the workspace; (2) a client attribute-value pair that consists of the attribute label "client" as a string and another string that contains the name given to the component model (e.g., "C3Trace1" or "PythonModel"); (3) a name attribute-value pair that consists of the attribute label "name" as a string and another string that contains a name given to the particular object itself; (4) a mode attribute-value pair that consists of the attribute label "mode" as a string and another string that is either "set" or "get"; (5) a type attribute-value pair that consists of the attribute label "type" as a string and another string that contains a name given to this kind of object (i.e., not the particular object, which is individuated by the name); (6) an argument attribute-value pair that consists of the attribute label "values" and another JSON object as the value.

In more detail: (1) The timestamp provides a way of sorting and retrieving objects as they are buffered in the workspace. (2) The client attribute-value pair is used to identify the socket connection on which the workspace reply should be delivered. (3) We have not yet used the name attribute-value pair, but we have included it in the anticipation of supporting more sophisticated workspace requests (e.g., an object that refers to itself, or circular reference graphs in general). (4) The mode attribute-value pair is used to determine whether the incoming request will result in the buffering of model output ("set") or in the retrieval of a previously buffered output ("get"). (5) The type attribute-value pair is used to individuate objects output by the component models in those cases where the component model produces or consumes different kinds of data. (6) The argument attribute-value pair contains the "payload" information shared between models and allows for complex data to be shared using any acceptable JSON value (i.e., another JSON object, array, number, string or Boolean constant).

4   Using the Workspace in the Not So Grand Challenge

The Not So Grand Challenge (NSGC) is a multi-partner research initiative sponsored by the Air Force Research Laboratory [7]. Like previous "model comparison" efforts [8], the goal of the NSGC is to extend the state of the art in agent modeling by providing a uniform interface to a constructive simulation environment and a set of required behaviors so that different approaches to behavior representation can be compared against a common baseline. In addition to understanding the models themselves, the NSGC also provides a context in which to assess development costs, the extent to which models are "data-driven" and the likelihood that a model can generalize to novel scenarios.

Each partner in the NSGC was tasked to develop a model of a pilot that would fly against another simulated pilot in a constructive fast-jet simulator. The goal was to produce tactical behaviors (e.g., changes in speed, heading altitude, etc.) that were both robust and flexible under different one-versus-one scenarios. Moreover, and by design, each partner came to the NSGC with different "architectural" approaches to human behavior representation; for example, one model was developed in Soar, another in Brahms, and yet another was implemented in a case-based reasoning system.

While the plurality of models served to highlight the differences among the various architectures, we viewed the NSGC primarily as an opportunity to demonstrate how our workspace would allow otherwise disparate human behavior representations to be combined to produce coherent behavior. Specifically, we integrated a task network model with an accumulator model to produce the required tactical behavior.

4.1   The Task Network Model

Task network modeling tools provide a framework for representing human behavior as a decomposition of operator goals or functions into their component tasks, which themselves can be further decomposed. This framework is visualized by way of an intuitive graphical representation of the process being modeled in which nodes represent tasks and directed edges represent the flow of control among the tasks. For the NSGC, we used C3Trace, a government-owned task network modeling tool, to implement a model of the red pilot performing a stern conversion. Figure 3 depicts the top-level view of the tasks that the pilot performs.

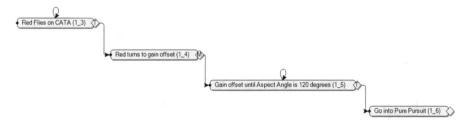

Fig. 3.   The task network model of a pilot performing a stern conversion.

The series of tasks that begins with "Red Flies on CATA (1_3)" represents the different states the red pilot occupies when performing a stern conversion—fly straight and level into the engagement until inside of 25 nm; maneuver to gain offset; go into pure pursuit at 120-degree aspect angle. The pilot remains in each state until certain conditions are met (described more fully later). This series of tasks implements a finite-state model of stern conversion; each state summarizes a specific set of actions that the pilot performs until the model transitions to another state.

4.2   The Accumulator Model

Although the task network model can serve as a complete representation of the red pilot, we found it useful to integrate the model with a higher-fidelity representation of change detection. The intuition here is that even if the red pilot is implemented as a simple finite state model, the transitions between states need not be driven only by strict rules or threshold conditions. Instead, we used an accumulator model to detect relevant changes in blue's tactical posture; specifically, when consistent changes in blue's aspect angle and heading rise above the level of noise. The accumulator model gets its name from the fact that evidence is accumulated over time (see Fig. 4). Incoming information, in this case kinematic information about the blue aircraft, drives two different accumulator models toward their respective decision bounds. One model detects changes in heading and the other detects changes in "aspect angle" (related but usefully different features of tactical maneuvering). The process terminates when accumulated evidence reaches a decision bound, at which time a categorization decision is made.

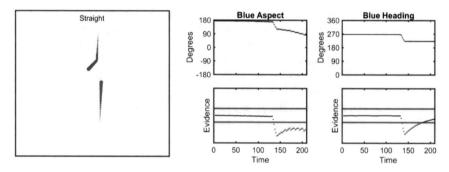

Fig. 4. An example of how evidence based on blue's aspect and heading are combined to detect a tactical change

Whereas the task network model bears a very clear correspondence to the tactical behavior it represents (i.e., the model recapitulates a real-world process flow), the accumulator model represents an implicit decision-making process that doesn't lend itself to an obvious or useful representation as a flow chart or finite state machine. Although domain knowledge is still needed for defining the inputs and decision boundaries, the inner workings of the model are determined by an off-line training process in which parameters are tuned to achieve the needed performance. In this case, that meant running the accumulator model against every one-versus-one baseline scenarios to determine whether the model could consistently and correctly identity tactical changes in blue given different intercept geometries. Having labeled blue behavior in each scenario by hand, and incrementally adjusting the parameters that control how much evidence is needed to reach a decision boundary and how long accumulated evidence persists, the model achieved reasonable performance after a few dozen training runs.

4.3   Integrating the Models

The initial development of the models occurred in parallel. Then, having satisfied ourselves that the task network model was providing at least a plausible representation of the stern conversion behavior and that the accumulator model was recognizing changes in blue's tactical posture, we integrated the models using the workspace. Conceptually the integration entailed that the conditions used in the task network to induce state transitions were extended to include output from the accumulator model. At a practical level, the integration entailed extending the transition conditions used in the task network model. For example, instead of using only range to target as the threshold conditions for transition from "Red Flies on CATA" to "Red turns to gain offset," we included a disjunct so that the transition would occur when the range to target was less than 25 nm *or* blue departed from straight and level flight, as determined by the accumulator model. We added a similar disjunct at two other key transitions in the task network model: the point at which the model goes into pure pursuits and inside a control loop that monitors the course of the pure pursuit.

4.4   Preliminary Results

The NSGC included a formative evaluation of the behaviors produced by each partner's model. The evaluation was conducted as an on-site, live demonstration of the partner models interacting with the Air Force's constructive air combat simulator. Each partner first explained their modeling approach in general before the models were run against a subset of the baseline scenarios. Subject matter experts observed the behavior in real time using a plan view display of the scenarios while simulation state data were also captured for later analysis. Our model, like the other partners' models, generated behavior that was plausible with respect to the limited scope of the baseline scenarios. Unlike the other partners, however, we demonstrated two different versions of our model, running the task network model with and without the accumulator model. While the accumulator model provided a degree of context sensitivity (e.g., reacting earlier to changes in the blue aircrafts tactical posture) than the more "ballistic" behavior of the task network model running in isolation, the combined model also introduced some unexpected and odd behaviors in some situations. Specifically, the accumulator induced some "metastable" behavior in which the red pilot model repeatedly engaged and broke off from the blue aircraft.

We did not try to patch that behavior for the formative evaluation but, rather, used it as example of how complex models of behavior might be developed using simpler component models. In fact, for us the NSGC formative evaluation was as much about establishing proof of principle for the workspace as demonstrating plausible behavior. In that respect, there were several useful results.

First, after having decomposed the required behavior representation into a process flow and a classification problem, it was possible to develop the task network model and accumulator model independently using different formalisms and development environments (i.e., a task network model implemented in C3Trace and an accumulator model implanted in Python). Even in this simple application, there were efficiencies gains that followed from dividing the labor between modelers who could use their preferred architecture to develop each model.

Second, the integration of the task network model and the accumulator model consisted of fixing inputs and outputs to each model, at a semantic level, and then ensuring that each model packaged and parsed the JSON-object syntax correctly. There was no need to develop or reconcile architecture-specific APIs or for each model to conform to a particular network protocol. This was a welcome departure from our previous experience in which significantly more effort had to be devoted to model integration than model development.

Third, to the extent that we were able to develop the models independently and define a clear interface between them, we were able to encapsulate the models. This, by itself, ensures a degree of modularity and points to the clear possibility of combining different models, a different classifier, for example, to produce the required behavior. The workspace thus exploits a natural coupling among modularity, composability and model reuse. Further, this same degree of encapsulation makes it easier to assign credit or blame in the verification and validation of a model's behavior. Indeed, when we included the output from the accumulator model in the task network model, there was no mystery as to what change led to the metastable behavior and, accordingly, we could immediately begin to experiment with possible solutions without first engaging in a painstaking review of our code to localize the problem. The efficiency gain here is both practical and conceptual.

5 Discussion

Hybrid architectures are nothing new to cognitive modelers. Likewise, federated models have long been used in the simulation of complex systems of systems. But somehow the need for tools that support the efficient integration of different modeling formalisms has gone largely unmet. All too often modelers face a Procrustean choice, either recasting a behavior to meet the requirements of their preferred architecture or modifying those architectures in one-off and *ad hoc* manners to represent a needed behavior. While this unfortunate choice might be unavoidable in the context of cognitive science, in more practical settings the ability to pick and choose among architectures and modeling formalisms has clear benefits.

The workspace we just described is intended to allow the modeler to 'pick the right tool for the job.' In this respect, it aligns human behavior representation with the best practices in federated simulation and its concomitant reflection of modern software engineering. Indeed, encapsulation, modularity, composability and reuse are all hallmarks of any sufficiently complex software engineering project. And while we have previously emphasized the practical benefits, we also see that the workspace might also impose a discipline on human behavior representation that brings with it theoretical benefit as well. At the very least, model re-use requires general functionality instead of specific implementations. At the representation level, this corresponds to adopting general ontological commitments that can be mapped to varying simulation implementations. Further, the workspace encourages interactions among component models to be automatic, direct, and transparent. Creating idiosyncratic protocols between pairs of models limits their potential for re-use. Instead, the input and output of each component model should be fixed regardless of which component models it interacts with. Science is best served when the models it produces are clear in their operation and their implications.

References

1. Bishop, M., et al.: Insider threat identification by process analysis. In: IEEE Security and Privacy Workshops, pp. 251–264. IEEE, San Josep (2014)
2. Anderson, J.R., Lebiere, C.: The Atomic Components of Thought. Erlbaum, Mahwah (1998)
3. Newel, A.: Unified Theories of Cognition. Harvard University Press, Cambridge (1994)
4. Gluck, K.A.: Cognitive architectures for human factors in aviation. In: Salas, E., Maurino, D. (eds.) Human Factors in Aviation, 2nd edn, pp. 375–400. Elsevier, New York (2010)
5. John, B.E., et al.: Predictive human performance modeling made easy. In: SIGCHI Conference on Human Factors in Computing Systems, pp. 455–462. Association for Computing Machinery (2004)
6. Pew, R., Mavor, A.S. (eds.): Modeling Human and Organizational Behavior: Applications to Military Simulations. National Academy Press, Washingto, D.C. (1998)
7. Stacy, W., et al.: Agents from the future. In: I/ITSEC. National Training and Simulation Association, Orlando, FL (2017)
8. Pew, R., Gluck, K.A. (eds.): Modeling Human Behavior with Integrated Cognitive Architectures: Comparison, Evaluation, and Validation. Lawrence Erlbaum Associates, Mahwah (2005)

Applications in Safety and Risk Perception

The Effect of Hazard Clustering and Risk Perception on Hazard Recognition

Timothy J. Orr$^{(\boxtimes)}$, Jennica L. Bellanca, Brianna M. Eiter,
William Helfrich, and Elaine N. Rubinstein

Pittsburgh Mining Research Division, National Institute for Occupational Safety
and Health, Centers for Disease Control and Prevention, Pittsburgh, PA, USA
{TOrr, JBellanca, BEiter, WHelfrich, ERubinstein}@cdc.gov

Abstract. Active mining operations are complex, dynamic environments that can present workers with an array of potential safety and health challenges. From missing fire extinguishers to large equipment and falling rocks, hazards exist that mineworkers must be cognizant of to keep themselves and their coworkers safe. While hazard identification is a key skill that mineworkers must possess to ensure workplace safety, the location and perceived risk of the hazards may alter this ability. To further explore these effects, NIOSH researchers conducted a study to characterize how mineworkers search for and identify hazards. Researchers asked participants to search 32 static panoramic scenes depicting typical locations at a surface stone mine—pit, plant, roadway, and shop—with each containing zero to seven hazards. Mineworkers tended to miss hazards when they were in clusters—i.e., where two or more hazards appeared within the worker's central field of view. This paper examines the relationship of clustered hazards, perceived risk and identification accuracy and how location and experience affect it. Based on the results, strategies will be suggested that mineworkers can use to help identify hazards in their workplace.

Keywords: Visual search · Simulator · Mineworker safety

1 Introduction

The mine environment poses significant challenges for operators to provide a safe and healthy workplace, as evidenced by the sharp increase in the number of fatalities in metal/nonmetal (M/NM) mines between 2013 and early 2015 [1]. In response, the Mine Safety and Health Administration (MSHA) provided a renewed focus on workplace exams. In 2017, MSHA amended the workplace exam rule with the intent to improve identification of dangerous conditions and promote corrective action before accidents occur [2]. The final rule requires examinations to occur in a timelier manner and increases accountability by requiring the reporting of hazards that are not imme-diately corrected in exam reports. The final rule also requires a "competent person" to conduct workplace exams, but there is no guidance as to the qualifications of this person. The Code of Federal Regulations (30 CFR 56.2) merely states that such a person must possess abilities and experience that fully qualify them for such assigned duties [3].

© Springer International Publishing AG, part of Springer Nature (outside the USA) 2019
D. N. Cassenti (Ed.): AHFE 2018, AISC 780, pp. 349–360, 2019.
https://doi.org/10.1007/978-3-319-94223-0_33

At the most basic level, to complete a workplace exam successfully, a competent person must be able to recognize and mitigate hazards in the environment. Hazard recognition generally involves the identification and assessment of hazards; this relies heavily on the visual perception and decision-making abilities of the mineworker preforming the exam. Unidentified mine site hazards pose a risk to workers at the site. Therefore, understanding the factors that affect visual search and decision-making performance is crucial.

Visual search errors are generally thought of as perception issues that can be caused by a lack of skill, knowledge, or attention in relation to what or how to search. Specifically, scanning and recognition visual search errors occur when an individual either never fixates on an object or fails to recognize an object as an important feature that requires further examination. Decision-making errors are related to recognition errors in that they are based on knowledge and skill, but are top-down errors associated with cognitive processing [4]. Both visual search and decision-making errors have been studied in many fields and include several overarching factors: scene complexity, attention, salience, expectations, and experience.

Overall, scene complexity reduces search and decision-making performance. By increasing the number of salient objects in a scene, clutter reduces the prominence and accessibility of search targets [5]. Furthermore, the mere presence of any foveal stimulus has been shown to reduce the useful field of view (UFOV)—the area of vision that can acquire information during a fixation [6]. Visual clutter enhances the reduction of UFOV and is particularly prominent when background object and target similarity is high [7–9]. Consequently, visual clutter is believed to increase search times as a result of the increased number of fixations required to locate a hazard among distractors [8, 10]. It has also been shown that background information influences search time more than image area and density, suggesting that overall scene complexity may be more detrimental to search performance than local clusters [11]. Practically, all of these effects have been demonstrated in hazard recognition while driving, where hazard recognition performance is worse on busy urban roads than rural roads [12].

Visual clutter also requires more attention to process, but attention is a limited resource. It would be impossible for a mineworker to fully scan every inch of a mine site. Therefore, hazard salience plays a critical role. In radiology, among other fields, research has shown that less salient targets are more likely to be missed [13]. For hazard recognition, more risky hazards may be more salient. Salience can also be related to expectations. Studies on inattentional blindness have shown that viewers frequently fail to recognize unexpected changes to aspects of a video scene if they are given a cognitive task related to another aspect of the display [14]. Researchers also demonstrated that even with prior knowledge of an unexpected event, viewers were no less likely to notice any other unexpected event [14].

Expectations have also been shown to reduce hazard identification performance through decision-making errors like satisfaction of search (SOS) errors. SOS errors are defined as a reduction in detection once one hazard has been identified. Because searchers believe that there are limited hazards, once one is found, searchers do not expect there to be any more and stop searching, leading to missed hazards. Just like visual clutter, SOS effects have been demonstrated over various domains including radiology, cognitive psychology, and security screening [13]. Additionally, it has been shown that when target salience is variable, SOS errors lead to the more salient targets being identified either in a cluster or in isolation from distractors [15].

In combination with complexity, attention, salience, and expectations, experience has been shown to have an interacting effect on hazard recognition ability. Experts benefit from unique perception skills related to visual search and decision-making that enable them to recognize what is typical in a situation, perceive subtle discriminations, and quickly perceive, understand, and project future events [16]. Recognizing typicality allows experts to sort through cluttered scenes and only focus on relevant information. Discrimination allows experts to focus directed attention to differentiate details that less experienced people may not recognize. Mental modeling and situational awareness allows experts to better understand and recognize hazards as well as assess risk associated with potential outcomes.

The perception skills of experts have been demonstrated across many domains. Primarily based on visual search, numerous automotive studies have demonstrated the experience effect, finding that inexperienced drivers tended be more inflexible with their visual search patterns and were unable to adapt their attention as scene layout and complexity changed, resulting in poorer hazard recognition possibly due to reduced awareness and attention [18–21]. However, studies have also demonstrated the importance of targeted experience. Perlman et al. found that experienced safety professionals in construction outperformed both the less and similarly experienced workers on hazard recognition, indicating that experience needs to be task-specific for the worker to reap the benefits [17]. Eiter et al. also found a similar result with safety professionals and mineworkers performing workplace exams [22].

In order to systematically characterize image context, risk perception, experience, and their effects on hazard recognition accuracy, the National Institute for Occupational Safety and Health (NIOSH) conducted a research study in a simulated environment. The goal of this study was to simulate performing workplace exams within a controlled laboratory setting using an immersive virtual reality (VR) simulator. Mineworkers of varying experience levels were asked to search a series of panoramic images depicting typical work environments at a surface limestone operation. For this study, realism was prioritized over homogeneous hazard distribution. Representative locations (i.e., pit, plant, roadway, and shop) were also chosen to account for natural context variability. However, some locations at mine sites are inherently more visually cluttered than others (i.e., shop and plant). As a result, hazards were clustered in many images, where one or more hazard was within the participant's near-peripheral field of view. The clustering of hazards exemplifies a special case of visual clutter. Therefore, this naturalistic setup provided the opportunity to further explore hazard recognition performance by investigating the relationship of hazard clusters with perceived risk, experience, and accuracy.

The first hypothesis is that mineworkers are less likely to identify hazards in visually cluttered environments, including both local clustering as well as an overall cluttered environment. The second hypothesis is that mineworkers are more likely to identify hazards that they perceive to be higher risk based on accident severity, accident probability, and overall risk. The third hypothesis is that mineworkers with more safety experience are more likely to identify hazards. By better understanding these factors,

NIOSH can begin to tailor safety interventions to make mine examinations more effective, and thus reduce the likelihood of future workplace injuries.

2 Methods

2.1 Participants

A total of 49 individuals completed the study after providing informed consent to participate in the institutional review board approved protocol. Participants were screened for visual acuity of 20/40 or better, vision disorders, and color blindness. All participants were also screened for full peripheral vision of −45 to +85° in both eyes. Study participants were volunteers with varying levels of experience working in surface mining operations. The participants were categorized by expected experience level ranging from student (n = 14), inexperienced miners (n = 12), experienced miners (n = 11), and safety professionals (n = 12), where safety professionals had at least 2 years of experience in an environmental, health, or safety position for a mine operator or government agency, experienced miners had more than 2 years of experience as a mineworker or supervisor, inexperienced miners had some but less than 2 years of experience as a mineworker or supervisor, and students were defined as a person enrolled in a mining-related program that is not otherwise classified. For more detailed demographics, see Eiter et al. [22].

2.2 Panoramic Images

Thirty-two panoramic photographic images of a surface limestone mining operation were used. The images were categorized by location with eight from each of the pit, plant, roadway, and shop (Fig. 1). The pit and roadway images were generally more open lacking additional salient objects aside from the hazards. The plant and shop locations were characterized by significantly more salient object within the images. The plant images feature large conveyors spanning across the area and the shop images were particularly cluttered with shelves packed with various tools and spare parts. Each image contained zero (control) to seven hazards. The hazards were selected to be a representative range of type and severity of hazard based on MSHA data and subject matter expert (SME) feedback from former MSHA inspectors, mine safety professionals, and other mining experts within NIOSH [22].

Fig. 1. Image examples from the pit, plant, roadway and shop locations (left to right).

2.3   Procedure

Data collection took place in the Pittsburgh Mining Research Division (PMRD) Virtual Immersion and Simulation Laboratory (VISLab). The main VISLab simulator uses a 360° panoramic projection screen that is 10 m in diameter by 3 m tall. Imagery is front-projected onto this screen from six high-definition projectors (Titan 1080p 3D, Digital Projection, Kennesaw, GA) to create a seamless image. The screen provides a visual field that is about 35° vertically from the center of the area.

Ten motion-tracking cameras (T20, Vicon, Oxford, UK) were used in concert with eye-tracking glasses (ETG 2.0, SensoMotoric Instruments, Teltow, Germany) to track participant movement and resolve their point of regard within the display space. The study image sequencing was controlled by an in-house Unity application (Unity Technologies, San Francisco, CA). See Bellanca et al. for a more detailed discussion of the development of the images, laboratory setup, and data collection methods [23].

Participants were fitted with eye-tracking glasses connected to a small laptop placed in a backpack worn by the participant. Participants were also given a hand grip button in their dominant hand that was connected to wireless data streaming hardware that was also placed in the backpack. The eye-tracking glasses have passive motion-tracking markers to track the participant's head position, and several additional markers were placed on the participant's torso to resolve head motion relative to the body frame. A series of calibration tests for the motion-tracking system and eye-tracking glasses were then conducted in the 360-degree simulator to ensure the data collection software was accurately capturing the participant's gaze within the screen space. Once the data collection instruments were calibrated, researchers presented two panoramic images to the participants to allow them to familiarize themselves with the 360-degree simulator and button press control.

Once acclimated, participants were presented with the 32 images in four sets of eight grouped by location category (pit, plant, roadway, and shop). Blocks were randomized across participant category, and images were randomized within location category. Participants were allowed two minutes to view each image and were instructed to press the hand grip button as quickly as possible when they identified a hazard. If they decided they had identified all the hazards in an image, participants could press a second button on the hand grip to end that trial early. Once all images were complete, the glasses, markers, and backpack were removed.

Following the hazard identification, participants were debriefed about all the images and the hazards. Regardless of identification, participants were asked to complete a risk assessment for each of the hazards. The risk assessment measure used for the study was adapted from Perlman et al. [17], where participants were asked to use a 5-point Likert scale to rate accident severity, accident probability, and overall risk as described in Table 1.

2.4   Eye-Movement Data Processing

The eye-tracking data was transformed into image space via motion-tracking data such that hit/miss accuracy could be calculated using a region of interest (ROI) and the button press timing. ROIs of hazards were pre-defined based on input from SMEs.

Table 1. Risk assessment instrument.

	Assessed risk value				
	1	2	3	4	5
Accident severity	No injury	Minor injury no leave	Injury ≥ 3 days leave	Non-fatal major injury	Fatal
Accident probability	Very unlikely	Fairly unlikely	Average likelihood	Fairly likely	Very likely
Risk level	Very low	Low	Medium	High	Very high

Fixations were calculated using a dispersion algorithm (minimum duration of 75 ms, maximum dispersion of 50 pixels) on the scan path data. Hits or misses on the button presses were based on the central gaze position of the fixation prior to the button press. Criteria for hit classification were developed to account for intention (requiring a prior fixation near the target), decision-making, and motor delay (for late hits) [24] as described previously by Eiter et al. [22]. All button press data was visually verified by a member of the research team and reviewed by a second team member if necessary.

Hazards were grouped in clusters if two or more hazards were present in an observer's typical UFOV based on the distance between the closest edges of ROI boundaries. Specifically, the polygonal ROI boundaries were simplified to rectangular bounding boxes by using minimum and maximum values in x and y pixel space (Fig. 2). Minimum horizontal angles between each ROI pair were calculated. Pairs of hazard targets were considered to belong to a cluster if the minimum angle was less than or equal to 30°, as is common in standard vision periphery screening tests and considered the typical UFOV [25]. Using this definition, there were 27 clusters in the 32 pictures. The number of clusters per image ranged from zero to three, and the number of hazards per cluster ranged from two to four, with an average of 2.3 hazards per cluster.

2.5   Statistical Analysis

The effect of clusters and risk were analyzed in two different ways. First, to capture the effects of the predictors on identification of individual hazards, hit/miss accuracy was modeled using generalized estimating equations (GEEs) with a binomial error distribution and a logit link function. GEE was chosen to account for the possible correlation of hazard identification by subject. A full factorial model was tested including cluster, risk, location, and group. Pairwise comparisons were completed for any significant effect using estimated marginal means. Second, severity, probability, and risk were explored as dependent variables to examine perception trends, especially since the risk assessments were performed on all hazards following identification, and the possibility that awareness of identification influenced perception could not be ruled out. Because hazard realism was given precedence over experimental control, it was not possible to achieve equal distributions of the risk of hazards across the locations. Therefore, location is not considered as a predictor of severity, probability, or risk. For analysis,

Fig. 2. Close-up view of a panoramic image from the mine plant showing an area with clustered hazards where only the highest-rated hazard was identified. Green ROIs indicate a hit; red ROIs indicate a miss. Each ROI is also identified with a unique ID number. The scan path line is color-coded from cool to warm colors before and after the button press. The click icons indicate fixations and the yellow dot represents the fixation location associated with the button press.

these variables were averaged over hit/miss and cluster assignment for each participant. Using the aggregated data, a maximum likelihood linear mixed model analysis was performed. Pairwise comparisons were also completed for any significant effect using estimated marginal means (SAS, Cary, NC). The alpha was set to 0.05 for all multivariate models and post-hoc comparisons.

3 Results

3.1 Accuracy

According to the first hypothesis, overall, hazards in clusters were less likely to be found (p < 0.001), where 56% of non-cluster hazards were correct and 48% of cluster hazards were correct. In fact, 14 of the top 20 missed hazards overall were in a cluster.

The GEE also resulted in differences in accuracy by risk (p < 0.001), group (p = 0.001), location (p = 0.001), and the interactions of cluster x location (p < 0.001) and risk x group (p = 0.017) effect. Risk as a predictor of accuracy supported the second hypothesis, all the levels were significantly different except 1 and 2 (p < 0.01, p = 0.101), with riskier ratings (5) having a higher likelihood of a hit. The estimated means from 5 to 1 were 67%, 60%, 51%, 43%, and 37%, respectively. The group effect supported the third hypothesis, where the results revealed that the safety professionals performed better than the students and inexperienced mineworkers (p < 0.01). The estimated means were 48% for the students, 48% for the inexperienced, 53% for the experienced, and 58% for the safety professionals. The post-hoc analysis of the location effect revealed that accuracy in the shop was significantly better that all other locations (p < 0.01), with means of 59%, 48%, 50%, and 49%, running counter to the visual clutter hypothesis. Looking closer at the location-cluster interaction effect revealed that accuracy was significantly better for non-clustered hazards in the more cluttered plant and shop locations (p < 0.001) and not significantly different for the pit (p = 0.840) and roadway (p = 0.419) as depicted in Fig. 3.

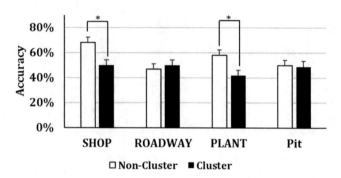

Fig. 3. Graph depicting the estimated marginal means of accuracy derived from the GEE model for clustered (black) and non-clustered (white) hazards by location. Bars connected with an (*) indicate significant differences at p < 0.01.

3.2 Perceived Risk

According to the second hypothesis, identified hazards on average were perceived as more severe, more probable, and overall riskier as displayed in Table 2. This is congruent with the significant correlation of risk and accuracy found in the GEE model.

Table 2. Estimated marginal means of average risk assessment ratings.

	Hazard Identification		
	Hit	Miss	Significance
Severity	3.542	3.313	<0.0001
Probability	3.248	2.899	<0.0001
Overall risk	3.482	3.142	<0.0001

Across the three risk variables, clusters were only found to be significant in the severity rating, where both the main cluster effect and cluster-accuracy interaction effect were found to be significant. Specifically, the results indicate that hazards in a cluster tended to be perceived as less severe (p = 0.015), with an average rating of 3.386 for clustered hazards and 3.469 for non-clustered hazards. Similarly, the cluster-accuracy interaction effect revealed that missed non-clustered hazards (3.238) tended to be perceived as less severe than missed clustered hazards (3.387) (p = 0.002), while identified hazards were not significantly different (p = 0.748).

Congruent with the GEE, there was a significant group-accuracy interaction for overall risk (p = 0.049). Analysis of the group-accuracy interaction effect for overall risk showed that group was significant for missed hazards (p = 0.030), but not for hit hazards (p = 0.280). Furthermore, for missed hazards, risk ratings of experts were significantly higher than ratings of safety professionals (p = 0.009) and ratings of students (p = 0.009), as depicted in Fig. 4.

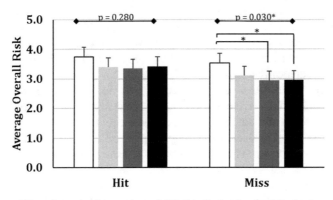

Fig. 4. Graph depicting the estimated marginal means derived from the mixed model of averaged overall risk ratings for hit and missed hazards across group.

4 Discussion and Conclusion

As hypothesized, hazards in clusters were less likely to be found. However, the detrimental effect of visual clutter was not supported in the effect of location. The more subjectively cluttered shop actually had the highest level of accuracy. Further investigation of the cluster-location interaction effect revealed that the two most cluttered environments (i.e. shop and plant) had higher accuracy rates for non-clustered hazards. Though this is counter to the visual search literature, this effect and overall cluster effect may be explained by SOS errors. The more sparse areas could have lead mineworkers to expect fewer hazards; this would lead to early termination in the pit and roadway locations. Similarly, given the overall cluttered nature of the shop once a hazard was found, the mineworker may have moved on to another area of the scene resulting in a low accuracy for cluster hazards as the secondary hazards were less likely to be found. Interestingly, the main effect of location from the GEE was slightly different from that previously estimated without clusters and risk in the model. The previous analysis found that roadway accuracy was significantly worse than the other locations [22]. While the estimated marginal means from the GEE still found the roadway accuracy to be lowest, though it was not significantly so. These differences underscore the importance of visual clutter and risk on accuracy.

Another explanation to the counter location effect could be the size and clarity of the hazards. Due to the naturalistic layout of the images, size and prominence were not evenly distributed. Furthermore, the contextual nature of the hazards makes the definition of size difficult and highly subjective. A more controlled stimulus may be better able to tease these effects apart, as these factors may be obscuring the expected effect.

The analysis supported the second hypothesis that identified hazards are perceived to be riskier. In the context of hazard recognition, risk can serve as a proxy for salience as it relates directly to the task objective of keeping mineworkers safe. In this study, risk was significantly related to accuracy on the level of individual hazards (GEE) as well as in aggregate across conditions (mixed model). However, it is important to note that the risk ratings may be biased because the participants provided these ratings as they reviewed their hazard recognition accuracy. The participants could simply have been justifying their selections by giving them a higher rating. Additional work should be done to systematically explore this relationship.

As alluded to above, a unifying explanation for the visual clutter and risk hypotheses could be SOS errors due to a number of factors (salience, time pressures, or preconceptions about the number of hazards present) [13]. In this case, the assessed severity, probability, and overall risk may be more salient to participants, leading to higher identification rates. Time may have also been a factor because many of the participants were forced to stop searching across the scenes when the two-minute time expired. Because the time-out occurrences were not sufficiently saturated across the groups and locations, this was not explicitly analyzed. A more directed study could confirm this hypothesis. Similarly, the inclusion of control images with no hazards may have also contributed to SOS errors. Although the number of hazards was unknown to the participants during data collection, they were informed that some scenes contained no hazards.

The effect of experience matches with previous work in that targeted experience leads to improved accuracy [16, 22]. The group effect may also speak to potential differences in attentional demands. As exemplified in driving studies, novices underperform experts partially because they have not internalized the search activities, thus requiring more attention and greater potential for errors [26, 27]. Similarly, the experience effect of risk suggests that experienced mineworkers may have more practice at performing risk assessments as they may be more likely to perform workplace exams on a daily basis as a part of their jobs. Experienced workers also have greater exposure to the hazards, and therefore may have a better understanding of the results and likelihood. Given these experience differences, more focus may need to be given to standardizing risk assessments to ensure that proper mitigation efforts are taken.

Mineworkers face a challenging and dynamic work environment that requires diligent attention to occupational risks so that they go home safely at the end of each shift. Improving the knowledge, skills, and abilities with which mineworkers scan their work environments for potential health and safety threats is key to reducing workplace illness and accidents. The important message from this research is that mineworkers, safety professionals, and trainers should be aware of the common shortcomings of human perception and decision-making. By understanding that many individuals will often miss key visual cues in their environment or lack knowledge about certain hazards, we can develop strategies to improve hazard identification performance.

Salience, attention, and experience should be considered in order to actually improve hazard recognition strategies to overcome detriments caused by visual clutter. Specifically, Ball et al. [7] demonstrated that with practice, error rates for identifying off-axis targets were reduced and the UFOV could be expanded by 10°. Therefore, practice can be used to combat visual clutter. In an effort to reduce attentional demands, task-specific risk checklists have also been shown to improve hazard detection, particularly in high-clutter environments. Checklists reduce the cognitive workload of a mineworker preforming a workplace exam as this information does not need to be retained in working memory [29]. More generally, object-based priming (previewing images of hazards that are likely to be encountered) has been shown to improve visual search performance, and thus exposure to example hazards through practice or otherwise may help improve knowledge and decision-making in identifying hazards [28].

Beyond these targeted efforts, a more detailed analysis of the scan path data might reveal behaviors used by participants who were more successful at identifying hazards. For example, it was observed anecdotally that in some cases participants were able to correctly identify additional hazards by making a second pass at the part of the scene containing the clustered hazards. While further analysis and research is needed to verify this observation, such strategies could also eventually be incorporated into training interventions.

Disclaimer. The findings and conclusions in this paper are those of the authors and do not necessarily represent the official position of the National Institute for Occupational Safety and Health, Centers for Disease Control and Prevention. Mention of company names or products does not constitute endorsement by NIOSH.

References

1. Mine Safety and Health Administration: Accident: Illness and Injury and Employment Self-Extracting Files (Part 50 Data). http://www.msha.gov/STATS/PART50/p50y2k/p50y2k.htm
2. Mine Safety and Health Administration: Examinations of Working Places in Metal and Nonmetal Mines. Final Rule. Federal Register, vol. 82, pp. 7680–7695 (2017)
3. Mine Safety and Health Administration: Definitions. In: Code of Federal Regulations. Title 30, Part 56.2 (2015)
4. Nodine, C.F., Kundel, H.L.: Using eye movements to study visual search and to improve tumor detection. Radiographics 7(6), 1241–1250 (1987)
5. Cole, B.L., Hughes, P.K.: A field trial of attention and search conspicuity. Hum. Factors 26(3), 299–313 (1984)
6. Leibowitz, H.W., Appelle, S.: The effect of a central task on luminance thresholds for peripherally presented stimuli. Hum. Factors 11(4), 387–391 (1969)
7. Ball, K.K., Beard, B.L., Roenker, D.L., Miller, R.L., Griggs, D.S.: Age and visual search: expanding the useful field of view. J. Opt. Soc. Am. A 5(12), 2210–2219 (1988)
8. Ho, G., Scialfa, C.T., Caird, J.K., Graw, T.: Visual search for traffic signs: the effects of clutter, luminance, and aging. Hum. Factors 43(2), 194–207 (2001)
9. Owsley, C., Ball, K., Sloane, M.E., Roenker, D.L., Bruni, J.R.: Visual/cognitive correlates of vehicle accidents in older drivers. Psychol. Aging 6(3), 403 (1991)

10. Shoptaugh, C.F., Whitaker, L.A.: Verbal response times to directional traffic signs embedded in photographic street scenes. Hum. Factors **26**(2), 235–244 (1984)
11. Drury, C.G., Clement, M.R.: The effect of area, density, and number of background characters on visual search. Hum. Factors **20**(5), 597–602 (1978)
12. Chapman, P.R., Underwood, G.: Visual search of driving situations: danger and experience. Perception **27**(8), 951–964 (1998)
13. Fleck, M.S., Samei, E., Mitroff, S.R.: Generalized "satisfaction of search": adverse influences on dual-target search accuracy. J. Ex. Psychol.-Appl. **16**(1), 60 (2010)
14. Simons, D.J.: Monkeying around with the gorillas in our midst: familiarity with an inattentional-blindness task does not improve the detection of unexpected events. i-Perception **1**(1), 3–6 (2010)
15. Berbaum, K.S., El-Khoury, G.Y., Franken, E.A., Kuehn, D.M., Meis, D.M., Dorfman, D.D., Kathol, M.H.: Missed fractures resulting from satisfaction of search effect. Emerg. Radiol. **1** (5), 242–249 (1994)
16. Klein, G.A., Hoffman, R.R.: Cognitive Science Foundations of Instruction, pp. 203–226. Lawrence Erlbaum Associates, New Jersey (1992)
17. Perlman, A., Sacks, R., Barak, R.: Hazard recognition and risk perception in construction. Saf. Sci. **64**, 22–31 (2014)
18. Mourant, R.R., Rockwell, T.H.: Strategies of visual search by novice and experienced drivers. Hum. Factors **14**(4), 325–335 (1972)
19. Evans, L.: Traffic Safety and the Driver. Van Nostrand Reinhold, New York (1991)
20. Crundall, D.E., Underwood, G.: Effects of experience and processing demands on visual information acquisition in drivers. Ergonomics **41**(4), 448–458 (1998)
21. Underwood, G., Crundall, D., Chapman, P.: Selective searching while driving: the role of experience in hazard detection and general surveillance. Ergonomics **45**(1), 1–12 (2002)
22. Eiter, B., Bellanca, J., Helfrich, W., Orr, T., Hrica, J., Macdonald, B., Navoyski, J.: Recognizing mine site hazards: identifying differences in hazard recognition ability for experienced and new mineworkers. In: International Conference on Applied Human Factors and Ergonomics, pp. 104–115 (2017)
23. Bellanca, J.L., Orr, T.J., Helfrich, W., Macdonald, B., Navoyski, J., Eiter, B.: Assessing hazard identification in surface stone mines in a virtual environment. In: Advances in Applied Digital Human Modeling and Simulation, pp. 217–230 (2016)
24. Hale, S., Myerson, J., Smith, G.A., Poon, L.W.: Age, variability, and speed: between-subjects diversity. Psycol. Aging **3**(4), 407–410 (1988)
25. Heijl, A., Lindgren, G., Olsson, J.: Normal variability of static perimetric threshold values across the central visual field. Arch. Ophthalmol. **105**(11), 1544–1549 (1987)
26. Underwood, G., Chapman, P., Brocklehurst, N., Underwood, J., Crundall, D.: Visual attention while driving: sequences of eye fixations made by experienced and novice drivers. Ergonomics **46**(6), 629–646 (2003)
27. Underwood, G.: Visual attention and the transition from novice to advanced driver. Ergonomics **50**(8), 1235–1249 (2007)
28. Kristjánsson, Á., Campana, G.: Where perception meets memory: a review of repetition priming in visual search tasks. Atten. Percept. Psycho. **72**(1), 5–18 (2010)
29. Liao, P.C., Sun, X., Liu, M., Shih, Y.N.: Influence of visual clutter on the effect of navigated safety inspection: a case study on elevator installation. Int. J. Occup. Saf. Ergo. 1–15 (2017)

From the Laboratory to the Field: Developing a Portable Workplace Examination Simulation Tool

Brianna M. Eiter[(⊠)], William Helfrich, Jonathan Hrica, and Jennica L. Bellanca

Pittsburgh Mining Research Division, National Institute for Occupational Safety and Health, Centers for Disease Control and Prevention, Pittsburgh, PA, USA
{BEiter,WHelfrich,JHrica,JBellanca}@cdc.gov

Abstract. To perform a successful workplace examination, mineworkers must be able to find and fix hazards at their workplace. NIOSH recently completed a laboratory study to identify differences in hazard recognition performance for mineworkers, safety professionals, and mining engineering students tasked with performing a simulated workplace examination in a virtual environment. The laboratory methodology and study results were used to develop a training product aimed at improving mineworker safety. The purpose of the current chapter is to describe the efforts that were taken to modify the laboratory workplace examination simulation into a portable software tool called EXAMiner, which can be used for data collection and training purposes in the field. This chapter provides an explanation of the literature and results from the NIOSH laboratory research studies used to inform and motivate development of EXAMiner. In addition, the software specifications are explained.

Keywords: Mine safety · Hazard recognition · Virtual reality

1 Introduction

Over sixteen months, from October 2013 to January 2015, 37 mineworkers were fatally injured at metal/nonmetal (M/NM) mine sites [1], doubling the number of fatalities that occurred in each of the previous two years. To begin to address this concern, the Mine Safety and Health Administration (MSHA) called for an increased focus on "daily and effective workplace exams to find and fix hazards" as one of several areas where improvement was necessary to promote the health and safety of the mineworker [2]. MSHA has also proposed an update to the regulation related to workplace examinations. The proposed rule requires examinations of the working place to be conducted before work begins or as a mineworker begins work in that location. It also requires examination records to include descriptions of adverse conditions that are not corrected immediately and the dates of corrective action for those conditions [3]. Practically, this means that more mineworkers will be conducting workplace examinations, and more mineworkers will be accountable for the results.

D. N. Cassenti (Ed.): AHFE 2018, AISC 780, pp. 361–372, 2019.
https://doi.org/10.1007/978-3-319-94223-0_34

Hazard recognition is the identification and understanding of a condition or behavior that can cause harm [4]. Hazard recognition is a cognitively complex process that mineworkers must be prepared to execute [5], and it represents a special challenge for mineworkers because of their diverse work activities that involve the use of complex machinery and processes that take place in a dynamic, challenging environment [6]. While hazard recognition is critical for mineworkers' health and safety, recent research shows that, when tested in the laboratory, mineworkers are not identifying a significant number of hazards [5, 7]. National Institute for Occupational Safety and Health (NIOSH) researchers further examined hazard identification through a laboratory study where subjects performed a simulated work place exam; this study is referred to as the laboratory study from here on. In this study, researchers asked experienced and inexperienced mineworkers, mine safety professionals, and mining engineering students to perform a simulated workplace examination [7]. Participants performed this hazard recognition task within a virtual environment, and were instructed to search true-to-life size panoramic images of typical locations at surface limestone mines (e.g., the pit, plant, roadway, and shop) for hazards. After completing the hazard recognition task all participants received feedback about their hazard recognition performance and were debriefed about the hazards they accurately identified and missed. Consistent with the findings from other studies, hazard recognition task results show that experienced and inexperienced mineworkers and mining engineering students, on average, identified only 53% of the hazards. In addition and importantly, the participating mine safety and health professionals—those people at the mine site tasked with ensuring that mineworkers are trained and capable of recognizing hazards—were only able to accurately identify 61% of the hazards. These identification rates are well below the previously established 90% standard for mastery [8].

As the industry awaits the final promulgation of the new workplace examination rule, NIOSH has focused their efforts on the development of practical solutions from research findings for industry stakeholders to use to improve hazard recognition ability so that mineworkers can perform more effective workplace examinations and work more safely. Specifically, NIOSH is adapting the laboratory simulated workplace exam methodology to create EXAMiner—a portable workplace examination simulation designed to address critical competencies associated with hazard recognition ability. EXAMiner is designed to give mineworkers the opportunity to perform a simulated workplace examination, much the same way participants in the laboratory study performed a workplace examination in a laboratory environment [7]. NIOSH researchers are using the laboratory study materials and methodology to develop EXAMiner, and incorporating critical competencies and training strategies. The initial development of EXAMiner is described in the following sections

2 Training Strategies to Improve Hazard Recognition Ability

As NIOSH researchers modify the laboratory materials into a training tool, it is important to adopt proven training strategies in EXAMiner. Training strategies are the tools, methods, and contexts that are combined and integrated to create an interesting and engaging delivery approach [9]. An effective training tool forms a cohesive strategy by incorporating information, demonstration, practice, and feedback [10]. Information is the concepts or facts that have to be learned, demonstration provides visual depictions of the information, practice is the application of learned information and competencies in a safe environment, and feedback about performance. Each of these concepts is incorporated into EXAMiner to address hazard recognition competencies and improve hazard recognition ability.

EXAMiner delivers information about general and site-specific hazards through descriptions, statistics, and mitigation strategies during session debriefs. Hazard recognition is demonstrated within EXAMiner through a computer-based simulation that allows interactive and immersive activity by recreating all or part of a work experience [11] and encourages experiential learning [12]. EXAMiner provides high-fidelity panoramic images that allow the mineworker to visually experience all of the details of the mining environment. Additionally, using EXAMiner, mineworkers can practice searching for and finding hazards through a simulated workplace exam. The realistic and relevant hazard recognition practice in EXAMiner makes the skills and concepts practiced more likely to transfer to the job [13]. Finally, EXAMiner provides feedback about performance on the hazard recognition search task through debrief. Research indicates that technology-based simulation is more likely to be impactful and practice is much more likely to result in learning when it is paired with a debrief session, where trainees are given a post-training review and provided feedback about their performance [14].

3   Hazard Recognition and Critical Competencies

A growing body of research suggests that successful hazard recognition requires an individual to possess a complex set of competencies. These competencies include stable attributes such as intelligence [15], risk perception [16], and general and site-specific hazard knowledge [17]. They also include hazard recognition skills such as visual search [18], situational awareness [19], and lastly, they include temporary states such as alertness and distraction [20]. NIOSH researchers focused on four competencies to improve hazard recognition ability while developing EXAMiner: general hazard recognition knowledge, site-specific hazard recognition knowledge, visual search, and pattern recognition. Researchers selected these competencies because they are important, basic competencies all mineworkers should have and they were easily visualized using panoramic images in the hazard recognition laboratory study. In the following, we present how these competencies transfer from the laboratory to implementation in EXAMiner.

3.1   General Hazard Recognition Knowledge

General knowledge of hazards better prepares mineworkers to recognize hazards when they are present, and therefore, it is presumed, results in fewer incidents and accidents [17, 21]. General hazard recognition knowledge in mining is the knowledge of basic hazards found at most mine sites, regardless of commodity (e.g., gold, limestone, or coal). Examples of general hazards include slip, trip, and fall (STF) hazards such as material on a walkway or stairwell, electrical hazards such as splits and splices in electrical cords and cables or open electrical boxes, and fire hazards such as flammable material in close proximity to an open flame, or missing or inaccessible fire extinguishers. General hazard recognition knowledge is essentially the minimum level of training because mineworkers must first know and understand what hazards are in order to be able to find them at the worksite [5].

Within the laboratory study, general hazard recognition knowledge was represented by the hazards that were included in the panoramic images. Examples of general hazards include STF (e.g., material on walkway) and electrical hazards (e.g., unattended open electrical box). These panoramic images are also included in EXAMiner. General hazard recognition knowledge is also reinforced during the session debrief with information about the specifics of the hazards. Hazard information includes a description of the hazard, statistics to support why the hazard is risky, and information from the Code of Federal Regulation (CFR Part 46.5 and 46.8 of Title 30 Mineral Resources) that can be used to mitigate the hazard.

3.2 Site-Specific Hazard Recognition Knowledge

While there are hazards that are common across and within commodities, there are also site-specific hazards. Surface M/NM mineworkers face unique challenges. For instance, mineworkers operating a dredge work around water and have to be knowledgeable of how to correctly don the appropriate personal protective equipment (PPE), such as a life vest. Mineworkers operating front-end loaders and haul trucks in pits and quarries have to be knowledgeable about the unique geological characteristics of the material being mined at their site. This knowledge is critical for identifying areas of instability during highwall inspections. It is critical that mineworkers know and understand the specific risks that may be present in their work environments so they can recognize hazards.

Site-specific hazard recognition knowledge was also represented during the laboratory study by the hazards in the panoramic images, for instance cracks in the highwall. Again, the panoramic images are included in EXAMiner and reinforced during the session debrief with a description of the hazard, statistics to support why the hazard is risky, and information from 30 CFR that can be used to mitigate the hazard.

3.3 Visual Search

In addition to hazard knowledge, miners have to be able to search their surroundings efficiently and effectively to actually identify hazards. There is ample research from a number of industries (e.g., aviation, mining, automobiles) suggesting that visual search skills are trainable [12]. Visual search performance changes both qualitatively and quantitatively with extended training [22], and visual search training that engages workers in exercises that requires them to actively detect hazards is more effective than training that only provides verbal information about the hazards [23].

The laboratory study was designed to identify differences in how mineworker experience affected how they search for and find hazards. To complete the hazard recognition task, participants were instructed to search the panoramic images for hazards. While they performed this visual search task, study participants wore eye-tracking glasses to monitor and record their eye-movements. EXAMiner gives mineworkers approximately the same visual experience as the participants in the laboratory study. Using EXAMiner, mineworkers can practice searching for and finding hazards in the panoramic images visualized through a projector or monitor instead of in theater virtual laboratory environment.

3.4   Pattern Recognition

Pattern recognition or pattern matching is another skill that influences hazard recognition abilities. Experienced decision makers use it to more efficiently identify whether a specific situation reflects either normal operations or an abnormality [24]. Examples of safety-related patterns in mining are acceptable locations of equipment or personnel during operations, how equipment typically operates, or the typical cycle of changes in a mine environment over time. Changes to these patterns can signify when something is out of place, not operating correctly, or "does not look right," which may trigger the mineworker to recognize a hazard. Loveday et al. [25] suggest that while experience may be a necessary precursor to pattern recognition it is not sufficient on its own.

The panoramic images used during the laboratory study were developed to include multiple examples of the same hazard. For instance, there are several hazards related to a fire extinguisher: the fire extinguisher is missing, the signage denoting a fire extinguisher is missing, a garbage blocks the fire extinguisher, etc. Including multiple examples of a hazard within EXAMiner should strengthen the representation of that hazard in the mineworker's memory. Consequently, the mineworker should be more likely to recognize a fire extinguisher hazard in the future, even if it was not specifically in the training [26].

4   Use Case Definition for EXAMiner

NIOSH researchers identified the most critical areas and common training limitations to guide the development and determine how EXAMiner could ultimately be used to deliver training to the mining industry.

To be Used by the M/NM Mining Sector. NIOSH researchers developed EXAMiner as a response to the increase in M/NM fatalities and the call for more effective workplace examinations [1, 2]. The first step toward developing this simulation tool was a laboratory study that was designed to identify differences in hazard recognition ability based on mineworker experience [7]. Materials for the study include thirty-two panoramic images of typical locations at a surface limestone mine, where there are 101 hazards represented. These materials address general and site-specific hazards and strengthen pattern recognition common to the M/NM mining sector.

To be Used During Part 46 Training. EXAMiner was specifically designed to be used during training. The Code of Federal Regulations (CFR Part 46.5 and 46.8 of Title 30 Mineral Resources) requires mine operators to provide safety and health training to all new mineworkers and refresher training to all of their mineworkers each year. MSHA regulates the content of training plans. Trainers must follow an MSHA-approved training plan and meet the following criteria during their training classes:

- **New Miner Training:** Includes an initial 24 h of safety training within the first 60 days of work and 4 h of site-specific training before beginning work. The training must address site-specific hazards and include an introduction to the work environment, electrical hazards, emergency medical procedures, the health and safety aspects of tasks to be assigned, and the rules and procedures for reporting hazards.
- **Annual Refresher Training:** 8 h refresher training typically covers similar topics to new miner training. Refresher training must include instruction on changes at the mine that could adversely affect the mineworker's health or safety. Refresher training should also include a discussion of the hazards associated with the equipment that has accounted for the most fatalities and serious injuries, including mobile equipment conveyor systems, cranes, crushers, excavators, and dredges.

To address content relevant to Part 46 new miner and annual refresher training, EXAMiner provides information about general and site-specific hazards. This information is presented as hazard descriptions, statistics, and mitigation strategies during the session debrief.

To be Used in Group, Classroom Training. Part 46 of the CFR does not include specific requirements for where new miner or annual refresher training should take place. However, it is common for mine operators to choose to hold training either in a classroom at the mine site or a meeting room at a hotel or convention center near the mine site. Therefore, the infrastructure and resources available for training can vary greatly from one mine operator to another. While some mine operators may have technologically advanced classroom spaces equipped with up-to-date computer and projector systems, the vast majority rely on older model laptops and portable projector systems. In addition, given the remote location of some mine sites, a reliable internet connection may not be available. These factors affect what materials trainers select and how trainers choose to conduct training classes.

To minimize the overall cost and impact of training, mine operators typically conduct new miner and annual refresher training in groups led by one or two instructor trainers. Notably, this approach is consistent with training across industries, where 49% of training happens in an instructor-led, classroom setting [27]. To accommodate this training approach, EXAMiner was developed for group training sessions.

To be Used with Customized Materials. Mine safety and health trainers working in the mining industry have a limited supply of mining-specific training materials. NIOSH previously produced hazard recognition training materials that were delivered using the View Master [5]. This training program was an innovative and effective approach to hazard recognition training in that it provided mineworkers with a 3-D experience of their work environment. There are, however, some limitations with this program. First, it was developed for single-person use, so trainers had to buy the View Masters for all the students to be able to use materials. Additionally, the training program only included a small set of materials, limiting their usefulness for subsequent trainings.

Finally, while the authors detail the process for developing 3-D materials, the process is complicated and time and material intensive.

EXAMiner was created to give trainers the opportunity to customize training scenarios either by creating their own training scenarios or by choosing to create random scenarios from thirty-one panoramic images included with the software. They can also use one of the NIOSH scenarios to focus on specific types of hazards (e.g., electrical or slip, trip, fall) during training.

5   EXAMiner Workflow

As a portable, simulation, training tool, EXAMiner addresses the aforementioned hazard recognition competencies and constraints, and as a research-to-practice product, EXAMiner closely mirrors the methodology used in NIOSH's laboratory study [7]. Consequently, the initial version EXAMiner is a stand-alone, downloadable software application designed for optimal use during an instructor-led Part 46 annual refresher or new miner training held in a classroom setting. Once installed, EXAMiner does not require an internet connection and can be run on a standard laptop with a projector system. While EXAMiner includes a general hazard recognition training intervention, it can also be coupled with other training packages, as it provides the framework to demonstrate, practice, and review hazardous scenes. The following sections outline the how the software currently operates, with the basic components including the session setup (red), search task (blue), session debrief (green), and training intervention (orange) as depicted in the software flowchart in Fig. 1.

5.1   Session Setup

In EXAMiner, a session is defined as a single instance that a scenario—a set of panoramic images—is used for the search task, and it encapsulates the entire training (e.g., pre- and post-intervention panoramas). The trainer can create a new scenario, randomly generate a scenario, or load a previously saved one. To create a unique scenario, EXAMiner allows trainers to select the panoramas they wish to include and save the scenario to be either used immediately in a session or loaded later. Currently, the included panoramas of the pit, plant, shop, and roadway can be added to a scenario in any order chosen by the user. A randomly generated scenario includes between two and six randomly selected panoramas from 31 included images. Lastly, EXAMiner also allows the trainer to load a NIOSH-created scenario or one of their previously created scenarios.

The current version of EXAMiner includes nine NIOSH-created scenarios, five of which focus on specific hazard types (e.g., electrical or slips and falls). The specific hazard scenes only include hazards of that type. For instance, the electrical scenario includes only panoramas with electrical hazards. In this case, mineworkers should be given instructions to search for and find the electrical hazards, and only the electrical hazards are identified as feedback in the debrief session.

The remaining four NIOSH scenarios were developed specifically for new miner training. These scenarios include the hazards that were missed most often by

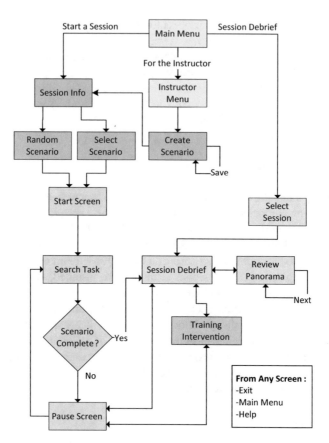

Fig. 1. A schematic depiction of the EXAMiner software workflow, including the navigation components (gold), session setup (red), search task (blue), session debrief (green) and training intervention (orange).

participants in the laboratory study with the least amount of mining experience—who were mining engineering students. These scenarios contain 23 hazards that had at least a 20% difference in hazard recognition accuracy as compared to the safety professionals, and are evenly distributed across the four panoramic locations.

.5.2   Search Task

Once a scenario is selected, EXAMiner progresses to the hazard recognition search task. During the search task, mineworkers practice searching for and finding hazards while performing a virtual workplace examination. Once the search task starts, one panorama is displayed at a time, and the trainer progresses through the panoramas in the scenario. Within a panorama, trainers can pan around the image in an unlimited loop, zoom in and out, and pan up and down (see Fig. 2). Hazards are identified by clicking. A two-minute time limit can also be set for each panorama in the session info. For timed sessions, a progress bar is displayed at the bottom of the screen, and after two minutes, the panorama disappears and EXAMiner automatically progresses to a pause screen. Alternatively, the trainer can end the panorama at any time. After the completion on one or more panoramas, the instructor can progress to the next panorama, load the training intervention, or stop the search task early and move on to the session debrief.

Fig. 2. The picture represents a current screenshot of the Search Task within EXAMiner with the navigation toolbar across the bottom of the screen. The green progress bar depicts the time remaining for a timed session. Red click icons represent locations that have been identified as hazards.

5.3 Session Debrief

Following the completion of one or more panoramas or upon loading a previously collected session, EXAMiner allows the trainer to review the results of the hazard recognition search task in a session debrief. The session debrief gives the trainer an opportunity to give mineworkers feedback on their performance and to spend time discussing the hazards that were found and missed. The software displays the results of the training session including: panorama thumbnails and identifiers (e.g., Shop 2), the number of hazards correctly identified (e.g., 3 of 3 for Pit 6), the total search times (MM:SS), and the total number of clicks (e.g., 7 for Pit 6) (see Fig. 3).

The trainer can choose to debrief any number of completed panoramas during the session debrief. A review of each of the panoramas includes correctly identified hazards (outlined in green), missed hazards (outlined in red), and click locations as well as more information about the hazards in the image (see Fig. 4). The hazard information details the "what," "when," and "how to fix" of each hazard; this addresses general and site-specific hazard knowledge. The hazard information is directly relevant to topics and material that have to be covered during Part 46 new miner and annual refresher training.

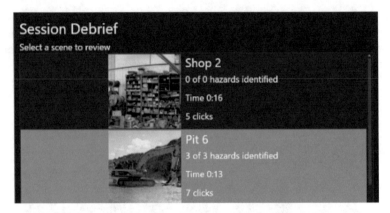

Fig. 3. The picture displays the current version of the EXAMiner's Session Debrief Screen, which provides the results of the loaded hazard recognition search task.

Fig. 4. The picture displays the current version of the EXAMiner's Panoramic Review Screen. Green boxed items indicate correctly identified hazards, and red boxes are unidentified hazards. The red click marks indicate where a trainer clicked during the search task. The upper left corner lists the hazards in the current panorama and the selected "Missing Midrail" hazard information is displayed in the upper right corner.

5.4 Training Intervention

Following the completion of at least one panorama in the search task, EXAMiner gives the trainer an opportunity to use the training intervention. Currently, the training intervention includes three modules: "Workplace EXAMinations and hazard recognition," "Are you searching like an EXAMiner," and "Taking it to the field." These training materials are designed to address general and site-specific hazard knowledge. More modules will become available through software updates as they are developed.

The intervention point also gives the trainer the opportunity to evaluate hazard recognition ability while using the training intervention. As depicted in Fig. 1, the current version of EXAMiner allows the trainer to move between the session debrief, search task, and training intervention to perform the search task on any number of panoramas and review them as they fit with the training modules in the training intervention. For example, in a pre/post format, the trainer can compare accuracy scores to determine whether there was a change in hazard recognition ability before and after the training intervention.

Acknowledgments/Disclaimer. NIOSH would like to thank Holly Tonini for her help in taking and editing the panoramic images, and Gregory Cole, Jonathan Fritz, and John Britton for programming the EXAMiner software. EXAMiner is currently being tested in the field and will be available to the public following final evaluation. The findings and conclusions are those of the authors and do not necessarily represent the official position of the National Institute for Occupational Safety and Health, Centers for Disease Control and Prevention.

References

1. Mine Safety and Health Administration: Accident: Illness and Injury and Employment Self-Extracting Files (Part 50 Data). http://www.msha.gov/STATS/PART50/p50y2k/p50y2k.htm
2. MSHA: Mine Safety and Health Administration. MSHA announces increased education, outreach and enforcement to combat increase in NMN mining deaths (2015b). http://arlweb.msha.gov/stats/review/2014/mnm-fatality-reduction-effort.asp
3. https://www.msha.gov/news-media/press-releases/2017/09/12/msha-proposes-changes-final-rule-workplace-examinations-metal
4. Bahn, S.: Workplace hazard identification and management: the case of an underground mining operation. Saf. Sci. **57**, 129–137 (2013)
5. Kowalski-Trakofler, K.M., Barrett, E.A.: The concept of degraded images applied to hazard recognition training in mining for reduction of lost-time injuries. J. Saf. Res. **34**(5), 515–525 (2003)
6. Scharf, T., Vaught, C., Kidd, P., Steiner, L., Kowalski, K., Wiehagen, B., Rethi, L., Cole, H.: Toward a typology of dynamic and hazardous work environments. Hum. Ecol. Risk Assess. **7**(7), 1827–1841 (2001)
7. Eiter, B.M., Bellanca, J.L., Helfrich, W., Orr, T.J., Hrica, J., Macdonald, B., Navoyski, J.: Recognizing mine site hazards: identifying differences in hazard recognition ability for experienced and new mineworkers. In International Conference on Applied Human Factors and Ergonomics, pp. 104–115. Springer, Cham (2017)
8. Barrett, E., Kowalski, K.: Effective Hazard Recognition Training Using a Latent-Image, Three-Dimensional Slide Simulation Exercise. Report of Investigation, Bureau of Mines (1995)
9. Salas, E., Tannenbaum, S.I., Kraiger, K., Smith-Jentsch, K.A.: The science of training and development in organizations: what matters in practice. Psychol. Sci. Public Interest **13**, 74–101 (2012)

10. Salas, E., Cannon-Bowers, J.A.: The science of training: a decade of progress. Ann. Rev. Psychol. **52**, 471–499 (2001)
11. Maran, N.J., Glavin, R.J.: Low-to high-fidelity simulation–a continuum of medical education? Med. Educ. **37**(s1), 22–28 (2003)
12. Kolb, D.: Experiential learning as the science of learning and development (1984)
13. Grossman, R., Salas, E.: The transfer of training: what really matters. Int. J. Train. Dev. **15**, 103–120 (2011)
14. Tannenbaum, S.I., Cerasoli, C.P.: Do team and individual debriefs enhance performance? A meta-analysis. Hum. Factors **55**(1), 231–245 (2013)
15. Wang, L., Song, M., Jiang, T., Zhang, Y., Yu, C.: Regional homogeneity of the resting-state brain activity correlates with individual intelligence. Neurosci. Lett. **488**(3), 275–278 (2011)
16. Hunter, D.R.: Risk perception and risk tolerance in aircraft pilots. Federal Aviation Administration Report no. PB2003100818. Federal Aviation Administration, Washington, D.C. (2002)
17. Albert, A., Hallowell, M.R., Kleiner, B., Chen, A., Golparvar-Fard, M.: Enhancing construction hazard recognition with high-fidelity augmented virtuality. J. Constr. Eng. Manag. **140**(7) (2014)
18. Gramopadhye, A.K., Drury, C.G., Prabhu, P.V.: Training strategies for visual inspection. Hum. Factors Ergon. Manuf. Ser. Ind. **7**(3), 171–196 (1997)
19. Endlsey, M.: Theoretical underpinning of situation awareness: a critical review. In: Endsley, M., Garland, D.J. (eds.) Situation Awareness Analysis and Measurement. Lawrence Earlbaum Associates, Mahwah, NJ (2000)
20. Wickens, C.D., Horrey, W.J.: Models of attention, distraction, and highway hazard avoidance. In: Driver Distraction: Theory, Effects and Mitigation, pp. 249–279. CRC Press, Boca Raton (2008)
21. Montgomery, J.F.: Safety and health training. In: Plog, B., Quinlan, P.J. (eds.) Fundamentals of Industrial Hygiene, pp. 680–681. National Safety Council, Itasca, IL (2002)
22. Shiffrin, R.M., Schneider, W.: Controlled and automatic human information processing: 11. Perceptual learning, automatic attending, and a general theory. Psychol. Rev. **84**(2), 127–190 (1977)
23. Blignaut, C.J.H.: The perception of hazard I. hazard analysis and the contribution of visual search to hazard perception. Ergonomics **22**, 991–999 (1979)
24. Kaempf, G.L., Wolf, S., Miller, T.E.: Decision making in the AEGIS combat information center. Proc. Hum. Factors Ergon. Soc. Ann. Meet. **37**(16), 1107–1111 (1993)
25. Loveday, T., Wiggins, M., Festa, M., Schell, D., Twigg, D.: Pattern recognition as an indicator of diagnostic expertise. In: Pattern Recognition-Applications and Methods, pp. 1–11. Springer, Heidelberg (2013)
26. Perdue, C.W., Kowalski, K.M., Barrett, E.A.: Hazard recognition in mining: a psychological perspective. Information circular/1995 (No. PB–95-220844/XAB; BUMINES-IC–9422). Bureau of Mines, Pittsburgh Research Center, Pittsburgh, PA, US (1995)
27. Association for Talent Development: Press Release: ATD Releases 2016 State of the Industry Report (2016). https://www.td.org/insights/atd-releases-2016-state-of-the-industry-report

Using High-Fidelity Physical Simulations of the Environment to Evaluate Risk Factors for Slips and Falls in the Mining Industry

Mahiyar Nasarwanji$^{(\boxtimes)}$, Jonisha Pollard, and Lydia Kocher

Pittsburgh Mining Research Division, National Institute for Occupational Safety and Health (NIOSH), Pittsburgh, PA 15236, USA
{MNasarwanji, JPollard, LKocher}@cdc.gov

Abstract. The shoe-floor interface is a key element in preventing slips and falls. The design of footwear and the floor surface should be considered to ensure worker safety. Testing various floor surface materials and boots in a real-world setting would impose unnecessary risk to participants and limit the extent of testing possible. Hence, two examples of high-fidelity physical simulation—an inclined grated metal walkway and a grated metal stairway—were built to evaluate risk factors for slips and falls associated with various walkway materials and boots with metatarsal guards. This paper discusses details and findings of the two studies. Also discussed are the advantages and disadvantages of using physical simulations of the environments, including decreased risk for participants and large space requirements for the experiment. Findings of the research can help select appropriate floor surface materials and boots for the mining industry and inform the use of future high-fidelity physical simulations.

Keywords: Physical simulation · Inclined walkway · Boots · Slip
Fall · Mining

1 Introduction

Slip, trip, and fall research has been vital to ascertain the root causes of slip, trip, and fall events, determine contributing factors, and develop guidelines for safe working conditions. Field research occurring in natural environments with slip, trip, and fall hazards can pose significant risks to participants and limit the extent of testing possible. However, simulations can be used to minimize participant risk while collecting the measures of interests. Different types of simulations have previously been used to create scenarios depicting falls from height, ladder falls, slips and falls on the same level, and to determine the influence of safety footwear in construction and the food service industries.

1.1 Use of Simulations in Slip and Fall Research

Virtual simulations of the environment in the laboratory allowed researchers to examine the human response to working at height [1]. The detection of elevation by humans occurs purely through visual stimulus. Virtual simulations allow researchers to induce the physiological and psychological effects of height exposure, without placing someone in a truly dangerous environment. Virtual simulations of elevation using virtual reality result in increased anxiety, increased perceived risk of falling, increased postural instability, and changes in skin conductance, which are similar to the reactions during exposure to real elevation [2]. Moreover, a virtual simulation of the environment significantly decreases perceptions of stability and balance confidence [3]. Virtual simulations of the environment with good visual fidelity may provide a safe and cost-effective means to assess working at heights and the risk for falls, evaluate fall prevention strategies, and study postural control in populations at risk for falls [2–4].

In contrast, physical simulations of the environment are used when the individual needs to physically interact with the environment. Physical simulations have been used in the past to research occupational ladder climbing and descending in laboratories. Ladders instrumented with force sensors allowed researchers to determine the force requirements for ladder climbing [5]. A ladder that simulates rung failure was used to examine the effect of ladder climbing pattern on fall severity [6]. In addition, a ladder equipped with a spinning, low-friction rung was used to examine the effect of hand position, age, and climbing dynamics on ladder slip outcomes [7, 8]. These studies utilized fall-arrest harnesses to ensure participants' safety. In Pliner and Beschorner [6], fall severity was evaluated by quantifying the peak weight supported by the harness that was attached to a load cell. These types of studies would not be feasible in a real-world setting due to the risk of falls and the need for custom-fabricated apparatuses and specialized instrumentation.

Level and inclined walkway slips have been examined using physical simulations of environment in the laboratory setting to determine performance of slip-resistant footwear, the effects of wearing personal protective equipment, slip recovery mechanisms, load carriage, age, and obesity. Participants have been subjected to dry, wet, and soapy conditions on carpeted and tiled floors, simulating slippery walkway conditions in restaurants [9]. Older and younger participants have been subjected to slippery floor conditions to determine the effects of aging on the slip initiation and fall frequency [10]. The objective of most of these studies is to improve the understanding of how humans slip, trip, or fall to help develop preventive measures. Prevention of slips, trips, and falls is most often multi-faceted and focuses on a number of areas including workplace design, personal protective equipment, maintenance of walking structures and equipment, and housekeeping.

1.2   Importance of the Shoe-Floor Interface

The footwear-floor interface is a key element in preventing slips, trips, and falls to the same level as it forms the interface between the human and the working environment. The type of floor surface used can be critical in preventing slips and falls and should be considered to ensure worker safety. Floor surfaces that do not provide adequate friction will pose slip hazards; and contaminants on the surface can modify the available friction. Grated metal is the material of choice for walkways at mines and is commonly used on level and inclined walkways, platforms, and stairs as it is rugged and prevents the accumulation of material on the walkway while offering some slip resistance. Level and inclined walkways that run along conveyers and equipment are often exposed to the environment; and, even though they are made of grated metal, they often have a buildup of contaminants on them (Fig. 1). In addition, anecdotal evidence indicates that these walkways along conveyers can have inclinations of over 20° to the horizontal. There is little guidance available on which grated metal flooring material can help reduce the risk of slips or on what inclinations may increase the likelihood of slips on grated metal walkways when contaminants are present.

Fig. 1. Examples of contaminants (*left*: snow and ice; *right*: rocks and accumulation) found on inclined grated metal walkways along conveyer belts at mines.

Footwear is the second component of the footwear-floor interface, and the critical aspects of footwear include tread design, sole flexibility, and shaft stiffness. A number of mines have recently been encouraging their miners to wear safety toe boots with metatarsal guards. Although boots with metatarsal guards may increase protection from falling objects, the slip or trip risks associated with wearing these boots is largely unknown. Boots with metatarsal guards may increase shaft stiffness, which has been shown to modify gait [11, 12]. In addition, for specific tasks such as stair ascent, where ankle range of motion and flexibility is necessary, footwear that reduces metatarsal motion could restrict ankle movement, thereby increasing slip or trip risks [12]. Safe footwear should have soles that allow for adequate traction and an overall design that does not affect personal walking characteristics [13].

Testing floor materials and footwear in the field or in a real-world setting would impose unnecessary risks to participants and limit the extent of testing that is possible, as seen in other slip and fall research. In addition, the use of high-precision, advanced measurement technologies to evaluate changes in gait or to detect the occurrence of a slip, such as motion capture and force plates, would not be viable in a real-world setting. In contrast, investigating these factors in a simulated environment are more viable. The objective of this work is to demonstrate the use of high-fidelity physical simulations of the environment to evaluate risk factors for slips and falls in the mining industry in two example studies.

2   Example Studies

2.1   Testing Walkway Materials on an Inclined Grated Metal Walkway

Simulation of Inclined Walkway Along Conveyers. A grated metal walkway, similar to what would be found beside a conveyer at a mine site (Fig. 2), was built with an adjustable inclination from the floor surface in 5° increments up to 20°. The walkway was 3.65 m long with a platform at the end that extended another 1.25 m. The width of the walkway and platform was 0.5 m. With the walkway raised to 20°, the platform was elevated 1.4 m above the ground. The walkway had handrails on both sides and the platform had handrails on the three exposed sides. The grated metal used as the flooring surface along the entire length of the walkway could be interchanged to use three different materials: (1) diamond weave material, (2) serrated, rectangular bar-type material, or (3) circular, perforated pattern material. The flooring materials and handrails used were those commonly encountered at mines and were installed per manufacturers' specifications.

Fig. 2. Adjustable inclined grated metal walkway built to simulate walkways along conveyers commonly found at mines.

Participants and Procedures. Detailed descriptions of the participants and experimental procedures can be found in Pollard et al. [14]. Twelve participants (including three women and nine men) participated in the study. In addition to the three-floor surface materials tested, the floor surface was tested in both a dry condition and in a contaminated condition. For the contaminated condition, the surface was coated with 99.99% Glycerol, using a brush and sprayed with a 2:1 Glycerol to water solution between trials. The contaminated condition simulated slippery conditions commonly encountered when these walkways are covered in ice, snow, grease, or oil. A ceiling-mounted fall arrest system was provided along the length of the walkway and two strain-based force plates were mounted in the center of the walkway to measure ground reaction forces that were used to calculate the required coefficient of friction. Participants were aware that the floor surface was contaminated and were instructed to walk across the walkway in both directions (up and down) without using the handrails.

Results. Figure 3 provides the number of slip events that occurred (out of 24 trials) for all participants for that condition. It is evident that a greater number of slips occurred at higher inclinations with slip events at inclinations as low as 10°. The circular, perforated material led to the most slips and the diamond weave led to the least number of slips. In addition, slips occurred more often on the contaminated metal gratings than on the dry metal grating. These findings can help make very specific recommendations to improve the workplace, including limiting walkway inclinations to less than 10° or utilizing the diamond weave materials for higher inclinations, up to 20°.

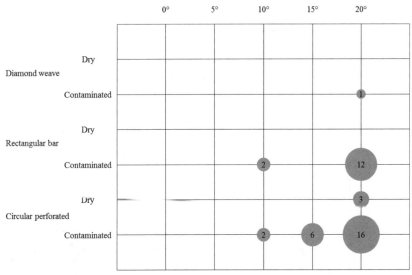

Number of slip events (out of 24 trials)

Fig. 3. Number of slips observed (out of 24 trials) for each combination of surface material, contaminant condition, and inclination of walkway.

2.2   Testing Boots with Metatarsal Guards on a Grated Metal Stairway

Simulation of Stairways at Mines. A stairway, similar to that traversed between levels at a mine, was constructed with steps made from the diamond-weave grated material. The stairway had five steps, each with a rise of approximately 0.2 m and a step depth of approximately 0.25 m, which led to a platform 1 m in length, for an overall rise of 1.22 m. The stairway was 0.6 m wide and had handrails on both sides of the stairs and on the three exposed sides of the platform.

Participants and Procedures. Detailed descriptions of the participants and experimental procedures can be found in Pollard et al. [15]. Five participants, wearing men's size 10 boots, participated in the study. Boots selected for the study were hiker-style Dr. Marten's Ironbridge Steel Toe boots with and without metatarsal guards. A motion capture system was used to track movement of the feet, using retro-reflective markers placed on the boots and the stairway. Toe clearance was measured when the toe crossed the front edge of the third step from the ground as the absolute horizontal distance (X), absolute vertical distance (Z), and the length of the resultant vector (shortest distance) between the reflective toe marker and the edge of the step in the direction of travel.

Results. Figure 4 shows the differences in toe clearance between the steel-toe boots with metatarsal guards and those without metatarsal guards in the horizontal (X) direction, the vertical (Z) direction, and the resultant vector (greater positive values indicate the metatarsal boots were closer to the edge of the stair tread). Although, there are minor differences between the boots, with metatarsal boots being on average 10 mm closer to the edge for the resultant vector, these differences are not statistically significant. However, due to the limited sample size and only a single pair of boots being tested, additional research is needed to evaluate the influence of metatarsal boots on toe clearance during ladder ascent, as well as heel clearance and foot placement during stair descent.

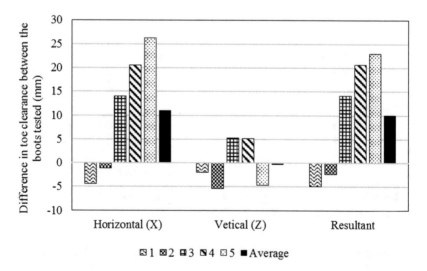

Fig. 4. Difference in toe clearance (in mm) between the two boots tested for the participants and the average in the horizontal direction, vertical direction, and the resultant vector (greater positive values indicate that the metatarsal boots were closer to the edge of the tread or less toe clearance).

3   Discussion

Slips, trips, and falls are a significant burden to the mining industry. Conducting detailed slip, trip, and fall research would be a challenge in the field due to hazards commonly encountered in the environment and the limited ability to use advanced measurement techniques. The objective of this work was to provide examples of how high-fidelity physical simulations of the environment can be used to evaluate risk factors for slips, trips, and falls in the mining industry. Research where participants are expected to slip, trip, or fall would not be possible without the use of safety precautions [6–10, 14]. Participant safety during slip and fall research is critical and can be ensured during simulations in laboratory environments by using systems that may not be practical in the field.

Laboratory simulations (both virtual and physical) also allow for the use of high-precision measurement techniques or systems. In our investigation of metatarsal boots [15], toe clearance was measured using an optical motion capture system. In our study on walkway materials [14], a similar system was used, along with force plates. It would be extremely difficult to set up this equipment in a field setting, therefore making laboratory simulation a preferred method.

The fidelity of the simulation should be selected based on the need for realism, study conditions tested, and the ability to generalize the findings beyond the laboratory simulation. For testing walkway materials, it was critical to ensure that there was an adequate length for participants to walk across and to test only those materials found in the field at inclinations observed at mine sites. Hence, the scale of the simulation, and materials used had to match the real environment closely, increasing the fidelity of the simulation. In contrast, as precise control of the contaminants was necessary to ensure conditions were repeatable between the walkway conditions and across all participants, glycerin was used as a contaminant to simulate slippery conditions like ice or snow. Hence, for the contaminant, the fidelity of the simulation did not match the real conditions encountered; but were instead selected to ensure repeatability and generalizability of the findings, while remaining practical for the researchers. Table 1 shows a list of advantages and disadvantages of using simulations in slip, trip, and fall research.

Table 1. A few advantages and disadvantages of using simulations in slip, trip, and fall research.

Advantages	Disadvantages
- Increased safety	- Space required for the simulation
- Can use advanced and precise measurement techniques	- Lack of participant access
- Can vary the fidelity	- Loss of context or realism

Using high-fidelity simulations in the laboratory also has some disadvantages. One potential limitation of this research was that the walkway materials tested were new. Grated metal walkaways at mines will wear over time due to exposure to the environment and pedestrian traffic and are often not replaced until they are significantly damaged. Our testing reflected a best-case scenario with new walkway materials. In our study on testing walkway materials [14], a minimum space of 8.2 m long by 4 m wide was required to house the simulated walkway with a minimum head clearance of 4.25 m (including accommodations for the fall harness). This space represents a large volume in an indoor laboratory. Similar studies, such as those carried out on ladders, would also require large volumes of space with additional head clearance to accommodate the ladders and required fall-arrest systems [6, 8]. Participant selection or access to participants is another concern when conducting laboratory simulations. In our two studies [14, 15], we did not use miners due to challenges with access to the population; however, there is unlikely to be a difference in gait between miners and the subject population tested. Other studies, such as in Pliner et al.'s research [7], recruited participants from populations that use ladders as part of their jobs, as there was likely to be an influence of expertise and comfort based on the frequency of ladder use. Participant selection needs to consider the necessity of having (or not having) prior knowledge or experience in the area of interest.

Simulations in the laboratory also have the potential to result in the loss of environmental context. In some scenarios, simulating the virtual environment alone is adequate because participants do not need to interact with objects within the environment to elicit a response [2]. In other scenarios, the physical interaction with objects within the environment is essential, requiring the simulation to mirror the physical interaction with the environment and work scenarios as closely as possible [7, 8, 14]. Combining high-fidelity physical simulations with virtual simulations could increase the realism of the simulation and are often used for motor vehicle research and training where both visual and physical interactions are necessary. In these scenarios, the physical interaction is a result of visual cues in the environment; hence, to study human responses both aspects of the environment need to be simulated.

4   Conclusions

Slip, trip, and fall research poses significant risks to participants, requiring safety precautions that are likely not present in field settings. These risks can often be minimized through the use of physical and virtual simulations in the laboratory. Simulations also allow the use of high-precision data collection systems, such as force plates and motion capture systems that are impractical for many field environments. Proper participant selection and the fidelity of the simulation is key to eliciting subject responses and ensuring the validity of the data.

Disclaimer: The findings and conclusions in this report are those of the authors and do not necessarily represent the official position of the National Institute for Occupational Safety and Health, Centers for Disease Control and Prevention.

References

1. Simeonov, P., Hsiao, H., Powers, J., Ammons, D., Kau, T., Amendola, A.: Postural stability effects of random vibration at the feet of construction workers in simulated elevation. Appl. Ergon. **42**(5), 672–681 (2011)
2. Simeonov, P., Hsiao, H., Dotson, B.W., Ammons, D.: Height effects in real and virtual environments. Hum. Factors **47**(2), 430–438 (2005)
3. Cleworth, T.W., Horslen, B.C., Carpenter, M.G.: Influence of real and virtual heights on standing balance. Gait Posture **36**, 172–176 (2012)
4. Wang, S., Liu, Y., Hoerter, M., Moreland, J., Page, G., Krotov, Y., Crites, S., Zhou, C.: Development of a safety training simulator for fall protection. In: AISTech-Iron and Steel Technology Conference Proceedings, United States, vol. 1, pp. 45–50, 8–11 May 2017
5. Bloswick, D.S., Chaffin, D.B.: An ergonomic analysis of the ladder climbing activity. Int. J. Ind. Ergon. **6**(1), 17–27 (1990)
6. Pliner, E.M., Beschorner, K.: Effects of ladder climbing patterns on fall severity. Proc. Hum. Factors Ergon. Soc. Annu. Meet. **61**(1), 940–944 (2017)
7. Pliner, E.M., Campbell-Kyureghyan, N.H., Beschorner, K.E.: Effects of foot placement, hand positioning, age and climbing biodynamics on ladder slip outcomes. Ergonomics **57** (11), 1739–1749 (2014)
8. Schnorenberg, A.J., Campbell-Kyureghyan, N.H., Beschorner, K.E.: Biomechanical response to ladder slipping events: effects of hand placement. J. Biomech. **48**(14), 3810–3815 (2015)
9. Hanson, J.P., Redfern, M.S., Mazumdar, M.: Predicting slips and falls considering required and available friction. Ergonomics **42**(12), 1619–1633 (1999)
10. Lockhart, T.E., Woldstad, J.C., Smith, J.L.: Effects of age-related gait changes on the biomechanics of slips and falls. Ergonomics **46**(12), 1136–1160 (2003)
11. Chiou, S.S., Turner, N., Weaver, D.L., Haskell, W.E.: Effect of boot weight and sole flexibility on gait and physiological responses of firefighters in stepping over obstacles. Hum. Factors **54**(3), 373–386 (2012)
12. Cikajlo, I., Matjacic, Z.: The influence of boot stiffness on gait kinematics and kinetics during stance phase. Ergonomics **50**, 2171–2182 (2007)
13. Gronqvist, R.: Mechanisms of friction and assessment of slip resistance of new and used footwear soles on contaminated floors. Ergonomics **38**(2), 224–241 (1995)
14. Pollard, J.P., Heberger, J.R., Dempsey, P.G.: Slip potential for commonly used grated metal walkways. IIE Trans. Occup. Ergon. Hum. Factors **3**(2), 115–126 (2015)
15. Pollard, J.P., Merrill, J., Nasarwanji, M.F.: Metatarsal boot safety when ascending stairs. In: IIE Annual Conference Proceedings, pp. 2064–2068 (2017)

The Effectiveness of Tactical Communication and Protection Systems (TCAPS) on Minimizing Hearing Hazard and Maintaining Auditory Situational Awareness

Jeremy Gaston$^{(\boxtimes)}$, Ashley Foots, Tim Mermagen,
and Angelique Scharine

United States Army Research Laboratory, Aberdeen, MD, USA
{jeremy.r.gaston.civ, ashley.n.foots.civ,
timothy.j.mermagen.civ,
angelique.a.scharine.civ}@mail.mil

Abstract. In military environments, maintaining Auditory Situational Aware-ness (ASA) and providing protection from hearing hazard are often dueling priorities. Traditional passive hearing protection devices (HPD's) can provide adequate protection to the soldier from impulsive and continuous noise hazards, but this can come at the cost of reduced ASA. Anecdotal reports indicate that many soldiers may forego HPD use entirely in an attempt to maximize ASA. However, unprotected ears can lead to temporary threshold shifts (TTS) and permanent threshold shifts (PTS) that can be much worse for ASA than HPD use. Tactical Communication and Hearing Protection Systems (TCAPS) are active electronic systems that can provide a potential solution to this problem by giving the soldier protection, environmental hearing and radio communications. This paper discusses evaluating the effectiveness of these systems through objective measures of attenuation and subject human sound localization.

Keywords: Sound localization · Sound attenuation
Tactical communications and hearing systems · Shock tube
Hearing protection device · Environmental hearing

1 Introduction

In military environments, maintaining Auditory Situational Awareness (ASA) and providing protection from hearing hazard are often dueling priorities. The need to provide adequate protection to warfighters is clear: hearing loss and tinnitus have been reported in the Veterans Benefits Administration Annual Benefits Report as the two most prevalent service-connected disabilities and have been for over a decade [1]. Although traditional passive hearing protection devices (HPD's) can provide adequate protection from impulsive and continuous noise, hazards soldiers frequently encounter, it can come at the cost of significantly reduced ASA [2], or at least perceived cost; anecdotal reports from soldiers indicate that many may forego HPD use entirely in an

© Springer International Publishing AG, part of Springer Nature (outside the USA) 2019
D. N. Cassenti (Ed.): AHFE 2018, AISC 780, pp. 382–391, 2019.
https://doi.org/10.1007/978-3-319-94223-0_36

attempt to maximize ASA [3]. However, unprotected ears can lead to temporary threshold shifts (TTS; [4]) and permanent threshold shifts (PTS; [4]) that can be much worse for ASA than HPD use would yield. A new class of protection systems called Tactical Communication and Protection Systems (TCAPS) offer a potential solution to these dueling priorities through systems that can provide adequate protection from military noise, while also maintaining ASA through an electronic environmental hearing restoration mechanism and integrated radio communications. This paper describes methods for characterizing TCAPS system using objective measures of sound attenuation performance and the subjective measure of single-source localization to assess ASA.

The form factor of hearing protection devices can be either in-ear (IE) or over-the-ear ((OE) see Fig. 1). Regardless of form factor, these can further be classified by whether the devices are passive or have active electronics. Passive HPD's are the most common types and can be linear and non-linear in the amount of attenuation provided. Passive linear HPD's provide a relatively fixed amount of attenuation across a range of sound levels. They provide passive protection by forming a physical barrier that prevents transmission of sound through the ear canal. When worn properly, these devices can be very effective at mitigating hearing hazard. However, depending on the amount of attenuation provided, they can make it difficult to hear lower levels of sounds and to understand speech [5, 6]. HPD's in general are also known to alter binaural and monaural spectral cues by covering, plugging, or obscuring the pinna which has the potential to negatively impact auditory localization. Like [7–10] traditional passive protectors, non-linear HPD's can provide a fixed amount of attenuation for continuous noise. Level dependent devices like the Combat Arms HPD contain a filter that mechanically prevents transmission of noise through the device at levels above a certain threshold [11]. Usually this filter can be blocked so that the device provides full passive attenuation, or left open so that the device only attenuates impulsive noise above a certain threshold.

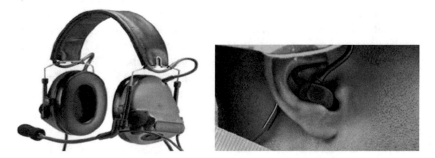

Fig. 1. Left: example of Over-the-Ear (OE) TCAPS system. Right: example of In-the-ear (IE) system.

Active HPD's use electronics to mitigate noise hazards reaching the user's ears. Noise canceling headphones are one type of device that uses active means to reduce noise by producing a signal that is 180° out of phase with the sound environment.

These devices can be effective at reducing exposure to continuous noise, but have little effect on impulsive noise and still have a largely negative impact on ASA. A type of active device that offsets the loss of ASA, is the class of active HPD systems called Tactical Communication and Hearing Protection Systems (TCAPS). These active electronic systems protect the warfighter from hazardous noise, provide communications, and restore some of the ambient hearing loss that affects ASA. TCAPS systems have an active environmental hearing restoration system that operates by filtering the sound environment and passing it through an IE or OE type headset. It is this "pass-thru" that helps to maintain ASA. The device creates a physical barrier, providing passive attenuation by default. The active environmental hearing restoration system filters sound, compressing high-level continuous noise and shutting off transmission of high-level impulsive noise. These systems typically also have the capability for integration into radio communications, but there are "lite" versions that do not have this capability.

2   Measuring TCAPS Performance

Here we describe the results of a test we conducted to evaluate the performance of several TCAPS candidates in measures of attenuation and auditory localization as a measure of ASA. The test consisted of one OE (OE-1) and three IE (IE-1, 2, 3) TCAPS systems and an IE passive nonlinear system (IE-P) as a benchmark for comparison.

2.1   TCAPS Attenuation

The established method for measuring the attenuation of hearing protectors is the real-ear attenuation standard outlined in ANSI/ASA S12.6 [12], and is commonly known as the REAT standard. To evaluate the effectiveness of hearing protection, one measures the level of sound at the listener's hearing threshold with and without the hearing protector. However, the REAT is designed to measure attenuation at threshold levels and most TCAPS are designed to provide normal "pass-thru" of sounds below approximately 85 dB(A). As a result, the attenuation measured at threshold is not representative of the attenuation at very-high impulse noise levels. For active devices like the TCAPS, it is preferable to use an auditory test fixture (ATF) that simulates the properties of the human outer ear to measure attenuation (see Fig. 2). The methods for assessments using ATF's are outlined in ANSI/ASA S12.42 [13]. Using this method, we measured the attenuation of steady-state and impulsive noise.

Steady-State Attenuation. Steady-state attenuation was measured using a G.R.A.S 45CB ATF in a diffuse, spectrum balanced sound field created at the U.S. Army Research Laboratory (ARL) in the Dome Room of the Environment for Auditory Research (EAR). Sound levels were collected through the ATF with noise levels (at the ears/center of room) of 105, 95, 85, and 75 dB. The sound data were processed into one-third octave bands and overall values were corrected to free field for analysis. Sound data for each system was collected in an active positive amplification mode, a unity mode (mode typically equal to normal hearing), a negative amplification mode,

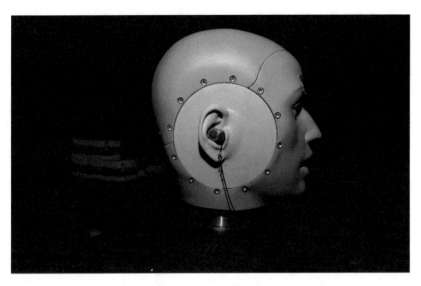

Fig. 2. G.R.A.S 45CB acoustic test fixture depicted with an in-ear TCAPS system inserted into the ear simulator.

and a passive mode (system off). The exact system setting in reference to the four modes of operation differed slightly from system-to-system. In passive conditions, the attenuation of the majority of the systems were high enough that the measurements were in the noise floor of the ATF for the lower background noise levels. As such, only the attenuation levels for the 105 dB noise level are presented for passive conditions. When the electronics were active for the systems, this was not a concern, as the levels were always much greater than the noise floor.

As seen in Fig. 3, the three IE systems provided a variable level of attenuation that resulted in a fixed level under the protector across 85 dB to 105 dB background levels. This is the expected outcome for the unity settings of these systems that serve to compress high-level audio signals to no more than 85 dB(A). In contrast, the OE-1 and IE-P systems provided a relatively fixed level of attenuation that resulted in a variable level under the protector. This pattern is the expected outcome for passive systems, such as would be expected for the IE-P system for continuous noise.

Impulse Attenuation. Impulse attenuation was also measured on a G.R.A.S. Sound and Vibration 45CB ATF. Impulse test sounds were generated by ARL's shock tube called the Pneumatic Impulse Noise System (PINS) as shown in Fig. 4. The generated peak levels at the test fixture recording location were measured to consistently be 165 dB (± 3 dB). Bare-head measurements at a free-field impulse level of 155 dBP were taken on the 45CB ATF in order to calculate the ATF to free-field transfer function. The transfer function was necessary to correct all measurements to free-field per ANSI 12.42 because there could be an 8- to 14-dB gain in peak pressure level because of the resonances of the ATF's simulated ear canal coupler, which also occurs in human ear canals.

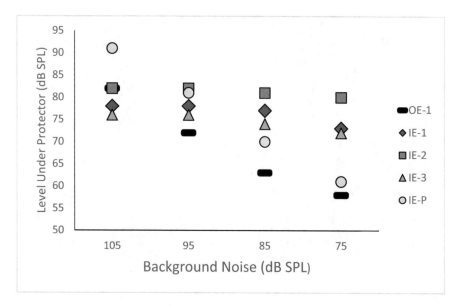

Fig. 3. Average attenuation level under the protector for unity system settings as a function of background noise level.

Fig. 4. Typical setup for impulse noise testing consisting of the PINS shock tube and 45CB ATF.

Measurements were taken on each system in four different modes. The modes used during testing were active, helmet worn-active, helmet-worn-passive, and passive. Helmet active was the candidate in a powered on state with the 45CA or 45CB ATF donning a standard issue Advanced Combat Helmet (ACH). Helmet passive was the candidate in a powered off state with the 45CA or 45CB ATF donning an ACH. All active measurements were taken at the highest pass through setting for each system. Figure 5 shows the average impulse peak attenuation for each of the TCAPS systems. All of the IE systems provided better attenuation than the OE system, with the IE-2 providing the best attenuation with an average peak attenuation of 59 dB.

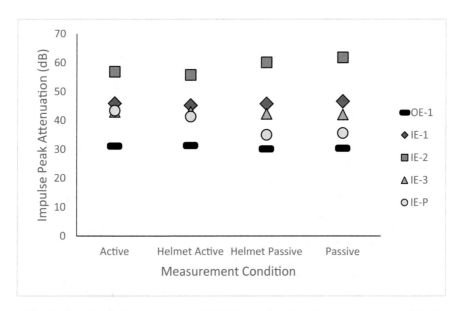

Fig. 5. Impulse peak attenuation for TCAPS as a function of measurement condition.

2.2 TCAPS Localization

TCAPS can affect ASA in different ways such as the ability to detect, recognize, identify, localize and communicate, [6] Here we focus on auditory localization. Although these devices offer environmental hearing restoration, a number of device properties can affect the degree to which auditory localization is negatively impacted. The physical changes to the profile of the ear caused by wearing TCAPS alters the spatial information available to the user and the may be the source of perceived loss of auditory situation awareness [3]. Studies consistently show that head-worn equipment impairs auditory localization performance [14, 15]. Consequently, the research and development of new helmets and TCAPS have included evaluations of their effects on auditory localization performance [11, 16–18]. Currently, there has been no officially established criteria for acceptable auditory localization performance. However, we have developed with our partners in the Test and Evaluation (T&E) community a method for evaluating the effects of TCAPS on auditory localization performance that has been used to down select fieldable TCAPS systems. For this particular test, the selecting authority defined a requirement that no system exceed greater than an average of 20° of error across loudspeaker locations.

Localization Methods. Six civilians with former military training served as test participants in a localization study on the four TCAPS systems described above, the IE-P system and a bare head condition. All testing was completed in a hemi-anechoic chamber that was equipped with a horizontal circular array of loudspeakers with a diameter of 6 m. The array contained 180 Meyer Sound MM-4XP loudspeakers arranged every 2° of angular separation. The present test used eight of the locations on the array, with the locations separated by approximately 45° (0, 44, 90, 134, 180, 224, 270, and 316°). In the center of the room, there was a listener station with a freely-rotating adjustable chair instrumented with a laser pointer and optical shaft encoder that determined the location of the chair when the participant's response was recorded.

Two types of stimuli were used during localization testing. The first was an AK47 muzzle blast impulse and the second was a 500-ms burst of white noise. Each signal was presented ten times from each speaker in each block of trials for a total of 160 trials. The white noise was presented at 70 dB and the impulse was presented at 100 dBP. Each participant was tested using each of the systems (four TCAPS and IE-P) as well as their bare head. Three samples of each of the five systems were tested. Thus, there were three blocks of trials for each system, in each of two system modes for a total of six blocks of trials per participant. Over the course of the test, this resulted in a total of 120 trials per loudspeaker position and a grand total of 960 trials per system for each test participant. All testing was completed with the participants wearing the Advanced Combat Helmet and the order of test blocks was randomized. The TCAPS were tested at two different operational modes for localization testing which included: unity mode and a positive amplification mode. Unity mode was an active (system on) mode typically equal to normal hearing. Positive amplification mode was an active mode of increased sound amplification relative to unity mode. The baseline IE-P was testing in both the open and closed positions. Following a stimulus presentation, test participants made responses by turning the participant chair and laser pointer to the location they thought the sound had come from. Responses were calculated as the angular error between the listener response and the actual location the sound came from.

Localization Results. The uncorrected localization error for both the AK47 and white noise stimuli are presented in Fig. 6. No system met the 20° requirement for either stimulus when evaluated for average localization error across sound source locations. In other words, no system or even the bare head was able to meet the 20° requirement for the AK47 stimuli. In contrast, for the white noise stimulus all but the OE-1 and the IE-3 systems were able to meet the 20° requirement. Across both signal types, the IE-1 was the best performing TCAPS and the OE-1 was the worst performing system. This makes sense since the IE-1 had the smallest profile and microphone baseline distance of any of the systems and the OE-1 the largest profile and baseline.

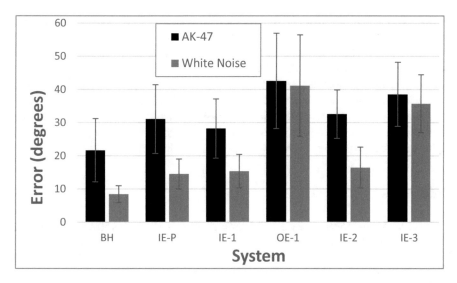

Fig. 6. Localization error for each TCAPS system, IE-P and bare head as a reference for the AK-47 and white noise stimuli.

3 Discussion

While the measures of attenuation presented here are formalized in a standard [13], there is no current "official" standard for assessing localization performance for TCAPS or HPD's in general. Recently, a working group has been formed to develop a standard method of measuring auditory localization as a function of head-worn devices [19]. There are four methods described in this standard. The first is a screening method that uses 8 loudspeakers arranged in four pairs placed orthogonally. The listener responds by naming the source, either verbally, or by selecting it on a tablet. By sitting in each of two orientations, localization can be measured for 16 locations. This test gives a measure of accuracy and allows for the quantification of errors due to blur (near misses), or front-back reversals. Method 2 uses a horizontal array of 36 equally spaced loudspeakers, and requires the use of a tracker and a pointing system. As in Method 1, listener responses are counted as correct or incorrect, and errors are quantified according to whether they are due to blur or reversals. Method 3 is an aurally guided visual detection task that uses the same array of loudspeakers as Method 2. It is intended to provide a measure of the functional impact of hearing protection on target detection time. Method 4 is a high resolution method using 180 loudspeakers spaced at 2° intervals. All four methods use the same sound stimuli: a brief 250 ms pulse consisting of pink noise with 10 ms onset and offset ramps; and a 7000 ms pulse train consisting of repetitions of the short stimulus. The 250 ms pulse is to be used in a 2:1 ratio with the longer pulse train. The standard also describes the acoustical standards for the test facility and listener hearing and performance requirements.

Finally, there is the obvious question of whether single-source localization is generalizable as a measure of ASA to everyday listening conditions. Meaningful sound events rarely occur in isolation. Rather, they are more often embedded in a complex sound context, and more complex localization contexts may be a better measure of real-world ASA. Basic research in our lab has looked at how increasing real-world relevant complexity can have significant effects on performance. For example, listeners can miss changes in spatial sound arrays composed of multiple environmental sounds even when the sound are clearly identifiable in isolation [20]. More directly related to localization, there is evidence that the sensory-level changes induced by TCAPS may be significant in single-source localization, but the perceptual-level effects of multi-sound contexts may swamp any sensory level differences [21].

References

1. D. o. V. Affairs: Veterans Benefits Administration. Annual Benefits Report - 2010 (2012)
2. Brown, A.D., Beemer, B.T., Greene, N.T., Argo, T., Meegan, G.D., Tollin, D.J.: Effects of active and passive hearing protection devices on sound source localization, speech recognition, and tone detection. PLoS ONE 10(8), e0136568 (2015)
3. Clasing, J., Casali, J.G.: Warfighter auditory situation awareness: effects of augmented hearing protection/enhancement devices and TCAPS for military ground combat applications. Int. J. Audiol. 53, S43–S52 (2014)
4. Henderson, D., Hamernik, R.P., Dosanjh, D.S., Mills, J.H.: Noise-Induced Hearing Loss, pp. 41–68. Raven, New York (1976)
5. Suter, A.H.: The ability of mildly hearing-impaired individuals to discriminate speech in noise (AMRL Report No. TR-78-4), Aerospace Medical Research Laboratory, Wright-Patterson Air Force Base, OH (1978)
6. Suter, A.H.: The effects of hearing protection on speech communication and the perception of warning signals (HEL-TM-2-89), US Army Human Engineering Laboratory, Aberdeen Proving Ground, MD (1989)
7. Abel, S., Hay, V.: Sound localization the interaction of aging, hearing loss and hearing protection. Scand. Audiol. 25(1), 3–12 (1996)
8. Atherley, G., Noble, W.: Effect of ear-defenders (ear-muffs) on the localization of sound. Occup. Environ. Med. 27(3), 260–265 (1970)
9. Noble, W.G., Russell, G.: Theoretical and practical implications of the effects of hearing protection devices on localization ability. Acta Otolaryngol. 74(1), 29–36 (1972)
10. Vause, N.L., Grantham, D.W.: Effects of earplugs and protective headgear on auditory localization ability in the horizontal plane. Hum. Factors 41(2), 282–294 (1999)
11. Scharine, A.A., Weatherless, R.A.: US Marine Corps level-dependent hearing protector assessment: objective measures of hearing protection devices (ARL-TR-7203. DTIC No.: ADA597684), Aberdeen Proving Ground (MD), Army Research Laboratory (US) (2014)
12. ANSI/ASA S12.6: American National Standard Methods for Measuring the Real-Ear Attenuation of Hearing Protectors. Acoustical Society of America, New York City (2016)
13. ANSI/ASA S12.42: American National Standard Methods for Measurement of Insertion Loss of Hearing Protection Devices in Continuous or Impulsive Noise Using Microphone-in-Real-Ear or Acoustic Test Fixture Procedures. Acoustical Society of America, New York City (2010)
14. Scharine, A.A.: Degradation of auditory localization performance due to helmet ear coverage: the effects of normal acoustic reverberation (ARL-TR-4879), Aberdeen Proving Ground, U.S. Army Research Laboratory (2009)

15. Scharine, A.A., Binseel, M.S., Mermagen, T., Letowski, T.R.: Sound localisation ability of soldiers wearing infantry ACH and PASGT helmets. Ergonomics **57**(8), 1222–1243 (2014)
16. Scharine, A.A., Domanico, M.C., Foots, A.N., Mermagen, T.J., Weatherless, R.A.: Auditory localization performance with Gamma Integrated Eye and Ear Protection. (ARL-TR-7914), U.S. Army Research Laboratory, Aberdeen Proving Ground, MD (2016)
17. Scharine, A.A., Weatherless, R.A.: Evaluation of variants of 3 M Peltor ComTAC tactical communication and protection system (TCAPS) headsets: measures of hearing protection and auditory performance. (ARL-TR-6667), U.S. Army Research Laboratory, Aberdeen Proving Ground, MD (2013)
18. Scharine, A.A., Weatherless, R.A.: Helmet Electronics & Display System-Upgradeable Protection (HEaDS-UP) Phase III Assessment: Headgear Effects on Auditory Perception (ARL-TR-6723), U.S. Army Research Laboratory, Aberdeen Proving Ground, MD (2013)
19. BSR/ASA S3.71: Draft American National Standard: methods for measuring the effect of head-worn devices on free-field directional sound localization in the horizontal plane. American National Standards Institute/Acoustical Society of America, New York City (201X)
20. Gaston, J., Dickerson, K., Hipp, D., Gerhardstein, P.: Change deafness for real spatialized environmental scenes. Cogn. Res. Princ. Implic. **2**(1), 29 (2017)
21. Gaston, J., Mermagen, T., Foots, A., Dickerson, K.: Auditory localization performance in the azimuth for tactical communication and protection systems. J. Acoust. Soc. Am. **127**(4), 1 (2015)

Improving Safety Training Through Gamification: An Analysis of Gaming Attributes and Design Prototypes

Leonard D. Brown$^{(\boxtimes)}$ and Mary M. Poulton

Lowell Institute for Mineral Resources, University of Arizona,
P.O. Box 210012 1235 E. James E. Rogers Way, Rm. 229,
Tucson, AZ 85721, USA
{ldbrown, mpoulton}@email.arizona.edu

Abstract. New approaches are needed to improve outcomes for safety training in hazardous industries. We use an evidence-driven approach to identify the key attributes of serious games that have the potential to improve safety training. Following a detailed needs assessment, we identified four major themes of usability problems which may be addressed through gamification: Limited accessibility, lack of context, lack of consequence, and absence of practicum. Based on our analysis, a series of application prototypes was developed to improve safety training in the mining industry. In particular, we discuss *Harry's Hard Choices*, a game for mining emergency response training. Pilot tests indicate high levels of user satisfaction and engagement and anecdotal evidence of training transfer.

Keywords: Serious games · Training · Mining · Usability · Contextual Inquiry

1 Introduction

Many industries worldwide rely increasingly on workplace training. Training was a $61.1B industry in the U.S. in 2014, with safety-oriented training comprising 38.7% of the global training market [1]. With respect to safety, the global mining industry is an excellent case study, as it works to eliminate fatalities and serious injuries. Despite the time and money spent on training, efforts are not as effective as they could be. In particular, the quantity and complexity of information covered in mine safety training, the unique technical vocabulary of the industry, the dynamic nature of the mining workplace, and the changing demographics of the workforce all suggest a need for major changes in the way mine safety training is performed.

Lutz and Lutz [2] analyzed U.S. mine fatalities in 2013, 2014, and 2015 and their relationship to training. In 2013, 12 fatality reports noted that the miners killed lacked some type of training. In two of the 12 fatalities, basic safety training was missing. In the remaining 10 fatalities, task training was either completely missing or significant portions of it were missing. In 2014 there were 11 fatalities in which investigators at the Mine Safety & Health Administration (MSHA) noted inadequate task training. Furthermore, MSHA investigation reports are often vague about what type of training had

© Springer International Publishing AG, part of Springer Nature 2019
D. N. Cassenti (Ed.): AHFE 2018, AISC 780, pp. 392–403, 2019.
https://doi.org/10.1007/978-3-319-94223-0_37

been performed. An analysis of surveillance data provided by the National Institute for Occupational Safety and Health (NIOSH) [3] showed that most trainers were still using didactic (i.e. lecture-based) methods, which research has shown to be of limited effectiveness [4]. Peters, Vaught, & Mallett summarize the situation:

"Collectively, our nation's miners sit through millions of hours of mandated S&H training each year, and mining companies spend millions of dollars to provide this training. Unless effective training materials and methods are used, miners are unlikely to learn what they need to know to actually help reduce their risk of suffering occupational injury and illness. Many miners sit through the same training lectures and films year after year in order to fulfill the requirements of the law. In these situations, their 'training' ends up being a very unfortunate waste of time and resources i.e., a wasted opportunity" Peters et al. [5].

Salas *et al.* [6] find that "properly designed training works and... the way training is designed, delivered and implemented can greatly influence its effectiveness." A large body of research has been published on strategies to improve learning for various types of occupational training [7–10]. Serious games are increasingly being used for training in fields ranging from medical science to homeland security. Building upon work by NIOSH to use gamification for underground mine map reading training [11], we have pursued an in-depth study of serious games for safety training. In this article, we identify major usability problems in mine safety training and discuss our preliminary work to address those problems using serious games.

2 Related Work

A major goal in designing serious games is to identify the specific attributes and capabilities that will enhance learning. An overwhelming variety of game attributes and design patterns have been identified. For example, Garris et al. [12] suggested that six key attributes are necessary: Fantasy, rules/goals, sensory stimuli, challenge, mystery, and control. Wilson et al. [13] expanded these six to include adaptation, assessment, conflict, interaction (equipment, interpersonal, social), language/communication, location, pieces/players, progress and surprise, representation, and safety. Other researchers have proposed similar collections of attributes [14–17]. Although an exhaustive survey is not possible due to space considerations, we outline some key attributes of serious games relevant to this work.

Challenge. Challenge relates to the level of difficulty inherent in a game. Hays [18] listed challenge as a major factor contributing to user enjoyment and argued that challenge results from uncertain outcomes. Uncertainty can be designed into a game by using multiple levels of goals, hidden or incomplete information, and randomness [18, 19]. Furthermore, research suggests a need for an optimal level of challenge– that is, a balance between game play that is too easy and game play that is difficult [12]. Wilson *et al.* [13] hypothesized that, "As the challenge feature in a game increases, so will declarative knowledge and learner's retention of that knowledge. However, a point will be reached when too much challenge will hinder and decrease learning (i.e., an inverted U relationship)."

Fantasy. Fantasy, as defined by Wilson *et al.* [13], is a "make-believe environment, scenario, or characters. It involves the user in mental imagery and imagination... and analogies for real-world processes. The user is required to take on various roles in which they are expected to identify." Activities in the game environment allow users to explore situations that are detached from daily realities [12]. The fantasy world can model specific parts of the real world as necessary to meet game objectives. Fantasies allow users to assume different roles and explore different mindsets [9]. Garris et al. [12] suggested that "material may be learned more readily when presented in an imagined context that is of interest than when presented in a generic or decontextualized form."

Flow. An appealing aspect of serious games rests in their ability to engage an audience. Csikszetmihalyi [20] describes a state in which "people become so involved in what they are doing that the activity becomes spontaneous, almost automatic; they stop being aware of themselves as separate from the actions they are performing." For this state, Csikszetmihalyi [20] suggested the notion of *flow* as an optimal experience that merges individuals' actions with activity. Furthermore, Hays [18] described flow as a sense of enjoyment, where "one 'loses' oneself in the activity." Research suggests that achieving a "flow state" may have significant advantages for learning in both the short term and long term [21, 22].

Rules. Game rules define the structure, boundaries, and goals of a game [12]. Rules may be aligned with reality or completely artificial, depending on the circumstances of the game. Garris et al. [12] noted that "goals should be clearly specified, yet the possibility of obtaining that goal should be uncertain. Games should employ progressive difficulty levels, multiple goals, and a certain amount of informational ambiguity to ensure an uncertain outcome."

Assessment. We define *assessment* as a measure of achievement within a game [13]. It is enabled by the tracking of in-game activities. A cycle of practice, re-teaching, and revision can be used to create embedded assessments in which students are engaged in the content and achievement is improved [23]. In particular, Gee [14] suggested that games allow users to engage in "reflective practice," through a four-step cycle of probing, hypothesizing, reprobing, and rethinking. Shute *et al.* [23] described an ideal situation that called "stealth assessment," in which the embedded assessments are "so seamlessly woven into the fabric of the learning environment that they are virtually invisible."

3   Field Studies

A major goal of this work was to identify the endemic usability problems of safety training. As a first step, we conducted a three-year, comprehensive field study of current training pedagogy. Our objectives were two-fold: First, we wanted to gain a holistic view of training – how the individual parts of the training regimen came together to shape the miners' understanding of safety; second, we wanted to examine the protocols, data sets, and flow of information that workers experienced.

Our field studies employed a data-driven method called Contextual Inquiry (CI), which is the first step in a usability engineering process called Contextual Design [24]. As a participatory design process, CI immerses a multi-disciplinary team of experts in the application domain, allowing them to closely engage with end users and see problems from the users' points of view. Each field study involved three stages. In the first stage, our research team participated in each of the training activities as a form of apprenticeship. The second stage involved observation, in which we watched the interactions between trainers and students in the classroom without disrupting the natural flow of their activities. Observation periods generally ran from 4 to 6 h at a time. The third stage was an open-ended debriefing, in which we interviewed both students and trainers. Many of our questions were prompted either by our personal experiences during apprenticeship and our observations in stage 2.

The mining industry is regulated by Title 30 of the U.S. Code of Federal Regulations (CFR), which directs MSHA to oversee all mining activities in the United States. The CFR Title 30, Parts 46 and 48, outlines safety training requirements, including courses for new miners, annual refreshers, and newly hired experienced miners. At least 27 topics are specified for each of these courses, although there is substantial overlap in content. Note that compliance is based on seat time alone. As it was impractical to perform Contextual Inquiry on all of the required training topics, we selected a subset in consultation with a panel of industry experts in mine safety. Representative activities were selected to include didactic training (e.g. lecture-based with slides), simulation software, and training videos, for the following:

- *Introduction to the Workplace.* New and experienced miner training courses must include an introduction to the work environment. Mine sites are complex and dynamic, so training activities must include tutorials on mine structure, methods, equipment, maps, and terminology.
- *Situational Awareness.* A survey of mine fatalities suggests that many accidents involve similar workplace hazards and risk-taking behaviors. As a result, there are numerous training topics which emphasize situational awareness. Specific training activities included surveying fatal accidents, recognizing hazards, and identifying mitigation strategies.
- *Emergency Preparedness.* Potential emergencies can result from a variety of events, such as explosions, fires, inundation, and ground falls. Emergency preparedness training is a two-part process that involves both generic classroom instruction and site-specific drills. Each mine must also maintain a first responder team which is required to conduct monthly refreshers.

4   Usability Themes

Contextual Inquiry [24] uses a data-driven discovery process to quantify the design space and identify significant usability problems. In our case, CI allowed us to transform a collection of field study observations and measurements into a hierarchical model, called an Affinity Diagram, which suggested some of the most significant obstacles to mine safety training. This analysis served as a starting point for designing more effective training through serious games.

4.1   Empirical Evidence

In our field studies, we collected a large body of empirical evidence. In particular, three types of evidence were collected:

- *Observations.* Observations were collected during the apprenticeship and "pure observation" phases and were based on events that were directly apparent and required no interpretation. Examples include "Text on slide is hard to see" and "No explanation given for technical terminology." Note that observations may or may not have related comments or inferences.
- *Comments.* Comments represent a direct form of user feedback. Comments consisted of verbal and written quotes obtained during the debriefing session. They illustrate users' opinions outright. Examples include "The instructions were confusing" and "We do it differently at my site." Where feasible, comments were collected from both trainers and trainees.
- *Inferences.* Inferences stem from observations that led to specific lines of questioning in the post-session debrief. They are based on the interpretations of our research team and further validated by user feedback. Inferences are valuable because they offer deeper insight into the user experience and can suggest causality. Examples include "The user is visibly frustrated that he cannot find his way" and "Dealing with computer interface distracted user from task goal."

Over one thousand pieces of evidence were collected in our three field studies. Through an aggregation process, we then reduced the evidence to 473 discrete data points. Aggregation involved eliminating duplicate data, such as observations that were recorded by different members of the team for the same application, coalescing similar inferences into one interpretation, and removing multiple instances of analogous user comments which might have slightly different wording. However, we did not eliminate observations, inferences, or comments that were similar if they occurred in multiple field studies, as this would have diminished the strength of the evidence.

4.2   Affinity Diagram

Based on the three types of evidence that we collected, an Affinity Diagram [24] was constructed to identify the major themes of usability problems. The diagram is built inductively from the ground up; similar evidence is recursively grouped together and labeled. The diagramming process results in a hierarchy of problem categories which at the top level identifies the principal themes of usability problems. Note that a piece of evidence may be placed into more than one grouping as necessary. A strength of the Affinity Diagram is that it does not use preconceived groups; the problem categories arise spontaneously based on the intrinsic structure of the evidence. The number of pieces of evidence in a category also suggests the frequency of the problem.

The abridged Affinity Diagram may be found in Fig. 1. For brevity, the diagram omits the individual pieces of evidence that were used to derive its structure; the abridged diagram represents the top three levels of the hierarchy. Our analysis suggests that there are 45 specific usability problems in safety training (*bottom level of chart*). The mean was 10.47 data points, with a standard deviation of 4.21, per usability

LIMITED ACCESSIBILITY	
Obstacles to acceptance	
	The age gap
	Aversion to (new) technology
	The time it takes
	Ergonomic risk factors
Beyond the user's skill set	
	Difficult tasks
	Cognitive overload
	Lack of remediation
Limited language & workplace literacy	
	A multi-lingual audience
	Limited language understanding
	Presumption of workplace literacy
	Variation across work sites
Difficult for novice users	
	Apparent learning curve
	The need for technology tutorials
	A multitude of standards
	Lack of concise mechanics
	Clumsy context switches
Overly limited interface mechanics	
	Artificial constraints
	Necessity of tactile feedback

LACK OF CONSEQUENCES	
No cause-and-effect	
	Unclear rules and boundaries
	Confusing timelines and events
Scenarios that are static or inflexible	
	No workplace dynamics
	Lack of hypotheticals
Confusion about task progress	
	Where do I go?
	What do I do?

LACK OF CONTEXT	
Media do not meet needs of task	
	Problems with computer media
	Problems with print & other media
Details do not meet needs of task	
	Missing or unavailable data
	Too little detail
	Too much detail
Lack of spatial awareness	
	Limited point of view
	Disorienting workplace structure
Users cannot relate	
	Stylized or abstract
	Lack of real world examples
	Unauthentic usage of equipment

ABSENCE OF PRACTICUM	
Inadequate story	
	Standardized content
	Believable delivery
Lack of motivation	
	Boredom among users
	Users as competitors
Few opportunities for crew interaction	
	Importance of team activities
	Few opportunities for teamwork
Hard to communicate	
	Verbal communication problems
	Lack of workplace-specific forms
Need for hands-on experimentation	
	Few chances to practice techniques
	Lack of substantive feedback
	Few ways to evaluate competency

Fig. 1. Abridged Affinity Diagram. A hierarchical clustering of the evidence from our field studies suggests that there are four principal themes usability problems in mine safety training.

problem. Together, these 45 specific problems illustrate 17 major categories (mid-level). We have further grouped the 17 major categories into 4 usability themes for safety training (top level):

1. *Limited Accessibility*. Promote acceptance across a range of worker demographics, engaging a new generation of miners in addition to experienced miners. Accommodations should be made to address computer and workplace literacy issues and second language learners.
2. *Lack of Context*. Provide a meaningful context and level of detail to illustrate important relationships. In particular, training must be grounded in the familiar, relating new concepts to each worker's experience.
3. *Lack of Consequences*. Illustrate the consequences of choices. Workers require a better understanding of timelines and causality if they are to recognize hazards and anticipate the outcomes of good and bad choices.
4. *Absence of Practicum*. Emphasize story-driven scenarios and evaluation. Hands-on activities reinforce learning and allow for the assessment of worker safety competencies.

5   Game Prototypes

Many of the usability problems outlined above can be resolved through active learning techniques [4] using well-designed serious games. In particular, we have developed a suite of computer-based video games for mine safety, including *Harry's Hard Choices*, which teaches mine emergency response. A thorough treatment of this game prototype and several others may be found in Brown [25].

5.1   Harry's Hard Choices

Our serious game *Harry's Hard Choices* (*HHC*) has its inspiration in a series of NIOSH paper exercises for mine emergency response, in which miners "role play" through a disaster story [26, 27]. This type of experiential training allows the miners to think about options in a relevant context. However, the paper exercises do have substantial limitations: They cannot give users a visceral experience that simulates the chaos of a real emergency; they are serialized based on the sanctioned answer to each question; they provide limited information about the emergency situation; and they cannot show cause-and-effect in response to users' free will.

In *HHC*, the user assumes the role of a foreman tasked with evacuating his crew in the face of a rapidly unfolding mine disaster (Fig. 2). Similar to the paper scenarios, our game is set in an underground coal mine and involves a mine fire. However, we have de-serialized the events of the paper scenarios and greatly extended them to create an immersive sandbox where users have free will to explore the environment, use equipment, and engage in a variety of crew-based interactions. Based on realistic situations that miners may face in an emergency, a diverse range of hazards has been added, including roof falls, expiring and defective respirators, inoperative refuge chambers, flammable gases, broken escape lines, and electrical faults. Story events can occur in parallel and there are many possible outcomes. They can also be randomized, so that a different sequence of events is presented each time the game is run.

Fig. 2. *Harry's Hard Choices*. Save a crew of miners (*top*) from an unfolding disaster featuring smoke, fire, lethal gases, ground falls, faulty equipment, and many other hazards (*bottom*).

5.2   Game Attributes for Safety Training

To address the usability themes discovered in our field studies, we sought motivation from best practices in games for entertainment. Entertainment games can be very effective at immersing users in complex virtual worlds where they must spend hours performing sophisticated tasks to achieve game goals. The best games sell millions of copies and maintain a user base for many years; consider The Elder Scrolls V: Skyrim. There are many design patterns that can make games more effective; a non-exhaustive survey may be found in [28]. We have cross-walked the usability themes for safety training with the game attributes that were found to be effective for learning (Sect. 2) and aligned them with design patterns for entertainment games. We then implemented a series of these design patterns in our prototype, HHC (see Table 1).

As an example, consider the "Limited Accessibility" theme of Fig. 1. To address problems with user acceptance, we have applied design patterns that promote *challenge* and *flow* as outlined in Sect. 2. A set of game metrics was implemented, including a scoring system that awards or deducts points based on user decision making. A fatigue metric models exposure and exertion, similar to "hit points" in games for entertainment. The crew of non-player characters (NPCs) is likewise affected; crew members may be injured, incapacitated, or even die. A time metric models the sense of urgency appropriate to mine disasters; as such, time penalties are enforced for poor decisions and procrastination, which can lead to unwinnable conditions as the situation deteriorates. Finally, a morale metric models the impact of users' decision making on the NPC crew. As morale decreases, non-player characters will become increasingly stressed and irritated, culminating in the NPCs abandoning the group. The intent of these design patterns is to draw users into the game and give them a vested interest in the welfare of the crew that they are trying to save. Our preliminary user study, discussed below, suggests that the game metrics substantially increased users' engagement, compelling them to try again and again to improve their outcomes.

Table 1. A sample of game attributes and design patterns for safety training.

Usability theme	Game attribute(s)	Design pattern(s)
1 Limited accessibility	Challenge, flow	Game metrics
2 Lack of context	Fantasy	Emergent stories, microworlds
3 Lack of consequences	Rules	Game traps
4 Absence of practicum	Assessment	Game metrics

6 Pilot Test

Our field studies suggest that user acceptance and engagement are core requirements for training transfer (see Fig. 1, "Limited Accessibility"), a conclusion that is supported by prior research [12, 14, 22]. A pilot study of user acceptance was conducted using a "beta" version of HHC. This was the first time that domain users participated in rigorously play-testing the game with open-ended interaction and full gamification (Table 1). As such, the objective of the study was to get feedback on the experience, including story design, game metrics, and user interface. Our primary evaluation instruments included pre- and post-session questionnaires. The pre-session questionnaire covered user demographics and prior experience with gaming. Our post-session questionnaire adapted questions from standard usability instruments for the testing of interactive systems [29, 30]. Seven-point Likert scales were used.

A total of 12 users ($N = 12$) participated in our 90-min pilot study. In addition to 8 students from the Dept. of Mining and Geological Engineering at the University of Arizona, 4 professional miners were solicited from a mine site in Arizona. Seven of our users indicated that they had completed a new miner training course. Eight users were male and 4 were female, with a mean age of 28 years (range 18–56). The users reported a diverse range of technology and gaming experience; 25% reported themselves as "gamers." As such, the users were given a 25 min tutorial on the game objectives and mechanics and then allowed to play for up to 45 min. Their objective was to rescue as many crew members (up to 9) and obtain the highest score possible. There were no restrictions on game play, and users could restart at any time.

A summary of user feedback is shown in Fig. 3. Given the low sample size, we did not run a statistical test for significance. Users' opinions of the game were very favorable on all but one metric, "Satisfying vs. Frustrating." This result was attributed to a bug in the game's path-following algorithm that has since been corrected. Users also left favorable comments, such as "Great simulation" and "Fun learning experience." Users mentioned that they were "Surprised by the level of accuracy," and added that the game was "Very real," with "lots of options." A notable question posed to users was, "Would you be interested in playing this game again?" All 12 answered "Yes." In earlier testing, user feedback had been lukewarm, with none of our earlier users indicating the affirmative. The difference in outcomes rests primarily on the addition of game design patterns outlined in Table 1. In our previous test, *HHC* had been implemented as a vanilla simulation, instead of a full-fledged gaming experience. Furthermore, game play observations suggested that users' task performance and decision making improved through subsequent runs of the game. For instance, users' results improved from 0 or 1 crew members evacuated in the first iteration to a mean of 3.80 (out of 9) on the best attempt, which was typically the last. Note that the scenario parameters changed randomly on each run of the game to prevent cheating.

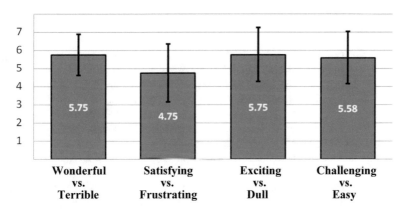

Fig. 3. Pilot study results for user satisfaction, $N = 12$ (*higher is better*).

7   Conclusions

This work represents a multi-disciplinary effort to improve training in safety-focused industries. We have taken an in-depth look at the mining industry, and, based on three years of field studies, identified four major themes of usability problems. In particular, our studies suggest that user acceptance and engagement are of utmost concern in safety training. To address these usability problems, we propose to draw from lessons learned in games for entertainment; that is, better use of gamification can lead to more effective training and better outcomes. Consider our prototype, *Harry's Hard Choices*, which draws users into mine emergency response with a challenging game experience full of realistic happenings, temperamental coworkers, and imperfect outcomes. Pilot tests suggest a high level of user satisfaction, which is a pre-requisite for high acceptance and engagement.

We are continuing our analysis of field study data with the goal of identifying design guidelines. We are also developing a suite of serious games that will provide effective training and competency assessment, in a medium that encourages discussion of safety culture. *Harry's Hard Choices* is now being deployed in industry training courses, and efficacy tests are under way. Although this work focuses on the mining industry, we hope that the insights and technologies developed here will be transferable to other industries where workplace safety is of principal concern.

Acknowledgements. We thank Michael Peltier for his significant contributions to the serious games development. We also thank Eric Lutz, Aly Waibel, Michelle Lutz, and the many industry experts who participated in our field studies. This work was supported by NIOSH award 1U60-OH010014, Science Foundation Arizona award SRG-0330-08, and MSHA Brookwood-Sago awards BS-22468-11-60-R-4, BS-23833-12-60-R-4, and BS-26353-14-60-R-4. The authors have disclosed a financial interest in Desert Saber, LLC to the University of Arizona. These interests have been reviewed and are being managed by the University of Arizona in accordance with its policies on outside interests.

References

1. Pappas, C.: Top 10 e-Learning Statistics for 2014 You Need to Know (2014). http://elearningindustry.com/top-10-e-learning-statistics-for-2014-you-need-to-know
2. Lutz, M., Lutz, E.A.: The future of training in a data-driven mining industry. In: SME Annual Meeting, 24 February 2016. SME, Phoenix (2016)
3. McWilliams, L.J., Lenart, P.J., Lancaster, J.L., Zeiner Jr, J.R.: National Survey of the Mining Population: Part II Mines. U.S. Department of Health & Human Services, CDC/NIOSH IC 9528, NIOSH Pub. 2012-153 (2012)
4. Dale, E.: Audio-Visual Methods in Teaching, 3rd edn. Dryden Press, New York (1969)
5. Peters, R.H., Vaught, C., Mallett, L.: A Review of NIOSH & US Bureau of Mines research to improve miner's health & safety training. In: Brune, J. (ed.) Extracting the Science: A Century of Mining Research, pp. 501–509. Society for Mining, Metallurgy & Exploration, Littleton (2010)
6. Salas, E., Tannenbaum, S., Kraiger, K., Smith-Jentsch, K.: The science of training and development in organizations: what matters in practice. Psychol. Sci. Public Interest **13**(2), 74–101 (2012)
7. Burke, L., Hutchins, H.: Training transfer: an integrative literature review. Hum. Resour. Dev. Rev. **6**(3), 263–296 (2007)
8. Juul, J.: The game, the player, the world: looking for a heart of gameness. In: Copier, M., Raessens, J. (eds.) Proceedings of Level Up: Digital Games Research Conference, Utrecht, Netherlands, pp. 30–45 (2003)
9. Malone, T.W., Lepper, M.R.: Making learning fun: a taxonomy of intrinsic motivations for learning. In: Snow, R.E., Farr, M.J. (eds.) Aptitude, Learning and Instruction III: Conative and Affective Process Analyses, vol. 3, pp. 223–253. Hillsdale, NJ, Erlbaum (1987)
10. Thiagarajan, S.: Team activities for learning and performance. In: Stolovitch, H.D., Keeps, E.J. (eds.) Handbook of Human Performance Technology, pp. 518–544. Jossey-Bass/Pfeiffer, San Francisco (1999)

11. Mallett, L.G., Orr, T.J., May, I., Cole, G.P., Lenart, P.J., Unger, R.L., Beshero, D.C., Hall, E. E., Vaught, C., Brune, J., Klein, K.: Underground Coal Mine Map Reading Training. U.S. Department of Health & Human Services, NIOSH/CDC, Pub. No. 2009-143c (2009)

12. Garris, R., Ahlers, R., Driskell, J.E.: Games, motivation, and learning: a research and practice model. Simul. Gaming 33(4), 441–467 (2002)

13. Wilson, K.A., Bedwell, W.L., Lazzara, E.H., Salas, E., Burke, C.S., Estock, J.L., Orvis, K. L., Conkey, C.: Relationships between game attributes and learning outcomes. Simul. Gaming 40(2), 217–266 (2009)

14. Gee, J.P.: What Video Games Have to Teach Us About Learning and Literacy, 2nd edn. Palgrave/Macmillan, New York (2007)

15. Pavlas, D., Bedwell, W., Wooten II, S.R., Heyne, K., Salas, E.: Investigating the attributes in serious games that contribute to learning: flow and grow. In: Proceedings of Human Factors and Ergonomics Society 53rd Annual Meeting, vol. 53, no. 27, pp. 1999–2003 (2009)

16. Van Staalduinen, J.P., de Freitas, S.: A game-based learning framework: linking game design and learning outcome. In: Khyne, M.S. (ed.) Learning to Play: Exploring the Future of Education with Video Games, pp. 29–45. Peter Lang, New York (2011)

17. Yusoff, A.: Conceptual framework for serious games and its validation. Ph.D. dissertation, University of Southampton, Southampton, UK (2010)

18. Hays, R.T.: The effectiveness of instructional games: a literature review and discussion. Naval Air Warfare Center Training Systems Division Pub. 2005-004 (2005)

19. Malone, T.W.: Towards a theory of intrinsically motivating instruction. Cogn. Sci. 5(4), 333–369 (1981)

20. Csikszentmihalyi, M.: Flow: The Psychology of Optimal Experience. Harper, New York (1990)

21. Lieberman, D.A.: What can we learn from playing interactive games? In: Vorderer, P., Bryant, J. (eds.) Playing Video Games: Motives, Responses, and Consequences, pp. 379–397. Erlbaum, Mahwah (2006)

22. Squire, K., Jenkins, H.: Harnessing the power of games in education. Insight 3(1), 5–33 (2003)

23. Shute, V., Ventura, M., Bauer, M., Zapata-Rivera, D.: Melding the power of serious games and embedded assessment to monitor and foster learning: flow and grow. In: Ritterfeld, U., Cody, M.J., Vorderer, P. (eds.) Serious Games: Mechanisms and Effects, 295–321. Routledge, New York (2009)

24. Beyer, H., Holtzblatt, K.: Contextual Design: Defining Customer-Centered Systems. Morgan Kaufmann, San Francisco (1998)

25. Brown, L.D.: Design, evaluation, and extension of serious games for training in mine safety. Ph.D. dissertation, University of Arizona, Tucson, USA (2015)

26. Vaught, C., Hall, E.E., Klein, K.: Harry's hard choices: mine refuge chamber training. U.S. Department of Health & Human Services, NIOSH/CDC, IC 9511 (2009)

27. Brnich, M.J., Vaught, C., Kowalski-Trakofler, K.M.: Man's mountain refuge: refuge chamber training. Report of Investigations RI 9685, U.S. Department of Health & Human Services, NIOSH/CDC (2011)

28. Björk, S., Holopainen, J.: Patterns in Game Design. Charles River, Rockland (2004)

29. Chin, J.P., Diehl, V.A., Norman, K.L.: Development of an instrument for measuring user satisfaction of the human computer interface. In: Proceedings of Human Factors in Computing Systems (CHI), pp. 213–218. ACM Press (1988)

30. Davis, F.D.: Perceived usefulness, ease of use, and user acceptance of information technology. MIS Q. 13(3), 319–340 (1989)

The Factors Affecting the Quality of Learning Process and Outcome in Virtual Reality Environment for Safety Training in the Context of Mining Industry

Shiva Pedram[1]([⊠]), Pascal Perez[1], Stephen Palmisano[1], and Matthew Farrelly[2]

[1] University of Wollongong, Wollongong, NSW, Australia
spedram@uow.edu.au
[2] Mines Rescue, Coal Services, Woonona, Wollongong, NSW, Australia

Abstract. The ultimate aim of training is to improve task performance towards expert level. Novices and experts differ in their capability to understand and make sense of sensory information (for example, perception on environmental hazard). Computer-aided training, from online course to immersive simulation such as Virtual Reality (VR) [1]. van Wyk and de Villiers [2] define VR-based training environments as "*real-time computer simulations of the real world, in which visual realism, object behavior and user interaction are essential elements*". The use of VR-based training environments assumes that Human-Machine interaction stimulates learning processes through better experiencing and improved memorization, leading to a more effective transfer of the learning outcomes into workplace environments. However, there are many human factors (internally and externally), which have impact on the quality of the training and learning process which need to be identified and investigated. The present study was conducted with Coal Services Pty Ltd, a pioneering training provider for the coal mining industry in NSW, Australia. The research focussed on 288 rescuers and the specific training programs developed for them. In this article, initially factors affecting the quality of the training and learning process for underground mine rescuers have been identified and then measured by using pre- and post-training questionnaires. We attempted to determine how much of the trainees' perceived learning could be explained by pre-training (9 in total) and post-training (16 in total) factors. The relatively small size of the sample (288 observations for 17 predictors) and the high level of correlation between variables led us to Principal Component Analysis (PCA). Principle Component Analysis (PCA) has been used to investigate the underlying relationship among different variables. This technique results in factor reduction based on hidden relationships. Based on the nature of the pre-training factors mostly contributing to each component we have used the first 3 Components to create 3 new aggregated variables: "Positive State of Mind" (Component 1), "Negative State of Mind" (Component 2) and "Technology Experience" (Component 3). Similarly, based on the nature of the post-training factors mostly contributing to each component we have used the first 3 Components to create 3 new aggregated variables: "Positive Learning Experience" (Component 1), "Negative Learning Experience" (Component 2) and "Learning Context" (Component 3).

© Springer International Publishing AG, part of Springer Nature 2019
D. N. Cassenti (Ed.): AHFE 2018, AISC 780, pp. 404–411, 2019.
https://doi.org/10.1007/978-3-319-94223-0_38

Keywords: Human factors · Virtual Reality · Training · Evaluation
Mining industry

1   Introduction

Learning can be defined as a psychological process involving a change in the way a person responds to a situation based on experience [3]. This change might be reflected in the person's behaviour, for instance the development of new operational skills. It might also result in knowledge acquisition and attitude formation [4]. Ideally, these changes should be long-lasting in order to be recalled whenever a relevant situation arises. Therefore, effective learning is not only about the mere acquisition of knowledge or skills, but it lies also in the ability to transfer them into real-life situations that can be associated to the initial training context [3]. The ultimate aim of training is to improve task performance towards expert level. Ericsson, Krampe [5] conducted studies across various activity domains and concluded that, on average, it required nearly 10 years (or 10,000 h) for a person to achieve expert level performance in his/her domain of activity. Therefore, one of the fundamentals for training design is reducing the gap of knowledge between novice and expert. [6], studying learning differences between novices and experts, identified the following sources of differentiation: perception, decision making, actions and attention. Novices and experts differ in their capability to understand and make sense of sensory information (for example, perception on environmental hazard). Based on the same sensory information, experts better recognise patterns, predict future, anticipate problems and take appropriate decisions. Experts are more capable at discriminating perceptual events [7] such as errors or hazards. However, evidence suggests that perceptual skills can be acquired through training [8, 9]. Experts are generally faster at making decisions and mobilising relevant knowledge and procedures [6]. Moreover, experts tend to react and move faster than novices due to their higher perceptual abilities, body control and focus [6]. However, Chapman [10], studying novice and expert drivers concluded that novice drivers could significantly modify their behaviour after appropriate training.

Nowadays, the use of computer simulations as learning environments has progressively embraced technological innovations from chart-based interfaces to fully immersive environments [11, 12]. VR technology has brought immersive and interactive features allowing users to 'feel' the experiment [13]. VR technology has been used to train for various operations and dangerous circumstances where it is believed that training objectives cannot be achieved easily or the cost will prohibitive. van Wyk and de Villiers [2] define VR-based training environments as "real-time computer simulations of the real world, in which visual realism, object behaviour and user interaction are essential elements". The use of VR-based training environments assumes that Human-Machine interaction stimulates learning processes through better experiencing and improved memorization, leading to a more effective transfer of the learning outcomes into workplace environments [14]. As stated by Meadows [15]: "When I hear, I forget; when I see, I remember; when I do, I understand". In order to investigate "how" the attribute of immersive-VR technology is able to support and enhance learning, the pedagogical benefits of VR as a learning tool need to be examined in a comprehensive way.

2 Participants and Study Context

2.1 Technology-in-Use

The research was conducted in collaboration with Mines Rescue Pty Ltd, a training provider for the coal mining industry in Australia that operates four training stations in New South Wales (Woonona, Lithgow, Newcastle, Singleton and Woonona). Each centre delivers classroom, onsite and VR-based training programs. Our study focussed on training programs developed for the mine rescue brigades. These brigades are made of five to seven highly specialized volunteers who act as primary responders in case of major mining incidents or accidents. The methodological framework was designed and tested at Woonona station. This paper focuses exclusively on the training programs developed for the 360° immersive theatre (360-VR). The 360-VR is a 10 m diameter, 4 m high cylindrical screen that displays a 3D stereo, 360° virtual environment, providing a fully immersive experience to participants equipped with 3D glasses.

2.2 Training Scenario

Before rescue brigades entering the 360-VR, they wear their BG4 suits and they go under oxygen. Then, rescue brigades were debriefed by trainer on the accident: "The mine has been operating under its Level 1 trigger. This allows mining to continue on a normal basis. At the end of the last shift at 6.57 am an alarm was activated on the surface monitoring system at the monitoring point M. A CO reading of between 30 and 40 PPM has been recorded since that time from the real time monitoring point. Results show no traces of hydrogen or ethylene from the sampling point and this has been confirmed. The control officer was unable to contact the longwall deputy who had already departed for the surface. The crew has now been held back awaiting an inspection to verify the source of the elevated readings." The large area within the theatre allows for a mixed reality experience, with small groups of trainees (5 to 7) able to interact both with props (virtual gas detectors) and with each other, in order to ensure that appropriate responses, activities and reflexes are included as part of the training experience. The trainer guides the trainees through successive stages of the scenario, prompting them for appropriate actions or responses.

2.3 Participants

Between March and July 2015, 94 trainees interviewed for this study and all of the participants in the study were male, aged between 24 and 64 years, with their time spent in mining and mines rescue ranging from between 5 and 40 years. The participants in this study were 94 experienced underground miners who had volunteered to join the rescue brigade. Also, 25 trainers have been interviewed who were in running the training courses in 360-VR.

3   Methodology

We have chosen to use users' opinion technique where users are asked to give their opinions on the conducted training, even though this technique does not reflect on knowledge creation and training transfer [16] but it allows us to understand the factors affecting the training process in 360-VR. The researcher attended all the 360-VR training sessions to observe trainees experiences. She also distributed the question-naires directly before and after these training sessions. The pre-training questionnaire was distributed to participants prior to attending the 360-VR training. The aim of this questionnaire is to measure the trainees' state of mind and experience with technology prior the training. After the VR training, the post-training questionnaire was distributed to measure the participants' learning and experiences as a user. Other key questions asked about the participants' perception of the perceived level of realism, the success and the usefulness of the VR training. Primary data was obtained using Likert Scale based Questionnaires. Our pre- and post-training questionnaires were based on items taken from established questionnaires.

Pre-training questionnaire included items from: Game Experience Measure (GEM): trainee's prior experience with computers and video games based on [17], Immersive Tendencies (ITQ): sense of focus, involvement and alert prior training session based on [18], Simulator Sickness (SSQ): history of motion or simulator sickness (nausea, disorientation and oculomotor symptoms) and self-assessment prior training based on [19], Dundee Stress State (DSSQ): sense of engagement, distress or worry prior training session based on [20] and Intrinsic Motivation Inventory (IMI): sense of motivation, confidence and competition prior training session based on [21].

And post-training questionnaire included items from: Simulator Sickness (SSQ): sense of sickness during session [19], User Interface (UIQ): easiness to use and per-ceived realism based on [17], Game Engagement (GEQ): sense of engagement with scenario and environment based on [17], Involvement and Presence (IPQ): sense of presence and involvement during session based on [18], Intrinsic Motivation Inventory (IMI): enjoyment and motivation during session based on [18], Immersive Tendencies (ITQ): sense of focus and immersion during session based on [22], Dundee Stress State (DSSQ): sense of pressure or tension during session based on [20]. Due to limits on our testing time (which prevented use of the full questionnaires) key items taken from standard questionnaires to measure each factor of interest. These factors had been identified by previous studies as being important for the success of VR training. In order to check that each group of items was still measuring the same factor (i.e. as the original full questionnaires), ensured that the Cronbach's Alpha value for each factor was above 0.7.

4   Result

Tables 1 and 2 summarize the mean values for each of the pre-training and post-training factors where Likert scales ranging from highly disagree (0) to highly agree (5). Even though trainees had limited gaming experience (M = 1.4) but they have been motivated (M = 4.2) to attend the training. Moreover, they reported better than average scores for "ease of use" (M = 3.6), "enjoyment" (M = 3.8), "presence" (M = 3.3), "usefulness" (M = 4.09) and perceived learning (M = 3.5) (scores out of 5).

Table 1. Mean value for pre-training factors

Type	Factor	Lower	Mean	Upper
Pre training	Gaming experience	1.3137	1.4185	1.5179
Pre training	Sense of alert and presence	3.8348	3.9427	4.0505
Pre training	Sense of stress	2.5461	2.7312	2.9162
Pre training	Sense of motivation	4.1142	4.2007	4.2873
Pre training	Sense of confidence and competency	3.8454	3.9355	4.0256
Pre training	Sense of worry	2.6344	2.828	3.0215
Pre training	Sense of competition	3.021	3.2151	3.4091
Pre training	Sickness	1.3285	1.359	1.3895

Table 2. Mean value for post-training factors

Type	Factor	Lower	Mean	Upper
Post training	Sense of engagement and interaction	3.3292	3.4663	3.6034
Post training	Sense of ease of use	3.5325	3.6957	3.8588
Post training	Sense of fatigue	2.1646	2.3258	2.4871
Post training	Sense of enjoyment	3.803	3.8966	3.9902
Post training	Sense of stress, pressure and tension	2.1497	2.2317	2.3138
Post training	Sense of pressure	3.1904	3.3015	3.4126
Post training	Sense of realism	2.8941	3.1989	3.5036
Post training	Sense of usefulness	3.9517	4.092	4.2322
Post training	Success	3.201	3.401	4.102
Post training	Sickness	1.3237	1.3931	1.4626
Post training	Perceived learning	3.3467	3.5301	3.7136

We attempted to determine how much of the trainees' perceived learning could be explained by pre-training (9 in total) and post-training (16 in total) factors. The relatively small size of the sample (231 observations for 17 predictors) and the high level of correlation between variables led us to a two-stage modelling process: (1) Principal Component Analysis to reduce the number of predictors; and (2) linear regression between perceived learning and aggregated predictors.

Table 3 shows that the first Component, explaining 34% of the variance, is characterised by 5 factors: "Alertness", "Motivation", "Confidence", "Wellbeing" and "Competitiveness". The second Component, explaining 17% of the variance, is characterised by 2 strongly correlated factors: "Worry" and "Stress". The third Component, explaining 13% of the variance, is characterised by 2 strongly correlated factors: "Gaming Experience" and "Digital World Involvement". Together these 3 Components explained 64% of the total variance.

Table 3. Structure matrix – PCA on pre-training factors

Structure Matrix

	Component		
	1	2	3
Alert	.862	-.238	.044
Motivation	.772	-.231	-.103
Confidence	.742	-.304	.000
Wellbeing	.703	-.301	.004
Competition	.642	.346	-.049
Worry	-.270	.839	-.109
Stress	-.317	.745	-.073
Gaming Experience	.071	-.128	-.855
Digital World Involvment	-.035	.366	-.783

Extraction Method: Principal Component Analysis.
Rotation Method: Oblimin with Kaiser Normalization.

Based on the nature of the factors mostly contributing to each component we have used the first 3 Components to create 3 new aggregated variables: "Positive State of Mind" (Component 1), "Negative State of Mind" (Component 2) and "Technology Experience" (Component 3).

Table 4 shows that the first Component, explaining 56% of the variance, is characterised by 11 correlated variables: "Task-Technology Fit", "Functionality", "Usefulness", "Ease of use", "Attitude", "Presence", "Engagement", "Interaction", "Enjoyment", "Immersion" and "Realism". The second Component, explaining 9% of the variance, is characterised by 3 strongly correlated variables: "Task Characteristics", "Feedback" and "Trainer". The third Component, explaining 8% of the variance, is characterised by 2 strongly correlated variables: "Stress" and "Simulation Sickness". These 3 Components explain 73% of the total variance.

Based on the nature of the variables mostly contributing to each component we have used the first 3 Components to create 3 new aggregated variables: "Positive Learning Experience" (Component 1), "Negative Learning Experience" (Component 2) and "Learning Context" (Component 3).

Table 4. Structure matrix – PCA on post-training variables

Structure Matrix

	Component		
	1	2	3
TTF	.908	-.540	-.258
Tool Functionality	.899	-.454	-.142
Usefulness	.897	-.427	-.156
Ease Of Use	.890	-.371	-.200
Atittude Towards Use	.883	-.519	-.276
Presence	.874	-.512	-.091
Engagement	.871	-.478	.013
Interaction	.858	-.443	-.099
Enjoyment	.767	-.445	-.384
Immersion	.725	-.358	.353
Realism	.705	-.290	-.159
Task Characteristics	.567	-.877	-.143
Feedback	.448	-.862	-.011
Trainer	.356	-.859	-.165
Stress Worry Pressure	-.104	.172	.840
Simulator Sickness	-.346	.148	.623

Extraction Method: Principal Component Analysis.
Rotation Method: Oblimin with Kaiser Normalization.

5   Conclusion

Positive learning experience (including: presence, engagement, enjoyment), negative learning experience (including: stress, worry and pressure) are important constructs in learning process and outcome. Most of the factors measured in the post-training questionnaire were found to be significantly correlated with each other and having an influence on perceived learning. Thus, it was difficult to single out key primary factors driving learning outcomes. Henceforth, our modelling attempts aimed at grouping these factors into positive or negative learning experience in order to infer the global and intertwined effects of these factors. Our explanatory model shows that 71% of the variance associated with perceived learning can be attributed to 3 aggregated variables describing positive and negative experiences during the session and learning context.

References

1. Newton, D., Hase, S., Ellis, A.: Effective implementation of online learning: a case study of the Queensland mining industry. J. Workplace Learn. **14**(4), 156–165 (2002)
2. van Wyk, E., de Villiers, R.: Virtual reality training applications for the mining industry. In: Proceedings of the 6th International Conference on Computer Graphics, Virtual Reality, Visualisation and Interaction in Africa. ACM (2009)
3. Pithers, R.T.: Improving Learning Through Effective Training. Social Science Press, Katoomba (1998)
4. Dewey, J., Boydston, J.A.: Essays on Philosophy and Education: 1916–1917. Southern Illinois University Press, Carbondale (1985)
5. Ericsson, K.A., Krampe, R.T., Tesch-Römer, C.: The role of deliberate practice in the acquisition of expert performance. Psychol. Rev. **100**(3), 363 (1993)
6. Tichon, J., Burgess-Limerick, R.: A review of virtual reality as a medium for safety related training in mining. J. Health Saf. Res. Pract. **3**(1), 33–40 (2011)
7. Blignaut, C.: The perception of hazard II. the contribution of signal detection to hazard perception. Ergonomics **22**(11), 1177–1183 (1979)
8. Starkes, J.L., Lindley, S.: Can we hasten expertise by video simulations? Quest **46**(2), 211–222 (1994)
9. Williams, A.M., Grant, A.: Training perceptual skill in sport. Int. J. Sport Psychol. **30**(2), 194–220 (1999)
10. Chapman, P., Underwood, G., Roberts, K.: Visual search patterns in trained and untrained novice drivers. Transp. Res. Part F: Traffic Psychol. Behav. **5**(2), 157–167 (2002)
11. Bell, P.C., Taseen, A.A., Kirkpatrick, P.F.: Visual interactive simulation modeling in a decision support role. Comput. Oper. Res. **17**(5), 447–456 (1990)
12. Jou, M., Wang, J.: Investigation of effects of virtual reality environments on learning performance of technical skills. Comput. Hum. Behav. **29**, 433–438 (2012)
13. Raskind, M., Smedley, T.M., Higgins, K.: Virtual Technology Bringing the World Into the Special Education Classroom. Interv. School Clin. **41**(2), 114–119 (2005)
14. Chen, I.Y.L., Chen, N.-S.: Kinshuk: Examining the factors influencing participants' knowledge sharing behavior in virtual learning communities. Educ. Technol. Soc. **12**(1), 134 + (2009)
15. Meadows, D.L.: Tools for understanding the limits to growth: comparing a simulation and a game. Simul. Gaming **32**(4), 522–536 (2001)
16. Nutakor, D.: Design and Evaluation of a Virtual Reality Training System for New Underground Rockbolters. ProQuest, Ann Arbor (2008)
17. Taylor, G.S., Barnett, J.S.: Training Capabilities of Wearable and Desktop Simulator Interfaces (2011). DTIC Document
18. Witmer, B.G., Singer, M.J.: Measuring presence in virtual environments: a presence questionnaire. Presence: Teleoper. Virtual Environ. **7**(3), 225–240 (1998)
19. Kennedy, R.S., et al.: Simulator sickness questionnaire: an enhanced method for quantifying simulator sickness. Int. J. Aviat. Psychol. **3**(3), 203–220 (1993)
20. Matthews, G., et al.: Validation of a comprehensive stress state questionnaire: Towards a state big three. Personal. Psychol. Eur. **7**, 335–350 (1999)
21. McAuley, E., Duncan, T., Tammen, V.V.: Psychometric properties of the intrinsic motivation inventory in a competitive sport setting: a confirmatory factor analysis. Res. Q. Exerc. Sport **60**(1), 48–58 (1989)
22. Slater, M.: Measuring presence: a response to the Witmer and Singer presence questionnaire. Presence: Teleoper. Virtual Environ. **8**(5), 560–565 (1999)

Classification Algorithms in Adaptive Systems for Neuro-Ergonomic Applications

Grace Teo$^{(\boxtimes)}$ and Lauren Reinerman-Jones

Institute for Simulation and Training, University of Central Florida,
3100 Technology Parkway, Orlando, FL, USA
{gteo,lreinerm}@ist.ucf.edu

Abstract. Adaptive systems typically comprise components that interrelate and interact to enable the whole system to respond and adjust to changes in the environment, operator, and task in order to regulate or maintain a level of performance or homeostasis. In so doing, they enable a degree of individual-ization and customization for many technological innovations such as managing the use of automation. Adaptive systems often involve some kind of feedback or closed-loop which requires a criteria for determining invoking thresholds, as well as some type of classification algorithm that models the type of changes to which the system has to adapt. This paper outlines the issues, considerations, and challenges associated with classification in adaptive systems, and reviews several algorithms that implement the feedback loop in neuro-ergonomic applications. These include logistic regression, Naïve-Bayes, artificial neural networks (ANN), and support vector machines (SVM) techniques.

Keywords: Classification · Adaptive systems · Neuro-ergonomics

1 Introduction

Adaptive systems invoke changes in system behavior to meet the operator's changing condition or state. Some adaptive systems involve a negative feedback loop that per-form a regulatory function to maintain a level of performance or state. In neuro-ergonomic applications, they have been used to regulate operator workload [1–3], engagement [4], and performance [3, 5, 6] by adjusting task allocation [4], task modes [3, 5], automation aiding [6, 7], task difficulty [8], and training [1], etc.

For many of these neuro-ergonomic applications, adaptive systems respond to real-time changes. These could be events in the environment (i.e., environment-based triggers), specific time and location (i.e., spatiotemporal triggers) [9], changes in task phase such as occurrence of critical events, level of performance, models of perfor-mance, or operator functional state [10], or combinations of these [11]. Many systems that adapt to operator functional state [10] use physiological inputs as indicators as they offer non-invasive, continuous monitoring with high temporal resolution. Central to these systems is a classification task where the system determines if a specific category of operator functional state (e.g., unmanageable workload, high fatigue and stress) is reached based on certain inputs. For instance, an adaptive system may initiate aiding by automation after determining that an operator is experiencing high workload based on

© Springer International Publishing AG, part of Springer Nature 2019
D. N. Cassenti (Ed.): AHFE 2018, AISC 780, pp. 412–420, 2019.
https://doi.org/10.1007/978-3-319-94223-0_39

multiple physiological workload measures. Accuracy of the classification depends largely on which and how inputs are used in the determination of category, and algorithms that perform the classification essentially build a model of how the inputs are related to the categorical output. This model is built during the *training phase* where the machine is trained with data of the inputs and outcomes of multiple cases (i.e., 'supervised learning' where the categorical outcome is known for each of the cases). From the model, the machine then predicts the unknown outcome or category of new cases from their inputs in the *testing phase*.

1.1   Requirements for Implementation in Adaptive Systems

Feign and colleagues [9] proposed that adaptions to meet the needs of the operator can: (i) modify function allocation, (ii) modify task scheduling, (iii) modify interaction, or (iv) modify content. To accomplish these, adaptive systems need to be responsive so classification algorithms must be executed online. Since, in most cases, the number of inputs directly contribute to the computation demand and time taken to build and execute the model in real-time, input (or feature) selection and extraction may be necessary before the classification [12].

Feature selection techniques can include setting variance thresholds and examining correlations. Setting a variance threshold excludes input variables on which the cases or observations across all outcomes do not vary by much (e.g., almost all cases have the same value on that input variable). Input variables that do not vary even for different outcome values do not provide much information about how different outcomes are obtained. On the other hand, setting a correlation threshold eliminates inputs that are highly correlated as this indicates redundancy of information. Typically, if two input variables are highly correlated, the one with the larger mean absolute correlation with other inputs can be excluded from the classification that follows. While feature selection reduces the number of features or inputs by elimination, feature extractor decreases the number of features by forming aggregates of features [12]. Principal component analysis (PCA) and linear discriminant analysis (LDA) are two techniques that accomplish this. Both PCA and LDA create aggregates that are linear combinations of the original input variables. PCA tries to capture as much variance in the original inputs when creating aggregates and can be done without knowledge of the outcome variable (i.e., unsupervised learning). In contrast, LDA requires knowledge of the outcome variable (i.e., a supervised method) as it seeks to maximize the extent to which the outcome categories are separate when creating aggregates [12].

Apart from the need to process data in real-time, algorithms used to determine operator functional state should also provide information about the inputs that trigger the adaptive action. This is so that system-appropriate behavior such as aiding by automation or adjustment of training pace or content, can be better designed. For instance, by knowing that the high workload state is associated with increased ocular activity, a better aid to develop would be one that alleviates visual demand.

1.2 Examples of Classification Algorithms for Adaptive Systems

Table 1 lists several of the many classification algorithms that have been used in adaptive systems. They differ in the degree to which their classification rules are transparent, their assumptions about the data, and their ability to deal with real-time data. All of these components potentially impact the suitability of any one classification algorithm for use in adaptive systems. The following sections provide a simple and brief overview of four algorithms and describe the use of these algorithms for classifying cases into a binary outcome such as "high" vs. "low" workload.

Table 1. Examples of classification algorithms

(a) Logistic Regression
(b) Naïve Bayes
(c) Artificial Neural Networks (ANN)
(d) Support Vector Machines (SVM)

2 Logistic Regression

2.1 Overview

Binary logistic regression predicts the likelihood that a given case results in either of two possible outcomes or categories from a combination of the inputs. Input variables may be categorical data that are coded, or continuous data. The algorithm relies on the maximum likelihood estimator (MLE) rather than the ordinary least squares (OLS) method to determine the parameters of the predictive model. The MLE method selects model parameters that maximize the likelihood that the model's prediction would result in the observed data (i.e., training data). This is in contrast to OLS which selects model parameters that minimize the deviations between the outcomes that the model predict and that of the observed data [13]. The predictive ability of the model is evaluated by "goodness-of-fit" measures such as Chi-square tests showing the disparity between expected and observed outcomes, classification tables that reflect the model's sensitivity and specificity, Receiver Operator Curves (ROC curves), the R^2 statistic, and Hosmer-Lemeshow tests that compares observed and expected frequencies of the two outcomes [13, 14].

2.2 Application to Adaptive Systems

In addition to overall model fit, logistic regression - like linear regression - includes information about the model's parameters. This include the magnitude and direction of the coefficients which reflect the relative importance of the various inputs in the prediction. Such information allows relevant inputs to be identified and be more closely examined.

3 Naïve Bayes

3.1 Overview

The Naïve Bayes classifier is based on Bayes' theorem which addresses the probability of an outcome given knowledge of conditions (i.e., inputs) that could relate to that outcome (see Fig. 1) [17].

$$P(A \mid B) = \frac{P(B \mid A)}{P(B)} P(A)$$

Fig. 1. Bayes' theorem for calculating the probability of an outcome (A), given a prior input (B)

For instance, the Naïve Bayes classifier may compute the probability that an operator is experiencing "high workload" from inputs such as "hours on task", "hours of sleep in the previous day", "age", and "health status". The classification would address the question: "What is the probability that a new operator is experiencing high workload given that he had been on the task for 12 h, has had 3 h of sleep in the previous day, is 60 years old, and in poor health?" A probability threshold can be set such that the adaptation or aid will be introduced when the threshold is reached. Advantages of the Naïve Bayes classifier include not requiring large training datasets, and being able to accommodate both categorical and continuous input data. There is also no assumption about the distribution of the data, although there is the assumption that the cases on which the algorithm is trained are independent.

3.2 Application to Adaptive Systems

The Naïve Bayes classifier has the advantage of being able to fine-tune its classifications in real-time as new data is incorporated. The algorithm is also readily interpreted and can provide information about the relative importance of the inputs in the classification across multiple cases. Being simple to implement, it is also computationally economical. Nevertheless, its simplicity is also due to its inability to deal with interaction of inputs and data that are not linearly separable. The algorithm may also be less able to handle data that are "noisier" such as physiological data which tend to be idiosyncratic and show greater inter-individual variability [18, 19]. While this had been addressed by using a hierarchical Naïve Bayes algorithm that includes a hidden node to account for between subject variability [20], this hidden node obfuscates the relationship between the inputs and outcomes.

4 Artificial Neural Networks (ANN)

4.1 Overview

A powerful machine learning algorithm is the Artificial Neural Network (ANN), so named because, like the brain, it consists of a network of nodes and synapses. An ANN may assume various architectures, but the most common is that of the Multilayer Perceptron (MLP) with at least three layers of nodes. The input layer contains data of the input or predictor variables, the hidden layer contains processing nodes, and the output layer comprises data of the outcome variable. Adjacent layers are connected by synapses. In the synapses, data from multiple neurons activated in the preceding layer are combined with weights and pass through an activating function. The result, which is transmitted from the synapse, determine which neurons in the subsequent layer get activated. The output or prediction of the ANN is the final result in the output layer. Data flow in ANNs during training can be unidirectional (i.e., feed-forward), or bidirectional (i.e., back-propagation of errors) allowing weights and the activating function in the processing layers to be adjusted so that the predicted output will best match the observed values [15].

4.2   Application to Adaptive Systems

Artificial Neural Networks have been used in several studies involving the use of adaptive systems to manage workload. Studies that use an ANN in their adaptive systems typically report high classification accuracy rates. These may be due to several reasons. First, the ANNs were subject-specific. Separate ANNs were trained for each subject [1, 6, 7] and involved repeated training, requiring larger volumes of data per subject. Next, the high accuracy rates were obtained from offline saliency analysis [16]. While this enabled some information about the relative importance of inputs to the outcome to be derived [7], the ANNs were not used in real-time to adjust system-behavior. ANNs that were executed online had poorer classification accuracy rates. Another major issue with using ANNs for adaptive systems is the limited access to information about which predictors are important in the classification and how these were used in the prediction. This is because it is difficult to know what happens in the hidden layers. Hence while ANNs can help predict the outcomes, they are less useful in explaining the basis for the classification, and they are considered a "black box" technique. These can limit the utility of ANNs in adaptive systems that require real-time classification.

5   Support Vector Machines (SVM)

5.1   Overview

In classifying a binary outcome variable, the Support Vector Machines algorithm seeks out the hyperplane (i.e., decision boundary) that best separates the outcomes categories in a feature space [17]. In a classification problem with only two inputs, this hyperplane may be a line. For instance, data of 10 cases on two input features, x (hours on task) and y (health status), as well as their outcomes are plotted in Fig. 2. While both the dotted and dashed lines successfully separate the outcome categories, the dashed line does so by a smaller margin compared to the dotted line, indicating that the dashed line is more likely to misclassify a new case. Between these two, the better hyperplane is the one whose distance to the nearest case of each outcome category is the greatest (i.e., the best line has the widest "safety margin" between the cases nearest to it) [21].

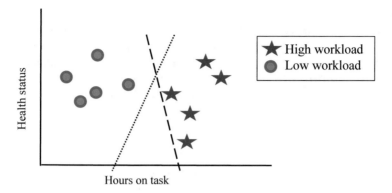

Fig. 2. Hyperplanes for linearly-separable data

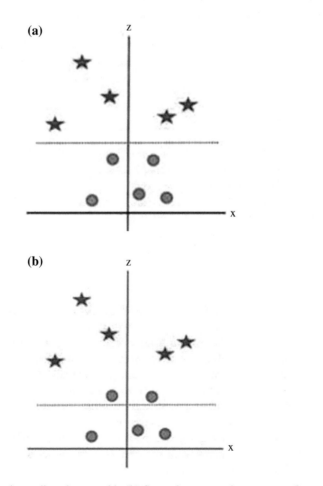

Fig. 3. (a) Data that is not linearly separable (b) Same data mapped onto a new feature space (not to scale) resulting in linear separability

When the data cannot be separated by a linear hyperplane (see Fig. 3a), the data may need to be mapped into a new feature space involving a higher dimension, e.g., dimension z, where $z = x^2 + y^2$ (see Fig. 3b). This mapping into the new feature space is specified by a kernel function. A hyperplane may then be found along these new dimensions. The output of SVMs are weights for each input feature, which, when applied to the new case, will result in a linear combination that predicts the outcome [17, 21].

5.2   Application to Adaptive Systems

SVMs have the advantage of being able to accommodate a large number of features without losing computational efficiency. However, like the ANN, SVMs are a "black box technique" because they are difficult to interpret which and how inputs are used in the classification since SVMs will classify even without the researcher specifying the kernel function that defined the feature space. In addition, SVMs are typically used to classify data off-line, and have been used to determine, from EEG data, when a driver is drowsy during a driving task performed before the classification analysis [22]. When there are new data, it would typically be necessary to retrain the SVM. These shortcomings may limit the use of SVMs in adaptive systems.

6   Summary

Compared to the logistic regression and the Naïve Bayes techniques, ANNs and SVMs are typically more capable of classifying "noisier" data. However their results are also more difficult to interpret and may not yield insights into the relationships between the inputs and outcomes. The algorithms also differ on the extent to which they can accommodate the streaming in of data in real-time which would impact the degree to which the adaptive systems are responsive.

7   Conclusion

Adaptive systems represent the next level of developments that would allow technology to be more contextualized and responsive to changing situations. When they are implemented to use operator-based triggers that are indicative of operator functional state, adaptive systems can customize system behavior and provide aid by technology to the individual. Although there are a variety of classification algorithms, the algorithms are not equally suitable for use in real-time adaptive systems which typically need classification rules to be updated with real-time data and information about how the inputs relate to the outcome to be accessible. These criteria may not always be met by the algorithm with the highest classification accuracy, implying that compromises and trade-offs, or a combination of algorithms, may be necessary in selecting algorithms for different adaptive systems.

Acknowledgments. This research was sponsored by the Army Research Laboratory and was accomplished under Cooperative Agreement Number W911 NF-14-2-0021. The views and conclusions contained in this document are those of the authors and should not be interpreted as representing the official policies, either expressed or implied of the Army Research Laboratory of or the U.S. Government. The U.S. Government is authorized to reproduce and distribute reprints for Government purposes notwithstanding any copyright notation herein.

References

1. Baldwin, C.L., Penaranda, B.N.: Adaptive training using an artificial neural network and EEG metrics for within- and cross-task workload classification. NeuroImage **59**, 48–56 (2012). https://doi.org/10.1016/j.neuroimage.2011.07.047
2. Hannula, M., Huttunen, K., Koskelo, J., Laitinen, T., Leino, T.: Comparison between artificial neural network and multilinear regression models in an evaluation of cognitive workload in a flight simulator. Comput. Biol. Med. **38**, 1163–1170 (2008). https://doi.org/10.1016/j.compbiomed.2008.09.007
3. Prinzel III, L.J., Freeman, F.G., Scerbo, M.W., Mikulka, P.J., Pope, A.T.: Effects of a psychophysiological system for adaptive automation on performance, workload, and the event-related potential P300 component. Hum. Factors **45**, 601–614 (2003)
4. Prinzel, L.J., Freeman, F.G., Scerbo, M.W., Mikulka, P.J., Pope, A.T.: A closed-loop system for examining psychophysiological measures for adaptive task allocation. Int. J. Aviat. Psychol. **10**, 393–410 (2000). https://doi.org/10.1207/S15327108IJAP1004_6
5. Freeman, F.G., Mikulka, P.J., Scerbo, M.W., Prinzel, L.J., Clouatre, K.: Evaluation of a psychophysiologically controlled adaptive automation system, using performance on a tracking task. Appl. Psychophysiol. Biofeedback **25**, 103–115 (2000)
6. Wilson, G.F., Russell, C.A.: Performance enhancement in an uninhabited air vehicle task using psychophysiologically determined adaptive aiding. Hum. Factors Soc. J. Hum. Factors Ergon. **49**, 1005–1018 (2007). https://doi.org/10.1518/001872007X249875
7. Wilson, G.F., Russell, C.A.: Real-time assessment of mental workload using psychophysiological measures and artificial neural networks. Hum. Factors **45**, 635–644 (2003)
8. Christensen, J.C., Estepp, J.R., Wilson, G.F., Russell, C.A.: The effects of day-to-day variability of physiological data on operator functional state classification. NeuroImage **59**, 57–63 (2012). https://doi.org/10.1016/j.neuroimage.2011.07.091
9. Feigh, K.M., Dorneich, M.C., Hayes, C.C.: Toward a characterization of adaptive systems: a framework for researchers and system designers. Hum. Factors J. Hum. Factors Ergon. Soc. **54**, 1008–1024 (2012). https://doi.org/10.1177/0018720812443983
10. Hockey, G.R.J.: Operator Functional State: The Assessment and Prediction of Human Performance Degradation in Complex Tasks. IOS Press, Amsterdam (2003)
11. Parasuraman, R., Mouloua, M., Molloy, R.: Effects of adaptive task allocation on monitoring of automated systems. Hum. Factors **38**, 665–679 (1996)
12. Dimensionality Reduction Algorithms: Strengths and Weaknesses. https://elitedatascience.com/dimensionality-reduction-algorithms
13. Field, A.: Discovering statistics using SPSS. Sage publications, Thousand Oaks (2009)
14. Goodness-of-fit tests for Binary Logistic Regression. http://support.minitab.com/en-us/minitab-express/1/help-and-how-to/modeling-statistics/regression/how-to/binary-logistic-regression/interpret-the-results/all-statistics-and-graphs/goodness-of-fit-tests/
15. Mehrotra, K., Mohan, C.K., Ranka, S.: Elements of Artificial Neural Networks (Complex Adaptive Systems). MIT Press, Cambridge, MA (1997)

16. Ruck, D.W., Rogers, S.K., Kabrisky, M.: Feature selection using a multilayer perceptron. J. Neural Netw. Comput. **2**, 40–48 (1990)
17. Provost, F., Fawcett, T.: Data Science for Business: What You Need to Know About Data Mining and Data-Analytic Thinking. O'Reilly Media Inc, Sebastopol (2013)
18. Besson, P., Dousset, E., Bourdin, C., Bringoux, L., Marqueste, T., Mestre, D.R., Vercher, J.-L.: Bayesian network classifiers inferring workload from physiological features: compared performance. In: 2012 IEEE Intelligent Vehicles Symposium (IV), pp. 282–287. IEEE (2012)
19. Johannes, B., Gaillard, A.W.K.: A methodology to compensate for individual differences in psychophysiological assessment. Biol. Psychol. **96**, 77–85 (2014). https://doi.org/10.1016/j.biopsycho.2013.11.004
20. Wang, Z., Hope, R.M., Wang, Z., Ji, Q., Gray, W.D.: Cross-subject workload classification with a hierarchical Bayes model. NeuroImage **59**, 64–69 (2012). https://doi.org/10.1016/j.neuroimage.2011.07.094
21. Understanding Support Vector Machine algorithm from examples (along with code). https://www.analyticsvidhya.com/blog/2017/09/understaing-support-vector-machine-example-code/
22. Yeo, M.V.M., Li, X., Shen, K., Wilder-Smith, E.P.V.: Can SVM be used for automatic EEG detection of drowsiness during car driving? Saf. Sci. **47**, 115–124 (2009). https://doi.org/10.1016/j.ssci.2008.01.007

Digital Modeling and Biomechanics

Repetitive-Task Ankle Joint Injury Assessment Using Artificial Neural Network

Sultan Sultan[✉], Karim Abdel-Malek, Jasbir Arora, and Rajan Bhatt

Center for Computer-Aided Design (CCAD), Virtual Soldier Research
(VSR) Program, The University of Iowa, Iowa City, USA
{asultan, jsarora}@engineering.uiowa.edu,
{karim-abdel-malek, rajan-bhatt}@uiowa.edu

Abstract. This research effort is to develop human simulation methods to predict and assess injuries due to the fatigue of a repetitive loading. Over the past few years, we sought to integrate high-fidelity computational methods for stress/strain analysis, namely finite element analysis (FEA), in combination with biomechanics predictions through digital human modeling and simulation (DHMS). Our previous work for the past 12 years has culminated with the development of a simulation environment called Santos™ that enables the prediction of human motion including many aspects of its biomechanics and physiological systems. The Santos environment provides a joint- and physics-based predictive simulation environment. Cumulative load theory states that repetitive activities precipitate musculoskeletal injury and suggests that cyclic load application may result in cumulative fatigue, reducing their stress-bearing capacity. Such changes may reduce the threshold stress at which the tissues of the joint components fail. Repetitiveness of the work activity has shown to be a strong risk factor for cumulative trauma disorders (repetitive strain injury). Hence, repetitive load cycling is a leading factor in the propensity for injury. Santos, and the developed method, are able to characterize the motion using an optimization algorithm that calculates the motion profiles (i.e., the kinematics of the motion across time for each degree of freedom for the body) and external forces during the task. An OpenSim model uses motion profiles and external forces to calculate forces of the muscles that articulate the joint. A multi-scale FEA system uses external forces and muscle forces for each task over a repetitive cycle as input. The FEA model of the selected joint (ankle) computes the stresses of all joint components at limited and random frames of the motion cycle. The Artificial Neural Network program (ANN) estimates the stress over the full motion cycle and compares current stresses of the components with the newly calculated yield strength that been affected by cyclic loading. As a result, the program indicates the injury status of the joint components. This paper presents promising results for this approach to predict and quantify injury in the ankle joint that is undergoing a cyclic repetitive motion. This integrated system is capable of showing the effects of various motions and task-parameters on the selected joint and by this we can modify tasks, save analysis time, and reduce the likelihood of injury.

Keywords: Digital human modeling · Multi-scale modeling · FEA modeling
Injury prevention · Fatigue loading

© Springer International Publishing AG, part of Springer Nature (outside the USA) 2019
D. N. Cassenti (Ed.): AHFE 2018, AISC 780, pp. 423–432, 2019.
https://doi.org/10.1007/978-3-319-94223-0_40

1 Introduction

Ankle joint injury is the third most common musculoskeletal injury in the human body [1]. The impact of this injury is a loss of manpower, an increase in healthcare costs, and a high probability of disabilities and fatalities [2, 3]. In the United States, there are 27,000 injuries each day due to sprains of the ankle and knee joints [4]. Joint injury occurs as result of tissue failure, particularly in articular cartilage [5]. This is likely due either to a single application of a heavy load or to cumulative cyclic application of a light load.

Two theories explain the failure of tissue due to the repetitive cycle of the applied load: the differential fatigue theory and the cumulative load theory. The first theory states that repetitive loading can alter muscle kinetics and that this changes the joint kinematics pattern and results in stress concentration in the tissue, causing failure [6]. The second theory focuses on the reduction of the stress-bearing capacity of the tissue [7, 8]. It explains that the repetitive cycle of the load will decline the power capacity of the muscle to a level of threshold stress at which the tissues fail.

The work in this study focuses on the cumulative load theory. Sultan and Marler [9] presented a multi-scale model for predicting a measure for knee and ankle joint injury based on analysis involving motion data and external ground reaction forces predicted by the digital human model (DHM) Santos in conjunction with muscle force computation by the OpenSim software.

The work of this study used a finite element model for the ankle joint to compute the stress field of all components of the ankle for a few selected frames of the motion. A developed program computed the injury index of all components for those selected frames. The neural network technique computed the injury index of the components over whole frames of the motion.

This study provides a method to measure ankle joint component injury for both soft tissues and bones due to a repetitive task cycle. The measuring system is capable of quantitatively assessing the level of injury of ankle cartilage. The importance of the proposed method is the ability to predict a quantitative scale of the risk of the injury. This can lead for proper action for ankle injury prevention.

2 Method

2.1 Dynamic Prediction System

The predictive dynamics approach usually simulates human motion with numerous degrees of freedom (DOFs) [10], where the joint angles are the DOFs. The Denavit-Hartenberg (DH) [11] formulation clearly defined the relation between the joint angles and the position of the points on a series of links in an optimization problem. The problem of optimization includes: (1) determining joint angles as a function of time, (2) optimizing the cost function, and (3) imposing constraints on the motion [12, 13].

At the University of Iowa, the Virtual Solider Research program developed a predictive dynamics system named Santos to predict the motion with 109 DOF [10].

Santos predicts the motion of land-markers in the form of x-y-z positions and three components of ground reaction force for a stair-ascending task in three different loading cases: with no backpack, with a 17 kg backpack, and with a 33 kg backpack, as shown in Fig. 1. The study conducted five different repetitive cycles—1 cycle, 10 cycles, 100 cycles, 250 cycles, and 500 cycles—for each load case. We developed program to feed the predicted data; the motion and ground reaction force into a computing muscle model for further analysis.

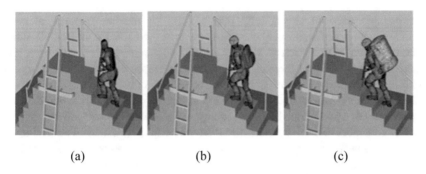

(a)   (b)   (c)

Fig. 1. Stair ascending: (a) no backpack, (b) 17 kg backpack, and (c) 33 kg backpack

2.2   Finite Element Model

The present work used the 3D FEA ankle model previously used by Sultan et al. [14], as shown in Fig. 2. There are seven components in the ankle joint: talus bone (TAB), tibia bone (TBA), talus cartilage (TCA), tibia cartilage (TAC), anterior talus ligament (ATL), posterior talus ligament (PTL) and deltoid ligament (DL). The finite element method meshed all joint components using a hexahedral element (Hex8) with 17674

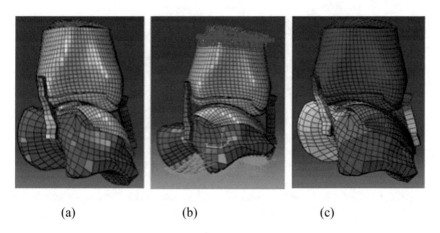

(a)   (b)   (c)

Fig. 2. Ankle joint FEA model: (a) hexahedral mesh element, (b) loading, and (c) displacement solution

nodes and 13849 elements of type C3D8. The study modeled the bone as an elastic material and modeled the soft tissue as a Noe-Hookean hyper-elastic material [9]. In the ankle joint, the contact problem worked in a pair of components, between the tibia cartilage and the talus cartilage. The coefficient of friction for the contact problems was 0.01 [15].

3 Injury Propensity and Repetitive Load

Enormous repetition of any activity places stress on the human physical system. Without enough time for recovery, repeating the load increases the stress concentration on the organism, which decreases the stress tolerance and finally results in a higher chance of injury. Different theories explain the occurrence of injury in any organism; among those are the differential fatigue theory and the cumulative load theory [6].

The present study focuses on cumulative load theory and experimental data taken from Kumar [6, 16]. Cumulative load theory states that repeated load may result in cumulative fatigue, and this reduces stress-bearing capacity. The decline in the maximal force or power capacity of muscle [8] may reduce the threshold stress at which the tissues fail. Using experimental data, we plotted the relationship between the strength of the tissue with the number of repeating cycles as shown in Fig. 3. We made data regression to fit the graph with a second-order polynomial equation as illustrated in Eq. (1):

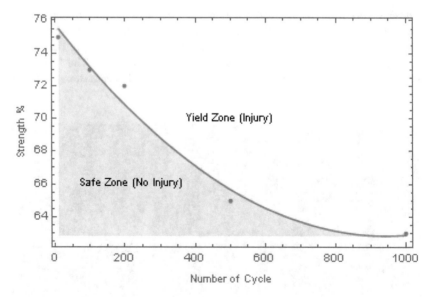

Fig. 3. Strength declination due to number of repetitive task cycles (data taken from reference [6], plotted with second-order polynomial fit).

$$R_f = 75.7395 - 0.0274 * N_r + 1.5446 * 10^{-6} * N_r^2 \tag{1}$$

$$S_c = R_f * S \tag{2}$$

where S_c is the current strength, S is the original strength, R_f is the reduction factor, and N_r is the number of task cycles.

This study used the same algorithm presented by Sultan et al. [14] to assess the level of injury of each component of the joint. The algorithm compares current stress with the yield stress of a component over the time of the stride. Due to continuous repeating of the cycle of the load, the yield stress declines based on Eq. (1), as illustrated in Fig. 3. For the stair-ascending task, we computed the injury index for all ankle joint components. We conducted the stair-ascending task for three different loadings: no backpack, 17 kg backpack load, and 33 kg backpack load, each at five different numbers of cycles: one cycle, ten cycles, 100 cycles, 250 cycles, and 500 cycles.

4   Result and Discussion

For the stair-ascending task, the dynamic prediction system provides data for motion and external forces such as ground reaction forces (GRF). A program in OPENSIM muscle modeling uses predicted data to compute muscle force. A developed program uses muscle forces and the external forces (GRF) to calculate compression and shear forces in the joint. A finite element model conducts analysis to compute the field stress at the joint for a few selected frames of the motion (10–20 frames). We used the neural network approach to compute the stress field for the entire motion of the stair-ascending task (140 frames). The value of Root Mean Squared Error (RMSE) and Mean Absolute Error (MAE) of the results of the Neural Network (NNT) are reasonable and shown in Table 1. The work of this study focuses on the cartilage part of the joint, where direct contact takes place during ankle articulation. The program determines and evaluates the ratio of current Von Mises stress [17] to the yield stress of all joint components.

A healthy joint means the value of the ratio for all components is less than unity. A value of unity or more of the ratio means a risky joint. A diagram illustrates the

Table 1. Neural network RMSE & MAE for 33 kg backpack, right ankle, stair-ascending results

Number of cycles	1	500
Number of input frames (Input points)	18	18
Number of output frames (Output points)	140	140
Number of variables (Ankle components)	7	7
RMSE - on grid (18 points)	0.063	0.085
RMSE - off grid (3 points)	0.065	0.088
MAE - on grid (18 points)	0.051	0.064
MAE - off grid (3 points)	0.054	0.067

complete method of injury assessment in Fig. 4. Results of the injury index for the right and left tibia cartilage and the talus cartilage of the ankle are shown in Tables 2 and 3.

$$\text{RMSE} = \sqrt{\frac{1}{n}\sum_{i=1}^{n}(y_i - y_i^*)^2} \qquad \text{MAE} = \frac{1}{n}\sum_{i=1}^{n}|y_i - y_i^*| \qquad (3)$$

where y_i and y_i^* are the actual value predicted value of the variable, respectively.

Table 2. Comparison of injury index of tibia cartilage and talus cartilage at different loadings for right ankle at 66% of stair-ascending stride

Number of Cycles	Right Tibia Cartilage			Right Talus Cartilage		
	Load-0	Load-1	Load-2	Load-0	Load-1	Load-2
1	0.187	0.227	0.263	0.169	0.169	0.172
10	0.188	0.227	0.264	0.170	0.171	0.172
100	0.191	0.231	0.268	0.173	0.174	0.175
250	0.198	0.239	0.279	0.174	0.176	0.181
500	0.216	0.261	0.303	0.194	0.195	0.198

Load-0 = no backpack, Load-1 = 17 kg backpack, Load-2 = 33 kg backpack.
The injury index at yield point is a unity.

Table 3. Comparison of injury index of tibia cartilage and talus cartilage at different loadings for left ankle at 12.66% of stair-ascending stride.

Number of Cycles	Left Tibia Cartilage			Left Talus Cartilage		
	Load-0	Load-1	Load-2	Load-0	Load-1	Load-2
1	0.177	0.185	0.261	0.166	0.171	0.172
10	0.178	0.185	0.261	0.166	0.171	0.172
100	0.188	0.218	0.266	0.169	0.176	0.178
250	0.241	0.255	0.276	0.175	0.181	0.182
500	0.262	0.273	0.299	0.191	0.197	0.198

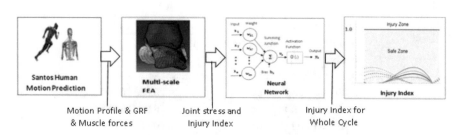

Fig. 4. Injury assessment approach

The results showed that the maximum injury index values for the tibia cartilage and talus cartilage of the right ankle were at 66% of the stair-ascending stride for three loading cases (see Fig. 5). The injury index values of the right tibia cartilage after 500 cycles are 0.216, 0.261, and 0.303, as presented in Table 2. For the same joint under the same task, the values of the injury index after 10 cycles are 0.188, 0.227 and 0.264. The injury index values of the right talus cartilage after 500 cycles are 0.194, 0.195, and 0.198, as presented in Table 2. For the same joint under the same task, the values of the injury index after 10 cycles are 0.170, 0.171, and 0.172.

The results also showed that the maximum injury index values for the tibia cartilage and talus cartilage of the left ankle were at 12.66% of the stair-ascending stride for three loading cases (see Fig. 5). The injury index values of the left tibia cartilage after 500 cycles are 0.262, 0.273, and 0.299, as presented in Table 3. For the same joint under the same task, the values of the injury index after 10 cycles are 0.178, 0.185, and 0.261. The injury index of the left talus cartilage after 500 cycles are 0.191, 0.197, and 0.198, as presented in Table 3. For the same joint under the same task, the values of the injury index after 10 cycles are 0.169, 0.176, and 0.178.

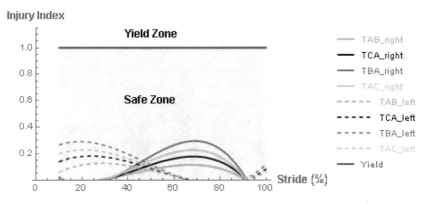

Fig. 5. A combination of right and left ankle components for 33 kg backpack after 500 cycles

Figure 6 shows plots of the injury index of tibia cartilage at 66% of the stair-ascending stride for the three load cases under different cycle numbers, respectively. Figure 7 shows a bar chart comparison of the injury index values for the tibia cartilage and talus cartilage under the 33 kg backpack for different numbers of cycles and at different percent of stride cycle.

Generally, from the injury index values given in Tables 2, 3 and the graphs, it is evident that there is an increase of 6% in the injury index value of the tibia cartilage and talus cartilage when repeating the task 250 cycles and an increase of 15% when repeating the task 500 cycles. This means that the probability of risk of injury of the cartilage of the ankle joint is higher by 6% when repeating the task 250 cycles and by 15% when repeating the task 500 cycles.

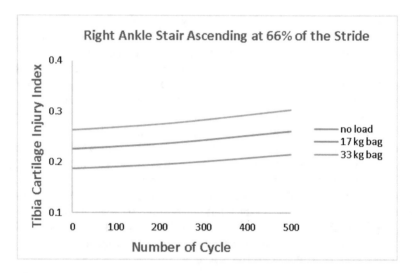

Fig. 6. Comparison of injury index at different cycle numbers for tibia cartilage

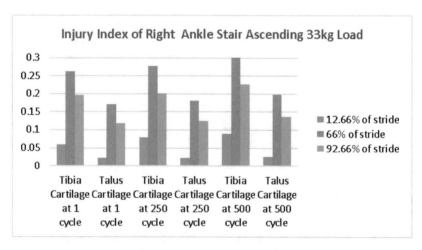

Fig. 7. Injury index of tibia cartilage and talus cartilage at 33 kg backpack for right ankle stair ascending.

5 Conclusions

This study has presented a method for injury risk assessment of ankle components due to a repetitive loading task. The DHM provides dynamic prediction data about motion and external forces for a stair-ascending task. The study involved muscle forces that articulate the ankle. An FEA model of the ankle using ABAQUS software computed stress fields for all joint components under different types of loading. A neural network technique (NNT) calculated the stress field of the components over the whole cycle of the motion of the load and computed an injury index for all ankle components for different numbers of repetitive loading cycles. The approach of measuring injury index of the ankle under repetitive loading is a powerful method and provides a quantitative measure of the probability of injury occurrence, as it measures the level of the risk under different load cycles. This, definitely, can indicate actions to consider component injury prevention.

References

1. Danny, L.T., Hollingsworth, J.: The prevalence and impact of musculoskeletal injuries during a pre-deployment workup cycle: survey of a Marine Corps special operations company. J. Spec. Oper. Med. Fall 9(4), 11–15 (2009)
2. Zambraski, E.J., Yancosek, K.E.: Prevention and rehabilitation of musculoskeletal injuries during military operations and training. J. Strength Cond. Res. 26(7), 101 (2012). National Strength and Conditioning Association
3. Army Report TB MED 592: Prevention and control of musculoskeletal injuries associated with physical training. Technical Bulletin, May 2011
4. Renstrom, P.A.F.H., Konradsen, L.: Ankle ligament injuries. Br. J. Sports Med. 31(1), 1–20 (1997)
5. Seedhom, B.B.: Conditioning of cartilage during normal activities is an important factor in the development of osteoarthritis. Rheumatology 45, 146–149 (2006)
6. Kumar, S.: Theories of musculoskeletal injury causation. Ergonomics 44(1), 17–47 (2001)
7. Gates, D.H., Dingwell, J.B.: The effects of neuromuscular fatigue on task performance during repetitive goal-directed movements. Exp. Brain Res. 187(4), 573–585 (2008)
8. Enoka, R.M., Duchateau, J.: Muscle fatigue: what, why and how it influences muscle function. J. Physiol. 586(1), 11–23 (2008)
9. Sultan, S., Marler, R.T.: Multi-scale predictive human model for preventing injuries in the ankle and knee In: 6th International Conference on Applied Digital Human Modeling, Las Vegas, July 2015
10. Abdel-Malek, K., Yang, J., Kim, J., Marler, R.T., Beck, S., Nebel, K., Santos: a virtual human environment for human factors assessment. In: 24th Army Science Conference, November, FL, Assistant Secretary of the Army, Research, Development and Acquisition, Department of the Army, Washington, DC (2004)
11. Denavit, J., Hartenberg, R.S.: A kinematic notation for lower-pair mechanisms based on matrices. J. Appl. Mech. 77, 215–221 (1995)
12. Marler, T., Knake, L., Johnson, R.: Optimization-Based Posture Prediction for Analysis of Box-Lifting Tasks. In: 3rd International Conference on Digital Human Modeling, Orlando, FL, July 2011
13. Marler, R.T.: A Study of Multi-objective Optimization Methods for Engineering Applications. Ph.D. Dissertation. University of Iowa, Iowa City (2005)
14. Sultan, S., Abel-Malek, K., Arora, J., Bhatt, R., Marler, T.: An integrated computational simulation system for injury assessment. In: 7th International Conference on Applied Digital Human Modeling, Florida, July 2016
15. Anderson, D.D., Goldsworthy, J.K., Shivanna, K., Grosland, N.M., Pedersen, D.R., Thomas, T.P., Tochigi, Y., Marsh, J.L., Brown, T.D.: Biomech. Model. Mechanobiol. 5(2–3), 82–89 (2006)

16. Sultan, S., Abdel-Malek, K., Arora, J., Bhatt, R.: Human simulation system for injury assessment due to repetitive loading. In: Cassenti, D. (ed.) Advances in Human Factors in Simulation and Modeling. AHFE 2017. Advances in Intelligent Systems and Computing, vol. 591. Springer, Cham (2018)
17. Andriyana, A.: Failure Criteria for Yielding, CEMEF UMR CNRS 7635, Sophia Antipolis (2008)

An Articulating Statistical Shape Model of the Human Hand

Jeroen Van Houtte[1]([✉]), Kristina Stanković[1], Brian G. Booth[1],
Femke Danckaers[1], Véronique Bertrand[2], Frederik Verstreken[2],
Jan Sijbers[1], and Toon Huysmans[1,3]

[1] imec-Vision Lab, University of Antwerp, 2610 Antwerp, Belgium
Jeroen.vanhoutte@uantwerp.be
[2] Orthopedic Department, AZ Monica Hospital, 2000 Antwerp, Belgium
[3] Applied Ergonomics and Design, Delft University of Technology, 2628 Delft,
The Netherlands

Abstract. This paper presents a registration framework for the construction of a statistical shape model of the human hand in a standard pose. It brings a skeletonized reference model of an individual human hand into correspondence with optical 3D surface scans of hands by sequentially applying articulation-based registration and elastic surface registration. Registered surfaces are then fed into a statistical shape modelling algorithm based on principal component analysis. The model-building technique has been evaluated on a dataset of optical scans from 100 healthy individuals, acquired with a 3dMD scanning system. It is shown that our registration framework provides accurate geometric and anatomical alignment, and that the shape basis of the resulting statistical model provides a compact representation of the specific population. The model also provides insight into the anatomical variation of the lower arm and hand, which is useful information for the design of well-fitting products.

Keywords: Articulation model · Registration · Human hand
Statistical shape modelling

1 Introduction

Shape models of faces and full-bodies have become valuable for many commercial applications of computer vision and graphics, ranging from customized design to motion tracking [1, 2]. Their potential for noise and artifact reduction, hole filling, and resolution improvement, have aided to employ low-budget scanners with low mesh quality [3, 4]. Recently, the popularity of these techniques has led to their consideration for modelling the human hand, most often for the task of hand tracking [5, 6].

In the context of hand tracking, shape models have been used with the primary goal of improving pose estimation [5–9]. In general, these techniques consist of a fixed prior rigged template model which can be aligned to person-specific depth images or 3D meshes. The alignment is often achieved by solving for the articulation and anthropometric parameters of the template that optimally match the subject's depth image or

3D mesh. The registration is regularized by principal component analysis (PCA) [5] or by an "as rigid as possible" (ARAP)-regularization [6].

As the focus of these techniques has been on obtaining accurate pose information, the level of geometric detail can vary significantly between models. Many models are composed of primitives like spheres and cylinders of fixed size, with the registration step simply articulating these primitives [5, 7]. Others use a more realistic skin geometry, but only allow the model to articulate [8]. Accommodating variations in hand shape and size has only recently been explored [6, 9], and those variations have not been restricted to a range of "natural" hand shapes and sizes.

It has been argued that detailed personalized hand models improve the accuracy of both model registration and pose estimation [10]. This argument was furthered by Khamis et al. who regularized possible hand shapes with a low-dimensional parametric shape model that included statistical shape variations of a population [11]. Ideally, this shape model would be based on a dataset of high-quality surface scans in the same pose, but Khamis et al. constructed their shape basis on low-quality depth scans which contained self-occlusions. To address the low quality, their statistical shape model was estimated simultaneously with each individual's hand shape and pose parameters.

Meanwhile, recent advances in optical scanning technology, such as the 3dMD-system [12], have enabled the acquisition of high-quality (<0.5 mm error) 3D surface scans, even from highly articulating objects like hands. It is expected that a statistical shape model based on these high-quality scans would reveal more geometric details and it is therefore the interest of this paper to build a high-resolution geometric shape basis that, to the best of our knowledge, has not been seen in the literature.

Building such as statistical model requires bringing the 3D scans of different subjects' hands into anatomical correspondence. Having an anatomical correspondence for all points of all meshes is critical to build an accurate and interpretable model. However, this is an especially difficult task for a complex articulating shape like the hand.

The aim of this study is to obtain reliable anatomical correspondence for building a statistical human hand model. We propose a registration algorithm, similar to the technique in [9], that aligns a template articulation model to a database containing 100 high-quality 3D scans of human hands. We hypothesize that the addition of the articulation model, and its corresponding registration algorithm, will allow us to more accurately obtain shape correspondences, and normalize for pose, in 3D scans of human hands. We further hypothesize that these advances in shape correspondences and pose normalization will facilitate the use of standard statistical shape modelling algorithms, like PCA, on 3D scans of human hands.

2 Methods

At a high level, our proposed shape modelling technique works as follows. An articulating reference of the human hand, with anatomically correct rotation axes, angles, and constraints, is constructed to act as prior in an articulation-based registration method. This reference hand is then registered to each optical surface scan of a database in order to make anatomically correct correspondences between them. Person-specific deviations that cannot be captured by the reference are accommodated through a subsequent elastic surface registration step. Finally, the registered surfaces are articulated to the same pose and PCA is used to derive the statistical shape model of the human hand. The following subsections discuss these steps in further detail.

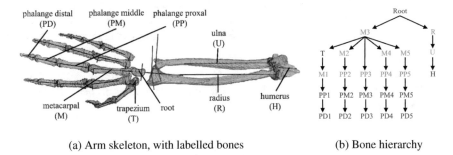

(a) Arm skeleton, with labelled bones   (b) Bone hierarchy

Fig. 1. Our reference articulating hand model is defined by the skeleton in (a). The bones in this skeleton are ordered in the hierarchical tree structure in (b) with an artificial root bone at the wrist. Arrows indicate the parent-child relationship. Colors indicate the corresponding articulation parameters: α(blue), α and β(green), γ(red), δ(orange). See text for further details.

2.1   Reference Articulating Hand Model

Reference Surface Geometry and Skeleton. Our reference hand is based on a single Magnetic Resonance Image (MRI) scan of the first author's right hand (repetition time [TR]: 4220 ms; echo time [TE]: 1560 ms; field of view [FOV]: 192 mm $\times$ 520 mm; resolution: 1 mm^3; no gap). The outer skin surface and all relevant bones were manually segmented from the MR image.

To construct the surface mesh of the reference hand, the binary label field of each body part, obtained from the MRI scan, was then converted to a triangulated surface mesh using a discrete marching cube algorithm [13]. The extracted skin surface mesh was then removed of noisy outliers, smoothed in volume-preserving way [14] and remeshed uniformly [15]. The reference skin mesh is denoted by $\widehat{\mathrm{M}}_M = (\widehat{V}_M, \varepsilon_M)$, with $\widehat{V}_M \in \mathbb{R}^{3 \times N_M}$ a matrix containing the coordinates of the N_M vertices in a rest pose (the rest pose being denoted by the hat), and $\varepsilon_M \in \mathbb{R}^{N_M \times N_M}$ representing the connectivity, which remains constant at all times.

An abstract line-skeleton $\hat{S}$, defined using the set of segmented bones, is shown in Fig. 1(a). Each segmented bone b is represented by a local coordinate frame in the skeleton (i.e. an origin and orientation). The origin of the bone is located at its center-of-rotation h_b and the orientation of its coordinate frame is as described by the International Society of Biomechanics (ISB) [16]. The orientation of each bone with respect to the world reference frame is described by the world-to-bone rotation matrix $C_b \in SO(3)$.

Bone Hierarchy. The set of bones are ordered in the hierarchical tree structure shown in Fig. 1(b). This hierarchy represents the parent-child relationships between the coordinate frames of each bone in the skeleton. The root of the hierarchy is an artificial bone located at the wrist with the same orientation as the third metacarpal bone. A wrist-rooted armature allows us to describe arm and hand motion independently from each other, but with respect to a common root coordinate system at the wrist (this decoupling will be a benefit in our registration tasks). Global hand motion is described by the third metacarpal bone, which the ISB standard defines as the parent of all other carpal bones [16].

Articulation. The articulation of the hand is defined by the state of its joints. Using our skeleton, the state of each joint can be described as a rotation between the local coordinate frames of adjacent bones:

$$R_b = C_b C_{p(b)}^{-1},   \qquad (1)$$

where $p(b)$ is the parent of bone b as defined by the tree structure in Fig. 1(b). The rotation matrix R_b captures how bone b is articulated with respect to its parent. This matrix can be decomposed into three rotation angles - $\alpha_b, \beta_b, \gamma_b$ - which match the ISB's joint angle descriptions [16]. The angle α_b is the primary angle of articulation and describes the bending of the fingers and flexion/extension of the wrist. The angle β is the secondary angle of articulation and describes ulnar/radial deviations of the wrist and the separation between the fingers. The angle γ is a roll angle around the bone's longitudinal axis. An additional angle, δ, is used to define pronation-supination of the arm. This motion is modelled as a rotation around an axis connecting the ulna at the wrist to the radius at the elbow. In the wrist-centered armature the ulna rotates around this axis over the radius. The degrees of freedom for each bone have been indicated by the color in Fig. 1(b): α(blue), α and β(green), γ(red), δ(orange).

When articulating a bone with a new set of angles, we recalculate the parent to bone rotation matrix as a concatenation of these rotation angles. From Eq. (1) it is possible to update the rotation matrix R_b since its parent maintained the same position in space. Relating the rotation matrix of the rest pose with this of the articulated pose, provides the rest-to-pose rotation matrix:

$$T_b = C_b \hat{C}_b^{-1}   \qquad (2)$$

from which we can update the head position of the bone and update the bones further down in the tree hierarchy.

Finally, we confine, by visual inspection, all joint articulation angles to remain within a natural range of motion. To accomplish this, we introduce a mapping from these constrained physical angles to "dummy" unconstrained variables as described in [17]. The benefit of the "dummy" unconstrained variable is that it can be optimized in the registration algorithm without any changes to the optimizer.

Anthropometric Scaling. Besides the articulation of the skeleton, the reference model also accommodates the anthropometric variations related to bone length and body part thickness. The model therefore adopts an affine scaling of each bone defined by a longitudinal scaling factor $s^{//}$ and a transversal scaling factor $s^{\perp}$. The scaling matrix in world coordinates can be written as follows:

$$S_b = C_b \, diag\left(s_b^{\perp}, s_b^{//}, s_b^{\perp}\right) C_b^{-1}. \tag{3}$$

The world to bone transformation including both articulation and scaling is therefore defined as:

$$F_b = S_b T_b. \tag{4}$$

In the reference hand, we apply longitudinal and transversal scaling on the lower arm, hand palm, and each finger separately. For the fingers, a single longitudinal scaling factor is used for all phalanges of the same digit; this is justified by the fact that the ratio of bone lengths between phalanges of a single digit obey closely the golden ratio rule [18]. Nevertheless, we allow the metacarpals to change in length independently from the phalanges in order to maintain flexibility of the reference during the registration task.

Skinning. To deform the reference skin mesh $\hat{M}_M$ to a new skin mesh $M_{M(\Phi)}$ in line with the articulation parameters ϕ of the skeleton, we employ Linear Blend Skinning (LBS) [19]. LBS updates vertices based on the skeleton's articulation via:

$$v_i = \sum_b w_{i,b} F_b \hat{v}_i + t_b, \tag{5}$$

with $t_b = h_b - \Gamma_b \hat{h}_b$ being a translation vector. The skinning weights $w_{i,b}$ capture how much vertex v_i is influenced by articulating bone b. They are obtained by solving a heat equilibrium analogy as described in [20].

During the pronation-supination movement of the lower arm, the amount of skin sliding gradually increases over the elongation axis of the arm. This twisting behavior cannot be explained with standard linear blend skinning since the expected skin deformation does not follow the transformation of its underlying bone. Instead, we model the skin deformation during pronation-supination by applying spherical linear interpolation (SLERP) [21] between C_{ulna} and C_{radius}, where the interpolation parameter t linearly increases from the ulna's head at the wrist to the radius' base at the elbow [9]. A vertex is then rotated with the interpolated rotation matrix depending on its location along the connection axis.

$$v_i = \left[w_{i,ulna} \, t \, T_{ulna} + w_{i,radius} \, (1-t) T_{radius} \right] \hat{v}_i. \tag{6}$$

2.2   Articulation-Based Registration

Hierarchical Optimization. The aim of this section is to fit the articulation model, described in the previous section, to a 3D surface scan, denoted by $M_T = (V_T, \varepsilon_T)$. This registration is done by optimizing a set of model parameters Ω via a non-linear Levenberg-Marquardt (LM) optimization scheme. The parameters include the articulation parameters, anthropometric scaling parameters, and rigid transformation parameters, summarized in Table 1. To avoid the optimizer ending in local minima, we subdivide the set Ω in several smaller groups of parameters $\Omega = \left\{ \Phi_j = \{\phi_i\}_j \right\}$ and order the parameters in each group Φ_j in a hierarchical structure which is optimized iteratively (e.g. A, A-B, A-B-C) by the LM optimizer. By using a wrist-centered armature, we can decouple the hand and arm related parameters and do their registration steps independently. The order in which we optimize the defined parameter groups are: "hand", "arm", "rigid", "scaling", "rigid", "hand", and "arm". Furthermore, we optimize each finger independently.

Landmark-Based Initialization. Before starting the hierarchical iterative optimization protocol, we initialize the registration by globally scaling and aligning the reference hand based on three landmarks: two at opposite sides of the wrist and one at the middle fingertip. Additionally, the length of the arm is set based on the distance between landmarks at the wrist and an additional landmark at the elbow pit. This second step was performed due to missing elbow geometry in our scan dataset, and would not be required if the elbow is thoroughly scanned.

Table 1. Parameter hierarchy. The parameter set is divided in independent parameter groups Φ_j. Parameters in each group are organised in different levels, where each level is optimised at a time. Optimisation is done iteratively between levels within each group.

Group Φ_j	Level $\{\phi_i\}$	Degrees of freedom	Relevant bones $B \subset S$
Hand	A	α_{M3}, β_{M3}	M2–5, PP2–5
	B	$\alpha_{PP2-5}, \beta_{PP2-5}$	PP2–5
	C	$\gamma_{root}, \alpha_{M3}, \beta_{M3}$	M2–5, PP2–5
Arm	A	α_R, β_R	U, R
	B	δ_U	H
	C	α_H	H
Scaling	A	$s^{\perp}_{U,R,H}, s^{\perp}_{M1-5,PP1-5,PM2-5,PD1-5}$	U, R, M2–5, PP2–5
Rigid	A	Global translation and rotation	Root, M5
Thumb	A	$\alpha_{M1}, \beta_{M1}, \alpha_{PP1}$	PP1, PD1
	B	α_{PD1}	PP1, PD1
	C	$\alpha_{M1}, \beta_{M1}, \alpha_{PP1}, \alpha_{PD1}, s^{//}_{M1,PP1,PD1}, s^{\perp}_{PP1,PD1}$	PP1, PD1
Finger *	A	$\alpha_{M*}, \beta_{M*}, \alpha_{PP*}, \beta_{PP*}, \alpha_{PM*}, \alpha_{PD*}$	PP*, PD*
	B	$\alpha_{PP*}, \beta_{PP*}, \alpha_{PM*}, \alpha_{PD*}, s^{//}_{M*}, s^{//}_{PP*,PM*,PD*}, s^{\perp}_{PP*,PM*,PD*}$	PP*, PD*

Energy Function. At each hierarchy level, we apply a LM optimization to minimize

$$\phi = \arg\min_{\phi} \left(\sum_{i=1}^{|V_M|} w_a(i, B) \left| \min_j \left(d \left(V_{M(\phi)}(i), V_T^{//}(j) \right) \right) \right|^2 \right), \tag{7}$$

where w_a is a binary weight used to turn on and off the contribution of vertices, depending on whether its corresponding bone is in the set of bones B considered to be relevant for the optimisation (see Table 1). $V_T^{//}$ is the subset of V_T consisting of vertices whose normals are within $72°$ from the normal at $V_M(i)$ (a more strict threshold of $37°$ is used for the scaling and arm optimisation steps). Rather than excluding points based on their normals, we search for the closest point that meets this normal angle condition. By doing so, we ensure that all points on the mesh will have a corresponding point (as long as the mesh is not too sparse). Points for which a counterpart was not found are excluded from the energy function.

The distance measure $d(p, q)$ used is the point-to-plane distance introduced by Park and Subbarao [22]. This is beneficial over point-to-point distance when using low resolution mesh, but cannot be used for optimizing arm supination since corresponding reference and target vertices lie in the same plane. In that situation, we replace the distance measure by its point-to-point variant.

2.3 Shape Correspondences

Initially, the vertices in our 3D meshes are randomly ordered, meaning that, say, vertex v_i in our reference mesh does not anatomically correspond to vertex v_i in another hand mesh. The number of vertices may also be different for every mesh. Before performing statistical analysis on these meshes, we must first establish an anatomical correspondence between them. This correspondence is achieved in two steps. First, the articulation-based registration, described above, is performed to align our reference hand to the target mesh. Second, an elastic registration algorithm is applied to provide a more precise anatomical correspondence between the reference mesh and the target mesh [23]. The final result is that the reference surface is deformed to have its shape as similar as possible to the shape of the target surface. At this point, the target mesh is replaced by the deformed reference, ensuring that each hand mesh has the same number of vertices ordered in the same fashion. This consistent vertex order ensures that every hand mesh has the same vertices in the same anatomical positions.

2.4 Pose Normalization

In the statistical model, we are only interested in anthropometric variations and want to normalize as much as possible for any variation due to pose and articulation differences. Therefore, we apply a pose normalization on the elastically deformed mesh, using the skeleton estimated by articulation-based registration. Pose normalization can easily be achieved by interchanging the rest and pose articulations, i.e. inverting the rest to pose transformation matrix in Eq. (2). Finally, all pose-normalized scans are centered around their center-of-gravity position.

2.5   Shape Modelling

To investigate the principal modes of variations present in the population, we apply a linear dimensionality reduction algorithm on the pose normalized registered scans. A popular choice for statistical shape modelling is a principal component analysis (PCA) [24]. In our context, PCA converts the vertex sets from all meshes into smaller sets of values through the definition of linearly uncorrelated variables called principal components. These principal components are defined by applying an orthogonal transformation on the original vertex coordinates. The position of vertex v_i in the statistical model is modelled as its average position μ_i plus a linear combination of principal components $P_{i,j}$:

$$v_i = \mu_i + \sum_j w_j P_{i,j}. \tag{8}$$

The weights w_j give the contribution of each principal component (PC) to the model instance. The calculated PCs describe orthogonal directions of variance and they are ordered based on the fraction of variance found along the direction.

3   Results

In this section, we provide the results of the proposed registration and model-building techniques after testing them on a set of 100 static optical surface scans acquired with a 3dMD system. For comparison purposes, we also applied the elastic registration on the dataset as described in Sect. 2.3 but without the articulation-based initialization of Sect. 2.2.

3.1   Articulation-Based Registration

Anatomical Correspondence. To quantify the anatomical accuracy of the registration method, we annotated 22 anatomical landmarks on the reference mesh and on each target scan. Landmarks were annotated at anatomical feature locations: at the elbow pit, at two opposite points around the wrist, at each fingertip and at all finger joints. We calculated the distance between the landmark positions on the moving mesh and their ground-truth counterpart on the target mesh. These distances were computed after our articulation-based registration, after our elastic registration, and for the result of a purely elastic registration, without articulation-based initialization.

The landmark correspondence results are shown in Fig. 2(a). The average distance between joint landmarks after articulation and elastic registration was 5.7 mm, compared to 6.8 mm without the articulation based initialization step. Anatomical alignment is the best at the fingertips and distal joints because its estimation relies on clear geometric features. The accuracy on the elbow pit alignment is low due to missing data at the elbow and limited geometry information at the upper arm. Given the improved landmark correspondence of our algorithm, we can conclude that the articulation-based registration - as an initialization step - improves the anatomical correspondence of the elastic registration.

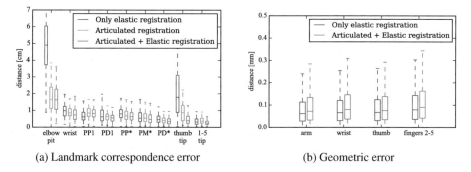

(a) Landmark correspondence error   (b) Geometric error

Fig. 2. The anatomical and geometrical correspondence results for our registration method and a purely elastic method. Anatomical correspondence error, using expert-denoted landmarks, is shown in (a) while geometric errors in the hand shapes are shown in (b).

Geometric Correspondence. To create shape correspondence, we replace a target mesh by the registered result. This step may introduce geometric error where the surfaces do not match exactly. We quantify this geometric correspondence accuracy by calculating the average distance between the target and the elastically registered mesh, in the normal direction on the registered mesh. The results are shown in Fig. 2(b), with the average distance between surfaces grouped by anatomical region. The average geometric accuracy of our algorithm was 0.12 mm.

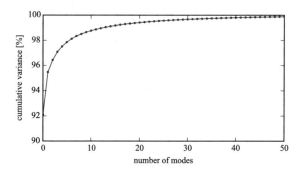

Fig. 3. Normalized compactness graph of the statistical hand shape model.

3.2   Statistical Model

Model Performance/Compactness. The compactness of a statistical model is a widely used measure to quantify how efficiently the model describes the total variance in the population [25, 26]. The compactness measure $C(m)$ is defined as the sum of the shape variance captured by the first m principal components:

$$C(m) = \sum_{i=1}^{m} \lambda_i, \qquad (9)$$

with λ_i the shape variance described by the ith PC. Figure 3 shows the normalized compactness results of our statistical hand shape model. The first principal component explains over 90% of the total variability in the dataset, while the first four PC account for over 97%.

Our model's first four principal components are visualized in Fig. 4. The average geometry is shown along with $\pm$ three standard deviations for each principal component. The first PC describes global scaling. The second PC describes variations in the length-to-thickness ratio of the arm, hand and fingers. The third and fourth PC are related to varying length and width of the fingers relative to arm size, respectively.

4   Discussion

We have presented a two-step registration method for 3D meshes of human hands. First, we matched an articulating prior model to a target scan, then we applied an elastic registration to obtain more precise shape correspondence. We demonstrated our method on a dataset of 100 optical 3D surface scans. We showed that the anatomical accuracy improves by 17% by initializing the elastic registration with the articulation-based registration result, while the average geometric accuracy stays around 0.12 mm. We further fed the registered surfaces into a statistical shape modelling algorithm and showed that the resulting model provides a compact representation of the population's variation. Only four principal components are needed to describe 97% of the shape variability in the dataset. We believe that our model is suitable for applications like hole-filling and resolution improvement, where pose estimation is an inevitable task. Our shape model could also be useful as a prior in a surface registration algorithm.

Nevertheless, our results did highlight a few limitations. We observed low accuracy on the estimation of the elbow pit location mainly due to missing data around the elbow and limited geometry at the upper arm. We also noted that the registration outcome highly depends on its settings (e.g. the ranges of motion, order of parameter optimizations, vertex normal thresholds). Finally, it is likely that some articulation information did make it into the shape model as a result of errors in the articulation-based registration. The source of these errors include the optimization settings, but also the limited degrees of freedom in the reference hand (e.g. the use of the golden ratio to scale finger bones). Our future work will look at addressing these limitations as well as extending the technique to the 4D modelling of hand motion.

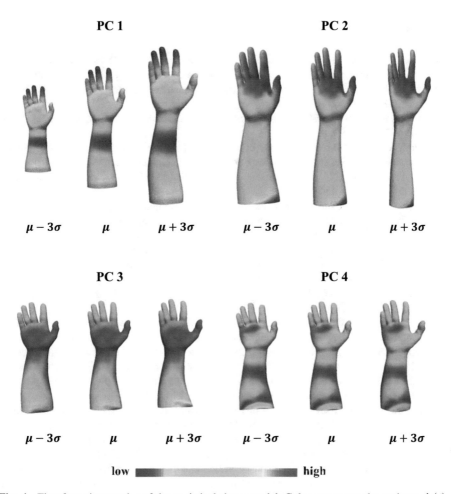

Fig. 4. First four eigenmodes of the statistical shape model. Color represents the variance $\lambda_i(j)$ for vertex j along the ith PC.

5 Conclusion

We presented herein a registration method for 3D meshes of human hands. It was based on the alignment of an articulating reference hand and elastic deformation. We demonstrated the registration's effectiveness by building a PCA shape model of the human hand. In the future, we will improve the anatomical accuracy of the methodology and to extend the method to model hand motion.

Acknowledgments. This work was supported by the Research Foundation in Flanders (FWO SB) and the VLAIO PLATO-project. The authors would like to thank Vigo nv, More Institute vzw and Orfit Industries nv for their continued contribution to the project.

References

1. Blanz, V., Vetter, T.: A morphable model for the synthesis of 3D faces. In: 26th Annual Conference on Computer Graphics and Interactive Techniques, pp. 187–194 (1999)
2. Hasler, N., et al.: A statistical model of human pose and body shape. Comput. Graph. Forum **28**(2), 337–346 (2009)
3. Anguelov, D., et al.: SCAPE: shape completion and animation of people. ACM Trans. Graph. **24**(3), 408–416 (2005)
4. Harih, G., Tada, M.: Development of a finite element digital human hand model. In: 7th International Conference on 3D Body Scanning Technologies, pp. 208–213 (2016)
5. Tagliasacchi, A., et al.: Robust articulated-ICP for real-time hand tracking. Comput. Graph. Forum **34**(5), 101–114 (2015)
6. Taylor, J., et al.: User-specific hand modeling from monocular depth sequences. In: IEEE Conference on Computer Vision and Pattern Recognition, pp. 644–651 (2014)
7. Oikonomidis, I., et al.: Evolutionary quasi-random search for hand articulations tracking. In: IEEE Conference on Computer Vision and Pattern Recognition, pp. 3422–3429 (2014)
8. Sharp, T., et al.: Accurate, robust, and flexible real-time hand tracking. In: 33rd Annual ACM Conference on Human Factors in Computing Systems, pp. 3633–3642 (2015)
9. Zhu, L., et al.: Adaptable anatomical models for realistic bone motion reconstruction. Comput. Graph. Forum **34**(2), 459–471 (2015)
10. Tan, D.J., et al.: Fits like a glove: rapid and reliable hand shape personalization. In: IEEE Conference on Computer Vision and Pattern Recognition, pp. 5610–5619 (2016)
11. Khamis, S., et al.: Learning an efficient model of hand shape variation from depth images. In: IEEE Conference on Computer Vision and Pattern Recognition, pp. 2540–2548 (2015)
12. Lübbers, H.-T., et al.: Precision and accuracy of the 3dMD photogrammetric system in craniomaxillofacial application. J. Craniofac. Surg. **21**(3), 763–767 (2010)
13. Grothausmann, R.: Providing values of adjacent voxel with vtkDiscreteMarchingCubes (2016)
14. Visual Computing Lab ISTI CNR: MeshLab. http://www.meshlab.sourceforge.net
15. Valette, S., Chassery, J.-M.: Approximated centroidal voronoi diagrams for uniform polygonal mesh coarsening. Comput. Graph. Forum **23**(3), 381–389 (2004)
16. Wu, G., et al.: ISB recommendation on definitions of joint coordinate systems of various joints for the reporting of human joint motion—Part II: shoulder, elbow, wrist and hand. J. Biomech. **38**(5), 981–992 (2005)
17. James, F.: Minuit-a system for function minimization and analysis of the parameter errors and correlations. Comput. Phys. Commun. **10**(6), 343–367 (1975)
18. Park, A.E., et al.: The Fibonacci sequence: relationship to the human hand. J. Hand Surg. Am. **28**(1), 157–160 (2003)
19. Lewis, J.P., et al.: Pose space deformation. In: 27th Annual Conference on Computer Graphics and Interactive Techniques, pp. 165–172 (2000)
20. Baran, I., Popovi, J.: Automatic rigging and animation of 3D characters. ACM Trans. Graph. Artic. **26**(3), 72 (2007)
21. Shoemake, K., et al.: Animating rotation with quaternion curves. In: ACM SIGGRAPH Computer Graphics, pp. 245–254 (1985)
22. Park, S.-Y., Subbarao, M.: An accurate and fast point-to-plane registration technique. Pattern Recognit. Lett. **24**(16), 2967–2976 (2003)
23. Danckaers, F., et al.: Correspondence preserving elastic surface registration with shape model prior. In: 22nd International Conference on Pattern Recognition, pp. 2143–2148 (2014)

24. Cootes, T.F., et al.: Active shape models-their training and application. Comput. Vis. Image Underst. **61**(1), 38–59 (1995)
25. Davies, R.H., et al.: A minimum description length approach to statistical shape modelling. IEEE Trans. Med. Imaging **21**(5), 525–537 (2002)
26. Su, Z.: Statistical Shape Modelling: Automatic Shape Model Building. University College London (2011)

The Effect of Object Surfaces and Shapes on Hand Grip Function for Heavy Objects

Mario Garcia[1], Jazmin Cruz[1], Cecilia Garza[2], Patricia DeLucia[2], and James Yang[1($\boxtimes$)]

[1] Human-Centric Design Research Lab, Department of Mechanical Engineering, Texas Tech University, Lubbock, TX 79409, USA
{mario.f.garcia, jazmin.aguilar, james.yang}@ttu.edu
[2] Department of Psychological Sciences, Texas Tech University, Lubbock, TX 79409, USA
cecilia.r.garza@students.tamuk.edu, pat.delucia@ttu.edu

Abstract. Successful grasp, transfer, and release of objects with the hand are important movements for completing everyday tasks. This study's objective is to understand the effect of an object's surface and shape on hand grip function for heavy objects in a young age group. For their functional prevalence and significance, grasp and release movements have been incorporated into many clinical hand function assessments, such as the Box and Block Test (BBT), a common test used to assess people's hand grip function. In the BBT, subjects transport a block from one side of a box to the other while crossing a partition and repeat the procedure as fast as possible. In this study we measured performance on the BBT in 20 right handed subjects between ages 20 to 30 years old. There were no statistically significant effects of object surfaces and shapes on hand grip function for heavy objects.

Keywords: Box and Block Test (BBT) · Hand dexterity · T-test

1 Introduction

Everyday activities require basic hand movements to complete specific tasks. These movements range from moving a light pencil in the office to moving a heavy dumbbell in the gym. To carry any type of object, one is required to grasp it and eventually release it. Due to the frequency of this set of movements throughout our lives, there have been several clinical tests to monitor hand grip function such as the Box and Block Test (BBT).

The BBT has various functions in the medical field. The BBT has been used to measure hand dexterity in patients with diseases such as cerebral palsy and stroke [1, 2], multiple sclerosis, traumatic brain injury, fibromyalgia, and upper limb amputation [3]. The BBT also can be used by physical therapists to evaluate the progress of rehabilitation in patients with hand impairments. For example, the BBT was used to evaluate the progress of stroke patients who received advanced rehabilitation techniques to improve hand function [4].

The improvement of the use of the BBT will have potential benefits for diagnosis and rehabilitation of hand impairments. In the current study, we measured performance on the BBT to determine whether there are effects of heavy object surfaces and shapes on hand grip function. If we find significant differences in BBT scores between objects with different shapes and textures, objects can be designed (e.g., with coating) to improve hand grip and thus facilitate the daily lives of those who suffer hand impairments.

2   Methods

2.1   Subjects

Subjects consisted of 20 (10 female) young right-handed adults who were between 20 and 30 years old (Mean: 22.35 yrs.; STD: 2.30 yrs.). All individuals had no history of motor movement or visual impairment. In addition, participants had no history of previous strokes, cerebral palsy, multiple sclerosis, traumatic brain injury, neuromuscular disorder, spinal cord injury, and fibromyalgia. The BBT time was set as 15 s.

2.2   Procedure

After completing a screening for study requirements and informed consent, subjects were seated in front of the experimental set up. This consisted of two identical boxes lined with foam that were connected in the middle by a partition and filled on one side with objects of different shapes and textures. Refer to Figs. 1, 2 and 3. There were three metal cylinders and three metal cubes. The objects were considered heavy because they weighted more than what was used in prior studies [5]. The cubes weighed 126 g and the cylinders weighed 100 g. Each object was covered with either a non-slip coating, spray paint, or nothing (just the raw metal).

Fig. 1. Cubes with various textures

Fig. 2. Cylinders with various textures

Fig. 3. Experimental setup

Subjects performed a modified version of the Box and Block Test (BBT) while seated upright with their elbows level with the table and each hand on either side of the box (see Fig. 3). Before each trial, one side of the box was filled with 30 objects. The side that was filled alternated from trial to trial. Subjects were asked to move as many blocks as possible from one side of the box to the other and were given 15 s to do so. Although a BBT score is usually calculated with a time span of 60 s, we shortened our testing to 15 s in order to save time and material expenses.

The start and end of the trial was timed and denoted with an audible chime. Rest was provided between trials. Subjects were instructed to use only their non-dominant (left) hand, to drop rather than throw the object, and to make sure that their fingers crossed the partition before dropping the block. Blocks transferred without following these rules did not count in the BBT score. Aside from these constraints, subjects could transfer the objects in any order and with any type of grasp or movement. The non-dominant hand is the weakest and slowest hand when it comes to movement which allowed for improvement over trials, and arguably simulates the hand of a patient with a disability or impairment; this may assist us while drawing conclusions at the end of our study [5].

The order in which the block shapes and textures was presented was randomized for each participant. Each of the six unique objects were transferred twice by each subject, resulting in 12 trials. Subjects were given practice before starting. The number of blocks that were moved from one side of the box to the other was counted and defined the BBT score on a given trial. All trials were recorded with a video camera in order to review each video in slow motion for an accurate BBT score.

3 Results

We calculated the BBT score for each condition (shape x texture) and conducted t-tests for these BBT scores to determine whether there were significant differences among conditions. No significant differences were found. We then compared differences among the three textures, and the three shapes, separately. Results are shown in Figs. 4, 5, 6, 7, 8 and 9. There again were no significant differences among textures or shapes.

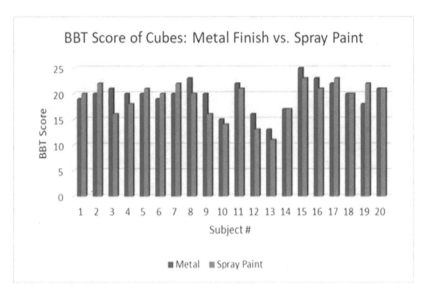

Fig. 4. BBT Score of Cubes: Metal Finish vs. Spray Paint (Cube Metal Finish Mean: 19.7, Cube Spray Paint Mean: 19.05)

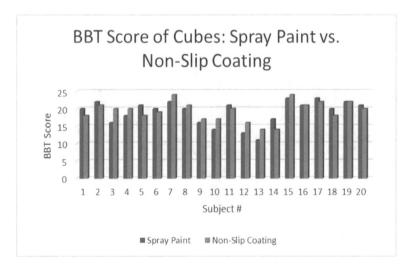

Fig. 5. BBT Score of Cubes: Spray Paint vs. Non-Slip Coating (Cube Spray Paint Mean: 19.05, Cube Non-Slip Coating Mean: 19.3)

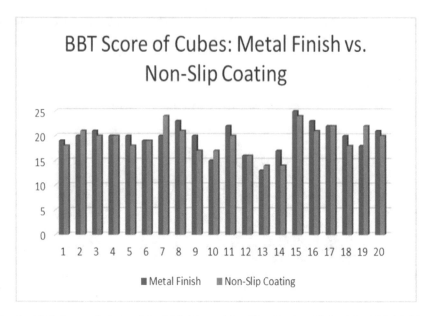

Fig. 6. BBT Score of Cubes: Metal Finish vs. Non-Slip Coating (Cube Metal Finish Mean: 19.7, Cube Non-Slip Coating Mean: 19.3)

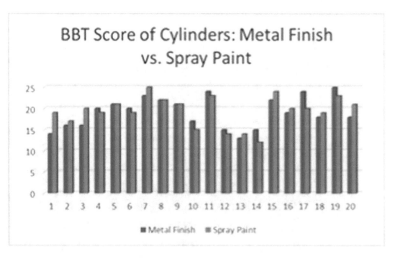

Fig. 7. BBT Score of Cylinders - Metal Finish vs. Spray Paint (Cylinder Metal Finish Mean: 19.15; Cylinder Spray Paint Mean: 19.4)

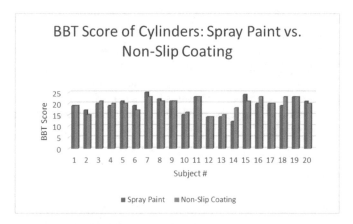

Fig. 8. BBT Score of Cylinders – Spray Paint vs. Non-Slip Coating (Cylinder Spray Paint Mean: 19.4; Cylinder Non-Slip Coating Mean: 19.65)

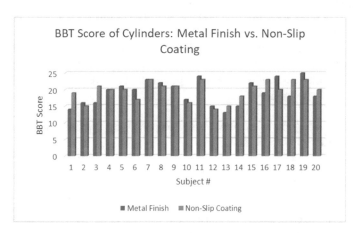

Fig. 9. BBT Score of Cylinders – Metal Finish vs. Non-Slip Coating (Cylinder Spray Paint Mean: 19.15; Cylinder Non-Slip Coating Mean: 19.65)

4   Discussion

Our finding did not indicate significant differences in BBT scores between shapes and textures of transferred objects, which is not consistent with previous studies for light objects. For example, a study using three different textures – paper, wood, and rubber, showed that subjects were able to move about 8% more rubber blocks than blocks covered with other materials (paper and wood). They concluded that object with a nonslip surface would have a higher BBT score than those with a slippery surface like paper or wood. [5] The fact that they used rubber as one of their textures, which has a significantly high Coefficient of friction (COF) than the rest of the materials, may have been the reason they saw a significant difference in their BBT score whereas we did not (we used paints instead). Another study measured the BBT in patients with cerebral palsy who transferred objects by using a glove based virtual reality system, rather than changing the textures in each shape. Results indicated a small improvement in BBT scores when using the VR-glove; this may be useful for physical therapists [6]. The discrepancy between our current results and prior studies may be due to the differences in textures that were used. Another reason is that most prior studies tested object that weighed less than our heavy objects. In addition, prior studies used only cubes whereas we used cylinders also.

5   Conclusion

We did not obtain significant differences in BBT scores between heavy objects of different textures and shapes. This may be due to the heavier weight we used compared to prior studies of light objects. Heavy objects require a greater amount of normal force on the cube or cylinder to successfully carry it over a partition. Failure to do this may result in dropping the object before it crosses the wooden barrier. In addition, the textures used in this experiment may not have differed sufficiently in their coefficient of friction (COF).

Future studies should attempt to replicate this study using textures with a higher COF. It would also be of interest to test people who actually have a hand impairment as it may provide different results from our experiment and have greater generalizability to people with such impairments.

Acknowledgements. This work was partly supported by National Science Foundation (#1559393) and Texas Tech University.

References

1. Cromwell, F.: Occupational Therapists Manual for Basic Skill Assessment: Primary Prevocational Evaluation. Fair Oaks Printing, Altadena (1976)
2. Slota, G.P., Enders, L.R., Seo, N.J.: Improvement of hand function using different surfaces and identification of difficult movement post stroke in the Box and Block Test. Appl. Ergon. **45**(4), 833–838 (2014)
3. Kontson, K., Marcus, I., Myklebust, B., Civillico, E.: Targeted box and blocks test: normative data and comparison to standard tests. PLoS ONE **12**(5), e0177965 (2017)
4. Knutson, J.S., Harley, M.Y., Hisel, T.Z., Chae, J.: Improving hand function in stroke survivors: a pilot study of contralaterally controlled functional electrical stimulation in chronic hemiplegia. Arch. Phys. Med. Rehabil. **88**(4), 513–520 (2007)
5. Seo, N.J., Enders, L.R.: Hand grip function assessed by the box and block test is affected by object surfaces. J. Hand Ther. **25**, 397–405 (2012)
6. van Hedel, H.J.A., Wick, K., Meyer-Heim, A., Eng, K.: Improving dexterity in children with cerebral palsy Preliminary results of a randomized trial evaluating a glove-based VR-system. In: International Conference on Virtual Rehabilitation (2011)

Approaches to Study Spine Biomechanics: A Literature Review

Jazmin Cruz[1], James Yang[1(✉)], and Yujiang Xiang[2]

[1] Human-Centric Design Research Lab, Department of Mechanical Engineering,
Texas Tech University, Lubbock, TX 79409, USA
{Jazmin.aguilar,james.yang}@ttu.edu
[2] Department of Mechanical Engineering, University of Alaska-Fairbanks,
Fairbanks, AK 99775, USA
yxiang@alaska.edu

Abstract. A large population will likely experience lower back pain during their lifetime. Severe cases of lower back pain can sometimes be caused by back conditions or diseases, eventually being alleviated through surgical procedure. Skilled surgeons can make educated decisions on the best procedure for their patients, but the development of a spine model that can estimate biomechanical properties of the spine could aid in surgical decision-making. This paper discusses the current state of the art of four approaches used to study spine biomechanics: *in vivo* experimentation, *in vitro* cadaveric testing, finite element analysis, and musculoskeletal modeling. It is concluded that using a combination of these methods can lead to more accurate spine models that could possibly lead to clinical use.

Keywords: Musculoskeletal modeling · Finite element analysis
Digital human modeling · Human spine

1 Introduction

Low back pain affects a large population of people at some point in their lifetime [1], although majority of these cases are not severe. In more serious cases, low back pain is typically associated with abnormalities within the lumbar portion of the spine. Some root causes of this low back pain include scoliosis and spondylolisthesis. Scoliosis is a spine disease that causes abnormal curvature of the spine. Spondylolisthesis is a spine condition where there is anterior slippage of the vertebra. A surgical solution to both conditions often result in spinal fusion, which involves realigning and fusing two vertebras together so that they can heal into a single bone. During this healing process, the spine is unstable, so rods and screws are used to increase stability.

As expected, invasive surgery is undesirable due to the economic and physical cost to the patient, however, it is sometimes necessary so that the patient can increase their quality of life. It is pertinent that the plan of action chosen by the doctor is the best surgical strategy for the patient. Experienced back surgeons can make educated decisions on what is the best procedure for a patient, but the ability to estimate the location of the problem within the spine or predict the performance of a surgery could aid the

D. N. Cassenti (Ed.): AHFE 2018, AISC 780, pp. 453–462, 2019.
https://doi.org/10.1007/978-3-319-94223-0_43

doctor in decision-making. With the use of various methods, this ability is made possible. The intention of this paper is to review the state of the art of methods for spine biomechanics research involving *in vivo* experimentation, *in vitro* cadaveric studies, finite element analysis and musculoskeletal modeling.

This paper is organized as follows. First, *in vivo* methods are described. Second, *in vitro* and cadaveric testing is examined. Next, finite element analysis models of the spine are explored. The next section discusses musculoskeletal models. The closing section includes a conclusion on the current state of methods for spine biomechanics and proposes direction for future research, such as a multi-scale method that has potential to create a more accurate and robust patient-specific spine model.

2 Methods Used to Study the Spine

2.1 In Vivo

In vivo experiments use living organisms, such as animals or human beings. While the use of animal and human testing is sometimes questioned, it can immensely aid in better understanding how the body would realistically react under the conditions that are being investigated. This type of experiment has been used to explore the biomechanical behavior of the spine and later serves as reference for other modeling methods that will be discussed later in this paper.

Animal testing can be helpful as that they sometimes have characteristics that are alike to humans. It has been found that the use of animal spines could be a comparable alternative to human spines, due to similarities between their axial loading [2]. However, the bone densities differ, which may make direct comparisons between the two difficult. Although there is regulation on the treatment of animal test subjects, such as providing them with anesthesia and relieving them with euthanasia, there are still arguments for alternative methods that do not require animal testing with the belief that animals do not deserve to be manipulated regardless of whether or not they feel pain [3]. Ethically speaking, it may not be a desirable route to pursue, especially since spine research is directly addressing humans. Besides the ethical dilemma, a major disadvantage to using animals is the economic cost of purchase [4] and the man-power needed for the maintenance of these animals, which makes this method less widely available to researchers.

Understandably, human testing is the most desirable form of *in vivo* experimentation. As opposed to animal testing, the use of a human allows for a more accurate representation of the human body and how it naturally moves. One type of *in vivo* experimentation involves motion capture tracking. Motion capture utilizes markers that are placed on participants, customized to their anatomy. These markers reflect light that is recognized by motion capture cameras, which track the trajectories of those markers to allow for kinematic calculation. Depending on the task of interest, the market placement can vary from experiment to experiment. Kinetics can also be obtained through motion capture when they are used in combination with force plates. Another kind of *in vivo* experimentation involves the use of medical imaging. It has been used to evaluate the center of rotation of vertebras [5] and measure the deformation of intervertebral discs [6]. The experiment has a participant perform various motions, like bending or twisting, while the medical imaging tool records the motion. One thing to keep in mind is that researchers, depending on their choice of medical imaging, must be aware of how much radiation their participants are experiencing to avoid dangerous exposure levels. Both motion capture and medical imaging have the potential to avoid invasive operations, while taking meaning measurements of the spine's performance under various loadings.

In addition to non-invasive studies, invasive *in vivo* studies involving human beings have found direction measurements, such as spinal loading, using surgically implanted transducers [7] and vertebral body replacements [8, 9]. Another research group performed an investigation that surgically implanted devices that were temporarily used to measure kinematics of the spine through motion capture [10].

As expected, the invasive procedure is not desirable due to the excessive cost of performing the surgical procedure, including the use of a surgeon, anesthesia, and a hospital room. In addition to cost, this method may not be the most practical approach since surgical procedures can be burdensome to the participants, such as experiencing pain post-surgery or requiring a considerable time commitment.

2.2 Cadaveric Testing (*In Vitro*)

In vivo experimentation would greatly advance the understanding of the human spine; however, the previous section demonstrates that it is not always ethical or practical. Cadavers can serve as a reasonable alternative. Cadavers have been used to perform *in-vitro* experiments to evaluate the performance of the spine.

The method begins with preparation of the cadaver specimens, including the separation of the spine into the segments of interest. The specimens are kept frozen and are thawed out before testing. The superior and inferior vertebras are potted and pinned so that they are fixed to the mechanical apparatus. The mechanical apparatus is used to apply various loading conditions on the cadavers that simulate motions such as bending and twisting. A variety of apparatuses have been used to test the biomechanical properties of the spine, as shown in Fig. 1.

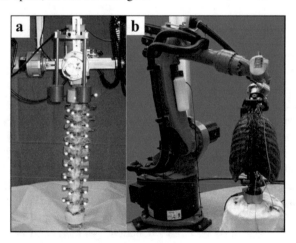

Fig. 1. (a) Cadaver set-up with motion capture [11]; (b) Cadaver set-up including entire ribcage [12]

The loading conditions that are applied to these cadavers are found through *in vivo* measurements from various literature, as mentioned previously. The information found through *in vitro* experiments includes the measurement of spinal disc pressure [13], range of motion [12], and others. This knowledge can be retrieved from a variety of test subjects. For example, healthy spine cadavers can be used to gain a better under-standing of how the healthy spine performs under various day-to-day loading condi-tions. In addition, surgical procedures, such as various methods of spinal fusion, can be tested to see how it would affect the range of motion post-surgery [14]. Although scoliotic spines are difficult to obtain, it might be possible to alter animal spines in such a way to simulate these diseased spines [11].

There are great advantages to using cadavers to simulate the human spine, including the ability to place pressure sensors within the specimen [13]. This allows a direct measurement of pressure that the vertebra disc is experiencing during a specific loading condition. Another advantage of this method includes the ability to test until an injury occurs without putting human life at risk. This method is the closest testing technique that mimics to the performance of a real human spine, which is why *in vitro* findings are often used for validation purposes in modeling methods that will be covered in later sections.

While cadavers allow researchers to ethically experiment with the spine, they come at a great economical cost and are one-time-use since it is possible to test until failure. Cadavers also lack true *in vivo* characteristics due to the preservation processes they endure [15], so the results that are found through this technique may not be an entirely accurate reflection of the spine *in vivo*. Also, the availability of human cadavers is not a major concern, but it does not allow for easy access to specific types of spines, such as adolescent human spines and diseased spines.

In summary, *in vitro* studies that use cadavers can be useful, but also have some limitations, including the unavailability of specific types of human cadavers. Despite its shortcomings, these *in vitro* studies are often used as reference for validation of pure computational simulation models [16].

2.3   Finite Element Analysis

Finite element analysis is a computer aided modeling technique that is used to evaluate a wide range of engineering problems, including structural analysis. It can simulate forces that are loaded onto structures and can visually display the expected deforma-tion, which makes it useful for modeling the structural behavior of the spine. Some common simulation outputs from these models include intradiscal pressure [17, 18], contact force on facet joints [19, 20], and intervertebral rotation [17, 21]. Healthy spines are often modeled since *in vivo* and *in vitro* information is readily available, but some studies have explored the biomechanics of diseased spines [22, 23].

For spine modeling, this method begins with creating the spine geometries through medical imaging, commonly a CT scan. The results found through finite element analysis have been found to be sensitive to geometry [21, 24], so it is important that the complex shapes are replicated accurately. Once the geometries of the bones and discs are built, they are smoothed and meshed appropriately. Once the pieces are meshed, they are imported into the finite element software and pieced together to create a complete spine model which also incorporates spinal ligaments. An example of a finite element model with a dissection of some spine pieces is shown in Fig. 2.

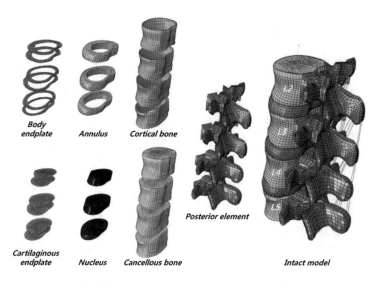

Fig. 2. Finite element model of the spine [19]

As shown in Fig. 2, there are several parts of the spine. These parts all have specific material properties, which are found through previous papers [25–27]. Even with these properties being available to researchers, generic values are the best that is available, meaning that patient-specific material properties are not easily accessible, likely due to the invasiveness of retrieving that information through *in vivo* methods. Fortunately, it has been found that material properties of the bone have little impact on the results found through finite element so it is possible that generic bone material properties will be close enough for prediction purposes [21]. The next step is validation. To validate this model, loading conditions, such as axial loading, lateral bending moments, and flexion/extension moments, are applied and are based on the loadings that occurred in *in vivo* or *in vitro* experiments.

Performing an analysis of a healthy or pre-surgery spine allows for an estimation of the current state of the patient, but finite element modeling also has the potential to aid in predicting the results of surgical procedures that might be used to alleviate lower back pain. It has been used to evaluate the effect of surgical procedures such as artificial disc replacement [28] and spinal fusion [29]. Through the use of finite element analysis, it could be possible to aid surgeons in decision-making concerning what would be the best choice for a specific patient.

As mentioned, current limitations of various finite element models include lack of material property values unique to each specific patient. While this shortcoming may not easily be resolved, improving the accuracy of more sensitive parameters could greatly enhance the performance of spine models. Another disadvantage is the use of generic loading conditions for motions. Again, these motions are based on the motions performed in *in vivo* or *in vitro* experiments that are referenced for validation purposes. These motions can be helpful with validating a model, but there is no patient-specificity to these motions.

2.4   Musculoskeletal Modeling

The purpose of musculoskeletal modeling is to predict muscle and joint forces that are produced during motion. This method has been used in a wide variety of research topics including total joint replacements, ankle sprains, ACL injuries, and many others [30]. Primarily, human-centered research most popularly utilizes this type of modeling, but it could also be used to simulate dynamic motion of animals. Through the power of musculoskeletal modeling, researchers have been able to investigate the internal biomechanics of the spine.

Musculoskeletal modeling begins with an input, which is motion data, and ends with an output of estimated muscle forces. The input, or motion, can be manually defined using the user interface, but motion capture data is commonly used when it is available. This aids in the specificity of the model at hand. As mentioned before, motion capture involves the use of reflective markers placed on the participant. In recent studies, marker placement has varied depending on what information is thought to be important. Two different marker placements can be seen in Figs. 3 and 4, but others have been used in the literature.

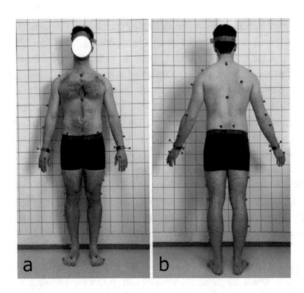

Fig. 3. (a) Front and (b) Back marker placement in Bassani et al. [31].

Prior to running a simulation, the models require critical information such as bone geometry, muscle placement, and muscle properties. The spine geometry, similar to finite element modeling, has been found to affect simulation results [32] so it is

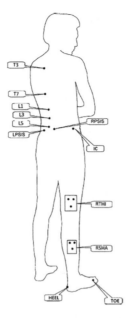

Fig. 4. Marker placement in Kuai et al. [35]

important to have accurate bone morphology. Once the rigid-body skeletal structure is set, each muscle must have a specified path where it lies on the skeletal body. When the muscle locations are set, the modeling system must also understand the way the muscle behaves, such as a direct point-to-point attachment as opposed to a muscle that wraps around bone. Lastly, the material properties of the muscle are defined. In various models, the material properties and location of muscles are based on *in vivo* or cadaveric studies [33, 34].

Depending on the software that is being used for musculoskeletal modeling, various models are readily available to be manipulated so researchers can scale a pre-built model so that the height and weight reflect the participant of interest. However, validation of the modified model is required to ensure that it is a reliable tool for simulation. The validation process involves a comparison of the musculoskeletal results to *in vivo* and *in vitro* results found from literature. Once the model has been validated, it can be used to estimate net muscle forces which can help in evaluating intradiscal pressure [31, 32]. The model can also evaluate range of motion of various skeletal segments during the completion of a variety of everyday tasks [35].

The advantage of musculoskeletal modeling is that it allows researchers to estimate the biomechanical activity that is happening within the body without the need of invasive measurement procedures. In addition, the lack of invasive procedures means that researchers who may not be able to afford *in vivo* testing can estimate *in vivo* muscle forces and evaluate spine biomechanics. Lastly, it can take *in vivo* motion of a human participant and use that data as a direct input, which is a better representation of how humans naturally move and can also make the model more patient-specific.

A major disadvantage to using musculoskeletal modeling is that it is often over-simplified. As mentioned before, there are several types of attachments defined within the model, including point-to-point and wrapping. It is desirable to use point-to-point more often than wrapping because it is a much simpler representation of the ligament and is less computationally expensive [36], although it may not be the most accurate reflection of the muscle. In addition, the ligament properties rely on those that were found in literature. This generality prevents the model from specifically modeling the participant and, instead, serves as a generic model.

Ultimately, musculoskeletal modeling is a promising, non-invasive method that can estimate internal muscle forces that occur within the body, which makes it a great alternative to invasive *in vivo* measurements. Musculoskeletal modeling has had many iterations of models that continue to gain robustness and accuracy [33, 34, 36–38].

3   Conclusion

There are many methods that have been used to gain a better understanding of the spine. The advantages and disadvantages of each method have been discussed, demonstrating that there is not a single best approach to analyze the biomechanics of the spine. For future work, multiscale modeling could be the next step in creating a more accurate and patient specific model of the spine. Multiscale modeling involves the simultaneous use of multiple modeling methods. For example, a study has used net forces found from musculoskeletal modeling as input values for their finite element model [39]. By combining multiple types of modeling, it may be possible to lessen the severity of some limitations that were mentioned in the previous sections, leading to a more powerful modeling method that could potentially be used clinically.

References

1. Woolf, A.D., Pfleger, B.: Burden of major musculoskeletal conditions. Bull. World Health Organ. **81**(9), 646–656 (2003)
2. Smit, T.H.: The use of a quadruped as an in vivo model for the study of the spine – biomechanical considerations. Eur. Spine J. **11**(2), 137–144 (2002)
3. Rollin, B.E.: Toxicology and new social ethics for animals. Toxicol. Pathol. **31**(1_suppl.), 128–131 (2003)
4. Costs of Animal and Non-Animal Testing : Humane Society International. https://www.hsi.org/issues/chemical_product_testing/facts/time_and_cost.html?, https://www.google.com/. Accessed 27 Feb 2018
5. Liu, Z., et al.: Sagittal plane rotation center of lower lumbar spine during a dynamic weight-lifting activity. J. Biomech. **49**(3), 371–375 (2016)
6. Wang, S., Xia, Q., Passias, P., Wood, K., Li, G.: Measurement of geometric deformation of lumbar intervertebral discs under in-vivo weightbearing condition. J. Biomech. **42**(6), 705–711 (2009)
7. Wilke, H.-J., Neef, P., Hinz, B., Seidel, H., Claes, L.: Intradiscal pressure together with anthropometric data – a data set for the validation of models. Clin. Biomech. **16**(1), S111–S126 (2001)

8. Dreischarf, M., et al.: In vivo implant forces acting on a vertebral body replacement during upper body flexion. J. Biomech. **48**(4), 560–565 (2015)
9. Rohlmann, A., Zander, T., Graichen, F., Bergmann, G.: Lifting up and laying down a weight causes high spinal loads. J. Biomech. **46**(3), 511–514 (2013)
10. Rozumalski, A., Schwartz, M.H., Wervey, R., Swanson, A., Dykes, D.C., Novacheck, T.: The in vivo three-dimensional motion of the human lumbar spine during gait. Gait Posture **29**(1), 165 (2009)
11. Wilke, H.J., Mathes, B., Midderhoff, S., Graf, N.: Development of a scoliotic spine model for biomechanical in vitro studies. Clin. Biomech. **30**(2), 182–187 (2015)
12. Lubelski, D., Healy, A.T., Mageswaran, P., Benzel, E.C., Mroz, T.E.: Biomechanics of the lower thoracic spine after decompression and fusion: a cadaveric analysis. Spine J. **14**(9), 2216–2223 (2014)
13. Doulgeris, J.J., et al.: Axial rotation mechanics in a cadaveric lumbar spine model: a biomechanical analysis. Spine J. **14**(7), 1272–1279 (2014)
14. Guo, S., et al.: A biomechanical stability study of extraforaminal lumbar interbody fusion on the cadaveric lumbar spine specimens. PLoS One **11**(12), e0168498 (2016)
15. Narici, M.: Human skeletal muscle architecture studied in vivo by non-invasive imaging techniques: functional significance and applications. J. Electromyogr. Kinesiol. **9**(2), 97–103 (1999)
16. Wang, S., et al.: A combined numerical and experimental technique for estimation of the forces and moments in the lumbar intervertebral disc. Comput. Methods Biomech. Biomed. Eng. **16**(12), 1278–1286 (2013)
17. Zhu, R., et al.: The effects of muscle weakness on degenerative spondylolisthesis: a finite element study. Clin. Biomech. **41**, 34–38 (2017)
18. Fan, W., Guo, L.X.: Influence of different frequencies of axial cyclic loading on time-domain vibration response of the lumbar spine: a finite element study. Comput. Biol. Med. **86**, 75–81 (2017)
19. Kang, K.T., Koh, Y.G., Son, J., Yeom, J.S., Park, J.H., Kim, H.J.: Biomechanical evaluation of pedicle screw fixation system in spinal adjacent levels using polyetheretherketone, carbon-fiber-reinforced polyetheretherketone, and traditional titanium as rod materials. Compos. Part B Eng. **130**, 248 256 (2017)
20. Xu, M., Yang, J., Lieberman, I.H., Haddas, R.: Lumbar spine finite element model for healthy subjects: development and validation. Comput. Methods Biomech. Biomed. Eng. **20** (1), 1–15 (2017)
21. Zander, T., Dreischarf, M., Timm, A.-K., Baumann, W.W., Schmidt, H.: Impact of material and morphological parameters on the mechanical response of the lumbar spine – a finite element sensitivity study. J. Biomech. **53**, 185–190 (2017)
22. Xu, M., Yang, J., Lieberman, I., Haddas, R.: Finite element method-based study for effect of adult degenerative scoliosis on the spinal vibration characteristics. Comput. Biol. Med. **84**, 53–58 (2017)
23. Wang, L., Zhang, B., Chen, S., Lu, X., Li, Z.-Y., Guo, Q.: A validated finite element analysis of facet joint stress in degenerative lumbar scoliosis. World Neurosurg. **95**, 126–133 (2016)
24. Niemeyer, F., Wilke, H.J., Schmidt, H.: Geometry strongly influences the response of numerical models of the lumbar spine-a probabilistic finite element analysis. J. Biomech. **45** (8), 1414–1423 (2012)
25. Schmidt, H., Heuer, F., Drumm, J., Klezl, Z., Claes, L., Wilke, H.-J.: Application of a calibration method provides more realistic results for a finite element model of a lumbar spinal segment. Clin. Biomech. **22**(4), 377–384 (2007)

26. Schmidt, H., et al.: Application of a new calibration method for a three-dimensional finite element model of a human lumbar annulus fibrosus. Clin. Biomech. **21**(4), 337–344 (2006)
27. Shirazi-Adl, A., Ahmed, A.M., Shrivastava, S.C.: A finite element study of a lumbar motion segment subjected to pure sagittal plane moments. J. Biomech. **19**(4), 331–350 (1986)
28. Zander, T., Rohlmann, A., Bergmann, G.: Influence of different artificial disc kinematics on spine biomechanics. Clin. Biomech. **24**(2), 135–142 (2009)
29. Xiao, Z., Wang, L., Gong, H., Zhu, D.: Biomechanical evaluation of three surgical scenarios of posterior lumbar interbody fusion by finite element analysis. Biomed. Eng. Online **11**(1), 31 (2012)
30. Delp, S.L., et al.: OpenSim: open source to create and analyze dynamic simulations of movement. IEEE Trans. Biomed. Eng. **54**(11), 1940–1950 (2007)
31. Bassani, T., Stucovitz, E., Qian, Z., Briguglio, M., Galbusera, F.: Validation of the AnyBody full body musculoskeletal model in computing lumbar spine loads at L4L5 level. J. Biomech. **58**, 89–96 (2017)
32. Putzer, M., Ehrlich, I., Rasmussen, J., Gebbeken, N., Dendorfer, S.: Sensitivity of lumbar spine loading to anatomical parameters. J. Biomech. **49**(6), 953–958 (2016)
33. Christophy, M., Senan, N.A.F., Lotz, J.C., O'Reilly, O.M.: A musculoskeletal model for the lumbar spine. Biomech. Model. Mechanobiol. **11**(1–2), 19–34 (2012)
34. Bruno, A.G., Bouxsein, M.L., Anderson, D.E.: Development and validation of a musculoskeletal model of the fully articulated thoracolumbar spine and rib cage. J. Biomech. Eng. **137**(8), 81003 (2015)
35. Kuai, S., et al.: Influences of lumbar disc herniation on the kinematics in multi-segmental spine, pelvis, and lower extremities during five activities of daily living. BMC Musculoskelet. Disord. **18**(1), 216 (2017)
36. de Zee, M., Hansen, L., Wong, C., Rasmussen, J., Simonsen, E.B.: A generic detailed rigid-body lumbar spine model. J. Biomech. **40**(6), 1219–1227 (2007)
37. Kim, Y., Ta, D., Jung, M., Koo, S.: A musculoskeletal lumbar and thoracic model for calculation of joint kinetics in the spine. J. Mech. Sci. Technol. **30**(6), 2891–2897 (2016)
38. Raabe, M.E., Chaudhari, A.M.W.: An investigation of jogging biomechanics using the full-body lumbar spine model: model development and validation. J. Biomech. **49**(7), 1238–1243 (2016)
39. Zhu, R., Zander, T., Dreischarf, M., Duda, G.N., Rohlmann, A., Schmidt, H.: Considerations when loading spinal finite element models with predicted muscle forces from inverse static analyses. J. Biomech. **46**(7), 1376–1378 (2013)

Development of a Tendon Driven Finger Joint Model Using Finite Element Method

Gregor Harih[✉]

Laboratory for Intelligent CAD Systems, Faculty for Mechanical Engineering,
University of Maribor, Maribor, Slovenia
gregor.harih@um.si

Abstract. Due to certain demanding manual tasks the loads on the human hand can be high, which can cause several disorders, among which are also tendon disorders. Many researchers tried to quantify the loads and provide mathematical models for the tendons of the hand. Since experiments and measurements in vivo are complex and usually not viable, we developed a finite element model of a finger joint, which utilizes tendon/muscle force for the joint movement. Initial simulations of the fingertip finite element model with the developed tendon joint model have shown accurate biomechanical behavior of finger movement and soft tissue deformation. We also compared the results in terms of relationship between tendon force and resulting fingertip (reaction) force from the simulation to an in vivo experiment and have confirmed that the results of the developed finite element model correspond well to the experimental results.

Keywords: Hand biomechanics · Finger joint · Simulation
Finite element method

1 Introduction

The mechanical behaviour of the biological materials of human hand is crucial during the human-product interaction, since forces and moments are transferred from the product to anatomical structures [1, 2]. It has been already shown that soft tissue deformation is crucial for stable manipulation of different object in hands [3]. To provide stability whilst holding an object in hands, the exerted normal finger forces must be reasonably high to prevent slippage and rotation of the object. With high exerted local and overall forces and contact pressures inside different anatomical structures can cause discomfort and also acute and cumulative traumatic disorders (ATD and CTD respectively) [4]. It has been also shown that tendon disorders are also among the most common CTDs at the workplace [5, 6].

The most important risk factors for tendon disorders have been shown to be the force in tendon, duration, dynamic loading and also sustained non-neutral postures [7]. Due to the high risk of tendon disorder development, many researchers have researched this topic to assess and quantify the mathematical models and risk factors. It has been shown that fingertip forces vary to huge extend by the occupational tasks. Low and hence non-hazardous tendon forces are usually obtained during low grasping forces, however high forces up to 190 N can be achieved during power grasps [8]. Several

© Springer International Publishing AG, part of Springer Nature 2019
D. N. Cassenti (Ed.): AHFE 2018, AISC 780, pp. 463–471, 2019.
https://doi.org/10.1007/978-3-319-94223-0_44

researches tried to measure and compare the finger forces to tendon forces where theoretical models have been developed along with in vivo measurements [7, 9–11]. Most of the theoretical models predicted the ratio between tendon force and finger pad force to be approximately 3, while the measurements from Kursa et al. [7] showed a ratio of 2.4 ± 0.7.

As presented, nature of the human hand and complex surfaces of the products usually prevent the direct measurements of forces, stresses, strains and contact pressures, therefore it has been already shown that a viable alternative method are computer simulations using finite element (FE) method [12, 13]. Several researchers have already utilised the FE method for modelling and simulating the biomechanical behaviour of the hand during different manual tasks and estimating the resulting loads. Researchers started approaching the problem using simplified 2D FE fingertip models during flat contact and analysed the mechanical responses to various loadings [14]. It has been shown that the soft tissue of the fingertip presents non-linear hyper-elastic material properties and experiences high local stresses and strains under dynamic loading. A system for simulation of human grasping using a simplified 2D finger model has been developed by previous researchers [15]. The model considered the deformability of the soft tissue using finite element method, however authors limited the simulation to real-time haptic systems. Chamoret et al. [16] developed a full hand 3D FE model with multiple nonlinearities: geometrical, material, frictional contact and impact. Anatomical correctness of the model was achieved using CT scan of an actual human hand. Authors presented a case study of an impact analysis of the hand model pushing against a wall, however they did not present a possibility to use the model for grasping simulations.

Despite several studies trying to develop an anatomically correct and numerically feasible and stable FE human hand model, they still leave a lot to be desired, especially in terms of joint definition. In previous research we already developed an angle driven joint definition, which has proven to be numerically feasible, stable and provides reasonable biomechanical behaviour during movement [17].

Human grasping is a dynamic process, where force/torque in joints is maintained while holding the object in hands. Therefore, angle driven joint definition is not adequate, since each joint is moved only to the extent to reach the predefined angle.

To overcome this limitation, we developed a tendon/muscle force driven joint definition presented in this paper, which allows the joint movement based on applied muscle/tendon finger force.

2 Methods

In our study we used finite element simulation software Abaqus/CAE 6.10 from Dassault Systems (France), which has been already used by many authors for simulating human tissue behaviour under mechanical stresses [1, 14].

Research methods mainly comprised of the accurate fingertip FE model geometry determination based on medical imaging [18]. After obtaining the DICOM images, manual segmentation of anatomical structures has been performed in medical imaging software and IGES models have been generated for the appropriate definition of FE model in Abaqus software. Special attention was given to material definition since soft tissue shows non-linear viscoelastic properties.

2.1 Material Properties

Fingertip bone and nail were assumed to be linear elastic with isotropic material parameters with Young's modulus of 17 GPa and 170 MPa respectively, with a Poisson ratio of 0.3 [14]. The material parameters of soft tissue were extracted from a uniaxial tensile test and were fitted to the Ogden hyper-elastic material model [19] (Table 1). Since soft tissue is almost incompressible, the Poisson ratio was determined to be 0.4 [14]. Steel as a quasi-rigid material was used for the tool handle material with Young's modulus of 210 Gpa and a Poisson ratio of 0.3.

Table 1. Material properties of the soft-tissue.

N	μ_i	α_i
1	−0.07594	4.941
2	0.01138	6.425
3	0.06572	4.712

2.2 Boundary Conditions

The biomechanical movement of human hand is determined based on geometry and material properties of the human hand, where bone link structure with ligaments and tendons is one of the most important [20]. Successful development of such biomechanical system requires simplifications, which is a compromise between accuracy of the biomechanical behaviour of the model and model complexity and calculation times. Since loads inside the soft tissue are the main interest, joints can be simplified to the extend they are numerically low-cost, but still provide accurate movement of the bones.

In this manner, the simplified angle driven joint definition has been already defined by us in previous research [17]. The definition is presented on the example of finger joint between distal and proximal phalange bone (Fig. 1). Firstly, the center of the rotation in proximal phalange bone is identified where a new local coordinate system is created. The coordinate system is then oriented in such manner that the "x" axis corresponds with the axis of the rotation of the joint axis. In the center of the new local coordinate system a new reference point (RP-PIP) is created. Another reference point (RP-DIP) is also created on the surface of the distal phalange bone. Both reference points are then connected to bones using constraints, which fix their translations and rotations relative to the bone. Between both reference points a rigid wire is then created, which is used to define a hinge connector that allows rotation only in "x" axis at the RP-PIP point. In this manner a simplified, numerically stable and biomechanically accurate joint can be defined.

Fig. 1. Definition of the tendon driven finger joint model.

Results have shown accurate joint movement and therefore also soft-tissue deformation (Fig. 2).

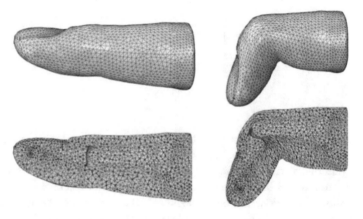

Fig. 2. Biomechanical behavior of the angle driven joint definition.

The angle driven joint definition has then been upgraded with an additional reference point RP-T$_1$ at the location where the tendon is attached to the distal phalange bone (Fig. 3). This reference point has been attached to the bone and all translations and rotations have been fixed. Another reference point RP-T$_2$ has been defined in the direction where the tendon is placed inside the sheats. Both reference points have been connected with a hinge wire connector, which allows rotation in the "x" axis at both reference points.

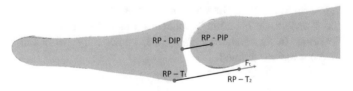

Fig. 3. Definition of the tendon driven finger joint model (upper: schematics, lower: FE definition).

To simulate the finger movement, concentrated force is applied at the $RP\text{-}T_2$ in the direction of the tendon, the "y" axis. This simulates the muscle force, which is transferred to the tendon and finally to the adjacent bone. Due to the moment arm, which is created based on the joint definition, a moment is created around the "x" axis of RP-PIP, which ultimately rotates the distal phalange bone around the center of rotation of the joint. The deformation of soft tissue is solely a result of material properties and joint definition.

2.3   Numerical Tests

Angle driven joint model has been already validated by us and it has been shown that the finger movement is accurate and the soft tissue deforms correctly [17]. In order to validate the newly developed tendon driven joint definition, the results from the simulation have been compared to the experimental results from Kursa et al. [7]. Authors have performed in vivo experiments to explore the relationship between external loading condition and internal tendon forces during an isometric task (Fig. 4).

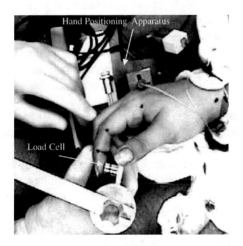

Fig. 4. Experiment performed by Kursa et al. [7] (reproduced figure).

To be able to compare our newly developed joint definition, we replicated the boundary conditions as close as possible to the boundary conditions of the experiment [7]. The load cell from the experiment has been modelled as a rigid appropriately sized cylinder and placed under the finger pad. Lower surface of the cylinder has been completely fixed. Afterwards tendon force of 50 N has been applied to the model, the same as in experiment. The simulation ran for 7 min on a cluster computer of 32 Intel Xeon processors (Fig. 5).

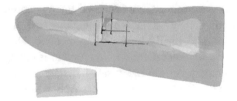

Fig. 5. Replicating the experiment boundary conditions inside Abaqus software.

3   Results and Discussion

Whole hand FE models have not been developed yet due complexity of the human hand anatomy and resulting complex biomechanics. Therefore, we developed a numerically feasible and stable tendon driven joint model, which allows finger movement based on muscle/tendon force. Successful development of such biomechanical system requires simplifications, which are a compromise between accuracy of the biomechanical behavior of the model and model complexity and calculation times. The tendon driven joint model is a continuation of our already developed angle driven joint model. Based on the biomechanics of the joint, the joint rotation center is set based on anatomical and topological features of the bones and does not consider the joint as two bones sliding on cartilage supported by ligaments, since such biomechanical system would be difficult to simulate using the finite element method. Used simplifications are also justified as our interest is in normal forces of the finger, tendon forces and contact pressures on the soft tissue. Based on our previous research, we have shown that such simplifications are reasonable and allow maintaining high level of accuracy of the simulated system.

The soft tissue deformation of the finger during simulation is a consequence of the bone link structure movement prescribed by the boundary conditions (i.e. muscle/tendon force). Therefore, after the simulation completed, we carefully investigated the bone movement during finger flexion. We observed that joint definition was set correctly, since the distance between bones was maintained during the finger flexion, which simulates the bones sliding on cartilage.

Compared to the angle driven joint model developed by us in previous research, the tendon driven joint model is comparably numerically low-cost. One additional rigid wire representing the tendon, which was introduced in this model, does not significantly increase the calculation time. Since the joint model shows correct biomechanical behavior, the movement and deformation of soft tissue can be also considered as biomechanically accurate, since the material model has been verified and validated by us in previous papers (Fig. 6).

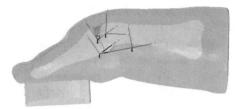

Fig. 6. Deformed fingertip after the applied 50 N of muscle/tendon force.

To validate the newly developed muscle/tendon driven joint model, we extracted the results of tendon force and reaction force on the cylinder from the simulation. We plotted the relationship between tendon force and resulting fingertip reaction force from our model and compared them to reproduced results from Kursa et al. [7]. Results from the simulations show good correspondence to the results from the experiment (Fig. 7).

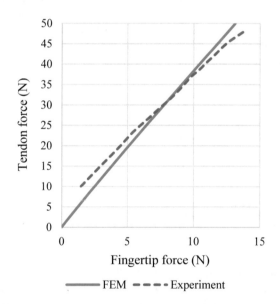

Fig. 7. Tendon force in comparison to the fingertip (reaction) force from the experiment and FEM.

Based on the results and comparison, we can assume that the developed muscle/tendon driven joint model is accurate and produces accurate fingertip reaction force when a tendon force is applied.

The slight difference in the incline of the relationship curve between experiment and numerical simulation could be due to the fact that boundary conditions could not be replicated fully due to lack of details of finger position in original paper of Kursa et al. [7]. Additionally, it has been shown by us in the past that fingertip geometry can vary to some extent between different subjects. Therefore, also the bone geometry varies and also tendon attachment to the bone. This influences the final biomechanics and finally the lever of the joint, which can hence produce different response.

Future work should explore the influence of different fingertip geometry on the response of fingertip force when tendon force is applied. Additionally, the joint model should be further developed to ultimately allow different inputs, such as joint angle rotation or tendon force and provide results in terms of joint angle rotations, tendon/muscle forces, fingertip (reaction) forces, contact pressure, contact pressure distribution, and finally stresses and strains inside soft tissue. Future work should also include the integration of this definition into a full hand finite element model, which would allow complex and comprehensive simulations of human hand movement and grasping.

4   Conclusion

In this paper we presented a newly developed FE finger joint model driven by a muscle/tendon force. The model has proven to be numerically feasible and stable and shows reasonable biomechanical behavior of movement and soft tissue deformation. Results of the model in terms of tendon force compared to fingertip force correspond well to the experiments, which shows the proposed model is accurate and biomechanically correct.

Acknowledgements. The authors acknowledge the project (Development of a computational human hand model for ergonomic product design, Project ID: Z2-8185) was financially supported by the Slovenian Research Agency.

References

1. Wu, J.Z., Dong, R.G.: Analysis of the contact interactions between fingertips and objects with different surface curvatures. Proc. Inst. Mech. Eng. Part H: J. Eng. Med. **219**, 89–103 (2005)
2. Wu, J.Z., et al.: A method for analyzing vibration power absorption density in human fingertip. J. Sound Vibr. **329**, 5600–5614 (2010)
3. Seo, N.J., Armstrong, T.J.: Investigation of grip force, normal force, contact area, hand size, and handle size for cylindrical handles. Hum. Factors **50**, 734–744 (2008)
4. Punnett, L., Wegman, D.H.: Work-related musculoskeletal disorders: the epidemiologic evidence and the debate. J Electromyogr. Kines. **14**, 13–23 (2004)
5. Moore, S., Garg, A.: Upper extremity disorders in a pork processing plant: relationships between job risk factors and morbidity. Am. Ind. Hyg. Assoc. J. **55**, 703–715 (2010)
6. Silverstein, B.A., Fine, L.J., Armstrong, T.J.: Hand wrist cumulative trauma disorders in industry. Br. J. Ind. Med. **43**, 779–784 (1986)
7. Kursa, K., et al.: In vivo forces generated by finger flexor muscles do not depend on the rate of fingertip loading during an isometric task. J. Biomech. **38**, 2288–2293 (2005)
8. Oh, S., Radwin, R.G.: Pistol grip power tool handle and trigger size effects on grip exertions and operator preference. Hum. Factors **35**, 551–569 (1993)
9. Schuind, F., et al.: Flexor tendon forces: in vivo measurements. J. Hand Surg. **17**, 291–298 (1992)

10. Dennerlein, J.T., et al.: Tensions of the flexor digitorum superficialis are higher than a current model predicts. J. Biomech.**31**, 295–301 (1998)
11. Chang, J., et al.: Investigation of index finger triggering force using a cadaver experiment: effects of trigger grip span, contact location, and internal tendon force. Appl. Ergon. **65**, 183–190 (2017)
12. Harih, G., Dolšak, B.: Recommendations for tool-handle material choice based on finite element analysis. Appl. Ergon. **45**, 577–585 (2014)
13. Harih, G., Kaljun, J., Dolšak, B.: Comparison of 2D and 3D finite element fingertip during grasping. In: 3rd International Digital Human Modeling Symposium, Tokyo, Japan (2014)
14. Wu, J.Z., et al.: Simulation of mechanical responses of fingertip to dynamic loading. Med. Eng. Phys. **24**, 253–264 (2002)
15. Pouliquen, M., et al.: Real-time finite element finger pinch grasp simulation. In: Eurohaptics Conference and Symposium on Haptic Interfaces for Virtual Environment and Teleoperator Systems, Fontenay-aux-Roses, France (2005)
16. Chamoret, D., et al.: A novel approach to modelling and simulating the contact behaviour between a human hand model and a deformable object. Comput. Methods Biomech. Biomed. Eng. **16**, 130–140 (2013)
17. Harih, G., Tada, M.: Development of a finite element digital human hand model (2016)
18. Yoshida, H., Tada, M., Mochimaru, M.: A study of frictional property of the human fingertip using three-dimensional finite element analysis. Mol. Cell. Biomech.: MCB **8**, 61–71 (2011)
19. Pan, L., Zan, L., Foster, F.S.: Ultrasonic and viscoelastic properties of skin under transverse mechanical stress in vitro. Ultrasound Med. Biol. **24**, 995–1007 (1998)
20. Brand, P.W., Hollister, A.: Clinical Mechanics of the Hand, 3rd edn. Mosby, St. Louis (1999)

Muscle Force Prediction Method Considering the Role of Antagonistic Muscle

Yuki Daijyu$^{(\boxtimes)}$, Isamu Nishida, and Keiichi Shirase

Department of Mechanical Engineering, Kobe University, Kobe, Japan
177t332t@stu.kobe-u.ac.jp,
{nishida,shirase}@mech.kobe-u.ac.jp

Abstract. Conventional musculoskeletal model often uses the optimization method to estimate muscle forces. However, the optimization method unfortunately does not usually consider the role of antagonistic muscles, which act in opposite direction to the prime motion or for restriction of rotational joint motion. Therefore, this study proposes a new method to estimate muscle forces considering the role of the antagonistic muscle during 3-dimensional motion. In this model, it is assumed that the agonist muscle is connected to the antagonistic muscle by a coupled spring. Joint torque is defined as the summation of the torques derived from the agonist muscles and the torques derived from the antagonistic muscles. Each muscle force can be estimated to keep balance among the torques generated by the agonist muscles and the antagonistic muscles respectively. The experiments were conducted to validate the proposed method to estimate muscle forces. Surface electromyograms (sEMG) were measured to compare with the estimated muscle forces. The experimental results showed that the estimated muscle forces had a good agreement with the sEMG of muscles.

Keywords: Computer human model · Musculoskeletal model
Redundant muscles · Antagonist muscle · sEMG

1 Introduction

Product design and work environment require the consideration of the physical properties of users and workers. Previously, a digital human model has been applied to evaluate the physical loads, such as joint forces, joint torques and muscle forces and to improve the adaptability, usability and safety of product design and work environment. Especially, the prediction of the muscle forces is very important to evaluate the physical loads during human motion.

Musculoskeletal models, which can predict muscle forces during motion, have been proposed. Conventional musculoskeletal model such as SIMM of Musculographics Inc and AnyBody of AnyBody technology Inc which are representative musculoskeletal model were developed. They use the optimization method [1], in which the sum of the predicted muscle forces is minimized [2]. However, the optimization method unfortunately usually does not consider the roles of an antagonistic muscle, which acts in opposition to the prime movers or restriction of a rotational motion about joint.

Actually, there are a few previous studies, which proposes the method considering the role of the antagonistic muscle based on the relationship between the output force distribution at the tip of a link and the electromyogram [3]. However, these previous models have limitation that only certain muscles in 2-dimensional motion can be predicted. Therefore, these previous models are difficult to evaluate muscles contributing to 3-dimensional motion due to its limitation.

This study proposes a new method to predict muscle forces considering the role of the antagonistic muscle during 3-dimensional motion. In order to validate the proposed method, surface electromyograms (sEMG) were measured to compare with the muscle forces predicted by the proposed method. In addition, the muscle forces predicted by the proposed method were compared with that predicted by the previous method.

2 Conventional Musculoskeletal Model

2.1 Optimization Method

Each muscle generates the torque to the attached joint. The total torque acting on a joint is added by the torque generated by each muscle. The torque generated by each muscle T is represented by muscle force F and pulley radius r as shown the Eq. (1). When the pulley radius is constant, the torque acting on a joint can be calculated according to the Eq. (1). However, each muscle force of the muscle attaching on the joint cannot be calculated linearly because the number of muscle attaching on the joint is larger than the number of the joint. Therefore, most conventional musculoskeletal models use an optimization method to predict each muscle force.

$$T = F \cdot r \tag{1}$$

The major optimization method is the method proposed by Crowninshield and Brand [2]. The method defines that each muscle force is distributed so as to extend the time, in which each muscle generates. In this method, the objective function described as the Eq. (2) is minimized. The constraint conditions of this method are described as the Eq. (3) and the Eq. (4). Here, the cross-sectional area is described as A and the maximum voluntary contraction is described as MVC. Unfortunately, in this method, the estimated muscle force is sometimes smaller than the actual one because this method ignores the function of the antagonistic muscles, which are the muscles that act in opposition to the prime movers or agonists of a movement.

$$I = \frac{1}{\sum_m \left(\frac{F_m}{A_m}\right)^{-n}} \tag{2}$$

$$T_i = \sum_m (r_m \cdot F_m) \tag{3}$$

$$0 \leq F_m \leq MVC \tag{4}$$

2.2   Functional Effective Muscle Method

The functional effective muscle method suggested by Oshima *et al.* [3] considers the roles of antagonistic muscles. The method includes three pairs of the antagonistic muscles in the upper limb as shown in Fig. 1. Oshima *et al.* [3, 4] also defined that these muscles act on the distal extremity and the maximum force of each muscle acts on the distal extremity as F'_{mf1}, F'_{me1}, F'_{mf2}, F'_{me2}, F'_{mf3}, F'_{me3} as shown in Fig. 1. Then, the maximum output force distribution on the distal extremity is geometrically a hexagon from maximum force of each muscle. They also verified the maximum output force distribution on the distal extremity is a hexagon in the experiment. They also investigated the vector of the output force on the distal extremity is related to the muscles activation pattern as shown in Fig. 2 [3, 4]. For example, when the vector of the output force is direction *a* as maximum in Fig. 1, the muscle forces are defined as 100% of maximum muscle force for muscles f1, e2 and e3 and 0% of maximum muscle force for muscles e1, f2 and f3. Therefore, the distribution of each muscle force is determined by the vector of the output force on the distal extremity and the muscle activation pattern as shown in Fig. 2. Unfortunately, this method applies only to the 2-dimensional motion. This method also estimates only three pairs of muscles. Therefore, this method cannot apply to the 3-dimensional motion and evaluate other muscles.

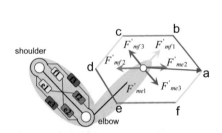

Fig. 1.   Force distribution by effective muscles

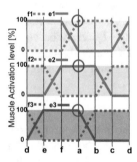

Fig. 2.   Muscle activity level in each force direction

3   Coupled Spring Model Considering Antagonistic Muscle

3.1   A Model Expressing Antagonistic Muscles with Coupled Spring

This study proposes a new model to estimate muscle forces by considering the antagonistic muscles. The feature of the proposed model is that it can estimate muscle force when the pair of the antagonistic muscles is defined. Therefore, this model can versatilely estimate muscle forces acting even on the 3-dimensional motion. This study expresses muscles considering the antagonistic muscle as a coupled spring as shown in Fig. 3. The agonist torque is defined as T_1. The antagonistic torque is defined as T_2. The relationship among the torque T acting on the joint, the agonist torque and the antagonistic torque is described as the Eq. (5). When the joint rotates by the angle $\Delta\theta$, the muscle force generated by the agonist muscle is expressed Eq. (6) with using spring rate k and pulley radius r as shown in Fig. 3.

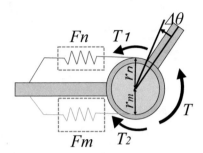

Fig. 3. Coupled spring model by considering the antagonistic muscle

$$T = T_1 - T_2 \tag{5}$$

$$F_n = k_n r_n |\Delta\theta| \tag{6}$$

Since the displacement of the spring of the agonist muscle is geometrically equal to that of the antagonistic muscle, the relationship of the displacement is described as the Eq. (7). The torque acting on the joint can be calculated with the human rigid segment model and Newton's equation. The pulley radius is considered constant. Then, if the spring rate k of the muscle is determined, it is possible to distribute the torque acting on the joint to the agonist torque T_1 and the antagonistic torque T_2 in accordance with the Eqs. (1) and (7)

$$\Delta\theta = \frac{T_1}{r_m^2 k_m} = \frac{T_2}{r_n^2 k_n} \tag{7}$$

However, it is impossible to distribute the torque acting on the joint only by the method described above in an upper limb because there is a biarticular muscle in the upper limb. The biarticular muscle generates the torque in two joint. So, this model defines the muscles existing on the upper limb as 6 representative muscles (f1-deltiod anterior, e1- deltoid posterior, f2- brachialis, e2- lateral head of triceps brachii, f3- long head of biceps, e3- long head of triceps brachii) as shown in Fig. 4. The torques acting on the shoulder joint and the elbow joint is described as the Eq. (8). Since the relationship between the displacement of the spring and the torque is described as the Eq. (7), the relationship between the torque acting on the joints and the displacement of the joint is described as the Eq. (9). In this case, it is also possible to distribute the torque acting on the joint to the agonist torque and the antagonistic torque if the spring rate k of the muscle is determined.

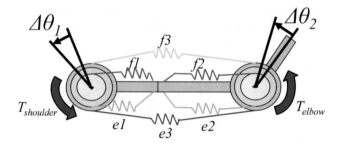

Fig. 4. Coupled spring model of upper arm considering biarticular muscles

$$T_{shoulder} = (|T_{f1}| + |T_{f3}|) - (|T_{e1}| + |T_{e3}|)$$
$$T_{elbow} = (|T_{f2}| + |T_{f3}|) - (|T_{e2}| + |T_{e3}|)$$

(8)

$$T_{shoulder} = \left(|r_{f1}^2 k_{f1} \Delta\theta_1| + |r_{f3}^2 k_{f3} (\Delta\theta_1 + \Delta\theta_2)|\right) - \left(|r_{e1}^2 k_{e1} \Delta\theta_1| + |r_{e3}^2 k_{e3} (\Delta\theta_1 + \Delta\theta_2)|\right)$$
$$T_{elbow} = \left(|r_{f2}^2 k_{f2} \Delta\theta_1| + |r_{f3}^2 k_{f3} (\Delta\theta_1 + \Delta\theta_2)|\right) - \left(|r_{e2}^2 k_{e2} \Delta\theta_1| + |r_{e3}^2 k_{e3} (\Delta\theta_1 + \Delta\theta_2)|\right)$$

(9)

3.2   Determination of Spring Rate

This study assumes that the muscle is a cylindrical material to determine the spring rate k of a muscle as shown in Fig. 5. The spring rate k is described as the Eq. (10) using the muscle length L, the muscle cross sectional area A, and the Young's modulus E in accordance with a Poisson effect.

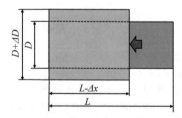

Fig. 5. Determination of spring rate in accordance with a Poisson effect

$$k = \frac{A E}{L}$$

(10)

When the cylindrical material expands and contracts, the cross-sectional area of the cylindrical material changes as shown in Fig. 5. This variation is described as the Eq. (11) using the longitudinal strain ε_1, the lateral strain ε_2, and the Poisson's ratio v. In accordance with the Eq. (11), the relationship between the cross-sectional area and the angle variation $\Delta\theta$ of the joint can be described as the Eq. (12). Therefore, by using the Eqs. (12) and (10), the spring rate of the muscle can be determined from the Eq. (13).

$$\varepsilon_1 = \frac{\Delta x}{L} \qquad \varepsilon_2 = \frac{\Delta D}{D} \qquad \upsilon = \frac{\varepsilon_1}{\varepsilon_2} \tag{11}$$

$$k = \frac{\pi D^2}{4}\left(1 + \frac{r \ \upsilon\Delta\theta^2}{L}\right) \tag{12}$$

$$k = \frac{A \ E}{L}\left(1 + \frac{r \ \upsilon\Delta\theta^2}{L}\right) \tag{13}$$

3.3 Muscle Force Estimated from Agonist Torque and Antagonistic Torque

It is possible to distribute the torque acting on the joint to the agonist torque and the antagonizing torque in accordance with the proposed coupled spring model considering antagonistic muscle. In this method, muscle force of each muscle can calculated by using the optimization method proposed by Crowninshield et al. from the distributed agonist torque and the antagonistic muscle. The calculated muscle force of each muscle considers the function of the redundant muscles because the torque acting on the joint is distributed to the agonist torque and the antagonistic torque before optimized.

4 Verification Experiment

In order to validate the proposed method, firstly, the result estimated by the proposed method was compared with the result by the optimization method and the result by the functional effective muscle method. Subsequently, the experiment, in which one participant output with the surface-EMG (sEMG) of each muscle measured by the electromyography, was conducted, and the obtained sEMG was compared with the result estimated by the proposed method.

4.1 Comparison with the Conventional Methods

In order to evaluate the difference of the estimated muscle force of the antagonistic muscle, the muscle force estimated by the proposed method was compared with that by the optimization method described in the Sect. 2.1 and that by the functional effective muscle method described in the Sect. 2.2. The estimation conditions were shown in Fig. 6, Table 1 [5] and Table 2 shows the muscle cross sectional area [6], muscle length and pulley radius, which is necessary for the muscle force estimation. Here, the length of the muscle and the pulley radius were determined using the Anatomy picture book [7]. The link length of the upper arm and the link length of the forearm were determined from the link length of the participant for sEMG experiment. Table 3 shows the conditions of the distal output.

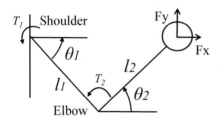

Fig. 6. Condition of the posture of the upper arm

Table 1. Parameters of the verification condition

Upper arm length $l1$ [m]	0.27
Forearm length $l2$ [m]	0.29
Shoulder joint angle $\theta1$ [deg]	45
Elbow joint angle $\theta2$ [deg]	45
Poisson's ration	0.49
Young's modulus [Gpa] [5]	0.102

Table 2. Condition of muscle properties of the upper arm

	Cross sectional area [mm²]	Length [mm]	Moment arm [mm]	Maximum muscle force [N]
Pectoralis major muscle	1590	177	52.0	783
Deltoid muscle front	1250	159	35.2	616
Deltoid muscle rear	1250	159	35.2	1194
Subscapularis muscle	1410	110	11.3	1347
Infraspinatus muscle	1190	123	11.3	1137
Teres major muscle	250	96	49.8	239
Teres minor muscle	370	101	11.3	353
Brachialis muscle	1440	268	68.2	485
Brachioradialis muscle	390	283	26.6	131
Flexor carpi radialis muscle	390	272	9.3	131
Flexor digitorum superficialis muscle	600	295	9.3	202
Pronator teres muscle	650	194	9.3	219
Anconeus muscle	130	100	25.5	23
Lateral head of triceps brachii muscle	2840	325	49.4	505
Biceps brachii muscle	820	237	35.0	739
Long head of triceps brachii muscle	1160	237	36.0	719

Table 3. Condition of the distal output

	Fx [N]	Fy [N]
Condition 1	−55	−10
Condition 2	−110	−20
Condition 3	−220	−40
Condition 4	−275	−50

Figure 7 shows the result of muscle force estimated by the optimization method, the functional effective muscle method and the proposed method.

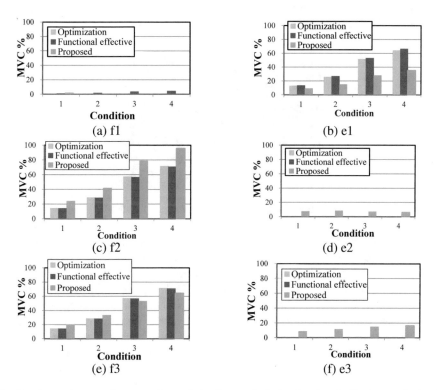

Fig. 7. Comparison result among the optimization method, the functional effective method and the proposed method

Muscles e1, f2, f3 are the agonist muscles and muscles f1, e2, e3 are antagonistic muscles under the conditions. The results estimated by the optimization method show that the antagonistic muscles are constantly zero. According to this result, the optimization method is considered not to estimate the role of the antagonistic muscle. The results estimated by the functional effective muscle method show that only muscle f1, which is one of the antagonistic muscle, is activated. Then, there is no antagonistic torque in the elbow joint. On the other hand, the results estimated by the proposed method show that the antagonistic muscle of the bipartite muscle e3 is activated.

Then it can be confirmed that the antagonistic muscles work in both joints of the shoulder joint and the elbow joint. Therefore, it is considered that the proposed method can express the influence of the antagonistic muscle, which could not be expressed by the conventional method.

4.2   Experiment with Surface-EMG (SEMG)

In order to validate the proposed method, the experiment, in which one participant (height: 167 cm, weight: 68 kg, age: 22 years old) output with the surface-EMG (sEMG) of each muscle measured by the electromyography after obtaining agreement about experiment contents, was conducted as shown in Fig. 8. The conditions of the posture and the muscle properties are shown in Fig. 6, Tables 2 and 3 as the same condition of the comparison with the conventional method.

Fig. 8. Measuring equipment of output force at the distal extremity

The condition of output force for verification was determined from the distal output force, which the participant output. The participant output four patterns of the distal extremity as shown in Table 5. In the experiments, the sEMG was measured using a myoelectric pad (Ambu Bluesensor P) and a myoelectric amplifier (Harada Industry Co., Ltd. EMG-021/025). Table 4 shows the measuring conditions of the sEMG. The sEMG was measured simultaneously with the distal output force on the three-component dynamometer. Measurement time was 10 s, with a 3 min break between each measurement. The obtained sEMG was smoothed by using a low pass filter of

Table 4. Measuring conditions of sEMG

Sampling frequency [Hz]	1000
Myoelectronic pad attachment position	Deltoid muscle front
	Deltoid muscle rear
	Biceps Brachii muscle
	Long head of triceps brachii muscle

Table 5. Condition of output force for verification

	Fx [N]	Fy [N]
Case1	−59	−14
Case2	−93	−14
Case3	−246	−50
Case4	−279	−48

4 Hz. After smoothed, the integrated sEMG was obtained and subtracted by the sEMG of the unloaded state.

Figure 9 shows the comparison of the result of %MVC estimated by the proposed method with that measured by the sEMG.

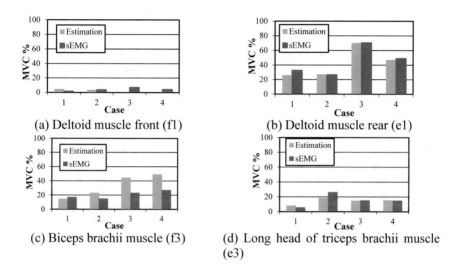

(a) Deltoid muscle front (f1)

(b) Deltoid muscle rear (e1)

(c) Biceps brachii muscle (f3)

(d) Long head of triceps brachii muscle (e3)

Fig. 9. Comparison of estimated muscle force and sEMG

In Fig. 9, muscle e1 and f3 are the agonist muscles and muscle f1 and e3 are the antagonistic muscles. The results show that the muscle force estimated by the proposed method is roughly in agreement with the result of the sEMG. Although the difference of the biceps brachii muscle between the estimated and the sEMG is larger as the distal output is larger, the reason of this is considered crosstalk of sEMG. Nevertheless, the proposed method was validated because the result shows that the estimated muscle force has good agreement with the measured sEMG.

5 Conclusion

This study proposed a new method to estimate muscle forces considering the role of antagonistic muscles. This study expressed muscles considering the antagonistic muscle as a coupled spring. The feature of the proposed model is that it can estimate the muscle force including the antagonistic muscles, which the conventional method cannot evaluate enough. The muscle forces estimated by the proposed method was compared with that by the conventional methods in order to validate the proposed method. Furthermore, Verification experiments using the sEMG were conducted.

- Compared with the conventional methods, it was confirmed that the estimated muscle force of the antagonistic muscle, which is the biarticular muscle acting at the shoulder joint and the elbow joint, was activated although that by the conventional methods was not activated.

• It was confirmed that the muscle force of the agonist and antagonistic muscles can be estimated by the proposed model. The muscle force estimated by the proposed method has good agreement with the measured sEMG.

In future work, it is verified whether the proposed method can be applied to the other pairs of antagonistic muscle and 3-dimensional motion.

Acknowledgements. This work has been partially supported by JSPS KAKENHI Grant Numbers JP17K17871 and Itochu Foundation.

References

1. Hase, K.: Motion analysis by SIMM, ARMO, any body. J. Soc. Biomech. **33**(3), 205–211 (2009). (in Japanese)
2. Crowninshield, R.D., Brand, R.A.: A physiologically based criterion of muscle force prediction in locomotion. J. Biomech. **14**(11), 793–801 (1981)
3. Oshima, T., Fujikawa, T., Kumamoto, M.: Functional evaluation of effective muscle strength based on a muscle coordinate system consisted of biarticular and monoarticular muscles - contractile forces and output forces of human limbs. J. Jpn. Soc. Precis. Eng. **65**(12), 1772–1777 (1999)
4. Oshima, T., Fujikawa, T., Kumamoto, M., Yokoi, N.: Functional coordination control of pairs of antagonistic muscles. Jpn. Soc. Mech. Eng. **63**(607), 769–776 (1997)
5. Akima, H., Kuno, A., Fukunaga, T., Katsuta, S.: Architectural properties and specific tension of human knee extensor and flexor muscles based on magnetic resonance imaging. Jpn. Soc. Phys. Fit. Sports Med. **44**(2), 267–278 (1995)
6. Holzbaur, K.R.S., Murray, W.M., Gold, G.E., Delp, S.L.: Upper limb muscle volumes in adult subjects. J. Biomech. **40**, 742–749 (2007)
7. Takahashi, A.: Exercise Anatomy Picture Book. Baseball Magazine Sha (1995). (in Japanese)

Automatic Learning of Climbing Configuration Space for Digital Human Children Model

Tsubasa Nose[1,2(✉)], Koji Kitamura[1], Mikiko Oono[1],
Yoshifumi Nishida[1], and Michiko Ohkura[2]

[1] Artificial Intelligence Research Center,
National Institute of Advanced Industrial Science and Technology,
2-4-7 Aomi, Koto, Tokyo 135-0064, Japan
{t.nose,k.kitamura,mikiko-oono,y.nishida}@aist.go.jp
[2] Shibaura Institute of Technology, 3-7-5 Toyosu, Koto, Tokyo 135-8548, Japan
ohkura@sic.shibaura-it.ac.jp

Abstract. Millions of children die from preventable injuries every year around the world. Environmental modification is one of the most effective ways to prevent these fatal injuries. The environment should be modified and products should be designed in ways that will reduce the risk of injury by taking child–environment and child–product interactions into account. However, it is still very difficult even for advanced simulation systems to predict how children interact with products in everyday life situations. In this study, we explored a data-driven method as a promising approach for simulating children's interaction with products in everyday life situations. We conducted an observational study to collect data on children's climbing behavior and developed a database on children's climbing behavior to clarify a climbing configuration space, which enables the prediction and simulation of the possible climbing postures of children.

Keywords: Climbing behavior · Configuration space
Digital human children model

1 Introduction

Millions of children are killed by accidental injuries annually around the world. According to a report by the United Nations Children's Fund [1], approximately 30% of the deaths involving children from 1 to 4 years of age can be attributed to unintentional injuries. Obviously, this is a major social problem, and there is a comparable trend in Japan. According to a report from the Tokyo Fire Department covering the years from 2011 to 2015, most of the accidents involving children below 5 years old result in fall injuries.

Currently, there have been various approaches to preventing accidental injuries in children. The World Health Organization has recommended the following three measures as effective methods for preventing child-related accidents [2]: product and

© Springer International Publishing AG, part of Springer Nature 2019
D. N. Cassenti (Ed.): AHFE 2018, AISC 780, pp. 483–490, 2019.
https://doi.org/10.1007/978-3-319-94223-0_46

environmental modification, education, and enforcement such as legislation and regulations. Of these three measures, product and environmental modification, which is often called a "passive strategy", is highlighted as one of the most effective methods for injury prevention.

The environment should be modified and products should be designed in ways that will reduce the risk of injury by taking the child–environment and child–product interactions into account.

Digital human models for simulating human behavior in a virtual environment are now available. These models are typically used for computer-aided ergonomic design (e.g., [3, 4]). However, it is still very difficult even for advanced digital human models, i.e., behavior simulation systems, to predict how children interact with products in everyday life situations. Because children significantly change their behaviors depending on their complex muscular and cognitive performance, consumer product's layout around the children in their homes. An effective means to simulate children's interaction with products in everyday life situations is to perform simulations using measured data obtained from behavioral observations.

On the other hand, image processing and machine learning technologies are advancing rapidly and can be used for data-driven approaches to developing a behavior simulation system [5].

In this paper, as a new methodology for a data-driven simulation system, we developed a system that collects three-dimensional posture data from color images and depth images by combining RGB-D cameras and pose recognition technology to acquire posture data. To construct a database of children's climbing behavior using our developed system, we conducted an observational study on how children interacted with a climbing apparatus.

2   Development of Auto-Generation System of Children's Climbing Behavior Database

2.1   Development of Three-Dimensional (3D) Posture Data Acquisition System

We developed a system that collects 3D posture data from color images and depth images using an RGB-D camera (Microsoft's Kinect). We subsequently used pose recognition software (OpenPose [6]) to acquire posture data, since the accuracy of the Kinect's posture detection algorithm decreases when a person touches an object. OpenPose allows us to extract the two-dimensional position of 18 parts of the body, such as the hands, feet, shoulders, and knees, on single images. Using the depth data from the Kinect and posture data detected by OpenPose, our developed system obtains 3D coordinate data of a detected person's posture (Fig. 1).

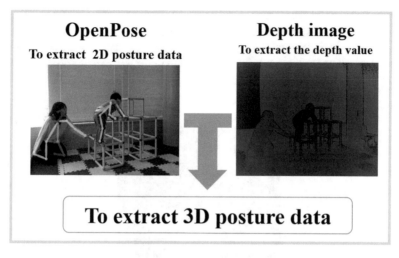

Fig. 1. System overview (Color figure online)

2.2 Behavior Observation

To construct a database of children's climbing behavior using our system, we conducted an observational study on how children interacted with a climbing apparatus (Fig. 2). Fourteen children aged 20 to 58 months participated in the study.

Fig. 2. Observation of climbing behavior

2.3 Development of Children's Climbing Behavior Database

Using the developed system, posture data when children climbed from behavior observation data was acquired. Additionally, we purposely extracted postures only when a child touched the climbing apparatus (Fig. 3) because children tend to react in response to the shape of an object. A total of 184 data points were extracted, and these posture data were compared with the color images to check whether the data were correct. It was confirmed that 141 posture data points corresponding to 76% of the total were valid. The other 43 data points were excluded because an object between the camera and the child (occlusion problem) prevented extraction of the correct depth value (Fig. 4). The 141 valid data points were manually corrected to be close to the actual posture, and the children's climbing behavior database was subsequently developed. Meanwhile, we think that the occlusion problem can be solved in the future by using multiple RGB-D cameras.

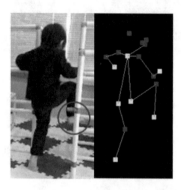

Fig. 3. Example of a child touching the climbing apparatus (Color figure online)

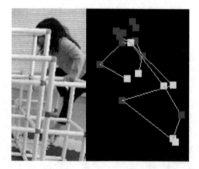

Fig. 4. Example of the occlusion problem

3   Analysis of Configuration Space Using Children's Climbing Behavior Database

3.1   Extraction of Feature Quantities

The coordinates of the acquired postures were converted into nine values to normalize the posture data. The nine values were the distances between: (1) the neck and the left hand, (2) the neck and the right hand, (3) the left and right hands, (4) the left hand and the left foot, (5) the neck and the hip, (6) the right hand and the right foot, (7) the hip and the left foot, (8) the hip and the right foot, and (9) the left and right feet (Fig. 5). The children's posture was determined based on these nine values.

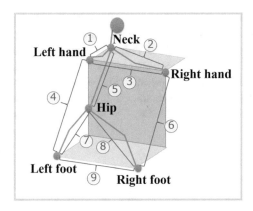

Fig. 5. Feature quantities

3.2   Analysis of Important Factors for Children's Climbing Behavior

To analyze factors important for children's climbing behavior, principal component (PC) analysis was performed on the extracted feature quantities. The results up to PC3 with a high proportion of variance are shown in Table 1. Focusing on the loading amount, PC1 is presumed to be a variable indicating "how close both feet are to the body," since the components relating to both feet are large in the negative direction. PC2 is presumed to be a variable indicating "how far both hands are from the body," since the components relating to both hands are large in the positive direction. PC3 is presumed to be a variable indicating "how far either of feet are from the body," since the components relating to the left foot are large in the negative direction and the components relating to the right hand are large in the positive direction.

Table 1. Results of principal component analysis

	Loading amount	PC1	PC2	PC3
Loading amount	(1) The neck and the left hand	-0.3133	0.4837	0.1662
	(2) The neck and the right hand	-0.2937	0.4973	0.0426
	(3) The left hand and the right hand	-0.1348	0.3925	0.0246
	(4) The left hand and the left foot	-0.4147	-0.1970	-0.4258
	(5) The neck and the hip	-0.2817	-0.4480	0.0686
	(6) The right hand and the right foot	-0.4398	-0.3194	0.3315
	(7) The hip and the left foot	-0.4125	-0.0469	-0.4244
	(8) The hip and the right foot	-0.4010	-0.0100	0.4833
	(9) The left foot and the right foot	-0.1473	0.1441	-0.5102
	Proportion of Variance	0.2792	0.1927	0.1640
	Cumulative Proportion	0.2792	0.4720	0.6359

PC: principal component

By plotting the analysis results in three dimensions, we were able to confirm a collection of points. This collection of points on the scatter plot can be treated as a data-driven configuration space, which indicates a space of posture that children are able to take when they climb. It can also be confirmed that the location of the configuration space differs depending on age (Fig. 6). Specifically, PC1 becomes more negative in older children. This is because older children are able to retain a posture of spreading and stretching their feet as they age, and the "degree of closing both feet to the body" becomes smaller.

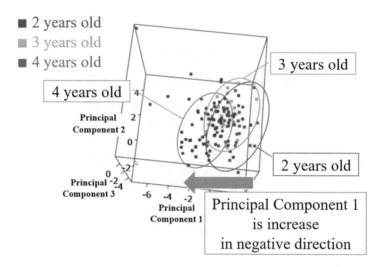

Fig. 6. Configuration space (generated using principal component analysis)

3.3   Analysis of Classification for Children's Climbing Behavior

Posture data were classified using the t-distributed Stochastic Neighbor Embedding method [7]. To investigate the classified data, we developed a system like that shown in Fig. 7. This system can confirm the color image and 3D posture data in the configuration space. Moreover, it allows users to check which posture data will be changed in the configuration space with the displayed posture data. Figure 8 shows that the system is useful for classifying typical climbing patterns. In future studies, we plan to accumulate a large amount of behavior data by installing 14 RGB-D camera into everyday life environments. We've already installed over into a child hospital to collect children's behavior data.

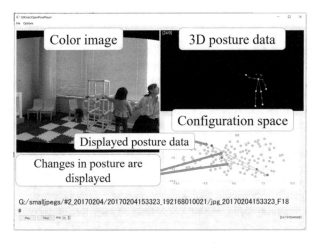

Fig. 7. Posture data analysis system (Color figure online)

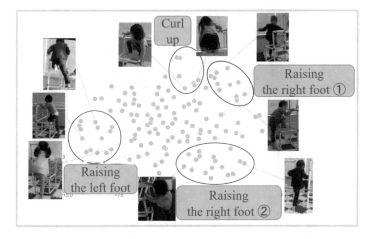

Fig. 8. Classification for children's climbing behavior

4   Conclusion

We developed a children's climbing behavior database using an RGB-D camera and posture detection technology and showed that it is possible to generate a configuration space using posture data of children's climbing behaviors. In the future, we plan to develop a risk assessment system using this configuration space that can determine whether children's climbing behavior could result in a serious injury.

Acknowledgements. This paper is partially supported by a project commissioned by the New Energy and Industrial Technology Development Organization (NEDO).

References

1. Heron, M.: Deaths: leading causes for 2014. Natl. Vital Stat. Rep. **65**(5), 17 (2016)
2. Sminkey, L.: World report on child injury prevention. Inj. Prev. **14**(1), 69 (2008)
3. Jung, K., Kwon, O., You, H.: Development of a digital human model generation method for ergonomic design in virtual environment. Int. J. Ind. Ergon. **39**(5), 744–748 (2009)
4. Yang, J., Kim, J.H., Abdel-Malek, K., Marler, T., Beck, S., Kopp, G.R.: A new digital human environment and assessment of vehicle interior design. Comput.-Aided Des. **39**(7), 548–558 (2007)
5. Farooq, A., Won, C.S.: A survey of human action recognition approaches that use an RGB-D sensor. EIE Trans. Smart Process. Comput. **4**(4), 281–290 (2015)
6. Cao, Z., Simon, T., Wei, S.-E., Sheikh, Y.: Realtime multi-person 2D pose estimation using part affinity fields. In: Computer Vision and Pattern Recognition (2017)
7. Bengio, Y.: Visualizing data using t-SNE. J. Mach. Learn. Res. **9**, 2579–2605 (2008)

Measurement System
of the Temporomandibulares Joint

André Solon de Carvalho$^{(\boxtimes)}$ and Eduardo Ferro dos Santos

Engineering School of Lorena, EEL, University of São Paulo,
Estr. Mun. do Campinho, s/nº, Lorena, São Paulo, Brazil
`andresolon@outlook.com.br`, `eduardo.ferro@usp.br`

Abstract. Opening and closing the mouth is one of the most important biomechanical movements of the human being, being one of the first to be performed even before birth. This movement is accomplished by a set joint called the temporomandibular Joint (TMJ). Their dysfunction causes a number of problems always accompanied by pain, in which much of the world population have disorders in this system, requiring the search for treatment. Their dysfunction causes a number of problems always accompanied by pain, in which much of the world population have disorders in this system, requiring the search for treatment. In this way, this work contributes to varieties of knowledge, clarification and care of patients suffering from temporomandibular dysfunction (TMD), with its main objective, the construction of a device able to measure and diagnose abnormalities during biomechanical movement of the opening and closing the mouth, through a low cost imaging system and easy to handle.

Keywords: Temporomandibular joint (TMJ) · Image capture
Mandibular deviation · Biomechanics

1 Introduction

TMJ is an abbreviation used to refer to the temporomandibular joint. It is considered the most complex joint in the human body [1]. Its complexity is due to the fact that it is connected, that is, one is dependent on the other, both left and right moves at the same time to perform the opening and closing of the mouth [2], another factor that explains its complexity is that it performs two different types of movements, the rotation and the translation during this process [3].

She is involved in swallowing, chewing, breathing and speech. The opening movement and closing the mouth is accomplished by a set musculoskeletal joint-called stomatognathic system in which highlights the most complex joint in the human body, called a temporomandibular joint (TMJ) [4].

Although several previous studies, there is still in the middle scientific lack of globally accepted parameters for evaluation and treatment of pathologies of the TMJ, compromising the objectives of the treatment, which currently depends more on the

© Springer International Publishing AG, part of Springer Nature 2019
D. N. Cassenti (Ed.): AHFE 2018, AISC 780, pp. 491–500, 2019.
https://doi.org/10.1007/978-3-319-94223-0_47

therapist's experience than scientific evidence [4]. This work has as its main goal, building a capable equipment to measure and diagnose the temporo mandibular joint dysfunction (TMD) during the biomechanical movement of the opening and closing of the mouth, through an imaging system.

We used low-cost and easy handling equipment, compared to systems on the market today. The acquired images are processed by a specific software for this purpose, where the data are processed and modeled, providing motion in the frontal plane and the lateral jaw by analyzing its respective track the movement of laterality, protrusion and maximum amplitude of the opening and closing of mouth, providing millimeter measured data.

Developed a good precision equipment using a new noninvasive technique and rapid implementation that does not cause discomfort to the patient and provides the healthcare professional a tool to aid the diagnosis of structural dysfunction of the masticatory system and determine important data for guidance and treatment planning of possible pathologies of the temporomandibular joint.

2   Experimental Development

To obtain the ideal volunteer position, you must use an ergonomically adjustable device, and both opted for a dental chair that allows the following settings: Seat height; Back back; and headrest (Fig. 1). The adjustment of the headrest is formed by a vertical prey stainless steel rod on the back of the back and the apparatus fixation head, this shank has a double system hinges and a lock that moves to the head of the patient, for best fit your ergonomic position.

This system is arranged orthogonally in order to provide a fixed reference system for the guidance of a professional.

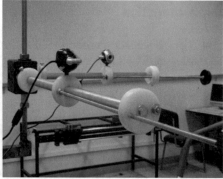

Fig. 1. Head restraint system.   **Fig. 2.** Support base.

2.1 Base of Support for Cameras

It is composed of metallic base supporting two aluminum bars allowing them to move in the vertical plane as required for the displacement of the shoot cameras (Fig. 2). This mechanical system allows manual positioning of the cameras in the three-dimensional shape space, allowing the professional to run the tests with ample opportunities in this position and individually for each patient.

2.2 Markers

To carry out the filming we used two types of markers, which after being idealized, were built to aid data collection.

2.3 Orthodontic Marker

The orthodontic marker (Fig. 3) was made of acrylic resin for a prosthetic dental arch where the volunteer was molded by a dentist. Through this mold was made a set of orthodontic appliance that is fixed in the anterior part of the lower dental arch and another in the upper. In this orthodontic appliance, a rigid steel wire 50 mm long and 1 mm in diameter and with a ball at its end red resin was stuck to facilitate their identification.

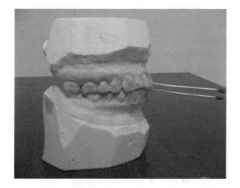

Fig. 3. Orthodontic marker (Color figure online)

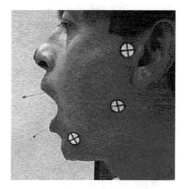

Fig. 4. Adhesive marker

2.4 Adhesive Marker

The adhesive marker (Fig. 4) was made in a circular format with a diameter of 15 mm and an x mark to mark its center. These were used in order to facilitate data collection procedure. Five adhesives were used for each data collection, three used to collect the lateral plane, placed as follows:

The first was placed above the mandibular condyle, which serves as a fixed point of collection; the second was placed on the lower jaw angle serving as the first moving point of collection; and the third was placed in the anterior mandible angle serving as the second moving point of collection.

The latter two adhesives are used for shooting in the frontal plane, wherein: one was placed on the nasal base serving as a fixed point; and the other was placed in the mental protuberance of the jaw, serving as a moving point.

2.5   Filming System

The system for capturing images consists of two type webcam cameras, which are processed by two computers, one for the camera that captures the footage in the frontal plane and one for the camera that captures the images in the lateral plane, the signals obtained by cameras are sent as data processing system for display in the preview screen with the help of the TMJ program, specific to the data conversion, this program was developed in MATLAB.

The system allows an analysis of the dynamics of jaw movements.

2.6   System Calibration

When evaluating the results of measurements made by means of an image capture system, it is necessary to establish a calibration method that provides appropriate parameters for the recovery of the distances traveled by the point of interest in the movement of the investigation.

To perform an accurate measurement through the proposed system, it is necessary to have a standard so that the computer can turn an analogue measurement (mm), digital (pixel), where this transformation is called calibration.

The TMJ program has a tool that when you start the program, it opens a window that asks the first photo to be analyzed, this should have a scale that is possible to compare and convert this measure of scale in millimeters, with the measure of the computer screen in pixel. For this scale, a caliper was used with a stipulated measurement of 50 mm. By choosing the space between point A and B, we have the image size in pixels, which divided by the scale value know the value of pixel/mm, as this procedure all of the following steps are processed pixel to millimeters, this is only necessary to be done for the horizontal axis, since the program calculates the calibration of the vertical axis automatically. Calibration is done for each of the collections, and for all volunteers. Studies [5] claim that this type of calibration allows for a lower precision 1%. 140 data collection were obtained, being 40 for voluntário1 and 2, and 30 volunteers for 3 and 4. Among these samples, 20 samples were selected for voluntary 1 and 2, and 15 for collecting volunteer 3 and 4, amounting 60 collected analyzed.

2.7   Subjects Evaluated

This work does not aim at a statistical study, nor a case study refers only to the determination of the operation of the equipment. Therefore, we used four volunteers, three female and one male, aged 35–45 years with no history of previous pathology of any kind. All the volunteers agreed to the disclosure of your data and pictures.

3   Methodology Used

Description of the sequence of operations during the evaluation of the volunteer:

1. Adjust the chair as the volunteer biotype so that he continue with his usual posture, as if we change your posture, the chewing muscles will be influenced in the evaluation;
2. Adjust the locking device, the patient's head;
3. Place the markers at predetermined locations;
4. Move the camera to the most favorable position, front or sideways, the patient's face by adjusting the desired distance, and the focus of the image;
5. Coupled to the head unit fastener, the volunteer is instructed to make movements of openings and closures mouth 10 times, where these movements are filmed in the frontal plane, the webcam 1 (Fig. 6) and the lateral plane on webcam 2 through a Webcam own program, this data is sent to the computers. Figure 5 presents a volunteer able to conduct the evaluation after follow the sequence of operations described.

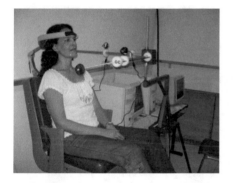

Fig. 5. Volunteer able to perform the evaluation.

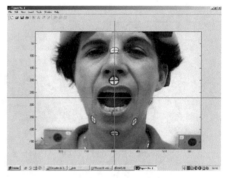

Fig. 6. Determination of points above the frame.

6. Transformed the AVI footage, picture frame by frame, in jpeg format, for each opening movement and closing the mouth.
7. Using the TMJ program developed for this work, the following steps were taken: (a) System Calibration (b) determine the points of movement, frame by frame through the "mouse" computer (c) Create a numeric table of coordinates points marked transformed into "mm" through calibration (d) Generate graph of plotted points of trajectories.
8. Analyze the results and give expert advice.

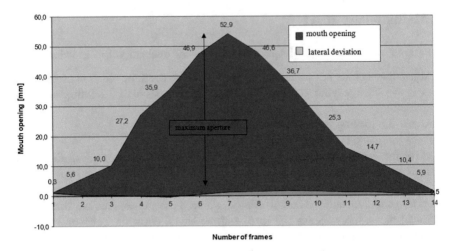

Fig. 7. Curve opening and closing movement of the mouth 4 of the volunteer.

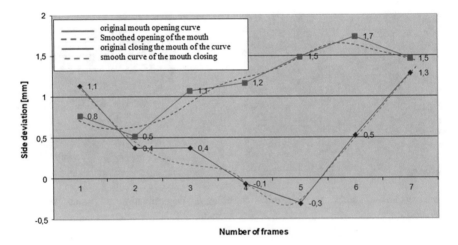

Fig. 8. Mandibular lateral deviation of the volunteer 4.

4 Results and Discussions

The results are shown in tables show the data collected and the mouth opening movement (AB) to the horizontal coordinate (x) and vertical (y) of four volunteers analyzed. For each volunteer were made five measured and calculated their average with the filming of the motion of: mouth opening in the frontal plane, opening and closing the frontal plane, opening and closing in a lateral plane, using this same sequence with the marker type apparatus orthodontic and then the adhesive for each volunteer was examined in which the efficiency of the orthodontic appliance type marker and marker type adhesive for all volunteers, these values are shown in Table 1. These results can also be presented in graphic form, such as they can be seen in the example of Figs. 7 and 8.

Mandibular deviations were calculated for the maximum openings of the mouth and jaw protrusions.

Table 1. Results of volunteers.

Results of volunteers								
Marker Type	Abscissa (X) [mm] Lateral deviation				Ordinate (Y) [mm] Max opening			
Opening orthodontic marker	Vol1	Vol2	Vol3	Vol4	Vol 1	Vol 2	Vol 3	Vol 4
	3.0	5.1	2.3	0.8	34.3	46.3	49.3	50.1
Opening and closing orthodontic marker	2.6	5.1	2.3	0.8	35.0	46.5	49.5	50.0
Opening and closing adhesive marker	2.8	5.2	2.3	0.5	34.7	46.5	49.4	49.9
Opening and closing orthodontic marker protrusion	5.1	8.3	9.0	9.7	334.5	446.6	449.0	551.6

With respect to the mandibular lateral deviation to be have a normal joint is necessary for it to move on its axis perfectly without any lateral deviation, but is described as within normal deviations of less than 2 mm [6]. The results in Table 1 could interpreted observer for voluntary 1, the higher mandibular lateral deviation was 3.0 mm and less than 2.6 mm between evaluations, these values being outside the normal standards recommended, but the patient can be treated for the temporo-mandibular joint.

The difference, of all reviews for this volunteer was 0.4 mm, which is considered quite satisfactory for the error comparison between the marker and the orthodontic adhesive.

For voluntary 2, it is observed that the higher mandibular lateral deviation was 5.2 mm and less than 5.1 mm between the evaluations and these values outside the normal standards recommended, but the patient can be treated to temporomandibular joint. The difference, of all reviews for this volunteer was 0.1 mm, which is considered quite satisfactory.

For voluntary 3, it is observed that the higher mandibular lateral deviation was 2.3 mm and less than 2.3 mm between the evaluations and these values outside the normal standards recommended, but the patient also can be treated for ear-jaw artic-ulation. There was no difference between the ratings for this voluntary and is con-sidered ideal for both markers.

For volunteer 4, it is observed that the higher mandibular lateral deviation was 0.8 mm and the smallest of 0.5 mm between the evaluations and these values within the recommended levels normal. The difference, of all reviews for this volunteer was 0.3 mm, which is considered quite satisfactory.

In the evaluation of maximum opening mouth it is known that this limitation is also characterized as a basic criterion for diagnosing TMJ disorders. It is considered limiting mouth opening when in its final phase measuring less than 35 mm [7]. For normal limits the maximum opening range of 40 to 60 mm when measuring the incisal edges of the upper and lower teeth [8].

For voluntary 1, the minimum value was 34.3 mm, and a maximum of 35.0 mm, and the difference between them of 0.7 mm and this is quite satisfactory. For this patient the normal ranges corresponds to a decrease in mouth opening, thereby taking the treatment indication joint.

For voluntary 2, the minimum value was 46.3 mm, and a maximum of 46.6 mm, and the difference between them of 0.3 mm, which is quite satisfactory. For this patient the normal ranges corresponds to a reduction of mouth opening, thereby taking the treatment indication joint.

For voluntary 3, the minimum value was 49.0 mm, and a maximum of 49.5 mm, and the difference between them of 0.5 mm, which is quite satisfactory, These values are within normal standard.

To volunteer 4, the minimum value was 49.9 mm, and a maximum of 51.6 mm, and the difference between them of 1.7 mm, which is quite satisfactory. These values are within the normal pattern.

As regards the evaluation of the movement protrusion [9] found val-or maximum protrusive on average from 9 to 10 mm, [10] found higher values (10.7 mm) in children from 10 to 16 years old. Determining the maximum length of the protrusive movement in adults [11] found a value of 9.0 mm, considering the methodology with a millimeter ruler. There average of 9.28 mm, the protrusive movement and protrusive movement when considered restricted less than 5 mm [12].

In Table 1 the results are shown on the abscissa axis (X).

In one volunteer, the maximum protrusion of 5.1 mm was observed, that by the standards, this value is within the normal range.

For volunteer 2, the maximum protrusion observed was 8.3 mm, which by the standards is within normal limits.

For the volunteer 3, the maximum protrusion observed was 9.0 mm, which by normality values is within normal limits.

For volunteer 4, the observed maximum protrusion was 9.7 mm, that for the normal standards, this value is within the normal range.

5 Conclusions

The TMJ dysfunctions are generally used to refer to the pathophysiology of joint and between them stands out the deviations of the trajectory of the mandibular movement in the opening and closing of the mouth, as well as their joint locks and hypermobility of this joint, in which these movements characterize the principal object of study of this work.

The values obtained in the measurements carefully follow the same pattern, being the maximum difference between them is mostly smaller than 0.5 mm, which for this study, this figure shows a very satisfactory precision, demonstrating that the data were collected and processed correctly, but it must be borne in mind that these values are identified as normal from studies of passive movements and compared with the data taken from active movements the value of normality can be considered even lower.

For the evaluation of mandibular movement, the data follow the same biomechanics of the standards described in the literature.

There were no significant differences between the use of markers like orthodontic appliance and the adhesive type, which proves that the use of adhesive on the skin does not suffer interference with the movement relative to the fixed marker in the bone of the dental arch.

Because of the facts presented it appears that the equipment developed using camcorders, is a very good method because it is simple and practical, non-invasive and very low cost, to show effective for what it is intended, which comes to meet your goal. Good results were obtained in a clear and with quite satisfactory precision, reinforcing its use in the day-to-day of the professional who does not have high quality measuring instruments.

References

1. Santos, E.C.A., Bertoz, F.A., Pignatta, L.M.B., Arantes, F.M.: Clinical signs and symptoms of temporomandibular disorders in children. Rev. Dent. Press Ortodon. Ortop. Facial **11**(2), 29–30 (2006)
2. De Leeuw, R.: Orofacial Pain: Evaluation Guide, Diagnosis and Treatment, 4th edn. Quintessence, São Paulo (2010)
3. Fiorelli, A., Arca, E.A., Fiorelli, C.M., Rodrigues, A.S., Furcin, A.C., De Vitta, A., et al.: The effects of a global postural exercise program on temporomandibular disorder. Motriz: Rev. Educ. Fis. **22**(4), 272–276 (2016). https://doi.org/10.1590/s1980-6574201600040009. http://www.scielo.br/scielo.php?script=sci_arttext&pid=S1980-65742016000400272&lng=en. Accessed 1 Mar 2018
4. de Chaves, P.J., de Oliveira, F.E.M., Damázio, L.C.M.: Incidence of postural changes and temporomandibular disorders in students. Acta Ortop. Bras. **25**(4), 162–164 (2017). https://doi.org/10.1590/1413-785220172504171249. http://www.scielo.br/scielo.php?script=sci_arttext&pid=S1413-78522017000400162&lng=en. Accessed 1 Mar 2018
5. Paulino, M.R., Moreira, V.G., Lemos, G.A., da Silva, P.L.P., Bonan, P.R.F., Batista, A.U.D.: Signs of prevalence and symptoms of temporomandibular disorders in pre-school students students: pool emotional factors, parafunctional habits and impact on quality of life. Ciênc. Collect. Health **23**(1), 173–186 (2018). https://doi.org/10.1590/1413-81232018231. 18952015. http://www.scielo.br/scielo.php?script=sci_arttext&pid=S1413-81232018000100173&lng = en. Accessed 1 Mar 2018
6. Barros, R.M.L., et al.: Development and evaluation of a system for three-dimensional analysis of human movements. Braz. J. Biomed. Eng. Camp. **15**(1–2), 79–86 (1999)
7. Lemos, G.A., Moreira, V.G., Forte, F.D.S., Beltrão, R.T.S., Batista, A.U.D.: Correlation between signs and symptoms of temporomandibular disorders (TMD) and the severity of the malocclusion. Rev. Odontol. UNESP **44**(3), 175–180 (2015)
8. Ingervall, B.: Range of movement of mandible in children. Scand. J. Dent. Res. **78**, 311–322 (1970)
9. Lemos, G.A., Paulino, M.R., Forte, F.D.S., Beltrão, R.T.S., Batista, A.U.D.: Influence of temporomandibular disorder presence and severity on oral health-related quality of life. Rev. Dor **16**(1), 10–14 (2015)

10. Gaviao, M.B.D., Chelotti, A., Silva, F.: Análise funcional da oclusão na dentadura decídua: avaliação dos movimentos mandibulares. Rev. Odontol. Univ. São Paulo **11**(Suppl. 1), 6–12 (1997). https://doi.org/10.1590/S0103-06631997000500010. ISSN 0103-0663
11. Heidsieck, D.S.P., et al.: Biomechanical effects of a mandibular advancement device on the temporomandibular joint. J. Cranio-Maxillo-Fac. Surg. **46**(2), 288–292 (2017)
12. Yamamoto, M.K., Luz, J.G.C.: Evaluation of mandibular maximum excursions in asymptomatic individuals. Rev. Assoc. Paul. Cir. Dent. **46**(3), 781–784 (1992)

Ergonomics Simulation and Evaluation Application for the Wheelhouse on Large Ships

Zhang Yumei[✉] and Wang Wugui

China Ship Development and Design Center, Wuhan 430064, China
zhangyumei821202@sina.com

Abstract. Background and objective: The wheelhouse on large ships is the place where personnel, equipment, environmental factor and navigation monitoring interaction concentrate. It's extremely easy to have human factor problems in the work area. The rationality of wheelhouse design scheme needs to be evaluated in design phase by ergonomics simulation. Methods: Firstly, the vitual cabin scene assembly models are built by using the ship 3D design software CADDS5. They're converted into intermediate format models of VRML and then imported into the simulation software DELMIA. Secondly the virtual crew human bodies are set up by 5%, 50% and 95% percentile according to the male body sizes of national standard GB/T 10000-1988. Thirdly the sailing commander, steersman and telegraph operator are selected to verify according to the typical tasks such as: (1) the sailing commander's riding comfort, convenience on up and down, lookout visibility; (2) the steersman's operability and reachability of the wheel, console panel, phone etc., as well as steering fatigue and lookout visibility; (3) the telegraph operator's operability and reachability of the telegraph, console panel, phone etc. Finally, via setting the crew of frequently-used walking path, the rationality of channel settings in wheelhouse was validated through collision and interference analysis. Results: Aiming at the typical position operators and space layout of wheelhouse on the large ships, the simulated validation was carried out with the human factors design requirements. The analysis and evaluation work includes the visibility and reachability, the working gesture of crew, the collision and interference checking, the rationality and comfort of cabin operation space. The concrete improvement suggestions are put forward to optimize design. Conclusion: The method adopted in this paper can be applied to ergonomics simulation analysis and evaluation for the wheelhouse design on large ships. This study has important significance in solving the relationship between the crew, equipment and cabin space in the ship design phase.

Keywords: Large ship · Wheelhouse · Ergonomics · Simulation
Evaluation

1 Introduction

The wheelhouse on large ships is critical place where personnel, equipment, environmental factor and navigation monitoring interaction concentrate. It's extremely easy to have human factor problems in the work area [1]. So, it is necessary to carry out

© Springer International Publishing AG, part of Springer Nature 2019
D. N. Cassenti (Ed.): AHFE 2018, AISC 780, pp. 501–511, 2019.
https://doi.org/10.1007/978-3-319-94223-0_48

ergonomics simulation and assessment of wheelhouse, solve relationship between the crews, equipment and cabin space in ship design phase. All these will be a great help to reduce human errors, ensure the safety navigation and overall efficiency performance of large vessel.

At present the digital ship is only geometric ship in the design phase. It's focus on form expression of the ship to mainly solve the problem of equipment layout. It doesn't consider the factor of the crew. By application of virtual simulation technology, the ship digital design is no longer bound to three-dimensional modeling level, but gradually developing towards virtual prototype and man-machine engineering. In order to make the designed ship better to be manipulated by the owner, the wheelhouse design scheme is carried out ergonomics check and evaluation using simulation software DELMIA. The specific work is as follows: 1. Building the virtual cabin scene of wheelhouse, through converting and importing the compartment structure and equipment assembly models. 2. Building the human models for key positions, such as the sailing commander, steersman and telegraph operator. 3. Simulating the human body postures and movements of typical task, carrying out ergonomics simulation verification, including lookout visibility, convenience on up and down of the sailing commander, operation accessibility and comfortableness of the integrated bridge equipment etc. The research of this article is useful to reduce design change and crew inconvenience, improve design quality, accelerate ship research and development efficiency, increase user satisfaction [2].

2   Building the Wheelhouse Models of Digital Ship

The compartment structure and equipment assembly models of wheelhouse are built by using the ship 3D design software CADDS5. They're converted into intermediate format model of VRML by lightweight processing and imported into the simulation software DELMIA. Thus, the virtual cabin scene of wheelhouse is completed.

In CADDS5 system, the digital ship consists of general parts and special parts. The special parts participate in ship module assembly directly, provide only the location coordinate of models, but exclude entity models. The assembly organization form of model nodes is formed, based on such digital ship construct ideas. According to the blocks and majors, this digital ship step-down classification is composed of preliminary configuration, hull structure, pipe, cable, equipment etc. Through decomposing step by step the ship assembly nodes present as a tree structure, depending upon the design requirements. The leaf nodes are all special parts [3]. Figure 1 shows the ship assembly structure tree in CADDS5.

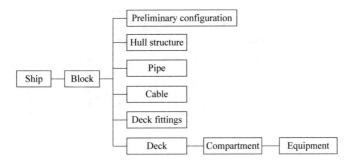

Fig. 1. Ship assembly structure tree in CADDS5

According to Fig. 2, it shows the data conversion process from CADDS5 to VRML. Firstly, the three-dimensional models of wheelhouse compartment are broken up by reference to different types of models, such as hull structure, pipe, cable, equipment etc. The assembly tree structure conversion script is compiled by setting the transformation interface separately. The hull structure adopts CADDSBATCH method. The other type models adopt GBFBATCH method. Through parallel transformation of special parts for each type model, the lightweight documents are generated, including *.ed and *.ol. Secondly the above-mentioned documents are manual processed as intermediate format model of VRML (*.wrl) by using Creo View software [4].

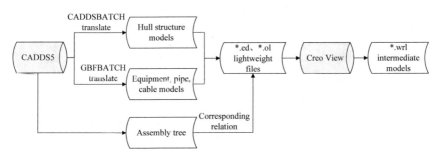

Fig. 2. Data conversion process from CADDS5 to VRML

The converted wheelhouse VRML models still retain assembling relation, geometric shape and other information of original CADDS5 models. In order to meet simulation requirements, the materials and textures of the following models such as the wheelhouse bulkhead, floor, integrated bridge equipment and command chair are built (see Fig. 3).

Fig. 3. Virtual scene of the wheelhouse in DELMIA

3 Building the Virtual Human Models

3.1 Analyzing Key Positions in the Wheelhouse

In wheelhouse design, the navigation officers should be considered first of all, including navigation information requirements, environmental information requirements, external view and means of communication. Next, the ship operators need to be supported reasonably, including manipulation information requirements, field of view, means of communication. In order to reduce personnel movement and information transmission, the equipment is properly configured and disposed, according to information and operating frequency in combination with post responsibility. It is beneficial to guarantee the safety of vessel navigation, increase shipping efficiency. Then the following aspects are concerned, including determining the working space of each position, reducing the lateral length of control console, enhancing the patency, openness and comfort of the wheelhouse. Finally, the long-term usage habit of users and layout aesthetics need to be taken into consideration measurably [5].

According to Fig. 4, the key positions in wheelhouse are set as follows:

1. Navigation command post

   Undertaker: the sailing commander
   Responsibility: The sailing commander is in charge of navigating and controlling the ship by giving instructions for the steering and telegraph.

2. Steering post

   Undertaker: the steersman
   Responsibility: The steersman is in charge of manipulating the steering console. He receives steering instructions from the sailing commander and executes correctly. When the steering angle meets command request, he feedbacks the result.

3. Operating telegraph post

   Undertaker: the telegraph operator
   Responsibility: The telegraph operator is in charge of manipulating the propeller monitoring station. He also receives operating telegraph instructions from the sailing commander and executes correctly. When the main engine speed meets command request, he feedbacks the result.

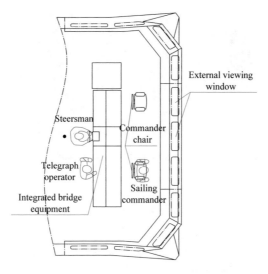

Fig. 4. Key positions setting in the wheelhouse

3.2   Setting the Crew Body Model Sizes

As shown in Table 1, the sizes of the crew body model are according to the average size of a Chinese adult with national standard GB10000-88. Then the virtual crew human bodies are set up by 5th percentile, 50th percentile and 95th percentile.

Table 1. Human body measurements of Chinese adults GB10000-88 (unit:mm)

Age and percent Measuring project	Male(18~60)			Female(18~55)		
	5%	50%	95%	5%	50%	95%
Height	1583	1678	1775	1484	1570	1659
Weight	48	59	75	42	52	66
Upper arm length	289	313	338	262	284	302
Forearm length	216	237	258	193	213	234
Thigh length	428	465	505	402	438	476
Leg length	338	369	403	313	344	375

The sailing commander, steersman and telegraph operator are all adult male. So in order to be more practical, the virtual human body sizes are modified and edited by reference to Chinese crew sizes [6]. Among them, the 5th percentile human body sizes of Chinese crew are: body height 1631 mm, sitting height 869 mm, eye height of sitting 761 mm, knee height of sitting 483 mm. Similarly, the 95th percentile sizes are: body height 1794 mm, sitting height 958 mm, eye height of sitting 858 mm, knee height of sitting 549 mm. The virtual human models of key positions in wheelhouse are built based on the above data (see Fig. 5).

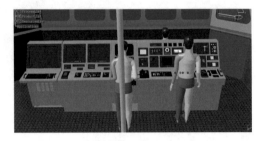

Fig. 5. Virtual human models of the sailing commander, steersman and telegraph operator

Usually the different percentiles are chosen on the basis of work nature and construction feature [3]. The basic principles are as follows:

1. The size of containing crew body uses 5th percentile.
2. The size which is contained in crew body uses 95th percentile.
3. The optimum operating range uses 50th percentile.

According to above principles, the human body size percentile of key positions in the wheelhouse is defined (see Table 2), aiming at ergonomic analysis for typical task operation.

Table 2. The human body size percentile of key positions

Key position	Operation analysis of typical tasks	Human body size percentile		
		5%	50%	95%
Sailing commander	Riding comfort	✓		✓
	Convenience on up and down		✓	
	Lookout visibility		✓	
	Channel accessibility			✓
Steersman	Operability and reachability of the wheel, console panel, phone etc.		✓	
	Steering fatigue		✓	
	Lookout visibility		✓	
Telegraph operator	Operability and reachability of the telegraph, console panel, phone etc.		✓	

4   Analyzing and Validating Typical Tasks

4.1   Riding Comfort of the Sailing Commander

Riding comfort means that the style and layout of seat minimize driving and riding fatigue as far as possible. So, the design of commander chair (see Fig. 4) should meet the human body comfortable posture requirements. Table 3 and Fig. 6 show the sitting posture analysis results of sailing commander.

Table 3. Sitting posture analysis of sailing commander

Size designation	Recommended range	Check results		Riding comfort evaluation		
		Male 5%	Male 95%	Good	General	Bad
Center line trunk and vertical angle	15°∼30°	19.8°	19.6°	✓		
Center line trunk and thigh angle	95°∼115°	106.4°	112.6°	✓		
Center line thigh and leg angle	100°∼135°	74.7°	83.9°			✓
Center line trunk and upper arm angle	0°∼30°	26.1°	22.6°		✓	

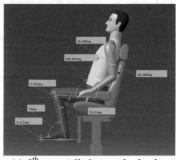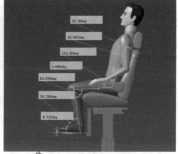

(a) 5th percentile human body size   (b) 95th percentile human body size

Fig. 6. The sitting posture of sailing commander

4.2   Convenience on up and Down of the Sailing Commander

The convenience on up and down of the sailing commander is important aspect of driving ergonomics research. The motion simulation of climbing the commander chair is displayed in Fig. 7. Judging from the results of analysis, the sailing commander has some deficiencies. It is recommended to adjust height and handrail of the commander chair, in combination with the above analysis of riding comfort.

Step 1   Step 2   Step 3   Step 4   Step 5   Step 6

Fig. 7. Analyzing convenience on up and down of the sailing commander

4.3   Lookout Visibility of the Sailing Commander

The command and outlook are the vital navigation task in wheelhouse of large vessel. So, there is high demand for lookout visibility of the commander. The excellence degree or grade of visibility is mainly influenced by chair height and placement. Figure 8 shows the lookout visibility analysis result by typical posture simulation of sailing commander. The helm indicator, intelligence composite display and terminal display control unit are all in sight. Meanwhile the vision field of window outside is broad.

| (a) Looking out the window | (b) Watching the helm indicator | (c) Watching the intelligence composite display leftside | (d) Watch the terminal display control unit rightside |

Fig. 8.   Analyzing lookout visibility of the sailing commander

4.4   Operation Accessibility of the Steersman

The operability of steersman is analyzed by dynamic motion simulation, according to personnel stance and hand reachability (see Fig. 9). The analysis results are as follows: The accessibility of wheel and rotary knob is ideal. But the operability of console panel rotary knob and button has a little trouble. The reachability of telephone which situates at the edge of console is not good enough.

| (a) Operating the wheel | (b) Operating the wheel rotary knob | (c) operating the rotary knob of console panel | (d) operating the button of console panel | (e) picking up the telephone |

Fig. 9.   Analyzing accessibility of the steersman

4.5 Operation Fatigue of the Steersman

The steersman needs to maintain same steering posture for a long time. Rapid upper limb assessment (RULA for short) is used to assess the risk of human upper limb musculoskeletal injury for work reason, through studying the posture, force and muscle use of human body part. Meanwhile the effect of neck, torso and leg on the upper limb is taken into account. The posture score is obtained finally by weighting all the factors. It indicates every posture score with different colors. Green represents level I (1~2 score), means best comfortable level. Yellow represents level II (3~4 score), means some comfortable level. Orange represents level III (5~6 score). It means some uncomfortable level and needs to be improved. Red represents level IV (7 score). It means uncomfortable level and needs to be adjusted immediately, otherwise causing human injury [7]. Figure 10 shows the fatigue analysis result of steering posture. The comprehensive score is 3. It indicates that the steering posture is some comfortable and can maintain a longer time.

Fig. 10. RULA analysis of the steersman (Color figure online)

4.6 Lookout Visibility of the Steersman

The excellence degree or grade of visibility of steersman is mainly influenced by the height and location of integrated bridge equipment, personnel stance. Figure 11 shows the lookout visibility analysis result by typical posture simulation of steersman. All the important parts and equipment are in sight, including the helm indicator, intelligence composite display, terminal display control unit and console panel etc. Meanwhile the vision field of window outside is broad.

(a) Looking out the window   (b) Watch the terminal display control unit rightside   (c) Watching the intelligence composite display leftside   (d) Watch the console panel

Fig. 11. Analyzing lookout visibility of the steersman

4.7   Operation Accessibility of the Telegraph Operator

The operability of telegraph operator is analyzed by dynamic motion simulation, according to personnel stance and hand reachability (see Fig. 12). The analysis results are as follows: The accessibility of telegraph and rotary knob is ideal. The reachability of telephone which situates at the edge of console is also ideal.

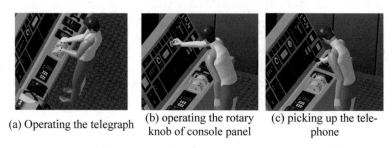

(a) Operating the telegraph   (b) operating the rotary knob of console panel   (c) picking up the telephone

Fig. 12. Analyzing accessibility of the telegraph operator

4.8   Spatial Rationality of the Wheelhouse

It can be seen that the wheelhouse is spacious from the Proportional relationship, between cabin environment and virtual humans, between virtual humans and equipment. The action and operation of key positions are convenient, including the sailing commander, steersman and telegraph operator. Via setting the frequently-used walking path of sailing commander, the rationality of channel settings in wheelhouse was validated through collision and interference analysis. The analysis result shows that command channel accessibility is good (see Fig. 13).

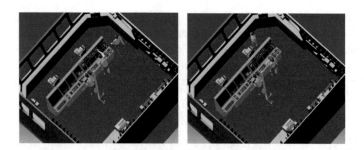

Fig. 13. Analyzing channel accessibility

5   Conclusion

The thesis has developed ergonomics simulation verification research, aiming at key positions and space layout of wheelhouse on large ships. The sailing commander, steersman and telegraph operator are selected to carry out the typical operation assignment human motion and posture simulation. The analysis and evaluation work include visibility and reachability, collision and interference checking, rationality and comfort of cabin operation space. The concrete improvement suggestions are put forward to optimize design according to the evaluation results.

The method adopted in this paper can be applied to ergonomics simulation analysis and evaluation for the wheelhouse design on large ships. This study has important significance in solving the relationship between the crew, equipment and cabin space in the ship design phase.

References

1. Qian, J., Yu, K., Yan, S.Y., et al.: Ergonomic design evaluation method for the wheelhouse on large ship. Chin. J. Ship Res. **9**(5), 15–21 (2014)
2. Zhang, Y.M.: A review of warship man-machine-environment system engineering. Chin. J. Ship Res. **12**(2), 41–48 (2017)
3. Fang, X.B., Chen, Y., Li, T.T., et al.: Design and implementation of the ship virtual maintenance simulation application system. Chin. J. Ship Res. **11**(6), 136–144 (2016)
4. Zhang, Y.M., Wei, Q.Q., Zeng, J.: Evaluation of overall warship operability and maintainability based on the technology of virtual reality. Chin. J. Ship Res. **8**(2), 6–12 (2013)
5. Li, Z.N., Wang, Y.Q., Xiang, G.Q., et al.: The human body data on naval vessel crew. Chin. J. Ergon. **2**(1), 53–55 (1996). 71
6. Xu, Q., Zhang, Y.M., Lv, J.F., et al.: Shape and Configuration Design of Navel Ships. National Defend Industry Press, Beijing (2016)
7. Sheng, X.Y., Sheng, X.J.: DELMIA Egonomics Tutorials. China Machine Press, Beijing (2009)

Using Digital Human Modeling to Evaluate Large Scale Retailers' Furniture: Two Case Studies

Carlo Emilio Standoli[1(✉)], Stefano Elio Lenzi[2],
Nicola Francesco Lopomo[2], Paolo Perego[1], and Giuseppe Andreoni[1]

[1] Politecnico di Milano - Dipartimento di Design, via Giovanni Durando, 38/A,
20158 Milan, Italy
{carloemilio.standoli, paolo.perego,
giuseppe.andreoni}@polimi.it
[2] Dipartimento di Ingegneria dell'Informazione,
Università degli Studi di Brescia, via Branze, 38, 25123 Brescia, Italy
{s.lenzi002, nicola.lopomo}@unibs.it

Abstract. The huge number of workers affected by musculoskeletal diseases and disorders demonstrates that prevention in working environment is an important issue. Large scale retail trade involves lots of workers, with a multiplicity of tasks and activities. The working tasks and environment condition the onset of these diseases. Digital Human Modeling (DHM) can support the design of the working environment, thus to avoid the risks of work-related musculoskeletal diseases and disorders (WRMSD). The aim of this study was to analyze and assess existing furniture by using DHM to evaluate the risk of developing WRMSD among supermarket clerks and cashiers. Two case studies are specifically presented. Assessment was realized in terms of reaching maps comparing different genders and height/weight percentiles. Preliminary findings suggested the use of dedicated guidelines to choose and set-up furniture in these specific applications, underling the variety of issues present in the large-scale retail trade.

Keywords: Digital Human Modeling · Human factors · Large scale retail trade
Work-related musculoskeletal disorders

1 Introduction

The identification of risks and the prevention of work-related musculoskeletal disorders and diseases (WRMSD) represents a very important issue, for both the epidemiology, the social burden and company costs. To prevent WRMSD, the employer is liable to check all the aspects related to the workplace and working conditions, such as environment, furniture, employees' activities and tasks. According to National and International Standards [1, 2] and these evaluations, employers must avoid all risks for employees in the best way possible: supplying aids, redesigning furniture, or changing the processes at the basis of workers' activities and tasks. This assessment should include a large population of workers and include the suitable related set of Standards, methods and tools.

© Springer International Publishing AG, part of Springer Nature 2019
D. N. Cassenti (Ed.): AHFE 2018, AISC 780, pp. 512–521, 2019.
https://doi.org/10.1007/978-3-319-94223-0_49

Not all the working conditions are properly fitted by the actual methods, also in relation to the rapid introduction of new workplaces and complex working tasks.

Large-scale retail trade is a huge market that involves a great number of skilled workers. In general, in each supermarket, there are lots departments (e.g., greengrocer, butcher, delicatessen, etc.), serving customers and selling a huge assortment of different goods. This variety of goods greatly influences the type of activities and tasks daily performed by workers, such as restocking activities. In this perspective, International [1] and National Standards [2] are necessary to identify workers' safeguard.

The aim of this paper was to present the benefits given by the use of an advanced Digital Human Model (DHM) in supporting the analysis of the existing solutions and the proactive design of novel working environment and furniture focusing on large-scale retail trade framework. This use of DHM was already applied in previous experiences related to different case studies [3–5]. In this paper, large scale retailers were analyzed and two specific working conditions were in particular addressed: the clerks (sorting and restocking activities) and the cashier workers. These are common activities for all the supermarkets.

The methodology comprises four different steps, from a starting observational phase with a detailed task analysis, to the digital and biomechanical modeling that is imported into a DHM software suite for the simulation of activities and the evaluation of postures and tasks. This paper describes the main preliminary outcomes of these activities.

2 Materials and Methods

To reach these goals, the research project was designed according to the following methodological steps: (1) conduction of an ethnographic analysis to understand the working activities, tasks and procedures specifically related to large scale retail distribution; (2) creation of the digital model of the working environment and in particular the existing furniture (shelves at different heights and a cash recorder desk); (3) simulation of the working postures in two specific case studies using a DHM system (Santos, Santos Human Inc.); (4) ergonomic assessment in the digital environment through a biomechanical software suite integrated in the DHM System. These ergonomic evaluations were performed considering 3 representative height/weight percentiles, according to the Standard ISO 3411 [6].

2.1 Ethnographic Research

This step represented a preparatory phase for the DHM evaluation. In order to understand the working methods and flows and to collect workplace information, an observational phase was needed. The analyzed large scale retailer represents a complex environment, in which different tasks and professional figures coexist. In most of the observed supermarkets, there are 8 different departments (i.e., fruit and vegetables, milk products, grocery and general merchandise, delicatessen, butchery, fishery, bakery and pastry). For each department, especially for those who required handmade activities (e.g., bakery), there are high skilled and qualified employees. These employees are exposed to a huge variety of risks and, to prevent WRMSD, the employer has to characterize in details all the employees' activities, tasks and environment.

The ethnographic phase involved 3 researchers and the observation of 6 stores of different sizes (small and medium size stores located in the city center, large size stores in the city suburbs). In each store, at least 1 employee coming from each of 8 departments was recruited. In addition, 6 cashiers were recruited. Researchers followed the employees for all their working hours, recording by using a video-camera; cashiers were recorded by using 2 fixed cameras, one placed on the frontal plane and one on the sagittal plane. The observational phase lasted 3 months. Figure 1 presents both the cashier and the restocking activities.

Fig. 1. On the left, a cashier waiting for a costumer. Cameras' set up is highlighted with green circles. On the right, an employee during a restocking activity. (Color figure online)

2.2   DHM Simulation Set-Up

In order to evaluate the workplaces and the related tasks and postures, a virtual environment was settled up. This section consisted in several steps:

1. Definition of the Avatars, both for gender and height/weight percentile, according to the Standards;
2. Preparation of the simulation environment and of the related furniture and aids creating the parametric digital models;
3. Definition of the parameter of interest, and exclusion of all the other variables that could influence the simulation.

Table 1. Height and weight of the selected percentiles, according to ISO 3411

Percentile	Gender	Height (cm)	Weight (kg)
5th	Female	156,6	62,8
50th	Male	173	78,7
95th	Male	190,5	94,6

Regarding the first point, these ergonomic evaluations were performed considering 3 representative avatars, belonging to the Caucasian population, covering 90% of anthropometric sizing: 5th percentile female, 50th percentile male and 95th percentile male. The avatars were anthropometrically differentiated by gender, somatotype and percentile, according to the Standard ISO 3411 (Table 1 and Fig. 2).

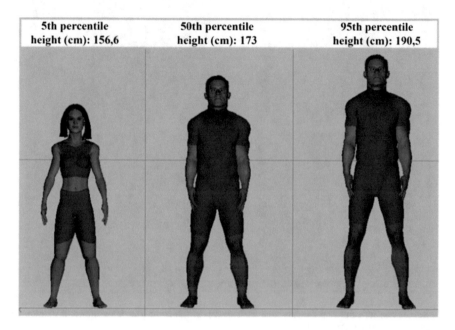

| 5th percentile
height (cm): 156,6 | 50th percentile
height (cm): 173 | 95th percentile
height (cm): 190,5 |

Fig. 2. From left to right: frontal picture of three Avatars, corresponding to the 5th, 50th and 95th percentiles, according to the ISO-3411.

Regarding the definition of the simulation environment, we decided - for this preliminary assessment - to simplify as much as possible the furniture 3D models (Fig. 3). In fact, the simulation algorithm includes several functions related to the 3D geometry of the environment (e.g. collision avoidance, vision cone, reaching areas); a geometry simplification significantly reduced the computational effort and time. In addition, we avoided the use of displayed goods that could be considered additional constraints. We are aware that displayed goods could influence the working activities, posture and strategies but, in this paper, we decided to investigate the simple reachability tasks. Furthermore, the main focus of this work was the general assessment of the workplace, considering posture as the most important parameter to be evaluated; for this reason, we didn't evaluate grasps.

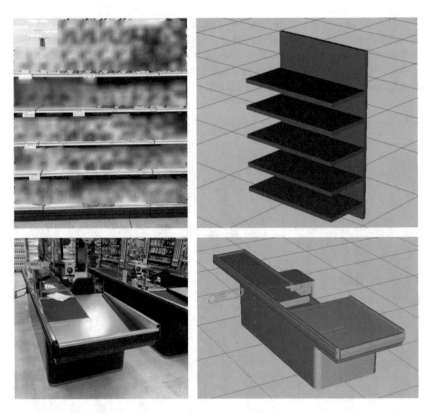

Fig. 3. On the left, the real display and cash desk; on the right, the 3D simplified display and cash desk, used for DHM evaluation.

3 Results and Discussion

The observational phase was useful for investigating the employees' activities, tasks and attitudes.

Cashiers spend most of their working hours at the cash desk. Sometimes, when there is a slight customers flow, they are asked to help clerks in restocking and sorting activities or to wait for new customers. Cashiers maintain a static seated posture at the cash desk, waiting goods carried by the conveyor belt. They have to pass goods from the right hand, through the scanner, to the left hand and release them to the customer. At the end of the process, they wait for the customer's payment. This is an iterative task but we're not interested in its repetitiveness.

Sorting and restocking activities take place in almost everywhere the supermarkets' area. There are different kind of furniture and displays, such as refrigerated displays, displays with shelves, displays for wine, tables and so on. The different kind of displays influenced the restocking activities. Operators have to manually load goods into the shelves. To ease this activity, a preparatory phase is needed: a dedicated employee placed goods' packages in a cart, taking into account their product group and the correct placement in the sales department. Clerks take the cart to the proper super-market area and they begin to sorting and restocking goods. Sometimes they use the cart as a prop, to avoid lifting heavy goods. In the store area, there are small store ladders, useful both for restocking clerks and customers.

A total of 24 simulations (6 for the cashiers and 18 for the restocking clerks) were carried out. As expected due to her limited anthropometrical dimensions, the 5th percentile avatar is the worst case in terms of reachability; for this reason results and discussion are focused on this situation. They are presented using reaching maps for both right and left hand. These maps define the comfort zone by means of a color gradient, that shifts from green (comfortable) to red (uncomfortable).

3.1   Cashier Desk

Each avatar was positioned in a sitting posture, in front of the cash desk, with one's arms on the cash-top. For the 5th percentile, a stool and a footrest were provided. Both the stool and the footrest 3D geometries were simplified. The stool was 54 cm high; the footrest was 17 cm high.

Each hand had a related End-Effector; the final target was the area in front of the user, where the cash-drawer and the scanner are placed.

Santos' Zone Differentiation tool was used to perform the evaluation. Using that tool, users can analyze information according to posture-based performance measures [7–9]. Zone Differentiation volume was created both for right and left hand; it was approximately $210 \times 170 \times 106$ cm, with a voxel resolution of $32 \times 32 \times 32$ cm (Fig. 4).

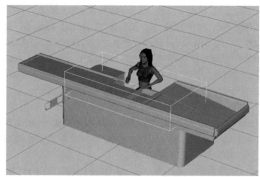

Fig. 4. On the left, the positioned avatar; on the right, the Zone Differentiation Volume used for DHM evaluation.

Results demonstrate that the 5th percentile avatar presents good values for both sides. 50th and 95th avatars didn't present issue in terms of reachability. Usually cashiers have to pass goods from the right hand, through the scanner area, to the left hand and release them to the customer. Both the right and left conveyor belts are easily reachable; the cash-drawer and the scanner are properly placed (Fig. 5).

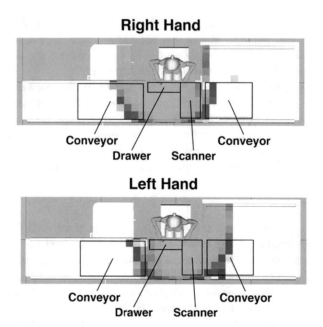

Fig. 5. Results of the comfort zone evaluation for the cash-desk. Results are expressed using reaching maps for both right and left hand.

3.2   Clerks

Each avatar was placed in different postures and positions, according to the different shelves height. Three case studies were analyzed, grouping shelves at different heights: First, we analyzed the lower shelves; second, the central shelves; third, the higher shelves. Each shelf was 5 cm thick. The lowest shelf was at 20 cm off the ground; there were approximately 30 cm between the other shelves. The highest was at 165 cm off the ground. For the last case, a ladder was provided. As the previous analysis, its 3D geometries were simplified, by using a cube 50 cm high.

The first avatar was in front of the display, in a crouch, with the arms close to the shelves. The second avatar was positioned in neutral standing posture, with the arms close to the shelves. The third avatar had the same posture of the previous one, but it was already on the ladder.

Each hand had a related End-Effector; the final target was the shelf, where goods are usually placed.

For each case study, Zone Differentiation volumes were created both for right and left hand. For the first case, they were approximately $150 \times 75 \times 85$ cm; for the second, $150 \times 75 \times 55$ cm; for the third, $150 \times 75 \times 30$ cm. For all them, we defined a voxel resolution of $32 \times 32 \times 32$ cm (Fig. 6).

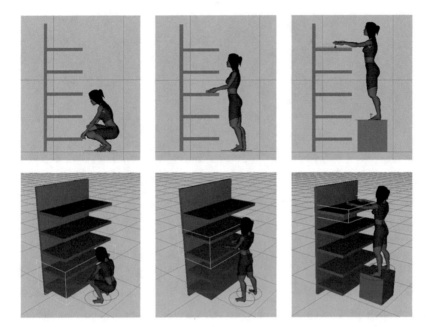

Fig. 6. The three positioned avatars and the Zone Differentiation Volume used for DHM evaluation.

Results underline a strong difference related to the reaching and comfort zones among the three groups of shelves. Central shelves present good results in terms of reachability. Lower shelves can be sorted and restocked hardly. Higher shelves, thanks to an aid such as a ladder, can be reached in a quite good way (Fig. 7).

50th and 95th avatars didn't present issue in terms of reachability of the central-higher shelves. As happened for the 5th percentile, they had problems both in postures and reachability for the lower shelves.

Aids could ease the working activities related to the medium-high level of shelves; shelves placed at the lowest level represent problems in terms of both posture and reachability.

Sometimes could happen to find more than 5 shelves, placed at different heights; it depends on the goods typology. DHM can ease the evaluation of the WRMSD risks of these cases thanks to virtual simulation.

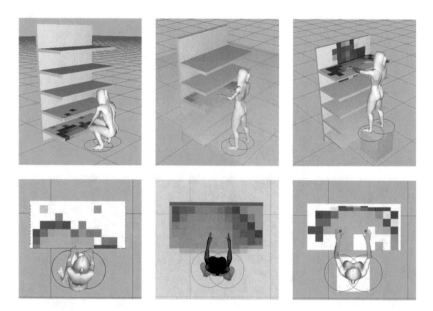

Fig. 7. Results of the comfort zone evaluation for the display, related to the three different case studies. Results are expressed using reaching maps from 2 different perspectives.

4   Conclusions

The methodology here described allows for the evaluation of working environments and furniture and the related WRMSD risks in a faster, accurate and detailed mode, compared to standard method i.e. recruiting a large population and making tests with real prototypes and furniture and collecting questionnaires and joint angles.

Moreover, the possibility to perform tests at the CAD-level means a considerable cost advantage; if needed, it allows manufacturers to modify and re-design furniture. Especially in Large Scale Retail Trade, where the shelves' configuration need to be frequently changed according to the displayed goods, the use of a DHM evaluation represents an effective approach.

Performing tests in the real-world environment on real users (of the three percentiles), means high time-consumption, high costs and reduced possibilities to modify and implement both environments and furniture.

In conclusion, this paper demonstrates that DHM could be a promising and useful tool, to be used since the early design phases of working environment and furniture.

Further developments and future works need to investigate how products and their features (i.e., typology, weight, dimension) can influence the working activities, postures and grasps.

References

1. International Standard ISO 11228: Manual handling-parts 1, 2, 3 (2007)
2. D.Lgs. 81/08: Testo Unico in materia di salute e sicurezza nei luoghi di lavoro – TUSL81 (2008)
3. Mazzola, M., Forzoni, L., D'Onofrio, S., Standoli, C.E., Andreoni, G.: Evaluation of professional ultrasound probes with Santos DHM. Handling comfort map generation and ergonomics assessment of different grasps. In: Proceedings of the 5th International Conference on Applied Human Factors and Ergonomics AHFE 2014, Kraków, Poland (2014)
4. Mazzola, M., Forzoni, L., D'Onofrio, S., Marler, T., Beck, S.: Using Santos DHM to design the working environment for sonographers in order to minimize the risks of musculoskeletal disorders and to satisfy the clinical recommendations. In: Proceedings of the 5th International Conference on Applied Human Factors and Ergonomics AHFE 2014, Kraków, Poland (2014)
5. Mazzola, M., Forzoni, L., D'Onofrio, S., Andreoni, G.: Use of digital human model for ultrasound system design: a case study to minimize the risks of musculoskeletal disorders. Int. J. Ind. Ergon. **60**, 35–46 (2016)
6. International Standard ISO 3411: Earth moving machinery – Physical Dimension of operators and minimum operator space envelope, 4th edn (2007)
7. Yang, J., Marler, T., Beck, S., Abdel-Malek, K., Kim, H.-J.: Real-time optimal reach-posture prediction in a new interactive virtual environment. J. Comput. Sci. Technol. **21**(2), 189–198 (2006)
8. Yang, J., Marler, T., Kim, H-J, Arora, J.S., Abdel-Malek, K.: Multi-objective optimization for upper body posture prediction. In: 10th AIAA/ISSMO Multidisciplinary Analysis and Optimization Conference, Albany, NY (2004)
9. Yang, J., Verna, U., Penmatsa, R., Marler, T., Beck, S., Rahmatalla, S., Abdel-Malek, K., Harrison, C.: Development of a zone differentiation tool for visualization of postural comfort. In: SAE 2008 World Congress, Detroit, MI (2008)

Author Index

© Springer International Publishing AG, part of Springer Nature 2019
D. N. Cassenti (Ed.): AHFE 2018, AISC 780, pp. 523–525, 2019.
https://doi.org/10.1007/978-3-319-94223-0

Printed in the United States
By Bookmasters